# Applied Fluid Mechanics

PRENTICE HALL INTERNATIONAL SERIES
IN THE PHYSICAL AND CHEMICAL ENGINEERING SCIENCES

Neal R. Amundson, Series Editor, *University of Houston*

Advisory Editors

Andreas Acrivos, *Stanford University*
John Dahler, *University of Minnesota*
H. Scott Fogler, *University of Michigan*
Thomas J. Hanratty, *University of Illinois*
John M. Prausnitz, *University of California*
L. E. Scriven, *University of Minnesota*

# Applied Fluid Mechanics

**Tasos C. Papanastasiou**

*Chemical Process Engineering Research Institute*
*Department of Chemical Engineering*
*ARISTOTLE UNIVERSITY OF THESSALONIKI*

P T R PRENTICE HALL

Englewood Cliffs, New Jersey 07632

**Library of Congress Cataloging-in-Publication Data**

Papanastasiou, Tasos C.
    Applied fluid mechanics/Tasos C. Papanastasiou.
    p.    cm.
    Includes bibliographical references and index.
    ISBN 0-13-060799-1
    1. Fluid mechanics.   I. Title.
TA357.P33   1993
620.1'06--dc20

92-34111
CIP

Editorial/production
 and interior design: *bookworks*
Acquisitions editor: *Betty Sun*
Managing editor: *Sophie Papanikolaou*

Artist: *Academy ArtWorks, Inc.*
Cover designer: *Ben Santora*
Copy editor: *Service to Publishers*
Buyer: *Alexis Heydt*

© 1994 by P T R Prentice Hall
Prentice-Hall, Inc.
A Paramount Communications Company
Englewood Cliffs, NJ 07632

The publisher offers discounts on this book when ordered
in bulk quantities. For more information, contact:

Corporate Sales Department
P T R Prentice Hall
113 Sylvan Avenue
Englewood Cliffs, NJ 07632

Phone: (201) 592-2863
FAX: (201) 592-2249

Printed in the United States of America

10  9  8  7  6  5  4  3  2  1

ISBN   0-13-060799-1

Prentice-Hall International (UK) Limited, *London*
Prentice-Hall of Australia Pty. Limited, *Sydney*
Prentice-Hall Canada Inc., *Toronto*
Prentice-Hall Hispanoamericana, S.A., *Mexico*
Prentice-Hall of India Private Limited, *New Delhi*
Prentice-Hall of Japan, Inc., *Tokyo*
Simon & Schuster Asia Pte. Ltd., *Singapore*
Editora Prentice-Hall do Brasil, Ltda., *Rio de Janeiro*

*Dedicated to my Parents*
*Haritos Papanastasiou*
*and*
*Yalatia Christofi*

# Contents

# CHAPTER 2   FLUID STATICS   46

# CHAPTER 3   MASS, ENERGY, AND MOMENTUM BALANCES   84

# CHAPTER 4    VISCOUS FLOW AND FRICTION: CONFINED, OPEN, FREE STREAM, AND POROUS MEDIA FLOWS    139

## CHAPTER 5    INTRODUCTION TO DIFFERENTIAL FLUID MECHANICS    229

## CHAPTER 6    UNIDIRECTIONAL FLOWS    274

## CHAPTER 7   TWO-DIMENSIONAL LAMINAR FLOWS: CREEPING, POTENTIAL, AND BOUNDARY LAYER FLOWS   330

## CHAPTER 8   NEARLY UNIDIRECTIONAL FLOWS: LUBRICATION AND STRETCHING FLOWS   375

# CHAPTER 9    RHEOLOGY AND FLOWS OF NON-NEWTONIAN LIQUIDS   414

# CHAPTER 10   TURBULENT FLOW AND MIXING   455

# *Preface*

*Applied Fluid Mechanics* is based on the author's extensive research and teaching experience in Newtonian and non-Newtonian process fluid mechanics at both the receiving and teaching ends. The book aspires to address both the undergraduate student's difficulties and often, frustration in mastering one of the toughest, yet most widely required courses in engineering education, and the teacher's time constraint in first building the required fundamentals and then proceeding to traditional as well as modern applications of fluid mechanics. Given the background of the author, the book is somewhat biased toward materials processing applications, which is vital to several emerging technologies. Fluid mechanics courses are indeed in a unique position to build the fundamentals needed to address not only related heat and mass transfer processes, but also the transfer, processing, shaping, and forming of materials in any fluid or fluidized state. Gas, liquid, granular solid, melt, suspension, emulsion, plasma, and mixtures of them are processed according to universal fluid mechanics and related transport phenomena principles of continuum. Thus the material of the book is developed around several realities:

1. Fluid mechanics is usually studied in one of the first "hard-core" engineering courses in the undergraduate curriculum, and must inevitably serve as an introduction to subsequent transport phenomena courses.
2. The subject of fluid mechanics is often—and justifiably—perceived by students as constituting a "tough" course. To pinpoint the areas of potential

difficulty, the book places special emphasis on the concepts of coordinate systems (which are unique to transport phenomena, especially to the subset of fluid mechanics), control volumes and changes within them, conventional and diffusional transport, and the origin of body and contact forces. This is done in a simplified manner, consistent with the general undergraduate level.

3. Most traditional fluid mechanics textbooks devote considerable attention to topics of primary interest to mechanical and civil engineering, but not necessarily to chemical, process, and materials engineering. In this book, traditional subjects of hydraulics, potential flows, and boundary layer and turbulent flows have been shortened to make space for topics related to small-scale laminar flows arising in materials processing.

4. Courses in fluid mechanics are becoming increasingly important in their own right as vehicles for addressing the processing of such materials as polymers and composites, coatings, thin films, and so on. Therefore, the book includes a chapter on lubrication and stretching forming flows, which are good prototypes for a diversity of fluid processing.

5. The effect of surface tension on fluid statics and small-scale laminar flows under free surface and/or interfaces is highlighted throughout the book, with emphasis on static equilibrium across interfaces and uneven thin-film flows.

6. Most industrial fluid processing includes non-Newtonian liquids. A separate chapter of the book deals with the physics, rheology, and rheometry of these liquids, primarily on their differences from Newtonian fluid in selected model flows.

7. The book is profusely illustrated with application-oriented problems for design and processing. Each chapter concludes with several problems of an applied nature, some solved and others suitable for homework, including applications involving transferring and handling fluids, polymer and coating processing, air-pollution and particle control, water circulation and irrigation, lubricant application, and other subjects.

8. To challenge advanced readers, some material of intermediate fluid mechanics level has been included selectively, such as fluid interfacial mechanics, transient flows, convering and diverging flows, concepts of vorticity dynamics and of linear stability analysis, and viscoelastic flow. These topics are to be viewed as optional, to be addressed based on time availability and level of audience.

9. To fit fluid mechanics theory into contemporary engineering and technology developments, topics from recent research activities are included in the form of application examples or problems, accompanied by extensive literature citations.

10. Because it is rather impossible to cover all the topics in a typical undergraduate course, the material is organized as described below, to allow alternative coverages, beyond six minimum required chapters.

Throughout the book, primary emphasis and care is given to the physics of flow, based on the properties and the mechanics of continuum, under the influence of driving forces and resisting cohesiveness. Mathematics are used to quantify a priori justified physical laws and conservation statements, and to solve the resulting mathematical problems. The former task of mathematics is considered and emphasized the most, since the resulting mathematical problems within the context of this book are easily solvable by means of tabulated methodologies in standard math handbooks.

The book is organized into chapters as follows:

**(a)** Chapters 1 to 4 are on *macroscopic fluid mechanics* and *hydraulics,* primarily with mechanical and civil engineering applications that include large-scale—often turbulent—flows and fluid storage and transportation equipment.

**(b)** Chapters 5 to 7 are on *microscopic* or *differential fluid mechanics,* with applications that include small-scale laminar flows arising mostly in fluid processing.

**(c)** Chapters 8 to 10, also based on differential fluid mechanics, are on *application*-oriented material. Chapter 8, deals explicitly with lubrication and forming flows that arise in materials processing; Chapter 9, deals with non-Newtonian flows also important in materials processing; and Chapter 10, is on turbulent flow applications and fluid mixing.

The material of each of the ten chapters is summarized below.

Chapter 1, on *Introduction to Fluid Mechanics,* deals with the continuum approximation of a fluid made up by molecules, and properties arising due to its continuum character, important to fluid flow: *Extensive,* such as mass, volume, and weight; and *intensive,* such as density, surface tension, pressure, viscosity, velocity, and stress. *Newton's law of motion* and other *conservation equations* and *equations of state* are introduced. It takes four to five hourly lectures to cover Chapter 1.

Chapter 2, on *Fluid Statics,* addresses the mechanics of fluid equilibrium, or *fluid statics,* under the action of *gravity* and *centrifugal* and *inertial* forces, that give rise to the *isotropic pressure,* responsible for *buoyancy, flotation, density stratification,* and *stability.* The principles of static equilibrium are generalized to *equilibrium across interfaces* of immiscible liquids, where additional *capillary pressure* results due to the action of *surface tension.* It takes four to five hourly lectures to cover Chapter 2 completely.

Chapter 3, on *Mass, Energy, and Momentum Conservation,* deals with the development of mass, momentum, energy, and solute conservation equations when a fluid system is removed from static equilibrium, and how to calculate *average spatially uniform properties* such as temperature, pressure, and concentration as if the system were under *well-mixed conditions.* The interactions of the system with its surroundings in the form of *exchange rates* of mass, momentum, heat, and solute and *body and contact forces,* as well as ongoing *internal changes* and *transformations,* are quantified. Applications are upgraded from tank filling which requires mass balance alone, to tank level control, requiring both mass and energy balances, and to liquid jets, which require simultaneous application of mass, energy and momentum balances. A special section is devoted to Bernoulli's equation and the circumstances under which it can apply. The chapter ends with the concept of *frictional forces* and *energy losses,* which leads to Chapter 4. It may take five or six hourly lectures to cover the entire chapter.

Chapter 4, on *Viscous Flow and Friction,* introduces the continuum mechanism, which gives rise to *shear stresses and friction,* which in turn leads to *mechanical energy losses* and *pressure drop.* Therefore, a supply of external power is required to move fluids through pipes and open channels. Frictional energy losses are quantified by exact equations in laminar flow, and by empirical equations and diagrams for turbulent flow. Confined flows in pipes and channels, flows in open channels, and free stream flows overtaking solid, liquid, and gas obstacles to flow are analyzed within this content. Flows in natural and synthetic porous media, gas compressible flows, mechanics of particulate motion, atomization of liquids two phase-flow and power supplying pumps are examined. It may take six to eight hourly lectures to cover Chapter 4.

Chapter 5, on *Introduction to Differential Fluid Mechanics,* emphasizes the necessity of *breaking into the system* introduced in Chapter 2 for macroscopic balances, in order to study the *spatial distribution* of velocity, pressure, stress, concentration, and temperature. The strategy adopted in Chapter 5 is application of the appropriate conservation equations to *differential control volumes,* well into the interior of the system, to derive the appropriate governing *differential equations of mass and momentum* for the fluid system at hand. *Solution steps* and *boundary conditions* needed to solve these equations are discussed, and ways to calculate useful quantities such as *flow rate, boundary stress and force, residence time, vorticity and rotation,* and *deformation rates* are shown. Chapter 5 concludes with the easy to study *radial frictionless flows* and *radial flows in porous media,* by means of the *continuity equation* to calculate the velocity profile, and then *Bernoulli's* or *Darcy's equations* to calculate the pressure given the velocity profile. It may take five to seven hourly lectures to cover this important introductory Chapter 5, in detail.

Chapter 6, on *Unidirectional Laminar Flows,* deals with the application of the differential principles developed in Chapter 5 to important *transfer* and *processing* flows that are amenable to analytic solutions. These flows include

(1) *rectilinear Cartesian flows* such as channel and thin-film flows under drag (Couette flow), pressure (Hagen–Poiseuille flow), gravity, and combinations of these;

(2) *axisymmetric rectilinear* (pipe, annulus, axisymmetric film), *radial* (sink and source), and *torsional* (rotating cylinder-type) flows; and

(3) *spherical radial* (sink/source and cavity growth/collapse) flows.

Chapter 6 continues with *multidimensional flows,* (spiral, helical, deflected source/sink) obtained by *linear superposition* of already studied unidirectional flows. The chapter concludes with elements of transient flows, where initiation of flow in cavity by wind and transient radial flows are studied. The former flow is a good prototype of *vorticity dynamics*, which may serve as an introduction to *boundary layer flows*, examined in Chapter 7. It takes five or six hourly lectures to complete Chapter 6.

Chapter 7, on *Two-dimensional Laminar Flows*, introduces the *Navier–Stokes equations* by generalizing the principles of Chapter 5 and 6. At this stage these equations are much easier for the student to understand and appreciate having gone through Chapters 5 and 6, where the underlying principles of the Navier–Stokes equations are built step by step. Several of the flows analyzed earlier are revisited to demonstrate the alternative approach of solving flow problems by means of the universally applied Navier–Stokes equations. The Navier–Stokes equations are then simplified by means of the *streamfunction* to study two-dimensional flows in converging/diverging channels which may include *creeping, lubrication, reverse,* and *radial flow,* depending on the *Reynolds number,* and associated *vorticity dynamics. Potential flows* are analyzed by means of the *potential function* and *Bernoulli's equation along streamlines,* to which the momentum equations degenerate. *Boundary layer flows* in the vicinity of solid boundaries surrounded by potential flow are analyzed by the *von Karman's appropriate method* and *Blasius's exact solution.* It takes five or six hourly lectures to cover Chatper 7.

Chapter 8, on *Nearly Unidirectional Laminar Flows,* extends both the differential control volume principles and the Navier–Stokes equations to *lubrication-type flows,* where a dominant velocity component varies drastically in vertical to flow direction(s) and slightly in the flow direction, and to *stretching flows,* where the dominant velocity component varies significantly only with respect to the flow direction. The first class of flows includes *hydrodynamic lubrication* of engines and bearings and *application of thin films or coating* under the influence of voscosity, gravity, and surface tension, studied by *Reynolds' lubrication equation. Forming processes,* such as *spinning of fibers, casting of films, film blowing, compression molding,* are studied by means of the *thin-sheet* or *beam approximation.* It takes five or six hourly lectures to complete Chapter 8.

Chapter 9, on *Flows of Non-Newtonian Liquids,* highlights important macroscopic differences between Newtonian and non-Newtonian liquids and explanations based on the nature of the involved molecules. *Constitutive equations* for *Newtonian, viscous inelastic, Bingham plastics, thixotropic, viscoelastic and elastomeric liquids* are summarized and discussed. Common *rheological instruments* to deduce or fit the constitutive equation are presented for both *shear and elongational measurements.* The second part of Chapter 9 deals with the combination of the conservation equations with selected non-Newtonian constitutive equations, to study selected confined and free surface *non-Newtonian flows.* It takes four to six hourly lectures to cover Chapter 9.

Chapter 10, on *Turbulent Flow and Mixing,* summarizes important phenomena and consequences on mixing, arising at high Reynolds numbers. *Onset of instabilities* that lead to turbulence, as predicted by the Navier–Stokes equations in laminar flows, are summarized. Beyond the onset of instability and turbulence, the same equations are *time-averaged* and the arising *Reynolds' stresses* are expressed by means of selected empirical correlations. The developed time-averaged equations for turbulent flow are then used to solve selected problems on turbulent flow in channels and pipes. One of the most important applications of turbulent flow and mixing is *agitation,* and selected topics and relative equipment are discussed. Chapter 10 concludes with *analogies among turbulent momentum, heat,* and *solute transfer.* It takes five or six hourly lectures to complete Chapter 10.

*Applied Fluid Mechanics* concludes with *Appendix A,* which summarizes *selection of a system of coordinates* and *vector and position representation,* the *significance of differential and integral calculus to fluid mechanics,* and the *relation between a physical flow problem and a differential equation and boundary conditions. Appendixes B, C, D, and E* provide information on *units and conversions,* useful *physical and mathematical constants,* and the important *dimensionless numbers* of fluid mechanics. Appendix F tabulates *Newton's law of viscosity,* which combines with the conservation equations to produce the *Navier–Stokes equations.* These equations are tabulated in Tables G.1 to G.3 of Appendix G, for the Cartesian, cylindrical, and spherical system of coordinates.

Most of the *solved examples* and *unsolved problems* are chosen to highlight the application of concepts to real-life processes, under reasonable assumptions and approximations. For several of the processes described in examples and problems, citation of key and accessible literature provides sources of further understanding and study.

A typical fluid mechanics course must cover the material of Chapters 1 to 4 and 5 to 7. An instructor may choose to cover first Chapters 5 and 6, based on differential control volume principles in detail, and to only summarize the first part of Chapter 7, on Navier–Stokes equations, before covering potential and boundary layer flows. Alternatively, one may choose to summarize needed features of the introductory Chapter 5, and then proceed to detailed coverage of the Navier–Stokes equations, by utilizing Chapter 6 as an application chapter to these equations. The author found the former approach suitable for second- to third-year student classes, and the latter approach more appealing for third- to fourth-year student classes. Chapters 8 and 9 are preferred for process- and materials-oriented classes, whereas Chapters 7 and 10 are essential to hydraulic and mechanical engineering applications.

I would like to thank several people in education and research that led to the conception, realization, improvement and completion of this book: My primary school teacher George Maratheftis; my high school physics instructor Andreas Stylianides; my transport phenomena instructor at the National Technical University of Athens, Prof. Nikolaos Koumoutsos; my graduate instructors and advisors at the University of Minnesota, Profs. L. E. Scriven and C. W. Macosko; my chairman Prof. H. S. Fogler and co-instructors, Profs. J. O. Wilkes and S. G. Bike, of the University of Michigan; my former students and now Prof. Andreas Alexandrou of Worcester Polytechnic, Prof. Rose Wesson-Williams of LSU, Dr. Zhao Chen of Eastern Michigan University, Dr. A. Nitin of Ford Motor

Company and adjunct Professor of the University of Michigan, Dr. K. Ellwood of Ford Motor Company, Dr. Mehdi Alaie of the University of Teheran, and Dr Nikolaos Malamataris of the Aristotle University of Thessaloniki; Ms. Ingrid Shriner-Ward and Dr. Androula Papanastasiou for the accurate and prompt completion of text and illustrations; all colleagues who kindly provided text and illustrations; the reviewers of drafts of the book, particularly Dr. Wasden (and Prof. N. Amundson) of the University of Houston for his detailed review and proofreading of the final text; and finally, the editors of the book, Mr. Michael Hays and Ms. Betty Sun of Prentice-Hall for attending to the entire project.

I am grateful to the entire faculty of the Chemical Engineering Department of the Aristotle University of Thessaloniki, particularly to Prof. C. Kiparissidis and former chairman A. Anagnostopoulos; and primarily to Prof. I. Vassalos, Director of the Chemical Process Engineering Research Institute at Thessaloniki, for securing the completion and publication of this project.

Tasos C. Papanastasiou

# *About the Author*

Tasos C. Papanastasiou is an Associate Professor at the Department of Chemical Engineering of the Aristotle University of Thessaloniki and the Chemical Process Engineering Research Institute at Thessaloniki, Greece. He holds a Diploma in Chemical Engineering from the National Technical University of Athens, and a M.Sc. degree in the area of environmental engineering and a Ph.D. degree in the area of fluid mechanics and rheology from the University of Minnesota. He is the author or co-author of over forty research publications in the area of fluid mechanics and coating flows, rheology and polymer and composite processing, computational mechanics and finite element methods, and environmental pollution transfer and control. He has developed and taught courses in fluid mechanics, polymer science and technology, computational mechanics and rheology, and engineering mathematics since 1985. The book *Applied Fluid Mechanics* combines his experience as an undergraduate and a graduate engineering student with early interests in fluid mechanics, as a production engineer and consultant to materials processing industries, and as a teacher of related courses to undergraduate and graduate students at the University of Michigan (1985–1992) and the Aristotle University of Thessaloniki (1992– ), and to industry engineers at Dow Chemical, Midland (1989–1992) and through the EEC 1993 COMETT "Flowtrend" Training Modules Project. His work is cited frequently in the literature and included in the 1992 "Who's Who in Science and Engineering in America".

# 1

# *Introduction to Fluid Mechanics*

## 1.1 CONTINUUM FLUIDS AND MECHANICS

Fluids, as the other forms of matter, such as solids, are discontinuous in nature, consisting of molecules. However, in contrast to solids, fluid molecules possess a high degree of motion freedom, such that they tend to assume the shape and occasionally occupy the volume of their containers. *Fluids* may include (1) *gases* of the highest degree of molecular motion freedom and lightest molecular weight such that they occupy the volume of their container; (2) *liquids* of retarded yet enough molecular motion to assume the shape but not occupy the entire volume of their container under gravity; (3) any other form of matter that can be made to behave as a fluid, for example, *molten metals and polymer melts* and *granular solids* (e.g., sand, wheat, etc.), in which case the making unit is the *grain*; (4) mixtures of two or more of the foregoing three. All of these forms of discontinuous matter are characterized by the ability of each of the constituting units to change its position with respect to its neighbors, which distinguishes them from plain solids. Due to these properties, *fluids are defined as matter than cannot support shear forces of any magnitude without continuous deformation and flow (i.e., relative motion).* Of course, fluids deform and flow under normal forces as

1

well, but so do solids, which, in contrast, can resist shear forces without any continuous deformation or flow, up to a certain level of shear force of structure collapse.

The vast majority of industrial and natural processes of interest to engineers involve length scales significantly larger than molecular dimensions and times significantly longer than molecular translation and vibration times. As a result, conclusions can be obtained by considering infinitesimally small *fluid particles* made up of enough molecules (or grains in case of granular matter) to allow statistical average properties for each fluid particle: The property of a fluid particle at a certain time (e.g., position in space, weight, temperature, velocity, and others) is the average of the properties of the molecules (or grains) making up the fluid particle at that time. Thus, each fluid particle is identified with a particular position in space at a particular time, which is about the geometric average of the positions of its molecules at that time. In this way, a fluid that is discontinuous on a molecular level is treated macroscopically as a continuum medium of fluid particles stacked together. Because different fluid particles occupy different parts of fluid space, each fluid property varies continuously in space and in time, from particle to particle. Under this *continuum hypothesis,* it is assumed that fluids are made up of particles stacked together without any vacuum or discontinuities in between. Each particle is a collection of different molecules and therefore has different properties, in such a way that a *continuous property distribution* in space and time results. By this continuous approximation of matter, formidable molecular calculations are avoided without sacrificing any practical accuracy. Indeed, for all practical purposes, the motion and position of individual molecules in a fluid container, for example, is of no interest, since the volume, weight, wall pressure, and other practical properties can be calculated from continuum principles.

*Properties* are conditions that characterize a state. Fluids exhibit extensive and intensive properties. *Extensive* properties depend on the amount of fluid: for example, mass, weight, volume, internal energy, and enthalpy. *Intensive* properties do not depend on the amount of fluid: for example, pressure, density, temperature, concentration, and specific heat. Pressure, temperature, and density are the most important intensive properties, and fully characterize equilibrium states of static fluids. In homogeneous fluids, pressure and temperature alone characterize equilibrium states. *Temperature,* symbolized by $T$, is a measure of thermal energy, and may vary with position and time. This dependence is expressed mathematically by $T(x, y, z, t)$. *Pressure,* symbolized by $p$, is the ratio of a *normal force,* $F_n$, acting on a surface by the area of the surface, $A$:

$$p = \frac{F_n}{A}. \tag{1.1}$$

Pressure is always directed toward the surface (i.e., it is a *normal stress*). A force that acts parallel to a surface induces *shear stress* on it, which is defined as equal to the magnitude of the force divided by the area of the surface, as explained later in this chapter. Under equilibrium conditions, pressure results from random molecular collisions with the surface and is also called *thermodynamic pressure.* Under flow conditions the resulting pressure, due to directed molecular collisions with the surface, is different from the thermodynamic pressure and is called *mechanical pressure.* The *thermodynamic* or *equilibrium pressure* can be determined from *equations of state,* such as the *ideal gas law* for gases and the *van der Waals equation* for liquids. However, the *mechanical pressure* can be determined only by means of energy-like *conservation equations* that take into account not just potential and thermal energy associated with equilibrium, but also kinetic energy associated with flow and deformation. The pressure is also, in general, a function of position and time [i.e., $p = p(x, y, z, t)$].

A case of position-dependent pressure is the *barotropic or hydrostatic pressure* in fluids at equilibrium under gravity, such as atmospheric air and ocean water, given by

$$p(z) = p_0 - \rho g(z - z_0), \tag{1.2}$$

where $p_0$ is the pressure at elevation $z_0$, $\rho$ the density of the fluid, $g$ ($= 981 \text{ cm/s}^2$) the acceleration of gravity, and $z$ the vertical distance over $z_0$. Equation (1.2) is utilized by nanometers to measure pressure, as shown in Fig. 2.4.

## 1.2 DENSITY AND EQUATIONS OF STATE

A fundamental property of continuum is the mass density. Density, symbolized by $\rho$, is the ratio of the mass $m$ to the volume $V$, occupied by a fluid when the volume shrinks to size $V_0$ significantly larger than that corresponding to molecular dimension, $l_m$:

$$\rho = \lim \frac{m}{V}, \qquad V \to V \gg O(l_m^3), \tag{1.3}$$

where $O(l^3)$ in general denotes volume *on the order of* $l_m^3$. The units of measure of density are those of mass divided by cubic length. For example, the density of water at 4°C and 1 atm is about $1 \text{ g/cm}^3$, or $1000 \text{ kg/m}^3$, or $62.4 \text{ lb/ft}^3$. Because molecules of different states and properties are included in different fluid particles of different positions at a given time, density, as any other property, is in general a function of time and position [i.e., $\rho = \rho(x, y, z, t)$]. Alternatively, since a particular temperature $T$, pressure $p$, and composition or concentration $c$ (in case of nonhomogeneous fluid) prevail at each position and time under conditions of local thermodynamic equilibrium, relations of the form $T = T(x, y, z, t)$, $p = p(x, y, z, t)$, and $c = (x, y, z, t)$ exist. Then the density of a fluid of molecular weight $M$ may also be considered as a unique function of temperature, pressure, and composition:

$$\rho = \rho(T, p, c, M), \tag{1.4}$$

and becomes

$$\rho = \rho(T, p, M), \tag{1.5}$$

for a purely homogeneous fluid. Equation (1.5) is an *equation of state at equilibrium, which relates deformation or strain or arrangement of matter in space defined by the density, to pressure stress at a given temperature.* Examples of specific equations are (1) the *ideal gas law,*

$$\rho = \frac{pM}{RT}; \tag{1.6}$$

(2) the *van der Waals equation* for liquids,

$$\left(p + \frac{\alpha}{V_m^2}\right)(V_m - b) = RT, \qquad \rho = \frac{m}{V} = \frac{nM}{V} = \frac{M}{V_m}, \tag{1.7}$$

where $R$ is the ideal gas constant equal to 10.73 $(lb_f \, ft^3)/(in^2 \, lbmol \, K)$, and equivalently, to 0.73 $(atm \, ft^3)/(lbmol \, K)$, or 0.083 $(lb \, atm)/(mol \, K)$, or 1.986 $Btu/(lbmol \, K)$, $M$ the molecular weight, $n$ the number of moles, $V_m$ the molar volume, $b$ the volume occupied by molecules, and $\alpha$ is a pressure correction factor; (3) the *linear nonisothermal liquid law,*

$$\rho = \rho_0 - \alpha(T - T_0), \tag{1.8}$$

where $\rho_0$ is density at a reference temperature $T_0$ and $\alpha$ is a volume expansion coefficient; and (4) the *incompressible liquid relation,*

$$\rho = \rho_0, \tag{1.9}$$

which holds for gases under isobaric and isothermal conditions and for liquids under isothermal conditions. *Incompressible fluids* that follow Eqn. (1.9) are the primary target of this book.

*Specific gravity, s,* is the ratio of density to that of a reference fluid. Such reference fluids are water at 4°C and 1 atm of density 1 $g/cm^3$ for liquids, and air at 1 atm and 60°F of density 1.22 $kg/m^3$ for gases.

*Compressibility* is a measure of the changes in volume, and therefore in density, of a certain mass of fluid, in response to changes in the normal forces acting on the volume. It is defined by

$$e_v = -\frac{1}{V}\left(\frac{\partial V}{\partial p}\right)_T. \tag{1.10}$$

Water, under ambient conditions, exhibits $e_v \simeq 5 \times 10^{-6}\,cm^2/N$, compared to that of steel, $e_v \simeq 5 \times 10^{-8}\,cm^2/N$. For air, $e_v \simeq 10\,cm^2/N$, as can be computed by

$$e_v = -\frac{1}{V}\left(\frac{\partial V}{\partial p}\right)_T = -\frac{1}{V}\frac{\partial}{\partial p}\left(\frac{nRT}{\partial p}\right)_T = -\frac{nRT}{V}\frac{\partial}{\partial p}\left(\frac{1}{p}\right)$$

(1.11)

$$= \frac{p}{p^2} = 1\,atm^{-1} = 10\,cm^2/Nt.$$

Thus water is 100 times as compressible as steel and 0.00005 times as atmospheric air. For these reasons, liquids under isothermal conditions are considered practically incompressible unless subjected to huge pressure changes. Compressibility effects do become important for long-distance gas (and gas–liquid) pipelines common to the petroleum industries (e.g., Section 4.11), and in petrochemical unit operations such as furnaces, crackers and reactors, where fluids encounter large pressure and temperature variations.

Compressibility is perhaps the most important fluid property, and in fact *fluid mechanics,* the science that studies the mechanics of equilibrium and motion of fluids, has been developed in several directions, depending on the density of the subject fluid: *hydromechanics* or *fluid mechanics* deals with incompressible liquids, at equilibrium (*hydrostatics*) or under flow (*fluid dynamics*); *gas mechanics* deals with compressible gases under flow (*gas dynamics*) and with compressible atmospheric air (*meteorology*). In general, there are not an apriori compressible or incompressible fluids. The very same fluid may behave as incompressible under certain pressure, temperature, and flow conditions, and as compressible under a different combination of these conditions, as Eqns (1.7) to (1.10) suggest. Figures 1.3 and 1.4 list values of density of air and water at selected conditions of temperature and pressure. Density values of other selected materials are tabulated in Tables 1.3 and 1.4.

**Example 1.1: Pressure Calculations**

Calculate the pressure in the following cases of equilibrium.

(a) In atmospheric air as a function of elevation $z$ measured in meters, if the temperature changes with elevation according to $T(z) = (25 - 0.01z)°C$, and the density according to $\rho(z) = (1.18 - 0.01z)\,kg/m^3$.

(b) On the walls of a balloon containing 0.25 kg of air at altitude $z = 100$ m.

(c) At the bottom of a cylindrical tank of radius $R = 1$ m containing $6.28 \times 10^3$ kg water at 4°C.

(d) At the cylindrical wall of the same tank.

(e) Under the 36-in$^2$ area of the shoe of a 170-lb man.

(f) Under the 3-in$^2$-area shoe of a hockey player of 170 lb weight, on ice.

**Solution**

**(a)** Air in this case can be taken as an ideal gas mixture of roughly 21% oxygen and 79% nitrogen, of molecular weight

$$M = (0.21 \times 32 + 0.79 \times 28) \text{ g/mol} = 29 \text{ g/mol} = 29 \text{ kg/mol}.$$

Thus the pressure is given by Eqn. (1.6):

$$p(z) = \frac{\rho(z)T(z)r}{M}$$

$$= \frac{(1.18 - 0.001z) \text{ kg/m}^3 \times (25 - 0.01z) \text{ K} \times 0.082 \text{ at L/(mol K)}}{29 \text{ g/mol}}$$

$$= (1.18 - 0.001z) \text{ g/L} \times (198 - 0.01) \text{ K} \times (0.082/29) \text{ atm L/(g K)}$$

$$= (1 - 0.0008z) \text{ atm}.$$

Thus the atmospheric pressure decreases with elevation from 1 atm at sea level of $z = 0$ to 0.16 atm at altitude $z = 2$ km.

**(b)** The external pressure from the surrounding air is

$$p(z = 100 \text{ m}) = (1 - 0.0008 \times 100) \text{ atm} = 0.92 \text{ atm}.$$

Since the balloon is at equilibrium, the internal pressure is everywhere identical to 0.92 atm. Under these conditions, the volume of the balloon is

$$V = \frac{nRT}{p} = \frac{mRT}{Mp}$$

$$= \frac{250 \text{ g} \times 0.082 \text{ L} \cdot \text{atm/mol} \cdot \text{K} (298 - 0.01 \times 100) \text{ K}}{29 \text{ g/mol} \times 0.92 \text{ atm}} = 228 \text{ L}.$$

**(c)** The weight of the water is

$$B = mg = 6280 \, kg \times 9.81 \text{ m/s}^2 = 61{,}606 \text{ N}.$$

This weight is supported by the bottom, of area

$$A = \pi R^2 = 3.14 \times 1 \text{ m}^2 = 3.14 \text{ m}^2,$$

and the resulting pressure on the bottom is

$$p = p_{\text{atm}} + \frac{B}{A} = 1 \text{ atm} + \frac{61{,}606 \text{ N}}{3.14 \text{ m}^2} = 1 \text{ atm} + 19{,}620 \text{ Pa} = 1.2 \text{ atm}.$$

**(d)** The pressure on the cylindrical wall depends on the depth of the water there, and it is identical to the weight force of the water above it. Thus the pressure right at the free surface line is identical to the atmospheric pressure. The pressure at half the depth of the water is

$$p = p_{\text{atm}} + \frac{0.5B}{A} = 1.1 \text{ atm},$$

and at two-thirds the depth is

$$p = p_{\text{atm}} + \frac{0.66B}{A} = 1.133 \text{ atm}.$$

**(e)** By the same arguments,

$$p = \frac{B}{A} = \frac{mg}{A} = \frac{170 \text{ lb} \times 32.17 \text{ ft/s}^2}{2 \times 36 \text{ in}^2} = 75.67 \text{ psi} = 5.34 \text{ atm}.$$

**(f)** Similarly,

$$p = \frac{B}{A} = \frac{mg}{2A} = \frac{170 \text{ lb} \times 32.17 \text{ ft/s}^2}{2 \times 3 \text{ in}^2} = 454 \text{ psi} = 30.9 \text{ atm}.$$

The huge concentration of pressure under the hockey shoe melts the ice underneath and a thin liquid water film is induced, which lubricates the player's motion on ice. This water film freezes back into ice upon departure of the shoe and thus eliminates the high pressure.

**Example 1.2: Density Calculations**

Calculate the density of the following substances at ambient conditions of $T = 20°C$ and $p = 1$ atm:

**(a)** steel;

**(b)** clean water;

**(c)** salt ocean water with 0.01 g/cm$^3$ of NaCl;

**(d)** Dry air;

**(e)** air of 30 g/m$^3$ humidity;

**(f)** granular iron of spherical grains of average diameter $d = 1$ mm packed in a 1-grain/mm$^3$ arrangement.

**Solution**

**(a)** From the tabulated data of Table 1.3, $\rho_s = 7.6 \text{ g/cm}^3$.

**(b)** From the tabulated data of Table 1.3, $\rho_\omega = 0.998 \text{ g/cm}^3$.

**(c)** Besides tabulated data, a simple expression could be

$$\rho_{sw} = \frac{m_{sw}}{V_{sw}} = \frac{m_w + m_s}{V_{sw}} = \frac{(1 + 0.01) \text{ g}}{1 \text{ cm}^2} = 1.01 \text{ g/cm}^3,$$

under the assumption of negligible volume change in mixing.

**(d)** Besides tabulated data, the ideal gas law may be used:

$$\rho = \frac{mp}{RT} = \frac{29 \text{ g/mol} \times 1 \text{ atm}}{0.082 \text{ L atm/mol K}) \times 293 \text{ K}} = 1.207 \text{ g/L} = 1.207 \text{ kg/m}^3.$$

**(e)** Along the lines of part (c):

$$\rho = \frac{m}{V} = \frac{m_{\text{air}} + m_w}{V_{\text{air}}} = (1.207 + 0.03) \text{ kg/m}^3 = 1.237 \text{ kg/m}^3.$$

**(f)** Along the lines of part (c), on a basis of $V = 1 \text{ mm}^3$:

$$\rho_{gi} = \frac{m}{V} \times \frac{m_{iron} + m_{air}}{V} = \frac{1.33\pi(d^3/8)\rho_{iron}}{1 \text{ mm}^3} + \frac{1 - 1.33\pi(d^3/8)\rho_{air}}{1 \text{ mm}^3}$$

$$= \frac{(0.52 \times 9 + 0.48 \times 0.001) \text{ mm}^3 \times \text{g/cm}^3}{1 \text{ mm}^3} = 4.68 \text{ g/cm}^3.$$

This is to be compared with the $9 \text{ g/cm}^3$ of the corresponding solid iron.

The conclusion here is that densities of solids and molten metals are much higher than those of liquids and the latter much higher than those of gases. This is a consequence of spacing of molecules in each of the three categories. The density of granular solids is reduced compared to that of the corresponding compact solids, the more so the larger the air void between grains.

## 1.3 VISCOSITY AND VISCOUS STRESSES

Most phenomena in nature behave according to the funadamental relation

$$\begin{pmatrix} \text{result,} \\ \text{or fiux} \end{pmatrix} = \begin{pmatrix} \text{transfer coefficient,} \\ \text{or inverse resistance} \end{pmatrix} \begin{pmatrix} \text{cause,} \\ \text{or gradient} \end{pmatrix}. \qquad (1.12)$$

Heat flow results from a temperature difference or gradient through a medium of a certain coefficient of conductivity or inverse resistance. The same relation governs electric current through a wire of high conductivity and low resistance under a voltage difference or gradient. In a slightly different manner, yet according to Eqn. (1.12), pieces of ice and sleds and hockey shoes slip on ice due to an intervening thin film of liquid water, as shown in Fig. 1.1 and analyzed below.

As summarized in Example 1.1(f), the thin liquid film forms due to

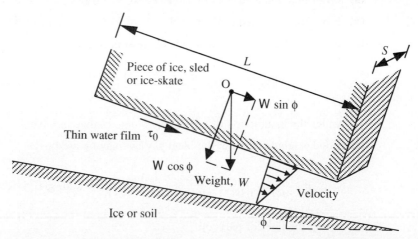

**Figure 1.1** Sliding devices slip on ice or soil by means of an intervening thin liquid water film formed under their weight, supported by the lower contact surface.

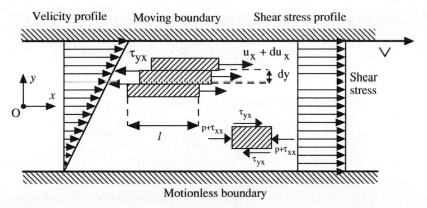

**Figure 1.2**  Flowing thin water film between stationary and moving boundaries.

excessive concentration of pressure or *normal stress,* which according to Eqn (1.1) is

$$p = \frac{B \cos \phi}{SL},$$  (1.13)

and flows in the direction of motion due to the *shear stress,*

$$\tau_0 = \frac{B \sin \phi}{LS},$$  (1.14)

induced simultaneously (see Example 1.3 for details). In a similar fashion, butter and honey spread thin on bread under the shearing action of a knife, and so does makeup on skin under the shearing action of fingers. Thus the application of an external shear stress, $\tau_0$, results in fluid flow, consistent with the fluid definition. In reference fo Eqn. (1.12), fluid flow is the result, the shear stress applied by the ice weight is the cause, and the transfer coefficient, or *fluidity* in this case, is a characteristic property of the fluid, identical to its inverse resistance to flow. The resistance to flow that inhibits fluidity is called *viscosity.* The induced *velocity profile* varies across the flowing fluid, as shown in Fig. 1.1 and magnified in Fig. 1.2.

The most important quantification of Eqn. (1.12) in fluid mechanics is *Newton's law of motion*[1]:

$$F = \frac{dJ}{dt} = \frac{d}{dt}(mu) = m\frac{du}{dt} + \frac{dm}{dt}u.$$  (1.15)

According to Eqn. (1.15), the action of any force $F$, on a mass $m$ traveling at velocity $u$, changes its *momentum, J,* defined by

$$J = mu,$$  (1.16)

at rate $dJ/dt$. In Eqn. (1.15), where $du$ is the induced velocity variation. For a constant mass, Newton's law of motion reduces to

$$F = \frac{d}{dt}(mu) = m\frac{du}{dt} + u\frac{dm}{dt} = m\frac{du}{dt}. \qquad (1.17)$$

If the unit of mass per volume, which is the density $\rho$, is used in Eqn. (1.15), the resulting Newton's law takes the form

$$\hat{F} = \frac{d\hat{J}}{dt} = \frac{d}{dt}(\rho u) = \rho\frac{du}{dt} + u\frac{d\rho}{dt}, \qquad (1.18)$$

where $\hat{F} = F/V = \rho F/m$ is force per unit volume. For incompressible fluids of $d\rho/dt = 0$, the last term of Eqn. (1.18) vanishes.

Thus when two fluid layers or particles moving with velocities that differ by $du = u_1 - u_2$ are brought in contact, which allows the exchange of $m$ mass between the two, a simultaneous momentum exchange results, at rate $m\,du/dt$. This exchange results in a contact force $F$ at the interface of contact, according to Eqn. (1.15), which is parallel to the geometrical velocity differential $du$. This interfacial force accelerates the slower moving layer and decelerates the faster, i.e., momentum is transmitted from high to low velocity regions. A relative example on a macroscopic scale is wind-induced circulation in the ocean, lakes, and cavities filled with water: Under no-wind conditions the water is in static equilibrium. Under wind blowing parallel to the free surface, mobile air molecules penetrate the free surface and initiate motion in a thin water layer adjacent to the free surface. Molecules from this mobile water layer penetrate the water layer underneath and set it in motion, and so on. The induced water flow is parallel to the wind direction. This is known as a drag-induced fluid motion, caused by a drag force from air to water along the interface of contact, and by the same mechanism from upper to lower water layers. Under a steady wind, this is an ongoing exchange of momentum that maintains a steady water flow. The induced velocity profile in the water is similar to that of Fig. 1.2, from the free surface where its velocity is identical to that of the wind (corresponding to the "moving boundary" of Fig. 1.2), to the bottom where the velocity is zero, identical to that of the "motionless boundary."

Consider now two parallel adjacent fluid layers of thickness $dy$, length $l$ and width $W$ in the normal to the paper direction moving along the $x$-direction with velocities $u_x$ and $u_x + du_x$, as shown in Fig. 1.2. The fluid molecules in each layer move randomly in all possible directions with average velocity $u_m$, calculated from molecular kinetic theory to be

$$u_m = \sqrt{\frac{8kT}{\pi m_m}}, \qquad (1.19)$$

where $k$ is the Boltzmann constant, $T$ the absolute temperature, and $m_m$ the molecular mass. This average velocity is in addition to the unidirectional velocities $u_x$ and $u_x + du_x$ that the molecules possess, riding each layer vehicle. The average velocity $u_m$ enables molecules to cross the interface of contact area $A = lW$ from one layer to another, at molecular mass flow rate $\dot{m}_m$ proportional to their average velocity $u_m$, the number of molecular collisions $N$ per unit area of contact, and the molecular mass $m_m$. In doing so, these molecules transfer net $x$-momentum across the contact area $A$ between the lower and upper layer given by

$$\dot{J}_x = A[m_m N(u_x + du_x)]u_m - A[m_m N u_x]u_m = A m_m N \, du_x u_m. \qquad (1.20)$$

This transfer changes the momentum of the layer of volume $A \, dy$ by

$$A \frac{dJ}{dt} dy = A m_m N \, du_x u_m \qquad (1.21)$$

and equivalently,

$$\frac{dJ}{dt} = m_m N u_m \frac{du_x}{dy}. \qquad (1.22)$$

In accordance with Eqn. (1.15), this momentum exchange results in the appearance of contact force $F$, per unit of contact area $A$, or *shear stress*, given by

$$\tau_{yx} = \frac{F}{A} = k \frac{1}{A} \frac{dJ}{dt} = \eta \frac{du_x}{dy}, \qquad (1.23)$$

where $k$ is a proportionality constant that depends on temperature and molecular mass and diameter. Equation (1.23) yields the resulting shear stress $\tau_{yx}$ in the $x$-direction on a surface perpendicular to the $y$-axis, resulting from a transverse velocity gradient, or *shear rate*, $\dot{\gamma} = du_x/dy$, in a fluid of viscosity $\eta$. For gases at low density, the viscosity in Eqn. (1.23) is calculated by kinetic theory to be[2]

$$\eta = \frac{2}{3\pi^{1.5}} \frac{\sqrt{mkT}}{d^2}, \qquad (1.24)$$

where $d$ is the molecular diameter.

   In the flows represented by Fig. 1.2, the velocity $u_x$ changes only vertically with respect to the flow direction, which gives rise to shear stresses. In a generalized fluid flow, the velocity may change in the direction of flow as well. Then, adjacent fluid layers in series in the flow direction exchange $x$-momentum ($m_m N u_x$) across their contact areas—perpendicular to the flow direction—by molecules crossing between layers of different velocities. By following the same procedure and considering the resulting change of $x$-momentum in a layer of thickness $dx$ due to exchanges with its upstream and downstream neighbors, *normal viscous stresses* $\tau_{xx}$ are shown to appear, given by

$$\tau_{xx} = 2\eta \frac{du_x}{dx}. \qquad (1.25)$$

Equation (1.25) yields the normal stresses developed in a flowing fluid of viscosity $\eta$, due to a streamline velocity gradient, or *extensional rate*, $\dot{\varepsilon} = du_x/dx$. This type of normal stress, which arises under conditions of relative flow and deformation, is called *viscous* to distinguish it from the pressure normal stress that is present even under no flow at all. Of course, in flow situations the action of the two is additive, yielding the *total normal stress.* Since there is always one type of shear stress under flow conditions alone, its "viscous" characterizaion is omitted. The two subscripts of a viscous normal stress are identical, because the direction of its action is parallel to the normal direction to the surface of action. In contrast, the subscripts of a shear stress cannot be the same because the two directions they represent are not identical; in fact, they are perpendicular.

Due to the sign convention used in Eqns. (1.23) and (1.25), the calculated viscous stresses are those imposed by the farther to the closer of the origin layer: i.e., shear stress from the layer at $y + dy$ to the layer at $y$, and normal stress from the layer at $x + dx$ to the layer at $x$. When the opposite sign is used in Eqns. (1.23) and (1.25), the stress from the closer to the farther layer is calculated.[2]

Figure 1.2 also shows a fluid volume in a flow, with the stresses acting in the $x$-direction. Due to the highlighted mechanism of momentum exchange with all its adjacent layers, there are a pair of shear stresses $\tau_{yx}$, and a pair of total normal stresses $\tau_{xx} + p$, made up by viscous and pressure normal stresses.

Equations (1.23) and (1.25), both conform to the general *constitutive equation*

$$\tau_{ij} = \eta\left(\frac{\partial u_i}{\partial x_j} + \frac{\partial u_i}{\partial x_j}\right), \qquad i = x, y \tag{1.26}$$

for a two-dimensional flow of two velocity components $u_x$ and $u_y$, changing with respect to both directions $x$ and $y$. Equation (1.26) is the celebrated *Newton's law of viscosity*. If Eqn. (1.26) is reversed to the form

$$\left(\frac{\partial u_i}{\partial x_j} + \frac{\partial u_j}{\partial x_i}\right) = \frac{1}{\eta}\,\tau_{ij} \tag{1.27}$$

viscosity is shown to measure resistance of adjacent fluid parts to flow at different velocities; that is, *viscosity is resistance exhibited by a fluid to relative flow and deformation,* imposed by viscous stresses. Of course, the viscosity of fluids is finite and so is their resistance to flow, and therefore ultimately, fluids flow under both shear and normal stress. Along these lines, solids may be viewed as fluids of infinite viscosity and resistance to flow. In fact, there exists a class of materials, called Bingham plastics, that at low shear stresses behave like solids of infinite viscosity; however once a certain value of yield stress is exceeded, they start to flow and deform like fluids of finite viscosity. Thus viscosity is related directly to the freedom of motion of molecules and therefore to their ability to travel across interfaces, causing an exchange of momentum. Therefore, viscosities of gases are much smaller than those of liquids and of the latter much smaller than the practically infinite viscosities of solids. Another consequence of the momentum exchange mechanism that controls viscosity is the fact that the viscosities of liquids decrease with temperature, which enhances molecular motion according to Arrhenius-type equations,

$$\eta(T) = A \exp\left(-\frac{B}{RT}\right), \tag{1.28}$$

and increase with excessive pressure that retards molecular motion, according to equations of the form,

$$\eta(p) = C \exp(Dp),\qquad(1.29)$$

where $A$, $B$, $C$, and $D$ are characteristic constants of the liquid. Under common processing pressure conditions, $D = 0$ and the viscosity of liquids is practically independent of pressure. For example, the viscosity of toluene at 30°C changes from 5220 $\mu$P to 8120 $\mu$P when the pressure changes from 0.1 MPa to 63.5 MPa, and from 4215 $\mu$P to 6840 $\mu$P at 50°C. The corresponding changes for $n$-heptane are from 3700 to 6300 $\mu$P and from 3040 to 5310 $\mu$P, respectively.[3] The excessively high freedom of molecular motion in gases causes temperature and pressure to have the reverse effect on gas viscosity, since temperature leads to an increasing number of molecular collisions and contact, whereas pressure retards them. For example, air at 30°C changes its viscosity from 188 $\mu$P to 200 $\mu$P for pressures of 1 and 60 atm.

The ratio of viscosity to density is the *kinematic viscosity, v,* defined by

$$v = \frac{\eta}{\rho}.\qquad(1.30)$$

As explained in Chapter 5, the kinematic viscosity is also a vorticity transfer coefficient, which determines how fast a shear stress signal propagates into fluids. Figures 1.3 and 1.4 illustrate how the viscosity of water and air changes with temperature for three different values of pressure.

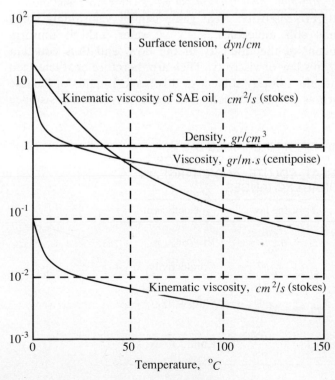

Figure 1.3  Viscosity, kinematic viscosity, and density of air at atmospheric pressure and different temperatures.

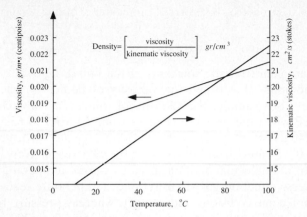

**Figure 1.4** Viscosity, density, kinematic viscosity, and surface tension (with air) of water at atmospheric pressure and different temperatures.

Newtons's law of viscosity is similar to Fourier's law of heat conduction and to Fick's law of mass diffusion, as illustrated in Table 1.1. The signs of these laws account for the fact that transfer of heat, solute, charge, and stress occurs (i.e., is positive) from regions of high to low temperature, concentration, charge density, and velocity, respectively. Of course, a positive stress from part A to B is necessarily accompanied by a negative stress from B to A, in accordance with *action–reaction* principles.

Viscosity is perhaps the most important property of flowing fluids, which are therefore characterized accordingly. *Ideal frictionless gases* offer no resistance to flow and therefore exhibit zero viscosity. Real gases exhibit vanishingly small viscosities. Liquids of small stiff molecules, such as water, exhibit uniform resistance to flow independent of the intensity of the flow, and thus constant viscosity according to Newton's law of viscosity. They are therefore characterized as *Newtonian liquids* and are the primary subject of this book. Liquids of flexible elastic macromolecules, such as polymeric melts and solutions, exhibit resistance

**TABLE 1.1**   TRANSFER MECHANICS BY DIFFUSION OR CONDUCTION OF STRESS OR MOMENTUM, HEAT, SOLUTE, AND CHARGE, IN ONE-DIMENSIONAL EXCHANGES

| Transfer of: | Transfer law | Coefficient |
|---|---|---|
| Stress | $\tau_{ij} = \eta \dfrac{\partial u_i}{\partial x_j}$ | Viscosity, $\eta$ |
| Heat | $\dot{H}_c = -k \dfrac{\partial T}{\partial x_i}$ | Conductivity, $k$ |
| Solute | $\dot{m}_c = -\mathscr{D} \dfrac{\partial c}{\partial x_i}$ | Diffusivity, $D$ |
| Charge | $I = -k_i \dfrac{\partial V}{\partial x_i}$ | Conductivity, $k_i$ |

**TABLE 1.2**  VISCOSITY CLASSES OF FLUIDS

| Newtonian | Shear thinning | Shear thickening | Bingham |
|---|---|---|---|
| Air | Polymer melts | Suspension | Ketchup |
| Water | Polymer solutions | Emulsions | Mayonnaise |
| Oils | Foams | Dispersions | Butter |
| Lubricants | Paint | Blends | Paint |
| Glycerin | Food sauces | | Mud, sludge |
| Monomers | | | Foams |
| Molten metal | | | Filled rubber |
| Honey | | | Waxes |
| Milk | | | Blood |

to flow and viscosity that change with the intensity of the deformation expressed by the velocity gradient or derivative, called the *shear rate*. This is due to their ability to orient themselves along the flow direction, reducing resistance to flow and viscosity, as illustrated in Fig. 9.1. These liquids are called *shear-thinning* liquids. On the other hand, suspensions and emulsions are made up by discontinuous phases in the form of solid or liquid sphere-like particles dispersed in a continuous fluid phase. Thin films of the continuous phase lubricate and facilitate relative motion of these particles, in general, therefore reducing viscosity. At high shear rates, however, this kind of lubrication breaks down, freedom of motion is highly retarded due to particle–particle contact and friction, and the viscosity increases. These liquids are characterized as *shear-thickening* liquids. Finally, a particular class of liquids, including mostly gell, sauce, and paint-like thick liquids called *Bingham fluids* or *viscoplastics*, flow only after being subjected to shear stresses that exceed a characteristic value, called *yield stress*.

Table 1.2 lists representative fluids belonging to each of these viscosity classes, and Fig. 1.5 illustrates how viscosity may change with the shear rate. Tables 1.3 and 1.4 list densities, viscosities, and kinematic viscosities of selected fluids at ambient or other selected conditions. *Non-Newtonian fluids* in general

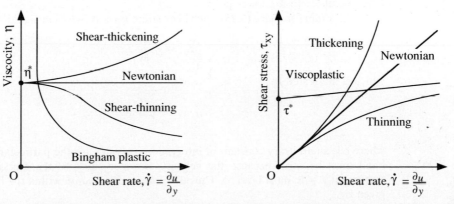

**Figure 1.5**   Viscosity and shear stress as functions of shear rate, $\dot{\gamma} = \partial u_x/\partial y$, of Newtonian liquid of viscosity $\eta^*$, of shear thinning and thickeneing liquids, and of Bingham plastic of yield stress $\tau^*$.

are the topic of Chapter 9, and their flow, deformation, and processing are examined exclusively in non-Newtonian fluid mechanics, rheology, and polymer processing publications (e.g., Refs. 4, 5 and 6).

### Example 1.3: Viscosity and Shear Stress

Analyze the process of Fig. 1.1 for an ice–water system by means of the equations of Section 1.1.2. Consider a 10-cm-long, 3-m-wide, 1-m-tall piece of ice, sliding over a slightly inclined ice surface, at angle $\phi = 2°$ with the horizontal, in the presence of a thin liquid water film of thickness 1 mm, at temperature 0°C.

**Solution**   At temperature 0°C the density of ice is $\rho_i = 0.915 \, \text{g/cm}^3$, and that of water is $\rho_w = 0.999 \, \text{g/cm}^3$. The viscosity of water is $\eta = 1.787 \, \text{cP}$. Thus the weight of the ice is

$$B = \rho g V = 915 \, \text{kg/m}^3 \times 9.81 \, \text{m/s}^2 \times 30 \, \text{m}^3 = 269{,}285 \, \text{N}.$$

The component of the weight perpendicular to the sliding plane is

$$B_y = B \cos 2° = (269{,}285 \times 0.999) \, \text{N} = 269{,}223 \, \text{N},$$

and therefore the prevailing pressure underneath the ice piece is

$$p = \frac{B_y}{A} = \frac{269{,}223 \, \text{N}}{30 \, \text{m}^2} = 8974 \, \text{Pa} = 0.088 \, \text{atm}.$$

The component of the weight that shears the liquid water film is

$$B_x = B \sin 2° = (269{,}285 \times 0.0349) \, \text{N} = 9395 \, \text{N},$$

and the resulting shear stress through the film is

$$\tau_{yx} = \frac{B_x}{A} = \frac{9395 \, \text{N}}{30 \, \text{m}^2} = 313 \, \text{Pa}$$

from the upper to the lower layer.

According to Eqn. (1.26), the shear stress for a Newtonian liquid is given by

$$\tau_{yx} = \eta \, \frac{\partial u_x}{\partial y},$$

which is integrated to

$$u_x = \frac{\tau_{yx}}{\eta} y + c,$$

where $c$ is an arbitrary constant of integration. Its value for the particular problem at hand must be zero, because the velocity of water stuck to the motionless lower surface, at $y = 0$, must be zero. Thus the *velocity distribution* within the liquid film is given by

$$u_x(y) = \frac{\tau_{yx}}{\eta} y = \frac{313}{0.001787} y = (175{,}000 y) \, \text{m/s}.$$

The sliding speed, $U$, of the iceberg, located a distance $H = 0.0005$ m over the motionless surface, is

$$U = u_x(y = 0.0005 \text{ m}) = (175{,}000 \times 0.0005 \text{ m}) \text{ s}^{-1} = 87.6 \text{ m/s}.$$

In reality, these results are altered by surface roughness and uneven liquid film thickness similar to those arising in *lubrication applications*[7] and by heat generation by *viscous dissipation* of the highly shearing sliding motion.[8] Internal velocity distribution in flowing fluids is the topic of Chapters 5 and 6, and lubrication-like thin films is the topic of Chapter 8.

## 1.4 SURFACE TENSION AND CAPILLARY PRESSURE

Surface tension is a thermodynamic boundary property, which relates the anisotropy of the interactions between liquid molecules at the boundary and in the bulk, as depicted by Fig. 1.6. For molecules well in the interior, interactions are isotropic, and the resulting net force on each liquid molecule vanishes. In contrast, molecules at the interface of a liquid with gas (or with another immiscible liquid) are attracted more by molecules in the interior of the liquid than by gas molecules across the interface, such that a nonzero net force from interface to the interior fluid results. Since this force comes from all possible directions of interface to the center, it turns the interface inward to nearly spherical shapes of interface. In the absence of any other form of appreciable force, a spherical interface results, a characteristic example being that of spontaneous formation of spherical droplets of mercury. This resulting inward force, combined with the volume-change resisting incompressibility, causes a membrane-like tension called *surface tension, $\sigma$,* in the tangential direction. Thus the surface tension acts as force per unit length of interfacial area.

From the thermodynamics point of view, if we multiply the *force/length* dimension of surface tension by *length/length*, surface tension is also *surface energy per unit interfacial area*. This energy is stored when an external force $F$ spends work $F\,dx$ to create a new amount of interfacial area, $dA$, according to the relation

$$F\,dx = \sigma\,dA. \tag{1.31}$$

It is known from thermodynamics that isolated systems at equilibrium tend toward states of minimum energy. Thus once the external force $F$ is eliminated, the system tends spontaneously to minimize its surface energy. Under isothermal conditions the surface tension is constant, and therefore minimization is achieved by minimizing the interfacial area $A$.

At equilibrium, there is a pressure jump, $p_A - p_B$, across the interface of two immiscible fluids, balanced by an equivalent *capillary pressure* due to the surface tension, according to

$$p_B - p_A = \frac{2\sigma \sin d\phi}{ds} = 2H\sigma, \tag{1.32}$$

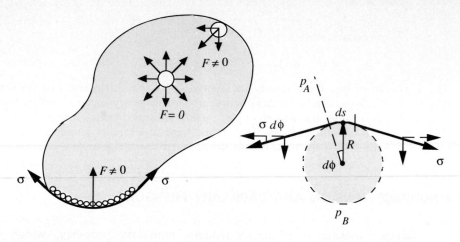

**Figure 1.6**  Physical mechanism of surface tension and the Young–Laplace equation.

as shown in Fig. 1.6. In Eqn. (1.32), $2H$ is the *curvature* of the interface, defined here as

$$2H = \frac{2 \sin \phi}{dS} = \frac{1}{R},$$  (1.33)

where $R$ is the radius of the maximum circle tangential to the line segment $dS$ at the point. These principles are easily generalized to two-dimensional surfaces (e.g., spherical droplet) to produce the celebrated Young–Laplace equation of capillarity,[9]

$$\Delta p = \sigma \left( \frac{1}{R_1} + \frac{1}{R_2} \right),$$  (1.34)

where $R_1$ and $R_2$ are the radii of the two mutually perpendicular maximum circles which are tangent to the two-dimensional surface at the point of contact. Equation (1.34) suggests that the units of surface tension are those of force per unit length: for example, dyn/cm, N/m, and others.

Curvatures of selected interfacial configurations include (1) plane, $R_1 = R_2 = \infty$ and $2H = 0$; (2) sphere, $R_1 = R_2 = R$ and $2H = 2/R$; and (3) cylinder, $R_1 = \infty$, $R_2 = R$, and $2H = 1/R$.

A good analog for surface tension and the Young–Laplace equation is an inflated balloon: $\Delta p$ is the pressure differential inside and outside the balloon, $R_1 = R_2 = R$ its radius of curvature, and $\sigma$ the tension of its thin elastic wall. Indeed, the situation is almost identical to a gas bubble in equilibrium within a liquid, discussed in Example 1.4. From the thermodynamics point of view, a force

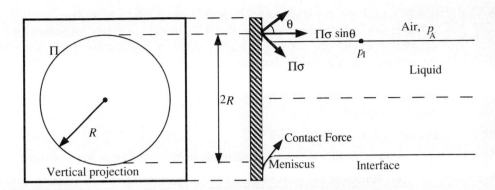

**Figure 1.7**  Contact angle $\theta$ and contact line $\Pi$ of a liquid cylinder to a solid plate.

$F = p_i S$, where $p_i$ is inflating pressure and $S$ the balloon neck's cross-sectional area, is spent to grow the balloon, creating a new quantity of balloon surface $dA$ and therefore storing $\sigma \, dA$ surface energy in the stretched wall. After elimination of inflation, the spontaneous tendency of the balloon is deflation, which reduces both its surface area and its surface energy.

Surface tension controls the angle of attachment of liquids to solids, known as the *contact angle*, as shown in Fig. 1.7. A liquid-free surface or interface, at *contact angle* $\theta$ defined between the two vertical lines to the solid and to the fluid interface there, results in a total force

$$F = \Pi \sigma \sin \theta = 2\pi R \sigma \sin \theta, \qquad (1.35)$$

from solid to fluid, where $\Pi$ is the *contact line* of the liquid–solid contact along which the membrane-like surface tension acts. This force causes liquids to rise or descend in narrow capillaries, for example, water to overcome gravity and advance upward through porous media (e.g., paper, cloth, sand, tree trunk, and branches).

Surface tension plays a significant role in a diversity of small-scale laminar flows, as well as in immiscible liquids under equilibrium. Liquid volumes tend to attain spherical shapes that exhibit the minimum surface-to-volume ratio, the more so the higher their surface tension. This finds applications in heat and solute exchange by means of liquid droplets and gas bubbles, in emulsion stabilization, in liquid spraying and atomization, and elsewhere. The tendency of surface tension to minimize interfacial area is of primary importance to the formation and stabilization of thin films (Chapter 8) with applications to coating[10] and to heat and solute exchange by thin liquid films.[11] Surface tension controls leveling and spreading of liquids on substrates with applications to spray coating or painting. There, gravity enhances spreading, which is opposed by surface tension and viscosity. The same mechanism prevails in enhanced oil recovery (Section 4.11.3), where significant amounts of residual oil remain trapped in the capillaries of the oil-reservoir, in the form of oil droplets under capillary pressure. Additional situations in which surface tension plays a significant role are examined throughout the remaining chapters.

Examples 1.4 and 1.5 highlight some applications of Eqns (1.31) to (1.34). Surface tension values of selected liquids are given in Tables 1.3 and 1.4, and for water in Fig. 1.4.

### Example 1.4: Capillary Pressure

A liquid droplet is in static equilibrium in stationary air at low pressure $p_G$.

(a) How does the pressure inside the droplet change for droplets of different radii $R$, for infinite, finite and zero surface tension $\sigma$?

(b) What are the corresponding expressions for the same droplet growing slowly with constant velocity $u$ in air?

### Solution

(a) The Young–Laplace equation in this case is

$$p - p_G = \sigma \left( \frac{1}{R_1} + \frac{1}{R_2} \right) = \frac{2\sigma}{R}, \tag{1.4.1}$$

and equivalently,

$$R = \frac{2\sigma}{p - p_G} = \frac{2\sigma}{\Delta p}. \tag{1.4.2}$$

According to Eqn. (1.4.1), the pressure inside droplets increases with surface tension and decreases with radius. For droplets of radius 0.02 cm, 0.2 cm, and 2 cm, the corresponding gauge pressures are: mercury, 0.048, 0.0048, and 0.00048 atm; water, 0.0073, 0.00073, and 0.000073 atm; and ethanol, 0.00225, 0.000225, and 0.0000225 atm. According to Eqn. (1.4.2), liquids of high surface tension may sustain pressure differences $\Delta p$ in the form of larger droplets than those of low surface tension, characteristic examples being those of mercury of surface tension 480 dyn/cm, against those of water of surface tension 73 dyn/cm, and ethanol of surface tension 22.5 dyn/cm, in the form of droplets of radii 0.096, 0.0146, and 0.0045 cm, respectively, at a pressure difference $\Delta p = 0.01$ atm.

Equation (1.4.1) shows that for $\sigma = 0$, the capillary pressure is $\Delta p = p - p_G = 0$ (i.e., capillary pressure arises only with finite values of surface tension). The latter is also zero across planar interfaces of $R = \infty$; that is, the pressure is continuous across planar interfaces even if $\sigma \neq 0$ according to Eqn. (1.4.2). In other words, fluids in contact along a planar interface have the same pressure at the interface, whereas fluids in contact along a curved interface exhibit different pressures by the amount of the capillary pressure.

(b) A mechanical energy balance between the droplet–air interface, where the pressure of the air is $p_g$ and its velocity identical to $u$, and far away, where the air is undisturbed at zero velocity and atmospheric pressure $p_\infty$, yields

$$p_G = p_\infty - \rho \frac{u^2}{2}. \tag{1.4.3}$$

Equation (1.4.3) combines with Eqn. (1.4.1) to yield

$$p - p_\infty + \rho \frac{u^2}{2} = \frac{2\sigma}{R}, \tag{1.4.4}$$

which relates pressure and kinetic energy of an inviscid liquid droplet growing in an inviscid fluid (air). Equation (1.4.4) is rearranged to the form

$$\frac{R}{4\sigma}(p - p_\infty) + \frac{\rho u^2 R}{\sigma} = 1. \tag{1.4.5}$$

In Eqn (1.4.5) the *Weber number* appears:

$$\text{We} = \frac{\rho u^2 R}{\sigma}, \tag{1.4.6}$$

which is a ratio of kinetic energy $\rho u$ to surface energy $\sigma/r$.

### Example 1.5: Measurement of Surface Tension

The Wilhelmy plate method (Fig. E1.5) is widely used to measure surface tension. A plate of known dimensions $S$, $L$, and $h$ and density $\rho_s$, is being pulled from a liquid of density $\rho_B$ and surface tension $\sigma$ in contact with air of density $\rho_A$. What is the relation between the force $F$ measured and the liquid surface tension $\sigma$?

**Solution**   The net force exerted by the air on the upper base surface according to Eqn. (1.2) is

$$F_A = (p_0 - \rho_A g h_A)SL, \tag{15.1}$$

where $p_0$ is the datum pressure at the interface. Similarly, the net force exerted by

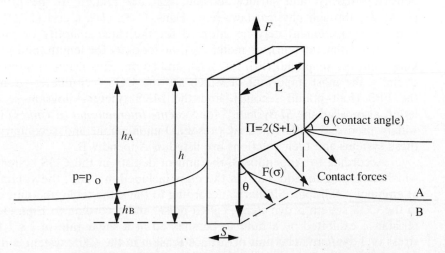

**Figure E1.5**   Wilhelmy plate method for measuring surface tension.

the fluid on the lower base surface is

$$F_B = (p_0 - p_B g h_B)SL. \tag{1.5.2}$$

The surface tension force on the plate in the vertical direction is given by Eqn. (1.35):

$$F(\sigma) = -\sigma \Pi \cos \theta. \tag{1.5.3}$$

The weight of the plate is

$$B = -\rho s g V, \qquad V = (h_A + h_B)SL. \tag{1.5.4}$$

A force balance in the vertical direction results in

$$F = gSL[h_A(\rho_S + \rho_A) + h_B(\rho_S + \rho_B)] + \sigma \Pi \cos \theta. \tag{1.5.5}$$

## 1.5 MEASUREMENT OF PROPERTIES

In the preceding sections, the definition and physical significance of properties important to fluid mechanics have been presented. These properties need to be quantified. This is done by defining how large or small the value of a property or quantity is, compared to characteristic, sensible, and/or experienced quantities. Water at temperature $-10°C$, which is lower than the characteristic temperature of $0°C$ assigned to its freezing, is understood to be in the solid state, at 1 atm pressure. Similarly, a liquid of density $0.1$ $g/cm^3$ is understood to be much heavier than air, of density on the order of $0.001$ $g/cm^3$, and much lighter than water, of density on the order of 1 $g/cm^3$, and therefore under equilibrium conditions the three will be stratified as air–liquid–water from top to bottom. The units °C, g, cm, and others are universally accepted *units of measurement*. There are primary or fundamental units of measurement, which for fluid mechanics are the length $L$, the mass $M$, and the time $T$, and secondary units of measurement, for example of density, viscosity, and surface tension, that are related to the primary ones, $[L, M, T]$, through physical laws [e.g., Eqns. (1.3), (1.23), and (1.31)]. Different units of measurement may be adopted for the same quantity or property, for example centimeter (cm), or meter (m), or foot (ft) for length, and gram (g), or kilogram (kg), or pound (lb), for mass, and so on. This leads to discrete systems of units, the most important of which are the CGS (centimeter–gram–second), the FPS (foot–pound–second), and the MKS (meter–kilogram–second). The latter, also called the *SI System*,[12] for *Système International (d'Unités)* is the most widely used and understood at present. Fundamental and secondary units for these systems and their relations are listed in Appendix B.

According to Appendix B, the unit of denisty in the CGS system is $g/cm^3$, which according to Eqn. (1.3) is the mass included in 1 $cm^3$, the volume of water at ambient conditions. Similarly, according to the same table, the unit of viscosity in the CGS system is dyn s/$cm^2$, called poise, and according to Eqn. (1.23), is the resistance exhibited by a fluid to be sheared at a shear rate of 1 $s^{-1}$ by a shear stress of 1 dyn/$cm^2$. The unit of surface tension in the CGS system is dyn/$cm^2$, as recorded in the same table, and according to Eqn. (1.34), for example, it is the

**TABLE 1.3** PROPERTIES OF PURE COMPOUNDS AT ATMOSPHERIC PRESSURE

| Fluid | Density (g/cm$^3$) | Viscosity (cP) | Kinematic viscosity (cSt) | Surface tension (against air) (dyn/cm) |
|---|---|---|---|---|
| Gases, $T = 20°C$ | | | | |
| Air | 0.0012 | 0.0181 | 15.05 | 0 (miscible) |
| NH$_3$ | 0.0007 | 0.0100 | 14.29 | 0 (miscible) |
| CO$_2$ | 0.0018 | 0.0146 | 8.11 | 0 (miscible) |
| CH$_4$ | 0.00066 | 0.0110 | 16.46 | 0 (miscible) |
| N$_2$ | 0.0011 | 0.0175 | 15.10 | 0 (miscible) |
| O$_2$ | 0.0013 | 0.0203 | 22.98 | 0 (miscible) |
| Liquids, $T = 20°C$ | | | | |
| Acetone | 0.792 | 0.32 | 0.404 | 26 |
| Benzene | 0.880 | 0.65 | 0.739 | 29 |
| Castor oil | 0.970 | 986 | 1016 | 35 |
| Ethyl alcohol | 0.791 | 1.2 | 1.518 | 22.5 |
| Light fuel oil | 0.927 | 148 | 159.6 | 25 |
| Medium fuel oil | 0.940 | 310 | 329.8 | 23.5 |
| Heavy fuel oil | 0.985 | 586 | 595 | 23.5 |
| Glycerin | 1.26 | 622 | 493.6 | 63 |
| SAE-10 oil | 0.869 | 70 | 80.55 | 36.5 |
| SAE-30 oil | 0.888 | 380 | 428 | 35 |
| Methyl alcohol | 0.805 | 0.6 | 0.745 | 23 |
| Olive oil | 0.916 | 84 | 91.7 | 32–35 |
| CCl$_3$ | 1.588 | 0.83 | 0.522 | 35 |
| Toluene | 0.866 | 0.6 | 0.692 | 28.5 |
| Seawater | 1.030 | 1.003 | 0.973 | 73 |
| Water | 0.988 | 1.002 | 1.014 | 73 |
| Water, $T = 100°C$ | 0.958 | 0.0217 | 0.023 | 59 |
| Mercury | 13.55 | 1.55 | 0.114 | 435 |
| Molten metals | | | | |
| Steel, $T = 165°C$ | 8 | 0.05 | 0.006 | 1600 |
| Lead, $T = 400°C$ | 11.33 | 1.7 | 0.150 | 4500 |
| Sodium, $T = 250°C$ | 0.97 | 0.381 | 0.393 | 198 |
| Potassium, $T = 250°C$ | 0.86 | 0.258 | 0.300 | 115 |

*Source:* Data from Refs. 2, 14, 15, and 16.

surface tension that gives rise to a pressure difference of $\Delta p = 1$ dyn/cm$^2$, across a spherical interface of radius $R = 1$ cm.

For any prediction of the flow behavior an apriori measurement of the density, the viscosity, and the surface tension of the liquid under consideration is required (Tables 1.3 and 1.4). The density is measured by means of *pycnometers,* which are based on Archimedes' principle of buoyancy. The viscosity is measured by means of *viscometers,* where the measured torque to drive a flow and the resulting deformation are related according to Newton's law of viscosity. The surface tension is measured by *tensiometers,* which record the force necessary to overcome the surface tension force, to form droplets and bubbles, or to break thin films. More sophisticated methods, usually based on optical techniques,[13] are

**TABLE 1.4** VISCOSITY AND SURFACE TENSION OF NEWTONIAN AND SLIGHTLY SHEAR-THINNING HOUSEHOLD LIQUIDS AT AMBIENT CONDITIONS, MEASURED TO WITHIN 5% EXPERIMENTAL ERROR[a]

| Fluid at 27°C | Shear rate $(s^{-1})$ | Viscosity (cP) | Surface tension (against air) (dyn/cm) |
|---|---|---|---|
| **Newtonian liquids** | | | |
| Water | 225–450 | 1.0 | 73 |
| Crisco vegetable oil | 22–45 | 42.5 | 33.5 |
| Mazola corn oil | 22–45 | 41.5 | 33.5 |
| Pompeian olive oil | 22–45 | 39 | 33.5 |
| La Choy soy sauce | 225–450 | 3.11 | 53 |
| Kroger Worcestershire sauce | 225–450 | 1.6 | 35 |
| Heinz white vinegar | 225–450 | 0.82 | 56 |
| Triaminic-DM cough relief | 90–450 | 17.5 | 47 |
| Scope mouthwash | 90–450 | 1.5 | 32 |
| Spic and Span pine cleaner | 90–225 | 7.2 | 31 |
| Coca-Cola Classic | 90–450 | 1.4 | 50 |
| Kroger whole milk | 225–450 | 2 | 48 |
| Kroger skim milk | 225–450 | 1.3 | 50 |
| **Shear-thinning liquids** | | | |
| Mylanta antacid | 22 | 50 | 29 |
| | 45 | 43 | |
| Dawn liquid dish soap | 2 | 185 | 24 |
| | 11 | 158 | |
| Old-Tyme coffee cream | 90 | 6.5 | |
| | 450 | 4.8 | |
| Mrs. Butterworth's syrup | 6 | 484 | 51 |
| | 12 | 473 | |
| | 24 | 439 | |
| Johnson & Johnson baby shampoo | 1 | 364 | 32 |
| | 2.3 | 321 | |
| | 5.75 | 310 | |
| Coppertone suntan oil | 4.5 | 75 | 33 |
| | 11 | 72 | |
| | 22 | 67 | |

[a] Measured in the Rheology Laboratory of the University of Michigan by a cone and plate rheometer.

employed when high accuracy is vital. The principles of operation of pycnometers, viscometers, and tensiometers are highlighted in several examples, beginning in this chapter.

## 1.6 LAMINAR AND TURBULENT FLOW

Fluids flow in pipes and channels under the action of external forces such as gravity (e.g., flow down an inclined plate), shear or drag force (e.g., butter spreading on toast by knife and paint by brush), and pressure difference between

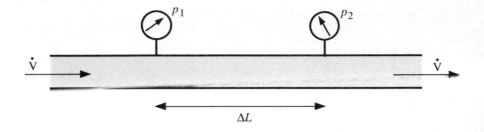

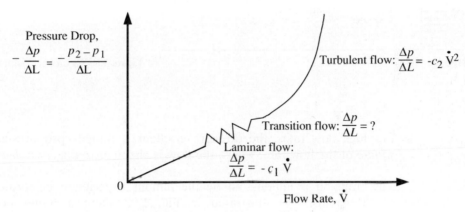

**Figure 1.8**  Pressure drop in a horizontal pipe.

inlet and outlet (e.g., medicine flow in hypodermic piston needle, filling and draining of a pipette, and pumping of water from a well). Depending on the magnitude of the external force, and the density and viscosity of the flowing fluid, the flow is smooth (or laminar) or chaotic (or turbulent), as explained below.

Figure 1.8 shows a pipeline where manometers record the pressures $p_1$ and $p_2$ at the beginning and the end of a test section of length $\Delta L$. The pipeline is kept horizontal, and therefore measured pressure differences are unaffected by hydrostatic pressure. In addition, a substantial length of straight pipe precedes the first pressure gauge, which allows the flow to *develop fully* at that location. For a given flow rate $\dot{V}$, it turns out that the pressure drop, $\Delta p = p_2 - p_1$, is directly proportional to $\Delta L$, and in fact the *pressure gradient*, $\Delta p / \Delta L$, is constant.

By varying the flow rate the measured pressure gradient is as shown in Fig. 1.8, which characterizes three distinct flow regimes:

1. For low flow rates, the pressure gradient is *linearly* proportional to the flow rate, and the flow is *laminar,* as characterized by the famous Reynolds experiment, described below.

2. For intermediate flow rates, there is no unique reproducible relation between flow rate and pressure drop, and the response *alternates* seemingly randomly between regimes 1 and 3; this is a *transition* flow regime.

dye injection capillary

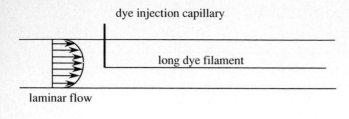

laminar flow

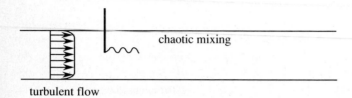

turbulent flow

**Figure 1.9**   Reynolds experiment.

**3.** For high flow rates, the pressure gradient is roughly proportional to the *square* of the flow rate, where the flow is characterized as *turbulent.*

The situation is illuminated by the famous experiment performed by Sir Osborne Reynolds,[17] as illustrated in Fig. 1.9, where a liquid flows in a transparent tube. A fine stream of a dye, or smoke in the case of gas flow, is introduced by a hypodermic needle into the center of the flowing stream and is observed as it is carried downstream by the fluid. Three distinct patterns are observed, corresponding to those characterized earlier and shown in Fig. 1.8 with distinct qualitative features:

**1.** For low flow rates, the injected dye or smoke jet maintains its integrity as a long filament that travels along with the fluid, under negligible diffusion and buoyancy forces, much like smoke lines left behind aircraft at high altitudes under no-wind conditions. The observed behavior is as shown in Fig. 1.9a, which characterizes laminar flow.

**2.** For intermediate flow rates, the results are irreproducible and the dye line departs from straight to wiggly in a random way.

**3.** For high flow rates (Fig. 1.9b), the jet of dye initially forms unstable wiggles and then mixes so rapidly with the surrounding fluid that its continuity breaks and therefore soon becomes invisible, much like smoke in a windy atmosphere. It appears that the fluid flow in the pipe exhibits time-dependent fluctuations of velocity that carries and diffuses the dye farther into the fluid.

Reynolds correlated these distinct flow regimes by means of a unique dimensionless number that incorporates the average fluid velocity $\bar{u}$, its density $\rho$

and viscosity $\eta$, and the diameter $D$ of the pipe. This number is defined by

$$\text{Re} = \frac{\rho \bar{u} D}{\eta} \tag{1.36}$$

and is called the *Reynolds number*. Table 10.1 shows which regime can be expected, depending on the kind of flow, for a given Reynolds number. The Reynolds number expresses the ratio of inertia to viscous forces. At vanishingly small Reynolds numbers, $\text{Re} \ll 1$, the flow is slow, called *creeping flow*, and viscous forces are dominant. Such flows occur primarily with highly viscous liquids flowing in geometries with constrictions and other obstacles to flow, profound examples being the processing of polymeric melts in molds and through dies. At finite but not large Reynolds numbers, $\text{Re} < 2100$ in pipes and channels and $\text{Re} < 10{,}000$ in free-stream flows overtaking submerged solids, inertia forces become dominant, and the fluid continues to flow smoothly, in layerwise parallel fashion, without intermixing between adjacent flowing layers. This is *laminar flow*. Beyond these Reynolds numbers, the layerwise motion starts to be disrupted by intermixing between layers in all possible directions, and eventually *turbulent flow* sets in. The detailed mechanics of laminar flow are the topic of Chapters 5 to 8 for Newtonian fluids and of Chapter 9 for non-Newtonian fluids, and turbulent flow is discussed in Chapter 10. Elements on both flows, necessary to address friction in pipes and other flow conduits as well as in free-stream flows, are summarized below and detailed in Chapter 4.

The patterns above characterize *fully developed flows,* where the velocity profile does not change with distance in the flow direction. This is not ture at the entrance of pipes and channels, under laminar or turbulent flow conditions. Solution of the two-dimensional flow equations between the sharp-edged entrance, where the velocity is plug, and far downstream, where the velocity becomes fully developed parabolic, yields the developing velocity profiles in between. Such calculations in this case show that the plug velocity profile before the entrance develops into its final fully developed parabolic profile within distance $l_s$ such that

$$\frac{l_s}{D} = 0.575 \, \text{Re} = 0.575 \frac{\rho \bar{u} D}{\eta} . \tag{1.37}$$

Within the distance $l_s$, the velocity is retarded to zero value at the solid wall and accelerated to its maximum value along the centerline, until it reaches its steady profile, which persists from there on, as a laminar or turbulent transition, characterized by the Reynolds number.

The expression for the Reynolds number can be rearranged to the form

$$\text{Re} = \frac{\rho \bar{u} d}{\eta} = \frac{\rho \bar{u}^2}{\eta (\bar{u}/D)} = \frac{\text{kinetic energy}}{\text{frictional losses}} = \frac{\text{inertia force}}{\text{viscous force}} ,$$

which suggests that frictional mechanical energy losses arise due to shear stresses, $\tau_{yx}$, parallel to the flow direction according to Eqn. (1.26):

$$[\tau_{yx}] = \eta\left[\frac{\partial u_x}{\partial y} + \frac{\partial u_y}{\partial x}\right] = \eta\left[\frac{\partial u_x}{\partial y} + 0\right] = \eta\frac{\bar{u}}{D},$$

where $[q]$ denotes the order of magnitude of $q$. These stresses depend on the Reynolds number according to

$$\tau_{yx} = f\left(\frac{\eta\bar{u}}{D}\right) = f\left(\frac{1}{\text{Re}}\right)\rho\bar{u}^2, \tag{1.38}$$

where $f$ is a *friction coefficient* that depends on the Reynolds number. The same equation can be arrived at by considering the mechanism of relative motion between adjacent fluid layers traveling at different velocities as shown in Fig. 1.2. The resulting *frictional* force $F_{yx}$, which resists the relative motion, must be overcome for such motion or flow to occur. The work required, which eventually is lost to frictional heat, to overcome $F_{yx}$ over distance $dx$ is

$$W = -F_{yx}\,\Delta L = -\tau_{yx}\Pi\,\Delta L, \tag{1.39}$$

where $\Pi$ is the area of action of $\tau_{yx}$. The corresponding required power to maintain such flow is

$$P = \frac{dW}{dt} = -\tau_{yx}\Pi\frac{dx}{dt} = -\tau_{yx}\Pi\bar{u}. \tag{1.40}$$

To force a fluid through a pipe of length $L$ and diameter $D$, the power required is

$$P = -\tau_{rx}^{W}(\pi DL)\bar{u}, \tag{1.41}$$

and for a channel of length $L$, width $W$, and gap $H$, the power required is,

$$P = -\tau_{yx}^{W}(2WL)\bar{u}. \tag{1.42}$$

In Eqns. (1.41) and (1.42), $\tau_{rx}^{W}$ and $\tau_{yx}^{W}$ are the shear stresses at the pipe and channel walls, respectively, and $\bar{u}$ is the average velocity, defined by

$$\bar{u} = \frac{\dot{V}}{\pi D^2/4}, \qquad \bar{u} = \frac{\dot{V}}{HW} \tag{1.43}$$

for pipe and channel, respectively, with $\dot{V}$ the flow rate (i.e., volume of flowing fluid per time unit). The required power is provided by gravity, by a shearing force, or by a pump, depending on the mechanism driving the flow. The quantification of Eqns. (1.40) to (1.42) for laminar and turbulent flows is the topic of Chapter 4, including designing and operation of power providing pumps.


## 1.7 DIMENSIONAL ANALYSIS AND DYNAMIC SIMILARITY

The Reynolds experiment[8] demonstrates that important conclusions on pipe and channel flows can be drawn by knowing the *dimensionless Reynolds number,* defined by Eqn. (1.36). More importantly, flows of fluids of different densities and viscosities, flowing under different velocity, in *geometrically similar* pipes even of different diameters are *dynamically similar,* for example laminar or turbulent, if their Reynolds numbers are identical.

   Pipe and channel flows do not depend separately on each of the variables $D$, $\rho$, $\bar{u}$, and $\eta$, but rather on their Reynolds number combination. In that respect, it would be meaningless to perform numerous experiments, each time varying one of the variables and keeping the rest constant, in order, for example, to determine transition to turbulent flow experimentally. Instead, fewer, meaningful experiments would involve changing the Reynolds number combination. Indeed, transition to turbulent flow in pipes occurs at around $\mathrm{Re} \simeq 2100$ no matter what the values of $D$, $\rho$, $\bar{u}$, and $\eta$ are.

   This means that a large-scale flow (e.g., oil transfer from tanker to refinery) of large viscosity $\eta$ and density $\rho$ in a long pipe of large diameter $D$ under flow rate $\dot{V} = (\pi D^2/4)\bar{u}$ and average velocity $\bar{u}$ can be studied in lab-scale equipment. This is done by means of a different liquid of different (smaller) viscosity $\eta_0$, different density $\rho_0$, in a much smaller pipe of diameter $D_0$, under higher average velocity $\bar{u}_0$ and flow rate $\dot{V}_0$, inasmuch as

$$(\mathrm{RE})_0 = \frac{\rho_0 D_0 \bar{u}_0}{\eta_0} = \mathrm{Re} = \frac{\rho D \bar{u}}{\eta}. \tag{1.44}$$

Therefore,

$$D_0 = \left( \frac{\rho \eta_0 \bar{u}}{\rho_0 \eta \bar{u}_0} \right) D = (\mathrm{SF}) D \ll D, \tag{1.45}$$

where (SF) stands for *scaling factor.* The SF can be chosen intelligently to be much smaller than unity, for example by choosing $\eta_0/\eta \ll 1$ and $\bar{u}/u_0 \ll 1$, which makes $D_0$ much smaller than $D$ and therefore lab reproduction possible. Then all conclusions drawn from the small-scale lab experiment apply to any larger-scale flow in a *similar geometry.* For example, the average velocity of the large-scale flow is

$$\bar{u} = \left( \frac{\eta \rho_0}{\eta_0 \rho} \right) \left( \frac{D_0}{D} \right) \bar{u}_0, \tag{1.46}$$

where the properties of the two fluids are known, the ratio $D_0/D$ of the two similar geometries is known, and $\bar{u}_0$ is measured experimentally in the lab. This kind of scaling-up is extremely important in aerodynamics, where large-scale flow around traveling vehicles is modeled by lab reproduction. This allows significant size reduction, and scaling of pilot-plant results to actual industrial-scale processes.

   The Reynolds number, although clearly the most common and important in fluid flow, is not the only dimensionless group that must be matched in scale tests. Like the Reynolds number in compressible flow, the *Mach number*,

$$\mathcal{M} = \frac{\bar{u}}{c}, \tag{1.47}$$

defined as the ratio of fluid velocity $\bar{u}$ to the speed of sound $c$ in the same fluid, distinguishes subsonic from supersonic compressible flow, and can be utilized and used along the lines discussed for the Reynolds number.

Other dimensionless numbers encountered so far include the Weber number encountered in Eqn. (1.45),

$$\mathrm{We} = \frac{Ru^2\rho}{\sigma} ;\tag{1.48}$$

*aspect ratios,* such as entrance length to diameter in Eqn. (1.37),

$$\mathrm{AR} = \frac{l_s}{D},\tag{1.49}$$

and the *friction factor,* encountered in Eq. (1.38),

$$f\left(\frac{1}{\mathrm{Re}}\right) = \frac{2\tau_{rz}^w}{\rho\bar{u}^2/2}\tag{1.50}$$

From the preceding discussion it is evident that:

1. These dimensionless numbers are extremely important in fluid flow.
2. These numbers appear naturally in the solution of the governing conservation equation, by appropriate rearrangement (e.g., Section 1.7.2).
3. It would be nice to have ways to deduce the dimensionless numbers involved without solving the governing conservation or other equations. This would handle even cases where either all the governing equations are not known, or even if they are known, they cannot be solved.

It turns out that the latter approach does exist, which allows determination of the controlling *dimensionless numbers.* There are, in fact, two approaches: the first, called *dimensional analysis,* is based on the fact that any dimensional equation, in terms of dimensional variables, must be dimensionally consistent, and therefore the kind of units in its left-hand side must be identical to those in its right-hand side; the second, called *dimensionless analysis,* is similarly based on the fact that all terms of any dimensionless equation (in terms of dimensionless variables), must be rid of any units. The two approaches are summarized below.

### 1.7.1 Dimensional Analysis

In fluid flow the fundamental units of mass, time, and length span all other units used in any flow situation. Therefore, if one of the involved variables, for example flow rate, is expressed in terms of the rest, the dimensionality or power of each of its three involved fundamental units must be identical to the dimensionality of the product of the corresponding ones in the rest of the variables (Bridgeman's principle[18]). The approach is presented in connection with the problem of fluid friction in a rough pipe. The expected important variables are the properties of the incompressible fluid $\rho$ and $\eta$. Geometrical characteristics of the pipe are its diameter $D$ and its *roughness* $\varepsilon$, which represent the height of hills or depth of valleys of a rough internal surface in contact with the flowing fluid. The roughness $\varepsilon$ has dimensions of length, and is zero for *smooth pipe*. The processing or flow conditions are the average velocity $\bar{u}$ and the pressure drop $\Delta p / \Delta L$ that drives the flow. If, indeed, these variables are enough to fully determine friction, their arrangement in any governing equation must be such that the equation is dimensionally consistent (Rayleigh's dimensionality principle[19]). Therefore, the group

$$\Pi = \left(\frac{\Delta p}{\Delta L}\right)^a \rho^b \eta^c D^d \varepsilon^j \bar{u}^k \tag{1.51}$$

must be dimensionless (Buckingham's pi theorem[20]). Thus the three fundamental units of mass $(M)$, length $(L)$, and time $(T)$ must have identical exponents in the two sides of Eqn. (1.51) according to

$$[\Pi] = [M^0, L^0, T^0]$$
$$= [M, L^{-2}, T^{-2}]^a [M, L^{-3}]^b [M, L^{-1}, T^{-1}]^c [L]^d [L]^j (L, T^{-1})^k, \tag{1.52}$$

which leads to the following system of equations:

$$[M]: 0 = a + b + c$$
$$[L]: 0 = 0 = -2a - 3b - c + d + j + k = 0 \tag{1.53}$$
$$[T]: 0 = -2a - c - k.$$

This system of three equations with six unknowns cannot be solved unless three of the involved unknowns are used as parameters or fixed arbitrarily. Any three of the six unknowns can be fixed; however, understanding of the physics of the flow allows the most optimum choice. The number of the linearly independent solutions to a linear homogeneous system is identical to the number of unknowns involved, six in this case, minus the number of the equations, three in this case. Three independent solutions are obtained in this case as follows.

Assume that $a = d = 1$ and $k = -2$. Then the rest of the unknowns from the solution of Eqn. (1.53) are $c = 0$, $b = -1$, and $j = 0$, and therefore Eqn. (1.51) is grouped to

$$\Pi_1 \left(\frac{\Delta p}{\Delta L}\right) D (\rho \bar{u}^2)^{-1} = \frac{\Delta p}{(\Delta L / D) \rho \bar{u}^2} \equiv Eu = 2 f_F, \tag{1.54}$$

and the *Euler number, Eu*, results, which is twice the friction factor. If now $c = 1$, $d = k = -1$ instead, the resulting values of the other unknowns are $a = 0$, $b = -1$, and $j = 0$, and therefore

$$\Pi_2 = \frac{\eta}{D \bar{u} \rho} = \frac{1}{Re}. \tag{1.55}$$

Finally, if $a = 0$, $d = -1$, and $j = 1$, the resulting values of the other unknowns are $b = 0$, $c = 0$, and $k = 0$, and therefore

$$\Pi_3 = \frac{\varepsilon}{D}. \tag{1.56}$$

Thus the arrangement of the dimensionless numbers takes the form

$$F\left(\frac{\Delta p}{(\Delta L/D)\rho \bar{u}^{-2}}, \frac{D\bar{u}\rho}{\eta}, \frac{\varepsilon}{D}\right) = 0, \tag{1.57}$$

and equivalently,

$$\frac{\Delta p}{(\Delta L/D)\rho \bar{u}^2} = 2f_F = G\left(\frac{D\bar{u}\rho}{\eta}, \frac{\varepsilon}{D}\right), \tag{1.58}$$

which means that the friction factor is expected to be a function of the Reynolds number and the *relative roughness*. The exact form of the functions $F(\cdot)$ or $G(\cdot)$ cannot be provided by the dimensional analysis. However, based on Eqn. (1.58) a minimum number of experiments can be designed to deduce these functions: First, the Re is kept constant, the relative roughness is varied, and the Eu is measured. Then $\varepsilon/D$ is kept constant, the Re is varied, and the Eu is measured. The results are plotted appropriately and the functional forms of $F$ or $G$ are deduced. This procedure is followed to construct friction factor diagrams (e.g., Fig. 4.3) and to validate them experimentally. Indeed, the equations of friction factor of Section 4.3 are represented by Eqn. (1.58).

In summary, the approach described is based on Rayleigh's principle on equation dimensionality,[19] expressed by Buckingham's theorem[20] that: *the m variables that can be expressed in terms of n fundamental units of measurement can be arranged in m − n nonredundant groups that include multiplications or divisions among them.* More details on dimensional analysis can be found in Refs. 21, 22, and 23.

### 1.7.2 Dimensionless Analysis

When the governing equations of flow or other process are known, there is an alternative more straightforward procedure to derive the involved dimensionless numbers. Consider a droplet of initial radius $R_0$ of a viscous liquid of viscosity $\eta$ growing slowly at constant speed $u_0$, in a second liquid of much smaller viscosity (inviscid fluid), density $\rho$, and interfacial tension $\sigma$ with the liquid inside the droplet. The droplet grows due to injection of more liquid to it by a hypodermic piston needle. It is required to find the evolution of the nearly uniform pressure inside the droplet that causes the expansion of the droplet. Obviously, Eqn. (1.4.4) applies here with the internal pressure force expanding the droplet augmented by the viscous normal stress, $\tau_{rr} = 2\eta(\partial u/\partial r)_{r=R}$. The corresponding external normal stress is negligible for an inviscid fluid. Thus force balance over the expanding spherical interface yields

$$[p^L + \tau_{rr}^L]_{r=R} = [p^I + \tau_{rr}^I]_{r=R} + \frac{2\sigma}{R} = \left[p_\infty - \frac{\rho u_0^2}{2} + 0\right] + \frac{2\sigma}{R} \tag{1.59}$$

with

$$R = R_0 + u_0 t = u_0\left(\frac{R_0}{u_0} + t\right) \tag{1.60}$$

and the normal stress approximated by

$$\tau_{rr}^{L} = 2\eta \left.\frac{\partial u}{\partial r}\right|_{r=R} \simeq 2\eta \frac{u_0}{R/2} \simeq 4\eta \frac{u_0}{R} = \frac{4n}{(R_0/u_0) + t}. \tag{1.61}$$

Equations (1.60) and (1.61) are substituted in Eq. (1.59) and the *dimensional governing equation* for the *dimensional pressure* evolution $p(t)$ with *dimensional time t* results:

$$p^{L}(t) - p_{\infty} = \frac{4\eta u_0}{R_0 + u_0 t} - \rho \frac{u_0^2}{2} + \frac{2\sigma}{R_0 + u_0 t}. \tag{1.62}$$

The dimensional variables $p(t)$ and $t$ in Eqn. (1.62) are turned into *dimensionless variables* $p^*(t^*)$ and $t^*$ by dividing each of them by a *characteristic known quantity* of the same dimensionality (i.e., having the same units of measurements). For the case considered, $p_{\infty}$ is a known characteristic pressure and $R_0/u_0$ known characteristic time. Thus

$$t^* = \frac{t}{R_0/u_0} = \frac{u_0 t}{R_0}, \qquad p^*(t^*) = \frac{p(t^*)}{p_{\infty}}. \tag{1.63}$$

These dimensionless variables are substituted in the dimensional equation, Eqn. (1.62), which results in

$$p_{\infty}[p^*(t^*) - 1] = \frac{4\eta u_0}{R_0(1 + t^*)} - \rho \frac{u_0^2}{2} + \frac{2\sigma}{R_0(1 + t^*)}, \tag{1.64}$$

which is further rearranged to the form

$$\frac{p_{\infty} R_0}{2\sigma} [p^*(t^*) - 1] = \frac{2\eta u_0}{\sigma} - \frac{\rho u_0^2 R_0}{4\sigma} + \frac{1}{1 + t^*}. \tag{1.65}$$

Equation (1.65) is a *dimensionsless equation* because each of its terms is dimensionless. The appearing *dimensionless numbers* are: (a) the Weber number, also called the Bond number, defined by Eqn. (1.48), which is a ratio of inertia to surface tension force; (b) the hydrostatic-to-capillary force ratio

$$\tilde{We} = \frac{p_{\infty}}{\sigma_0/R_0} = Eu \cdot We, \tag{1.66}$$

which is the product of the *Euler number,*

$$Eu = \frac{\Delta p}{\rho u_0^2} \tag{1.67}$$

with the Weber number; and (c) the *Capillary number*

$$Ca = \frac{\eta u_0}{\sigma}, \tag{1.68}$$

a ratio of viscous to capillary forces.

    Equations (1.65) to (1.68) state that the evolution of the dimensionless pressure depends on the We, $\tilde{\text{We}}$, and Ca dimensionless numbers, in addition to the dimensionless time $t^*$, and not on each of the six dimensional quantities $p_\infty$, $R_0$, $u_0$, $\sigma$, $\eta$, and $\rho$, separately. Thus three rather than six series of independent experiments would be sufficient to determine the behavior experimentally. Once the dimensionless pressure in terms of dimensionless time is determined, the corresponding dimensional pressure and times are recovered by Eqn. (1.63).

    The dimensionless governing equation, Eqn. (1.65), is reduced to simple forms in the appropriate limits: For infinitely large surface tension such that Ca $\ll$ 1, Wc $\ll$ 1, and $\tilde{\text{We}} \ll 1$, Eqn. (1.63) reduces to

$$p^*(t^*) = 1 = \frac{1}{\tilde{We}} \frac{1}{1 + t^*} \tag{1.69}$$

and equivalently,

$$p(t) = p_\infty + \frac{2\sigma}{R_0} \frac{1}{R_0 + U_0 t}, \tag{1.70}$$

which brings the bubble to static equilibrium. For dominant viscous forces (i.e., for Ca $\gg$ 1), Eqn. (1.65) predicts

$$\tilde{We}[p^*(t^*) - 1] = 4Ca \tag{1.71}$$

and equaivalently,

$$p(t) = p_\infty + 4\eta \frac{u_0}{R_0}. \tag{1.72}$$

For fast expansion We is dominant and therefore

$$\tilde{We}[p^*(t^*) - 1] = -\frac{We}{2} \tag{1.73}$$

and equivalently,

$$p(t) = p_\infty - \frac{\rho u^2}{2}. \tag{1.74}$$

These forms of simplified equations and resulting limiting solutions are important in cases where the complete governing equation is not solvable analytically. In many situations the limiting cases themselves are of primary interest, and in some cases conclusions for finite values of the dimensionless numbers can be guessed and checked by extrapolating opposite limiting cases (e.g., when the solution in the opposite limits Ca $\ll$ 1 and Ca $\gg$ 1 is known, conclusions about the solution for Ca $\simeq$ 1 may be drawn, but not always).

    As in the case of dimensional analysis, the exact form of the functions $f_1(\cdot)$, $f_2(\cdot)$, and $f_3(\cdot)$ cannot be deduced by dimensionless analysis. Dimensional analysis is appropriate when the system is very complex and the governing conservation and equations of state are not obvious, whereas dimensionless analysis is commonly performed on the known governing equations.

### Example 1.6: Dimensional and Dimensionless Analyses

A reactant at concentration $c_0$ enters a plug reactor of diameter $D$ and infinite length, carried by a fluid stream at velocity $u$, which is constant throughout the reactor. The reactant is consumed by a first-order reaction at rate $-kc$, where $c$ is its varying concentration along the reactor. Under these conditions, there is also a streamwise diffusion of the reactant, along the flow direction, at rate $\mathscr{D}(dc/dz)$, where $\mathscr{D}$ is a diffusion coefficient. The situation is shown in Fig. E1.6. Find the dimensionless numbers that control this process where convection (with velocity, $u$), reaction (with rate, $-kc$), and diffusion (with diffusivity, $\mathscr{D}$) coexist, as follows: (a) pretend that the governing conservation equation for the mass of the reactant is not known, and perform dimensional analysis; (b) derive the conservation equation of reactant mass and then perform dimensionless analysis; (c) compare the results of (a) and (b).

*Dimensional analysis.* The variables expected to enter in the description of the process are $c_0$, $D$, $u$, $k$, and $\mathscr{D}$. Therefore,

$$\Pi = c_0^a D^b u^c k^d \mathscr{D} \tag{1.6.1}$$

and

$$[M^0, L^0, T^0] = [M, L^{-3}]^d [L, T^{-1}]^c [T^{-1}]^d [L^2, T^{-1}]^e, \tag{1.6.2}$$

which leads to the following system of homogeneous equations:

$$[M]: a = 0$$
$$[L]: -3a + b + c + 2e = 0$$
$$[T]: -c - d - e = 0.$$

There are three equations with five unknowns and therefore two independent arrangements of variables in groups of dimensionless numbers, obtained as follows: $a = 0$ always and cannot be used as a parameter. Assume that $b = c = 1$, to obtain $e = 0$ and $d = -1$ by solving the system. Thus

$$\Pi_1 = \frac{uD}{\mathscr{D}} = \text{Pe} \tag{1.6.3}$$

is one dimensionless number, called the *Peclet number*, entering in solute transfer flows.[24] Now assume that $b = 1$ and $c = -1$, to obtain $e = 0$ and $d = 1$, and

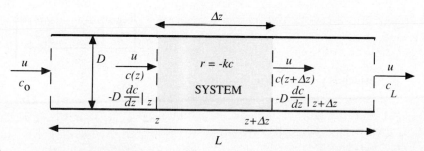

**Figure E1.6**   Convection–reaction–diffusion of reactant in plug reactor.

therefore

$$\Pi_2 = \frac{Dk}{u} = \text{Da}, \tag{1.6.4}$$

which is the *Damkohler number* entering in reacting flows.[25] Had the alternative choice $b = 2$ and $c = 0$ been made, the result would be $e = -1$ and $d = 1$, and

$$\Pi_3 = \frac{kD^2}{\mathscr{D}} = \text{Pe} \times \text{Da}, \tag{1.6.5}$$

which is not an independent group, being the product of $\Pi_1$ and $\Pi_2$.

*Dimensionless analysis.* First, the dimensional conservation equation of reactant mass is derived on the system of diameter $D$ and length $\Delta z$ shown, at steady state. It states that at steady state, convection by velocity $u$ and diffusion due to concentration gradient $dc/dz$, in the system through the cross section at $z$ and out of the system through the cross section at distance $z + \Delta z$, plus consumption due to the chemical reaction, are added to zero:

$$\underbrace{\left[ cu\pi \frac{D^2}{4} \right]_z - \left[ cu\pi \frac{D^2}{4} \right]_{z+\Delta_z}}_{\text{net convection}} + \underbrace{\left[ -\mathscr{D}\frac{dc}{dz}\pi \frac{D^2}{4} \right]_z - \left[ -\mathscr{D}\frac{dc}{dz}\pi \frac{D^2}{4} \right]_{z+\Delta z}}_{\text{net diffusion}}$$

$$\underbrace{- \left[ kc\pi \frac{D^2}{4}\Delta z \right]}_{\text{conversion}} = 0. \tag{1.6.6}$$

Under constant $u$, $D$, $k$, and $\mathscr{D}$, this equation is arranged to the form

$$\frac{u[c(z) - c(z + \Delta z)]}{\Delta z} + \frac{D\left[ \frac{dc}{dz}\Big|_{z+\Delta z} - \frac{dc}{dx}\Big|_z \right]}{\Delta z} - kc = 0. \tag{1.6.7}$$

This dimensional equation is now made dimensionless by substituting

$$z^* = \frac{z}{D} \quad \text{and} \quad c^* = \frac{c}{c_0} \tag{1.6.8}$$

in the form

$$\frac{uc_0}{\mathscr{D}}\left[ \frac{c^*(z^*) - c^*(z^* + \Delta z^*)}{\Delta z^*} \right]$$

$$+ \frac{\mathscr{D}c_0}{D^2}\left[ \frac{\frac{dc^*}{dz}\Big|_{z^*+\Delta z^*} - \frac{dc^*}{dz^*}\Big|_z}{\Delta z^*} \right] - (kc_0)c^* = 0, \tag{1.6.9}$$

which is rearranged in the form

$$\frac{uD}{\mathscr{D}}\left[ \frac{c^*(z)^* - c^*(z + \Delta z^*)}{\Delta z^*} \right] + \left[ \frac{dc^*}{dz^*}\Big|_{z^*+\Delta z^*} - \frac{dc^*}{dz^*} \right] - \frac{kD^2}{\mathscr{D}}c^* = 0. \tag{1.6.10}$$

In the limit of $dz \to 0$, the terms in the two parentheses become the corresponding derivatives. Thus

$$\frac{uD}{\mathcal{D}}\frac{dc^*}{dz^*} + \frac{d^2c^*}{dz^{*2}} - \frac{kD^2}{\mathcal{D}}c^* = 0. \tag{1.6.11}$$

The resulting dimensionless numbers are the *Peclet number*,

$$\text{Pe} = \frac{uD}{\mathcal{D}} = \frac{uc_0/D}{\mathcal{D}(c_0/D^2)} = \frac{\text{convection rate}}{\text{diffusion rate}}, \tag{1.6.12}$$

and the *Damkohler number*,

$$\text{Da} = \frac{kD}{u} = \frac{kc_0}{uc_0/D} = \frac{\text{reaction rate}}{\text{convection rate}}, \tag{1.6.13}$$

incorporated in the product

$$\text{Pe} \times \text{Da} = \frac{uD}{\mathcal{D}}\frac{kD}{u} = \frac{kD^2}{\mathcal{D}} = \frac{kc_0}{\mathcal{D}c_0/D^2} = \frac{\text{reaction rate}}{\text{diffusion rate}}. \tag{1.6.14}$$

Equation (1.6.11) can be solved easily for limiting cases of $\text{Pe} = 0$ or $\text{Pe} \to \infty$, $\text{Da} = 0$ or $\text{Da} \to \infty$, which represent cases of dominant diffusion, convection, or reaction.

### 1.7.3 Dimensionless Numbers in Fluid Mechanics

The most important dimensionless numbers of fluid mechanics are summarized below. Most of them are also discussed as they arise in connection with flows and processes examined in the following chapters. Under incompressible and isothermal conditions, the important dimensionless numbers are:

**1.** The *Reynolds number*,

$$\text{Re} = \frac{D\bar{u}\rho}{\eta}, \tag{1.75}$$

where $\bar{u}$ is an average velocity and $D$ a flow conduit dimension (e.g., diameter of pipe or depth of channel) which reveals the relative magnitude of inertia to viscous forces.

**2.** The *Froude number*,

$$\text{Fr} = \frac{gD}{\bar{u}^2}, \tag{1.76}$$

where $g$ is the acceleration of gravity and $D$ is the Reynolds number, which reveals the relative magnitude of gravity to inertia forces.

**3.** The *capillary number,*

$$Ca = \frac{\eta \bar{u}}{\sigma},\tag{1.77}$$

which reveals the relative magnitude of viscous to surface tension forces in flow situations[9];

**4.** The *Weber number,*

$$We = \frac{\rho \bar{u}^2 D}{\sigma},\tag{1.78}$$

which expresses the ratio of inertia forces to capillary pressure $\sigma/R$ forces.

Flows of highly viscous liquids are characterized by a vanishingly small Reynolds number and therefore called *creeping flows.* Most flows of polymers of large viscosities are creeping flows. The Reynolds number also serves to distinguish between *laminar* and *turbulent* flow; As demonstrated in Section 1.6, laminar flow in pipes and channels, characterized by parallel sliding motion of adjacent fluid layers without intermixing, persists for Reynolds numbers below 2100. Beyond that value, eddies start to develop within the fluid layers that cause intermixing and chaotic, oscillatory fluid motion, which characterizes turbulent flow. The Froude number is zero in strictly horizontal flows and maximum in vertical flows. The Weber and capillary numbers appear in flows under free surfaces and interfaces. The surface tension decreases, and thus the two numbers increase, by the addition of surfactants to the flowing liquids.

As we will see in Chapters 6 to 10, fluid flow is modeled by universally applied equations that include the common dimensionless numbers. These equations are nonlinear and therefore difficult to solve analytically. Knowledge of the order of magnitude of the involved dimensionless numbers, however, may allow the simplification and solution of these equations, for limiting values of the dimensionless numbers. Consider the equation

$$Re\left(u \frac{\partial u}{\partial x}\right) = \frac{\partial^2 u}{\partial y^2},\tag{1.79}$$

which may approximate the flow of liquid with velocity $u$ in the $x$-direction, over a submerged plate placed parallel to the flow direction. This is a nonlinear equation with no apparent analytic solution. However, for high Reynolds numbers and large distances $y$ over the plate, the left-hand term dominates and the solution to Eqn. (1.79) is well approximated by solution of the simplified equation

$$Re\left(u \frac{\partial u}{\partial x}\right) = 0,\tag{1.80}$$

which is simply $u = V$ (constant). For the low Reynolds number prevailing near the motionless plate, the right-hand term dominates, in which case the approximate solution is

$$u = c_1 y.\tag{1.81}$$

Thus a linear velocity profile is found to prevail near the plate, and a plug-like profile away from the plate, *without solving the complete governing equation,* Eqn. (1.80). Although Eqns. (1.79) to (1.81) are crude simplifications of what are known as *boundary layer equations,* the same procedure is used to study actual boundary layer flows, as discussed in Section 7.8. Dimensionless numbers important in fluid mechanics and in related transfer of heat and solute phenomena are tabulated in Appendix E.

## PROBLEMS

1. *Densities of fluids.* Estimate the density of the following fluids:
   (a) water of weight 20 kg that occupies a volume of 0.02 m³;
   (b) glycerin of weitht 200 kg that occupies a volume of 0.16 m³;
   (c) unknown liquid $X$ that fills a 1 m × 1 m × 1 m volume and weighs 200 kg;
   (d) unknown gas $Y$ that fills the same volume and weighs 2 kg;
   (e) air at temperature 0°C and pressure 2 atm;
   (f) unknown gas $Z$ that fills a 10-m³ volume at temperature 25°C, whose gauge pressure reads 1.5 atm. What are their corresponding specific weight and specific gravity?

2. *Newton's law of motion.* A body of mass 10 kg and volume 2 m³ is accelerated at 2 m/s². What is the magnitude of the force applied? If the distance of translation is 100 m, what is the work done by the force? If the initial speed of the body is zero, what is the traveling time?

3. *Energy and momentum.* Calculate the kinetic energy and momentum of a 2-kg mass of volume 0.1 m³ traveling at speed 10 m/s. What are the corresponding kinetic energy and momentum densities (i.e., energy and momentum per volume)? What amount of force is required to bring the body to a complete stop in a 10-m distance within 5 s? What are the work and power spent by such a force? How do the kinetic energy and momentum change within that time inverval? Plot the kinetic energy, momentum, force, and power for a constant-speed motion persisting for 1 min, followed by application of the brake force. How is this study related to the speed limit and distancing of vehicles on highways in preventing rear-end collisions?

4. *Pressure cases.* Estimate the resulting pressure in the following cases:
   (a) on earth by a standing man of 184 $lb_f$ who wears shoes of 0.3-ft² contact area each;
   (b) on earth by a four-wheeled vehicle weighing 10 tons with 160 cm² of tire flatness;
   (c) by air on a 2-m² surface of metal, paper, and plastic material, placed vertically, at sea-level altitude;
   (d) on the same surfaces placed horizontally at the same altitude;
   (e) by air on the eardrum of an ejected fighter pilot at 4000 m altitude;
   (f) by seawater on the eardrum of a diver at 100 m depth;
   (g) on an astronaut's body in space.

5. *Air-density tricks.* A device used frequently in "hands-on" science museums involves a vertical airstream and an inflated air balloon that can be left by the spectator at several spots adjacent to the stream. Describe and justify the motion of a balloon released at the center and then to the side of the stream, for cases of airstream temperature

lower, identical, and higher than that of the balloon. Will the same patterns be
observed for nitrogen and oxygen balloons?

6. *Atmospheric equilibrium.* The pressure, density, and temperature distributions under
   no-wind conditions are governed by the equations

$$\frac{dp}{dz} = -\rho g, \qquad p = \frac{\rho R T}{M}, \qquad T = T_0 - \beta z.$$

(a) Find the atmospheric pressure and temperature distributions, $p(z) = p_0 + \cdots$
    and $T(z) = T_0 + \cdots$, conditions, $p_0$, and $T_0$.
(b) Show that

$$\frac{p(z)}{p_0} = \left[ \frac{\rho(z)}{\rho_0} \right]^n,$$

with

$$\frac{n}{n-1} = \frac{gM}{\beta R}.$$

What is the value of $\beta$ (called the *lapse rate* in meteorology[26])?
(c) How could the exponent $n$ be related to atmospheric thermodynamics?

7. *Pressure and density stratification.* The temperature in a saltwater lake may vary nearly
   linearly from 15°C at depth 100 m to 20°C at the surface. Salt concentration $c$ may vary
   similarly from 0.01 to 0.001 g/cm³ between the two spots. The saltwater density can be
   approximated by

$$\rho(c, T) = \rho_0 - b_T(T - T_0) + b_c c,$$

where $b_T = -0.00011$ g/cm³°C and $\rho_0 = 1$ g/cm³ at 4°C and 1 atm.
(a) Plot the density distribution with depth from the bottom to the surface.
(b) Repeat for the pressure distribution.
(c) Repeat (a) and (b) for clean water at the reference conditions and compare the
    two results.

8. *Kinematic viscosity.* Calculate the kinematic viscosity of water and air at 1 atm and
   temperatures of 4°C, 10°C, 50°C, 70°C, and 100°C. Judging from your results, what is
   the physical significance of the kinematic viscosity, as opposed to the (dynamic)
   viscosity? Based on the momentum euation, for the velocity $u_x(y, t)$

$$\rho \frac{\partial u_x}{\partial t} = -\frac{dp}{dx} + \eta \frac{\partial u^2 x}{\partial y^2}$$

for an unsteady one-dimensional flow in a channel of length $L$ in the $x$-direction and
width $D$ in the $y$-direction, in what type of fluid motion is the kinematic rather than
the (dynamic) viscosity important? Define the Reynolds number based on the
kinematic viscosity and discuss its physical significance. (You may need to use
dimensionless analysis for the last two questions.)

9. *Sliding on thin film.* A rectangular plate of 50 N weight slides on its 0.04-m² area
   surface on top of a thin film of grease of viscosity 10 cP and thickness 2 mm, along an
   inclined plane at a 30° angle with the horizontal.
   (a) What is the resulting velocity of the plate motion?
   (b) What are the resulting velocity and shear stress distribution within the grease film?
   (c) Investigate how the plate velocity is affected by grease viscosity, film thickness,
       inclination angle, and plate weight-to-sliding surface ratio.

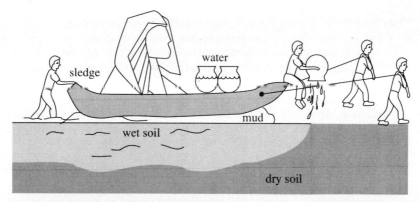

**Figure P1.12**   Lubrication-assisted transfer of heavy statue in ancient Egypt.

    **(d)** Greases and oils are used extensively as *lubricants*. How could your results in (c) apply to these situations?

10. *Ice sliding.* The piece of ice shown in Fig. 1.1 slides a distance of 100 m within 10 min, at an inclination of 5°. The ice can be considered to be a regular rectangle in shape, of 3 m width, 6 m length, and 2 m height, at 1 atm and −5°C. What are the resulting liquid water film thickness that *lubricates* the motion, the resulting velocity distribution within the film, the shear stress induced by the sliding ice, and the pressure within the liquid water film?

11. *Liquid spreading.* What shear force is required on a 2 cm × 10 cm knife blade to apply a 1-mm layer of butter viscosity 100 cP on toast, and then a 0.5-mm layer of honey of viscosity 10 cP on top of it? Recall one or more situations, for example paint rolling, where shearing forces are applied to spread liquids, and calculate the forces required accordingly.

12. *Lubrication at its inception.* Assyrians and Egyptians were the first to appreciate the merits of lubrication in moving sledges carrying heavy construction materials, such as the statue shown in Fig. P1.12. The lubrication technician, positioned at the leading edge of the sledge, pours water to lubricate its motion. For a statue weighing 10 tons positioned on a 2 m × 5 m wooden base, calculate the labor savings under the following rough assumptions:
    **(a)** The motion is slow, 5 m/min.
    **(b)** Under dry soil–wood contact conditions, the resisting shear stress is

$$\tau_d = \frac{kW}{A},$$

    where $k_d = 0.1$ is the friction coefficient, $W$ the total weight, and $A$ the area of contact.
    **(c)** The rate of water penetration to soil is such that a surface mud layer of viscosity 100 P and average thickness 3 cm is formed.
    **(d)** A man could apply an average 0.1 hp. What are the effects of base area $A$, soil quality parameter $k$, and rate of water supply?

13. *Sliding efficiency.* Explain the differences and similarities in sliding of (a) a hockey shoe vs. a roller-skating shoe; (b) a hockey puck vs. a soccer ball. Explain the use of salt on roadways during icy winter weather. Would a similar application enhance hockey and ski performance?

**14.** *Viscosity characterization.* Think of and list experiments or observations to distinguish among Newtonian, shear-thinning, shear-thickening, and viscoplastic liquids. Describe at least one method to construct the curves of Fig. 1.5, for fluids of unknown origin (i.e., *perform an elementary rheological characterization* of an unknown fluid).

**15.** *Viscosity measurement.* The following data have been obtained, by means of a *cone-and-plate viscometer,* in performing rheological characterization of five unknown fluids at *shear rates* of 0.1, 1, 10, 100, and $100\,\mathrm{s}^{-1}$. The recorded values of *shear stress* in $\mathrm{dyn/cm^2}$ for each shear rate are fluid A: 0.98, 7, 65, 500, and 2000; fluid B: 1, 18, 300, 6000, 30,000; fluid C: 0.31, 2.9, 31, 262, 3050; fluid D: 2, 21, 195, 1000, 12,000; fluid E: 12, 21, 110, 1050, 9800.

    **(a)** Construct plots of shear rate vs. shear stress for each fluid and characterize it as Newtonian, shear thinning, shear thickening, or viscoplastic.

    **(b)** What is the significance of the data in specifying fluid transport conditions for each fluid (e.g., flow rate achieved for given applied shear stress or pressure gradient, shear stress required for target flow rate, and others)?

**16.** *Elemenatary viscometer.* You are given two thin plates of 0.5 m length and 0.2 m width, a dynameter that measures force, a watch, and a scaled cylindrical container. With these tools at hands, how would you assemble an elementary viscometer (while waiting for a commercial one to be shipped to your company), and how is the viscosity of an unknown fluid determined by such a device? What diagrams could conveniently give the viscosity directly, by utilizing curve slope and intersection?

**17.** *Couette and plug flow.* In Couette flow the fluid velocity varies linearly from one boundary to the other, as shown in Fig. 1.2, for example. In plug flow, the fluid velocity is the same everywhere. Given the fact that a viscous fluid layer or particle in contact with a boundary has to move with the velocity of the boundary and that still air imposes negligible stress on adjacent viscous layers, show how such flows can be achieved by means of two parallel plates, by examining the following cases:

    **(a)** both plates of length 50 cm and width 20 cm, placed a distance 2 cm apart, traveling with speed 10 m/s;

    **(b)** lower plate stationary, upper moving at 10 m/s;

    **(c)** upper plate removed, lower plate traveling at 10 m/s;

    **(d)** both plates stationary. For each case determine the velocity profile from the lower to the upper plate, the flow rate obtained, and the *shear* or *drag force* required for each of the following fluids:

      (i) air of viscosity 0.01 cP;

      (ii) water of viscosity 1 cP;

      (iii) polymer melt of viscosity 1000 P.

      (*Hint:* First sketch the velocity profiles.)

**18.** *Film thickness by dimensional analysis.* Relate the thickness $h$ of a film of liquid of density $\rho$, viscosity $\eta$, and surface tension $\sigma$, falling under gravity $g$ at flow rate $\dot{V}$ adjacent to a vertical plate, to these variables and to length $L$ downstream from the leading edge of the plate. Identify the resulting dimensionless numbers. What ratios do they represent?

**19.** *Boundary layer thickness by dimensional analysis.* Liquid of viscosity $\eta$ and density $\rho$ approaches a plate at uniform velocity $u$, and forms a boundary layer of thickness $\delta(x)$ of parabolic velocity profile over the plate, which increases with downstream distance $x$ from the leading edge of the plate. Find the form of the relation $\delta = \delta(\rho, \eta, u, x)$ in terms of dimensionless numbers.

**20.** *Dimensionless analysis.* The exact equation that governs the flow of the film of

Problem 1.18 is[27]

$$\left(\frac{d^3h}{dx^3}\sigma - \rho g\right)\frac{h^3}{2\eta} + \dot{V} = 0,$$

where $x$ is the downstream distance (i.e., $0 \leq x \leq L$). Make the equation dimensionless, identify the resulting dimensionless numbers, and compare to those obtained by the dimensional analysis of Problem 1.18.

21. *Dimensionless analysis.* A convection–diffusion–reaction process in a plug reactor is described by the equation and boundary conditions,

$$u\frac{dc}{dx} = \mathscr{D}\frac{d^2c}{dx^2} - kc,$$

$$c(x = 0) = c_0, \qquad \frac{dc}{dx}\bigg|_x = L = 0$$

where $u$ is the plug velocity, $c$ the reacting species concentration, $x$ the downstream distance, $\mathscr{D}$ a diffusion coefficient, $k$ the specific raction rate, $c_0$ the entering reactant concentration, and $L$ the reactor length. Make the equation and boundary conditions dimensionless and identify and explain the resulting dimensionless numbers. You may also proceed with the solution to these equations.

22. *Scaleup of lab results.* Assume that the large-scale equivalent of Problem 4.19 is an air boundary layer adjacent to an aircraft wing surface, traveling at velocity $V$. How would you utilize small-scale experiments and dimensional analysis to predict the boundary layer thickness and related quantities around the large-scale wing?

23. *Friction by dimensional analysis.* Derive the functional form of the friction factor for flow in rough pipe and the form of drag coefficient for flow around a spheroid, by dimensional analysis considerations.

24. *Dimensional analysis of gravity waves.*[22] Figure P1.24 shows typical wave shapes along the free surface of a deep and wide water reservoir (e.g., a lake or the ocean).
    (a) Given that the speed of wave propagation, $c$, may depend on $h$, $\alpha$, and $\lambda$ as well as on density $\rho$ and gravity acceleration $g$, show that

$$\frac{c}{\sqrt{g\lambda}} = \begin{cases} F\left(\dfrac{\alpha}{\lambda}, \dfrac{h}{\lambda}\right) \\ \text{const.,} \end{cases}$$

for waves over deep water, where $h \gg \lambda \gg \alpha$.

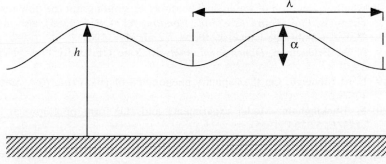

**Figure P1.24** Free-surface gravity waves. (From Ref. 22, by permission.)

**(b)** In the other extreme case, of long waves on shallow water, for which $\lambda \gg h \gg \alpha$, show that

$$\frac{c}{\sqrt{gh}} = \text{const.}$$

**(c)** How would you estimate the constants in these two equations?

# REFERENCES

1. I. Newton, *Principia* (*by S. Pepys, London, 1886*), Vol. 1, *The Motion of Bodies* (transl. F. Cajori), University of California Press, Berkeley, Calif., 1966.
2. R. B. Bird, W. E. Stewart, and E. N. Lightfoot, *Transport Phenomena,* Wiley, New York, 1960.
3. M. J. Assael, M. Pepadaki, S. M. Richardson, C. Oliveira, and W. A. Wakeham, Vibrating wire viscometry on liquid hydrocarbons at high pressure, *High Temp. High Resolut.* 23, 561 (1991).
4. R. B. Bird, R. C. Armstrong, and O. Hassager, *Dynamics of Polymeric Liquids,* Vol. 1, *Fluid Mechanics,* Wiley, New York, 1990.
5. R. I. Tanner, *Engineering Rheology,* London Press, Oxford, 1985.
6. J. R. A. Pearson, *Mechanics of Polymer Process,* Elsevier Applied Science Publisher, London, 1985.
7. N. Tipei, *Theory of Lubrication,* Stanford University Press, Stanford, Calif., 1962.
8. O. Pinkus, *Thermal Aspects of Fluid Film Tribology,* ASME Press, New York, 1990.
9. L. E. Scriven, Dynamics of a fluid interface, *Chem. Engrg. Sci.,* 12, 98 (1960).
10. B. V. Deryagin and S. M. Levi, *Film Coating Theory,* Focal Press, New York, 1964.
11. D. A. Edwards, H. Brenner and D. T. Wasen, *Interfacial Transport Processes and Rheology,* Butterworth Heinemann, Boston, 1991.
12. *Metrification in Scientific Journals,* Royal Society of London, London, 1968.
13. R. J. Goldstein, *Fluid Mechanics Measurements,* Hemisphere, New York, 1983.
14. J. H. Perry, *Perry's Chemical Engineer's Handbook,* 6th ed., McGraw-Hill, New York, 1984.
15. R. C. Reid, J. M. Prausnitz, and T. K. Sherwood, *The Properties of Gases and Liquids,* McGraw-Hill, New York, 1977.
16. J. Shackelford and W. Alexander, *The CRC Materials Science and Engineering Handbook,* CRC Press, London, 1992.
17. Sir O. Reynolds, An experimental investigation of the circumstances which determine whether a motion of water shall be direct or sinuous and the flow resistance in parallel channels, *Phil. Trans. Roy. Soc., London,* A174, 935 (1883); also in *Scientific Papers,* Vol. 2, Cambridge University Press, Cambridge, 1901.
18. P. W. Bridgeman, *Dimensional Analysis,* Yale University Press, New Haven, Conn., 1931.
19. Lord Rayleigh, On the capillary phenomena of jets, *Proc. Roy. Soc., London,* 29, 71 (1879).
20. E. Buckingham, Model experiments and the form of empirical equations. *Trans. ASME,* 37 263 (1915).
21. S. L. Kline, *Similitude and Approximation Theory,* McGraw-Hill, New York, 1965.
22. R. R. Long, *Engineering Science Mechanics,* Prentice Hall, Englewood Cliffs, N.J., 1963.

23. R. L. Panton, *Incompressible Flow,* Wiley, New York, 1984.

24. R. B. Bird, W. E. Stewart, and E. N. Lightfoot, *Transport Phenomena,* Wiley, New York, 1960.

25. H. S. Fogler, *Elements of Chemical Reaction Engineering,* Prentice Hall, Englewood Cliffs, N.J., 1986.

26. J. H. Seinfeld, *Atmospheric Chemistry and Physics of Air Pollution,* Wiley, New York, 1985.

27. T. C. Papanastasiou, Lubrication flows, *Chem. Engrg. Eoluc.,* 24, 50 (1989).

# 2

# *Fluid Statics*

## 2.1 STATIC EQUILIBRIUM OF FLUIDS

A fluid is in static equilibrium if *there is no relative motion of any part of the fluid with respect to the rest.* Thus, water in a lake and in an ocean, under conditions of no wind or other disturbances, is in static equlibrium, and so is water in a stationary reservoir and air in an inflated balloon. According to the definition given, the water in the container and the air in the balloon will continue to be in static equilibrium even if the container or the balloon are experiencing any translational or rotational motion that does not induce any relative motion within the fluid, i.e., even if the entire fluid and container move together as if a solid. These three cases of equilibrium are illustrated in Fig. 2.1. In none of these cases is there motion of any part of the liquid with respect to the rest, including the container.

A fluid in *static equilibrium* is characterized by the absence of shear or viscous normal forces that may arise under conditions of relative flow, as explained in Section 1.3. Thus *the only stress present at static equilibrium is the hydrostatic pressure, always normal and toward submerged surfaces.* According to Eqn. (1.2), to be derived subsequently, the value of hydrostatic pressure (stress) at a point depends only on the fluid's density and the depth of the point. Thus

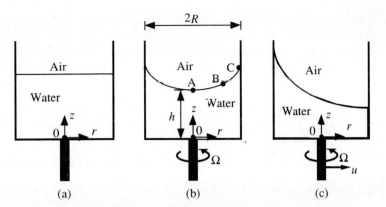

**Figure 2.1**  Cases of static equilibrium under no motion, under solid-body-like rotation, and under solid-body-like rotation and translation.

the magnitude of pressure on any differential surface $dA$ through the point, at that point, is independent of the orientation of the surface. Therefore, *pressure is an isotropic normal stress.* The equilibrium or thermodynamic pressure results from the collision of randomly moving fluid molecules with the surface.

### 2.1.1 Hydrostatic Equation

The simplest case of air at static equilibrium under gravity forces alone is shown in Fig. 2.2. Any volume of stationary air (under no-wind conditions), including the cubic one shown in Fig. 2.2, is at equilibrium, and according to classical mechanics laws, the summation of forces on it must be zero in any direction. Any direction is sufficiently represented by the three mutually perpendicular

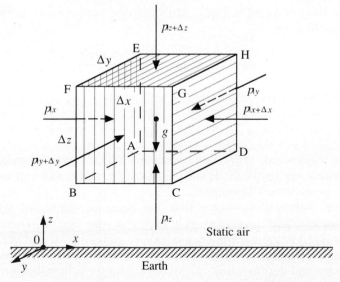

**Figure 2.2**  Static equilibrium of air under gravity.

directions $x$, $y$, and $z$ shown in Fig. 2.2. The forces acting on the volume by its surroundings are the isotropic pressure force toward each of its six square surfaces, and the gravity force at its mass center directed toward the center of the earth, parallel to the $z$-direction in Fig. 2.2. Thus, force balance in the $x$-direction results in

$$\Delta y\, \Delta z[p(x) - p(x + \Delta x)] = 0, \quad \text{thus } p(x) = p(x + \Delta x), \tag{2.1}$$

and similarly in the $y$-direction,

$$\Delta x\, \Delta z[p(y) + p(y + \Delta y)] = 0, \quad \text{thus } p(y) = p(y + \Delta y). \tag{2.2}$$

Force balance in the $z$-direction results in

$$\Delta x\, \Delta y[p(z) - p(z + \Delta z)] - pg\, \Delta x\, \Delta y\, \Delta z = 0, \tag{2.3}$$

which is rearranged to the form

$$p(z) - p(z + \Delta z) = \rho g\, \Delta z. \tag{2.4}$$

Equation (2.4) suggests that the difference in the barometric or hydrostatic pressure between two points in a fluid of density $\rho$, at elevation $\Delta z$ apart, is $\rho g\, \Delta z$. Here

$$g = 981 \text{ cm/s}^2 = 9.81 \text{ m/s}^2 = 32.17 \text{ ft/s}^2 \tag{2.5}$$

is the acceleration of gravity. Similarly, Eqns. (2.1) and (2.2) suggest that the same pressure does not change with horizontal distances. Equation (2.4) is rearranged in the form

$$\frac{p(z + \Delta z) - p(z)}{\Delta z} = -\rho g. \tag{2.6}$$

In the limit of $\Delta z \to 0$, Eqn. (2.6) yields the derivative of pressure with respect to $z$, by definition:

$$\frac{dp}{dz} = \lim_{\Delta z \to 0} \frac{p(z+) - p(z)}{\Delta z} = -\rho g. \tag{2.7}$$

Therefore,

$$\frac{dp}{dz} = -\rho g \tag{2.8}$$

is the celebrated *hydrostatic equation,* which governs pressure distribution in absolutely stationary fluids such as static air, static water in lakes and ocean, and fluids in stationary containers.

Equation (2.8) shows that the pressure decreases with altitude in the atmosphere and increases with depth in the ocean. These pressure changes transmitted to the eardrum cause pain (e.g., taking off and landing in a plane, and diving deep). Therefore, parachutists and divers must be protected or operate in altitudes and depths whose pressures guarantee a harmless environment.

Now imagine that the volume of Fig. 2.2 translates with velocity $u$ and acceleration $\alpha = du/dt$ in the $x$-direction. Then Newton's law of motion in the $x$-direction is

$$\Delta y\, \Delta z[p(x) - p(x + \Delta x)] = \alpha\rho\, \Delta y\, \Delta x\, \Delta z, \qquad (2.9)$$

and following the steps in Eqns. (2.6) to (2.9), it finally reduces to

$$\frac{dp}{dx} = -\rho\alpha. \qquad (2.10)$$

Equation (2.10) predicts that for an accelerating or decelerating solid-like motion with acceleration $\alpha$, there is a pressure change in the direction of motion given by the same equation.

If, in addition, there is solid-body-like rotation around an axis at a distance $r$ from a point in the fluid, with *angular speed* $\Omega$ such that a centrifugal acceleration toward the center is induced, given by

$$\alpha_r = -\Omega^2 r, \qquad (2.11)$$

then a pressure variation along the radius is induced, given by

$$\frac{dp}{dr} = \rho\Omega^2 r. \qquad (2.12)$$

For a general solid-body-like motion, Eqns. (2.8), (2.10), and (2.12) combine to yield

$$\frac{dp}{\rho} + g\, dz + \alpha\, dx - \Omega^2 r\, dr = 0. \qquad (2.13)$$

The equation relates the pressure of neighboring points in the fluid that are at vertical and radial distances, $dz$ and $dr$, apart, as shown in Fig. 2.3. Equation (2.13) is integrated between any two points, 1 and 2, in a fluid of density $\rho$ under gravity of acceleration $g$, solid-body-like translation in the $x$-direction with acceleration $\alpha$, and solid-body-like rotation with centrifugal acceleration $a_r$, in

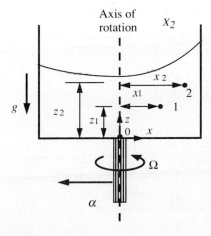
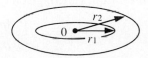

**Figure 2.3** Fluid in static equilibrium under gravity of acceleration $g$, linear translation of acceleration $\alpha$, and angular rotation of speed $\Omega$. The pressure difference between points 1 and 2 is given by Eqn. (2.14).

the form

$$\int_1^2 \frac{dp}{\rho} + g(z_2 - z_1) + \alpha(x_2 - x_1) - \frac{\Omega^2}{2}(r_2^2 - r_1^2) = 0. \qquad (2.14)$$

where $z_i$, $x_i$, and $r_i$ are finite distances or coordinates, as shown in Fig. 2.3. Equation (2.13) is the *generalized hydrostatic equation*, which reduces to Eqn. (2.8) in case of zero fluid acceleration and rotation.

### 2.1.2 Solution of Hydrostatic Equation

To integrate the first term of Eqn. (2.14), an *equation of state* between pressure and density is needed, as discussed in Section 1.2. Fluids whose density can be made to depend on either temperature or pressure alone are called *barotropic fluids*. For these fluids, the integral term of Eqn. (2.14) can be computed analytically, by eliminating the temperature, as follows.

*For incompressible liquids* under isothermal conditions, for which $\rho$ = constant, Eqn. (2.14) reduces to

$$p_2 - p_1 + \rho g(z_2 - z_1) + \alpha\rho(x_2 - x_1) - \rho\frac{\Omega^2}{2}(r_1^2 - r_1^2) = 0. \qquad (2.15)$$

In the case of an *ideal gas,* for which Eqn. (1.6) is appropriate, Eqn (2.14) reduces to

$$\frac{R}{M}\int_1^2 \frac{T\,dp\,\rho}{p} + g(z_2 - z_1) + \alpha(x_2 - x_1) - \frac{\Omega^2}{2}(r_2^2 - r_1^2) = 0, \qquad (2.16)$$

and under isothermal conditions, at temperature $T = T_0$, to the form

$$\frac{RT_0}{M}\ln\frac{p_2}{p_1} + g(z_2 - z_1) + \alpha(x_2 - x_1) - \frac{\Omega^2}{2}(r_2^2 - r_1^2) = 0, \qquad (2.17)$$

For *atmospheric air,* for which $\alpha = \Omega = 0$ with respect to the earth and of temperature that varies with elevation according to

$$T(z) = T_0 - \beta z, \qquad (2.18)$$

where $T_0$ is the sea-level temperature and $\beta$ a thermodynamic constant called the *lapse rate,*[1] Eqn. (2.13), given Eqns. (1.6) and (2.18), reduces to

$$\int_{p_0}^p \frac{dp}{p} + \frac{Mg}{R}\int_0^z \frac{dz}{T_0 - \beta z} = 0. \qquad (2.19)$$

Integration of Eqn. (2.19) leads to

$$\ln\frac{p(z)}{p_0} = \frac{Mg}{\beta R}\ln\frac{T_0 - \alpha z}{T_0}, \qquad (2.20)$$

and equivalently to

$$\frac{p(z)}{p_0} = \left(\frac{T_0 - \beta z}{T_0}\right)^{Mg/\beta R}. \qquad (2.21)$$

The corresponding air-density distribution is

$$\frac{\rho(z)}{\rho_0} = \left(\frac{T_0 - \beta z}{T_0}\right)^{(Mg/\beta R)-1}, \tag{2.22}$$

where $\rho_0$ is the air density at sea level under temperature $T_0$ and pressure $p_0$. For a typical lapse rate $\beta = 0.0066°C/m$, it is

$$\frac{Mg}{\beta R} = 5.2, \tag{2.33}$$

and therefore air-density variations up to 5000 m are less than $0.01\rho_0$.

Under *isothermal conditions* [i.e., for $\beta = 0$ in Eqn. (2.18)], the pressure and density variations with elevation are

$$\frac{p(z)}{p_0} = \frac{\rho(z)}{\rho_0} = \exp\left(-\frac{Mg}{T_0 R}z\right), \tag{2.24}$$

which are also the limits of Eqns. (2.21) and (2.22), respectively, in the limit of $\beta \rightarrow 0$.

### 2.1.3 Manometers

Manometers are devices utilized to measure pressure of static or flowing fluids, based on the hydrostatic equation. The prototype of a U-shaped manometer to measure the pressure of a flowing fluid is shown in Fig. 2.4. Depending on the expected magnitude of pressure $p$, the U-shaped tube is filled with mercury, oil, or water. U-shaped manometers with the upper end open as shown in Fig. 2.4 measure *gauge pressures*. *Absolute pressures* are measured with U-shaped manometers with the upper end closed, for example in vacuum installations. The absolute pressure (e.g. psi) is the gauge pressure (e.g., psig) plus the atmospheric (e.g., 14.696 psi) [i.e., $p(\text{psia}) = p(\text{psig}) + 14.696$].

Absolute pressures between 1 and 760 mmHg are measured with U-shaped

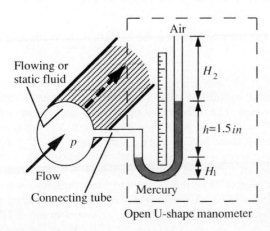

Open U-shape manometer

**Figure 2.4** Open U-shaped manometer that measures gauge pressure of flowing or stationary fluid. The measured gauge pressure is $p = \rho_{Hg}gh$.

manometers filled with mercury, whereas for pressures below 1 mmHg, oil is used. Pressures of 1 to 1000 $\mu$mHg are measured with the *McLeod manometer,* which is based on gas compressibility equations. High vacuum is measured by manometers of thermal conductivity or nuclear energy absorption. *Two-liquid manometers* are U-shaped tubes filled with two immiscible liquids of densities $\rho_A$ and $\rho_B$, used to measure *pressure differences* with high accuracy, as $\Delta p = g \, \Delta h(\rho_A - \rho_B)$. The U-shaped *inclined manometer* measures small pressure differences, as $\Delta p = g \, \Delta h(\rho_A - \rho_B) \sin \phi$. For pressures varying over a wide value range, metallic Bourdon-type manometers are employed. Details on diversity of manometers are tabulated in several engineering handbooks.[2]

The hydrostatic equations, Eqns. (2.14) to (2.17), are used to find the pressure distribution in closed and open liquid and gas systems, static or under rotation and/or translation, as demonstrated in Examples 2.1 and 2.2. Equations (2.20) to (2.24) are used to find pressure and density distribution in the atmosphere, as demonstrated in Examples 2.3 and 2.4. Equation (2.8) is used to find pressure distributions in stationary liquid reservoirs such as tanks, lakes, oceans and elsewhere, and to calculate resulting forces on submerged surfaces as detailed in Section 2.2 and Examples 2.5 and 2.6 and on submerged volumes according to Archimedes' principle of buoyancy,[3] as detailed in Section 2.2.1 and Examples 2.7 and 2.8, and in Section 2.2.2 and Examples 2.9 and 2.10.

The hydrostatic equation combined with the Young–Laplace equation of capillarity allows the determination of pressure distribution across free surfaces and interfaces of immiscible fluids and often the shape assumed by these surfaces, as detailed in Section 2.3 and in Examples 2.11 and 2.12.

**Example 2.1: Pressure Distribution in Stratified Liquids**

Stir 10 kg of oil of density $\rho_0 = 0.85 \, \text{g/cm}^3$ and 10 kg of water of density $\rho_w = 1 \, \text{g/cm}^3$ and mix in an open cylindrical container of radius $R = 10$ cm; then allowed to come to equilibrium.

**(a)** Show that the oil will always be on top of the water.

**(b)** What are the resulting pressures at the planar interface of the two liquids, and at the bottom of the container?

**(c)** What will happen to the two liquids if the container starts rotating with angular speed $\Omega$ rad/s?

**Solution**

**(a)** Consider first a small cylindrical volume of water of Volume $V_w$ surrounded by oil. According to Eqn. (2.6), the resulting upward net pressure force on the water volume is $F_p = (\Delta p)A = \rho_0 g A = \rho_0 g V_w$, where $A$ is the area of its basis. The opposing gravity force is $F_g = \rho_w g V_w$, downard. Thus $F_p < F_g$ and oil will cause any momentarily dispersed water to precipitate. The opposite conclusions are drawn for an oil volume dispersed in water. The resulting net pressure force is greater than the oil weight, and therefore water will cause any dispersed oil to ascend. A combination of the two expelling actions leads to a discrete oil layer on top of a water layer, as shown in Fig. E2.1a.

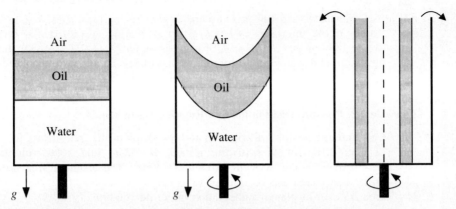

**Figure E2.1**   Stratification of oil on top of water: (a) under gravity, (b) under gravity and rotation, and (c) under rotation in the absence of gravity (e.g., in space).

**(b)**  The heights of each of the two layers are

$$h_o = \frac{V_o}{\pi R^2} = \frac{m_o}{\rho_o \pi R^2} = 37.4 \text{ cm}$$

$$h_w = \frac{V_w}{\pi R^2} = \frac{m_w}{\rho w \pi R^2} = 31.8 \text{ cm}.$$

It is convenient to measure distances up from the bottom, which is therefore identified with $z = 0$. Equation (2.8) applies in both layers. Consider first the oil layer, for which Eqn. (2.8) is integrated to

$$p_o(z) = -\rho_o g z + c_o, \qquad h_w \leq z \leq h_w + h_o, \qquad (2.1.1)$$

with $c_o$ an arbitrary constant of integration. Equation (2.1.1) holds at $z = h_w + h_o$, where the pressure is $p_o$, which substituted in Eqn. (2.1.1) determines $c_o = p_o + \rho_o g(h_w + h_o)$, and the pressure distribution in the oil layer is

$$p_o(z) = p_o + \rho_o g(h_w + h_o - z), \quad h_w \leq z \leq <h_w + h_o. \qquad (2.1.2)$$

The pressure at the interface predicted by this equation is

$$p_o(z = h_w) = p_o + \rho_o g(h_w + h_o - h_w) = p_o + \rho_o g h_o, \qquad (2.13)$$

identical to the supported columns of oil and air. Equation (2.8) applied to the water layer yields

$$p_w(z) = -\rho_o g z + c_w, \qquad 0 \leq z \leq h_w. \qquad (2.1.4)$$

At the interface at $z = h_w$, it is $p_o(z = h_w) = p_w(z = h_w) = p_o + \rho_o g h_o$, by Eqn. (2.1.3), which substituted in Eqn. (2.1.4) yields the pressure distribution in the water layer as

$$p_w(z) = p_o + \rho_o g h_o + \rho_o g(h_w - z). \qquad (2.1.5)$$

The pressure at the bottom is simply

$$p_w(z = 0) = p_o + \rho_o g h_o + \rho_w g h_w, \qquad (2.1.6)$$

identical to the supported columns of air, oil, and water.

Selected values of pressure are: at the free surface: $p_o(z = h_o + h_w) = p_o = 1\,\text{atm}$; at the interface: $p_o(z = h_w) = p_w(z = h_w) = p_o + \rho_o g h_o = 1.013\,\text{atm}$; and at the bottom: $p_w(z = 0) = p_o + \rho_o g h_o + \rho_w g h_w = (1.031 + 0.030)\,\text{atm}$. The configuration under gravity and rotation is that of Fig. E2.1b and is analyzed in Example 2.2. In space, in the absence of gravity and under rotation, the configuration is that of Fig. E2.1c (even in a bottomless container).

### Example 2.2: Pressure Distribution in a Rotating Open Liquid

Calculate the pressure distribution and the shape of the free surface for the situation of Fig. 2.1b, for $\Omega = 10\,\text{rpm}$, $\alpha = 0$, $d = 2\,\text{m}$, and total volume of water $V = 6.28\,\text{m}^3$, in contact with air of $p_{air} = 1\,\text{atm}$. The extended free surface is often utilized to increase mass transfer rates across. Calculate the increase of interfacial area $\Delta S$ over its planar configuration under no rotation.

**Solution**   For incompressible water, Eqn. (2.15) applies with $\alpha = 0$:

$$p_2 - p_1 + \rho g(z_2 - z_2) - \frac{\rho \Omega^2}{2}(r_2^2 - r_1^2) = 0. \tag{2.2.1}$$

One of the points is chosen conveniently to be the lower point at the interface, $A$, of $r_2 = 0$ and $z_2 = h$, where $p_{air} = 1\,\text{atm}$ in the absence of surface tension. Thus, Eqn. (2.2.1) becomes

$$p_{air} = p(r, z) + \rho g(h - z) - \rho \frac{\Omega^2}{2}(0 - r^2) = 0, \tag{2.2.2}$$

where $(z, r)$ is any point within the rotating liquid. Its pressure is therefore given by

$$p(r, z) = p_{air} + \rho g(h - z) + \rho \frac{\Omega^2 r^2}{2}. \tag{2.2.3}$$

The shape of the free surface is calculated as follows: Choose another point on the free surface, $B$, shown in Fig. 2.1b, of coordinates $r_B$ and $z_B$, where the pressure is also $p_{air}$. Then according to Eqn. (2.2.3), it is

$$p(r_B, z_B) = p_{air} + \rho g(h - z_B) + \rho \frac{\Omega^2 r_B^2}{2} = p_{air}, \tag{2.2.4}$$

which yields

$$z_B - h = \frac{\Omega^2 r_B^2}{2g} > 0. \tag{2.2.5}$$

Equation (2.2.5) suggests that for any point, $B$, on the free surface,

$$z_B > h, \tag{2.2.6}$$

and that the shape of the free surface is that of a paraboloid given by that equation.

The volume of a paraboloid $V_p$ is half that of the right cylinder that encloses it,[4] which for the case at hand is

$$V_p = \frac{1}{2}\pi R^2(H - h), \tag{2.2.7}$$

where $(H - h)$ is given by applying Eqn. (2.2.5) to point $C$, of $z_c = H$ and $r_c = R$, where $p = p_{air}$, that is,

$$H - h = \frac{\Omega^2 R^2}{2g}. \tag{2.2.8}$$

Equations (2.2.7) and (2.2.8) combine to yield the volume of the paraboloid occupied by air:

$$V_p = \frac{1}{2} \pi R^2 \frac{\Omega^2 R^2}{2g} = \frac{\pi R^4 \Omega^2}{4g}. \tag{2.2.9}$$

Then the volume $V$ occupied by the rotating liquid is

$$V = \pi R^2 H - \frac{\pi R^4 \Omega^2}{4g} = \pi R^2 \left( H - \frac{R^2 \Omega^2}{4g} \right), \tag{2.2.10}$$

and therefore

$$H = \frac{V}{\pi R^2} + \frac{R^2 \Omega^2}{4g}, \tag{2.2.11}$$

where $V$ is also the initial volume before rotation starts. Thus

$$h = \frac{V}{\pi R^2} - \frac{R^2 \Omega^2}{4g}. \tag{2.2.12}$$

The area of the free surface is computed by the revolution of the curve by Eqn. (2.2.5), around the $z$-axis:

$$S_p = \int_0^R 2\pi r \left[ 1 + \left( \frac{dz}{dr} \right)^2 \right]^{1/2} dr = \frac{2\pi g^2}{3\Omega^4} \left[ \left( 1 + \frac{\Omega^4 R^2}{g^2} \right)^{3/2} - 1 \right], \tag{2.2.13}$$

which represents an increase of

$$\frac{\Delta S}{\pi R^2} = \frac{S_p - \pi R^2}{\pi R^2} = \frac{2\Omega^2 R}{g} - 1 \simeq \frac{2\Omega^2 R^2}{3g}, \tag{2.2.14}$$

at high rotation, such that

$$\frac{2\Omega^2 R}{3g} \gg 1. \tag{2.2.15}$$

The increase in free surface rotation may be utilized to increase heat and solute transfer across rotating free surfaces and/or interfaces of fluids, since the amount of total rate of exchange, $\dot{H}$ or $\dot{m}_c$, is in general,

$$\dot{H} = KS \, \Delta T, \tag{2.2.16}$$

where $K$ is the appropriate exchange coefficient, $S$ the area of the surface across which the exchange takes place, and $\Delta T$ the driving temperature or concentration difference.

### Example 2.3: Temperature Distribution in the Atmosphere

The following elevation and pressure data have been collected:

| Elevation (m) | Pressure (atm) |
| --- | --- |
| 0 | 1.000 |
| 1000 | 0.895 |
| 2000 | 0.800 |
| 3000 | 0.710 |
| 4000 | 0.630 |
| 5000 | 0.560 |
| 6000 | 0.490 |
| 7000 | 0.435 |
| 8000 | 0.380 |

Calculate the temperature profile $T(z)$ for $0 \leq z \leq 8000$, with $T(z = 0) = 27°C$.

**Solution**    Equation (2.8) applies here:

$$\frac{dp}{dz} = -\rho g = -\left(\frac{pM}{RT}\right)z, \qquad (2.3.1)$$

which is integrated to

$$\ln\frac{p(z)}{p(z=0)} = -\frac{Mg}{R}\int_0^z \frac{dz}{T(z)} = -\frac{(29\ \text{g/mol})(981\ \text{cm/s}^2)}{840{,}000\ [\text{dyn}][\text{m}]/[\text{mol}][\text{K}]}\int_0^z \frac{dz}{T(z)} \qquad (2.3.2)$$

$$= -0.0327\int_z^0 \frac{dz}{T(z)}.$$

Since the temperature distribution $T(z)$ is not known, the integral cannot be evaluated analytically. However, discrete values of temperature with elevation are available, which allows numerical integration as follows.

Equation (2.3.2) is rearranged in the form

$$\ln\frac{p(z_{i+1})}{p(z_i)} = -0.0327\int_{z_i}^{z_{i+1}} \frac{dz}{T(z)} \approx -0.0327 \times \frac{z_{i+1} - z_i}{T[(z_{i+1} - z_i)/2]} \qquad (2.3.3)$$

and then to

$$T\left(\frac{z_{i+1} + z_i}{2}\right) = -\frac{0.327(z_{i+1} - z_1)}{\ln[p(z_{i+1})/p(z_i)]}. \qquad (2.3.4)$$

Based on the latter equation, the following table is constructed.

| Elevation $(z_i + z_{i+1})/2$ | Pressure, $z_i$ (atm) | $\ln[(p(z_{i+1})p(z_1)]$ | $T(°\text{K})$ |
|---|---|---|---|
| 0 | 1.000 | | 300 |
| 1000 | 0.895 | −0.1100 | 297 |
| 2000 | 0.800 | −0.1122 | 291 |
| 3000 | 0.710 | −0.1151 | 284 |
| 4000 | 0.630 | −0.1190 | 275 |
| 5000 | 0.560 | −0.1207 | 271 |
| 6000 | 0.490 | −0.1251 | 261 |
| 7000 | 0.435 | −0.1292 | 253 |
| 8000 | 0.380 | −0.1352 | 242 |

Thus the temperature distribution is almost linear, with gradient $dT/dZ = -0.007$ K and lapse rate $\beta = 0.007$ K/m:

$$T(z) = (300 - 0.007z)\ \text{K}. \qquad (2.3.5)$$

## Example 2.4: Pressure Distribution in the Atmosphere

Repeat Example 2.3 when the temperature is measured according to the following table, using a pressure of 1 atm at sea level.

| Elevation (m) | Temperature, T (°C) |
|---|---|
| 0 | 27 |
| 1000 | 24 |
| 2000 | 18 |
| 3000 | 11 |
| 4000 | 2 |
| 5000 | −2 |
| 6000 | −12 |
| 7000 | −20 |
| 8000 | −31 |

**Solution**    The given temperature distribution is used in Eqn. (2.3.3) to calculate the unknown pressure distribution, by

$$p(z_{i+1}) = p(z_i) \exp\left[-\frac{0.0327(z_{i+1} - z_i)}{T[(z_{i+1} + z_i)/2]}\right] \tag{2.4.1}$$

Starting from $z_i = z_0 = 0$, Eqn. (2.4.1) produces the following table:

| Elevation, $z_i$ (m) | $(z_i + z_{i+1})/2$ | $T$ (°C) | $T$ (K) | $p(z_i)$ (atm) |
|---|---|---|---|---|
| 1000 | 500 | 25.5 | 298.5 | 0.896 |
| 2000 | 1500 | 21.0 | 294.0 | 0.801 |
| 3000 | 2500 | 14.5 | 287.5 | 0.715 |
| 4000 | 3500 | 9.5 | 279.5 | 0.636 |
| 5000 | 4500 | 0.0 | 273.0 | 0.564 |
| 6000 | 5500 | −7.0 | 266.0 | 0.499 |
| 7000 | 6500 | −16.0 | 257.0 | 0.439 |
| 8000 | 7500 | −25.5 | 247.5 | 0.385 |

These results compare well within the data of Example 2.3, as they should.

## 2.2 PRESSURE FORCE

Any surface submerged in a fluid is subjected to an inward normal pressure at each point, depending on its location or depth within the fluid, but not on its orientation. Consider the inclined surface shown in Fig. 2.5, of total area $A$, submerged at an angle $\phi$ with the horizontal within a fluid of constant density $\rho$ in contact with air of atmospheric pressure $p(z = 0) = p_0$. The height dimension of

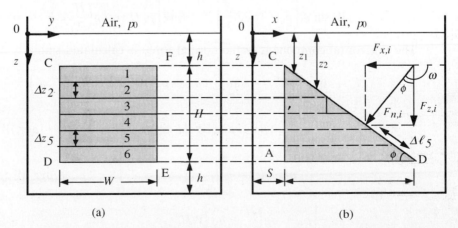

(a)                                          (b)

**Figure 2.5**  Force on surface of inclination $\phi$ and side $CD$ as shown in (b), projected horizontally as $CDEF$, as shown in (a).

the prism is perpendicular to the plane of the paper. The inclined surface is projected horizontally as shown in Fig. 2.5.

To calculate the total force on the inclined surface shown, we split the surface vertically into pieces 1 to 6, each of which has elementary area $\Delta A_i = W \, \Delta z_i$, over which the hydrostatic pressure can be considered constant given by $p_i = p_0 = \rho g z_i$, where $z_i$ is the distance of the elementary area $\Delta A_i$ from the free surface where the pressure is $p_0$, as shown in Fig. 2.5b. The normal force on each of the elementary surfaces is $F_{n,i} = p_i \, \Delta A_i = (p_0 + \rho g z_i) W \, \Delta l_i$. The resulting total normal force, $F_n$, is the summation of these individual forces:

$$F_n \simeq \sum_{i=1}^{k=6} F_{n,i} \simeq \sum_{i=1}^{k=6} p_i \, \Delta A_i \simeq \sum_{i=i}^{k=6} (p_0 + \rho g z_i) W \, \Delta l_i. \tag{2.25}$$

Equation (2.25) can be made more and more accurate as the number of subdivisions increases, such that the pressure becomes truly constant over each subdivision of vanishingly small height $\Delta z_i$. Then the approximation of Eqn. (2.25) can be replaced by an exact integral:

$$F = \lim_{k \to \infty} \sum_{i=1}^{k} F_{n,i} = \lim_{k \to \infty} \sum_{i=1}^{k} (p_0 + \rho g z_i) W \, \Delta l_i$$
$$= \int_A (p_0 + \rho g z) W \, dl = \int_A p(z) \, dA \tag{2.26}$$

over the the area $A$ of the submerged surface. This is the most general expression to calculate pressure forces over any submerged area: It is the integral of the pressure distribution $p(z)$ over the area $A$ of the submerged surface. Equation (2.26) is simplified and utilized as follows.

In case of a geometrically regular surface (e.g., square, triangle, circle, etc.), the integral can easily be calculated, as in the case of Fig. 2.5, where

$$F_n = \int_A p(z) \, dA = \int_0^l (p_0 + \rho g z) W \, dl = \int_h^{H+h} (p_0 + \rho g z) W \sin \phi \, dz$$
$$= W \sin \phi \left[ p_0 z + \rho g \frac{z^2}{2} \right]_h^{H+h} = W \left[ p_0 H + \rho g \left( \frac{H}{2} + h \right) \right] h \sin \phi. \tag{2.27}$$

The horizontal component of the normal force is calculated similarly:

$$F_x = \lim_{k \to \infty} \sum_{i=1}^{k} F_{x,i} = \lim_{k \to \infty} \sum_{i=1}^{k} (p_0 + \rho g z_i) W \, \Delta l_i \cos\left( \frac{\pi}{2} + \phi_i \right)$$

$$= -\lim_{k \to \infty} \sum_{i=1}^{k} (p_0 + \rho g z_i) W \, \Delta l_i \sin \phi = \lim_{k \to \infty} \sum_{i=1}^{k} (p_0 + \rho g z_i) W_i$$

$$= -\int_{A_x} p(z) \, dz = -\int_{A_x} p(z) \, dz = -\int_h^{H+h} (p_0 + \rho g z) W \, dz$$

$$= -\left[ p_0 + \rho g \left( \frac{H}{2} + h \right) \right] WH. \tag{2.28}$$

Here $\pi/2 + \phi$ is the angle between the normal direction *toward* the segment $\Delta l$ and the positive $x$-direction.

Thus, the important conclusions here are:

1. The horizontal component of the pressure force acting on an inclined surface is identical to the pressure force acting on its horizontal projection, as shown by Fig. 2.5 and Eqn. (2.28).

2. The resulting total pressure force on a vertical surface is the product of its area with the prevailing pressure at its geometrical centroid: In Eqn. (2.28), $WH$ is the area and $p_0 + \rho g(H/2 + h)$ the pressure at the centroid. When the area and the location of its centroid are easy to compute (e.g., triangular, square, circular). The approach is faster and easier to follow.

Similarly, the vertical component of the pressure force is

$$F_z = \lim_{k \to \infty} \sum_{i=1}^{k} F_{z,i} = \lim_{k \to \infty} \sum_{i=1}^{k} (p_0 + \rho g z_i) W \, \Delta l_i \cos \phi$$

$$= \lim_{k \to \infty} \sum_{i=1}^{k} (p_0 + \rho g z_i) W \, \Delta x_i = \int_{A_z} p(z) \, dx, \tag{2.29}$$

where $\phi$ is the angle between the normal direction *toward* the segment $\Delta l$ and the positive $z$-direction. To calculate the resulting integral, a relation between $z$ and $x$ along the surface is required, in order to eliminate $z$ in favor of $x$. This relation is the equation that represents the surface with respect to the origin $x = z = 0$. In the case of Fig. 2.5, this equation is that of a straight line of inclination $-\tan \phi$ (i.e., $z = h - \tan \phi (s - x)$), and therefore the transformed expression for the pressure is

$$p(z) = p[z = h - \tan \phi (s - z)] = p_0 + \rho g[h - (s - x) \tan \phi], \tag{2.30}$$

and the resulting vertical force is

$$F_z = \int_{s + H/\tan \phi}^{s} [p_0 + \rho g(h - (s - x) \tan \phi)] W \, dx$$

$$= \frac{WH}{\tan \phi} + \rho g W \left[ hx - \left( sx - \frac{x^2}{2} \right) \tan \phi \right]_{s + H/\tan \phi}^{s}$$

$$= p_0 \frac{WH}{\tan \phi} + \rho g W \left( \frac{hH}{\tan \phi} + \frac{H^2}{2 \tan \phi} \right) \tag{2.31}$$

$$= W \left[ \left( h + \frac{H}{2} \right) \rho g + p_0 \right] \frac{H}{\tan \phi}.$$

Thus the vertical component of the pressure force on an inclined surface is the product of the pressure at its centroid with its projected area in the vertical direction: In Eqn. (2.31) $(h + H/2)\rho g + p_0$ is the pressure at the centroid and $WH/\tan\phi$ is the vertically projected area.

Knowing the components $F_x$ and $F_z$, the normal to the surface force, $F_n$, is calculated by

$$F_n = (F_x^2 + F_z^2)^{1/2} = W\left[\left(h + \frac{H}{2}\right)\rho g + p_0\right]\left[H^2 + \left(\frac{H}{\tan\phi}\right)^2\right]^{1/2}$$

$$= W\left[\left(h + \frac{H}{2}\right)\rho g + p_0\right]H\sin\phi,$$

(2.32)

which agrees with Eqn. (2.27), as it should. The direction of $F_n$ is given by

$$\tan\omega = \frac{F_z}{F_x},$$

(2.33)

where $\omega$ is the angle between the normal force $F_n$ and the positive $x$-direction, as shown in Fig. 2.5.

### Example 2.5: Pressure Forces on Submerged Surfaces in Fluids

Calculate the resulting force and its horizontal and vertical components on the surfaces shown in Fig. E2.5a, submerged in water of density $\rho = 1$ g/cm$^3$. Apply to the case of $h = 0.2$ m, $H = 0.5$ m, and $\phi = 30°$.

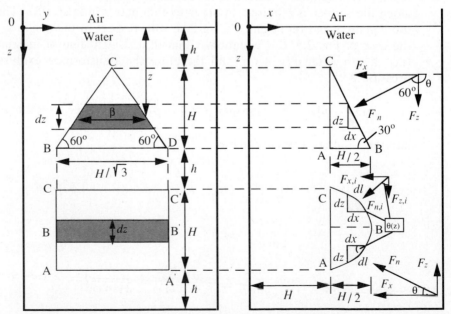

**Figure E2.5** (a) Pressure forces on submerged triangular and cylindrical surfaces, $BCD$ and $ABCC'B'A'$, respectively, oriented as shown in (b).

**Solution**    I. Force on triangular surface $BCD$, of inclination $\phi = 30°$:

$$G_n = \int_A p(z)\beta \, dl = \int_{A_z} p(z)\frac{2\sqrt{3}}{3}(z-h)\frac{dz}{\sin \phi} = \int_h^{h+H} (\rho g(z))\frac{2\sqrt{3}}{3}(z-h)\frac{dz}{\sin \phi}$$

$$= \frac{2\sqrt{3}\,\rho g}{3\sin \phi}\int_h^{h+H} z(z-h)\,dz = \frac{2\sqrt{3}\,\rho g}{3\sin \phi}\left[\frac{z^3}{3} - \frac{hz^2}{2}\right]_h^{h+H} \tag{2.5.1}$$

$$= \frac{2\sqrt{3}\,\rho g}{3\sin \phi}\frac{H^2}{2}\left(h + \frac{2}{3}H\right) = \left[\frac{H^2}{\sqrt{3}}\right]\left[\frac{\rho g}{\sin \phi}\left(h + \frac{2}{3}H\right)\right]$$

$$F_x = -\int_A p(z)\beta \, dl \sin 30° = -\left[\frac{H^2}{\sqrt{3}}\right]\left[\rho g\left(h + \frac{2}{3}H\right)\right] \tag{2.5.2}$$

$$F_z = \int_A p(z)\beta \, dl \cos \phi = \int_A p(z)\beta \frac{dz}{\sin \phi}\cos \phi = \left[\frac{H^2}{\sqrt{3}}\right]\left[\frac{\rho g}{\tan \phi}\left(h + \frac{2}{3}H\right)\right]. \tag{2.5.3}$$

Notice that Eqn. (2.5.1) shows that the resulting normal force is the product of the submerged area, $A = H^2\sqrt{3}\sin \phi$, times the pressure at the centroid of the surface, $p(z = h + 2H/3) = \rho g(h + 2H/3)$. Similarly, Eqn. (2.5.2) suggests that the horizontal force is also given by the projected area $A_z = H^2/\sqrt{3}$ times the pressure at the triangle's centroid, $p(z + h + \frac{2}{3}H) = \rho g(h + \frac{2}{3}H)$. Also, the vertical force, $F_z$, is the product of the projected area, $A_x = H^2/\sqrt{3}$, times the pressure at the centroid of $A$, at $z = h + 2H/3$. These relations could have been invoked in place of Eqns. (2.5.1) to (2.5.3) to attain the same results directly.

II. Force on cylindrical surface $ABCC'B'A'$ of varying inclination $\phi(z)$:

$$F_x = \sum_{i=1}^k F_{x,i} = \int_A \left[p(z)\frac{H}{\sqrt{3}}dl\right]\cos\left[\frac{\pi}{2} - \phi(z)\right] = -\int_{A_x} p(z)\frac{H}{\sqrt{3}}\frac{dz}{\sin \phi(z)}\sin \phi(z)$$

$$= -\int_{2(H+h)}^{H+2h} \rho g z \frac{H}{\sqrt{3}}dz = -\rho g\frac{H}{\sqrt{3}}\left[\frac{z^2}{2}\right]_{2(H+h)}^{H+2} \tag{2.5.4}$$

$$= -\frac{H^2}{\sqrt{3}}\rho g\left(2h + \frac{3}{2}H\right).$$

This is also in the form of projected area $A_x = H^2/\sqrt{3}$, times pressure at its centroid. According to Eqn. (2.29),

$$F_z = \sum_{i=1}^k F_{z,i} = \int_{A_1}\left[p(z)\frac{H}{\sqrt{3}}dl\right]\cos(\pi - \phi) + \int_{A_2} p(z)\frac{h}{\sqrt{3}}dl \cos \phi \tag{2.5.5}$$

$$= \int_{A_{1,z}} p(z)\frac{H}{\sqrt{3}}dx + \int_{A_{2,x}} p(z)\frac{H}{\sqrt{3}}dx,$$

where $A_1$ and $A_2$ are the areas of the lower and upper quarters of the semicircular surface, respectively, and $A_{1,z}$ and $A_{2,z}$ their projection areas in the $z$-direction. To proceed, the pressure $p(z)$ must be expressed in terms of $x$, as $p(x)$. This is done by means of the equation of the circular surface that is part of a circle of radius $H/2$ centered at $x = H$ and $z = 3H/2 + 2H$, which is

$$(x - H)^2 + \left(z - 2h - \frac{3H}{2}\right)^2 = \frac{H^2}{4}, \qquad x > 0, z > 0, \tag{2.5.6}$$

and therefore, along the semicircular surface profile, $z$ is related to $x$ by

$$z = \pm \left[ \frac{H^2}{4} - (x - H)^2 \right]^{1/2} + 2h + \frac{3H}{2},  \qquad (2.5.7)$$

where the minus sign applies to the upper quarter and the plus to the lower one. Therefore, the pressure along the semicircle can also be written in terms of $x$, as

$$p(x) = \rho g z = \rho g \left[ \pm \left( \frac{H^2}{4} - (x - H)^2 \right)^{1/2} + 2h + \frac{3H}{2} \right].  \qquad (2.5.8)$$

Then

$$
\begin{aligned}
F_z = {} & \int_H^{3H/2} \rho g \frac{H}{\sqrt{3}} \left[ \left( \frac{H^2}{4} - (x - H)^2 \right)^{1/2} + 2h + \frac{3H}{2} \right] dx \\
& - \int_H^{3H/2} \rho g \frac{H}{\sqrt{3}} \left[ -\left( \frac{H^2}{4} - (x - H)^2 \right)^{1/2} + 2h + \frac{3H}{2} \right] dx \qquad (2.5.9) \\
= {} & -\frac{\rho g H^3}{4}.
\end{aligned}
$$

The magnitude of the normal force is

$$F_n = (F_x^2 + F_z^2)^{1/2} = \rho g H^2 \left[ \left( 2h + \frac{3H}{2} \right)^2 + \frac{H^2}{16} \right]^{1/2},  \qquad (2.5.10)$$

and its angle $\theta$ with the horizontal is given by

$$\tan \theta = \frac{F_z}{F_x} = \frac{H}{4(2h + 3H/2)}.  \qquad (2.5.11)$$

### III. Application

*For the triangular surface:*

$$F_n = \frac{H^2}{\sqrt{3}} \frac{\rho g}{\sin 30} \left( h + \frac{2}{3} H \right) = \frac{0.25}{\sqrt{3}} \frac{1000 \times 9.81}{0.5} (0.2 + 0.33)\ \text{N/m}^2 = 1501\ \text{N}$$

$$F_x = -F_n \sin 30° = 751\ \text{N}$$

$$F_z = F_n \cos 30° = 1300\ \text{N}$$

$$\tan \theta° = -\frac{1270}{733} = -1.73, \quad \theta = \pi + 60°$$

*For the cylindrical surface:*

$$F_x = -0.25 \times 1000 \times 9.81 \times (0.4 + 0.75)\ \text{N/m}^2 = -2820.4\ \text{N}$$

$$F_z = -\frac{0.25/4}{2} \times 1000 \times 9.81\ \text{N/m}^2 = -306.6\ \text{N}$$

$$F_n = \sqrt{\frac{1820^2 + 613^2}{3}}\ \text{N} = 0000.0\ \text{N}$$

$$\tan \theta = \frac{613}{2820} = 0.21, \quad \text{thus} \quad \theta = 11.8°$$

### Example 2.6: Dam Construction

Consider the three possible constructions shown in Fig. E2.6 for a dam of width $W = 200\,\text{m}$ for a maximum water elevation $H = 30\,\text{m}$. Which would you suggest, based on the forces and their moments by the water on the dam wall?

**Solution**   *Construction (a):*

$$F_n = F_x = \left(\rho g \frac{H}{2}\right)(HW) = 1000 \times 9.81 \times \frac{900}{2} \times 200 = 8.83 \times 10^7\,\text{N}, \qquad F_z = 0.$$

*Construction (b):*   The angle between A'B' and the horizontal is

$$\tan \phi = \frac{2H}{W - S} = \frac{60}{6} = 10, \qquad \phi = 84°, \qquad \sin \phi = 0.994, \qquad \cos \phi = 0.1$$

$$F_n = \frac{WH}{\sin \phi}\left(\rho g \frac{H}{2}\right) = \frac{8.78 \times 10^7}{0.994} = 8.883 \times 10^7\,\text{N}$$

$$F_x = WH\left(\rho g \frac{H}{2}\right) = 8.83 \times 10^7\,\text{N}, \quad \cdot \quad F_z = (F_n^2 - F_x^2)^{1/2} = 9.69 \times 10^6\,\text{N}$$

*Construction (c):*

$$F_n = \frac{WH}{\sin \phi}\left(\rho g \frac{H}{2}\right) = 8.83 \times 10^7\,\text{N},$$

$$F_x = WH\left(\rho g \frac{H}{2}\right) = 8.78 \times 10^7\,\text{N}, \qquad F_z = (F_n^2 - F_x^2)^{1/2} = 6.2 \times 10^6\,\text{N}$$

Construction (b) is the worst because an upward vertical force, $F_z = 6.2 \times 10^6\,\text{N}$, results which will tend to lift the wall. Construction (c) is the best because the resulting vertical force, $F_z = 6.2 \times 10^6\,\text{N}$, can be supported by the ground. Construction (c) is also the best since ground reaction forces tend to eliminate the

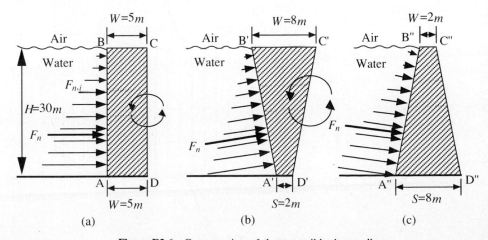

**Figure E2.6**   Cross section of three possible dam walls.

resulting moment of the distribution of hydrostatic forces that increase from top to bottom, which otherwise would tend to rotate the wall, as shown in cases (a) and (b). Case (b) is, of course, the worst from this point of view, too. Indeed, construction of dam walls is done along the lines of case (c).

## 2.2.1 Buoyancy Force

The total pressure force on a submerged body is the buoyancy force, acting in a direction opposite to that of gravity. A solid body or a part of fluid (e.g., liquid droplet or gas bubble) will float, equilibrate, or sink in a fluid bath, depending on the relative magnitude of buoyancy and gravity (or weight) forces. Examples 2.17 and 2.18 demonstrate relative applications.

### Example 2.7: Archimedes' Principle of Buoyancy

Prove Archimedes' principle of buoyancy[2] that the net force on a volume, submerged in a fluid, equals the product of its volume times the specific weight $\rho g$ of the fluid.

**Solution**    Figure E2.7 shows a generalized volume submerged in a fluid in static equilibrium. The force in the $z$-direction on each of the elementary cylindrical subdivisions of the volume, with respect to the system of coordinates shown, is

$$F_i = F_i^L + F_i^U = p_i^L(\Delta A_i^L)\cos\phi_i^L - p_i^U(\Delta A_i^U)\cos\phi_i^U, \qquad (2.7.1)$$

where $p_i$ is the hydrostatic pressure, given by

$$p_i = \rho g z_i, \qquad (2.7.2)$$

$\phi_i$ is the angle of the pressure force with the vertical $z$-direction, and $\Delta A_i$ is the elementary area of the basis of the elementary volume. The total vertical force is approximated by

$$F \simeq \sum_{i=1}^{n} F_i = \sum_{i=1}^{n} \Delta A_i \rho g(z_i^L \cos\phi_i^L - z_i^U \cos\phi_i^U). \qquad (2.7.3)$$

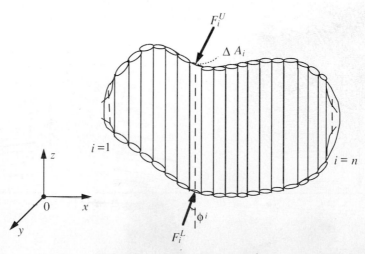

**Figure E2.7**   Origin of buoyancy force.

This approximation is made more accurate by increasing the subdivisions of the volume, in which case the summation becomes identical to the integral,

$$F_B = \lim_{\Delta A_i \to 0} F = \lim_{\Delta A_i \to 0} \left[ \rho g \sum_{i=1}^{n \to \infty} (\Delta A_i) \, \Delta z_i \right] = \rho g \int_V dV_i = -\rho g V. \qquad (2.7.4)$$

Equation (2.7.4) expresses Archimedes' principle of *buoyancy*—that the net pressure force on a volume submerged in a fluid, which he called *anosis* (i.e., the tendency of fluid to push up), is identical to the weight of fluid, $\rho g V$, displaced by the volume.

### Example 2.8: Measurement of Specific Gravity

The *pycnometer* shown in Fig. E2.8 is used to measure the density of liquids by comparing the difference $z$ of depths, when submerged and equilibrated in distilled water of known density $\rho_w$, and in a liquid of unknown density $\rho_x$. Derive the working equation that yields the unknown density $\rho_x$, in terms of $\rho_w$, $h$, the tube diameter $d$, and the known submerged volume up to indication $x$ in the water.

**Solution**  The pycnometer is at equilibrium in both cases and therefore its total weight, $B$, is balanced in each case by the developed buoyancy force. Thus

$$B = V_w \rho_w g = V_x \rho_x g, \qquad (2.8.1)$$

where $V_w$ and $V_x$ are the total submerged volumes in water and liquid of unknown density $\rho_x$, respectively. Thus

$$\left( V + \frac{\pi d^2 x}{4} \right) \rho_w g = \left[ V + \frac{\pi d^2 (x - z)}{4} \right] \rho_x g, \qquad (2.8.2)$$

where $V$ is the spherical volume. Equation (2.8.2) is rearranged in the form

$$s_x = \frac{\rho_x}{\rho_w} = \frac{1}{1 \pm \alpha h}, \qquad \alpha = \frac{\pi d^2}{4 V_w}, \qquad (2.8.3)$$

where $s_x$ is the specific gravity of the liquid of unknown density $\rho_x$. Based on Eqn. (2.8.3), pycnometers are graded and scaled appropriately to read the specific gravity and density of the measured liquid directly. Often, density is proportional to a certain valuable component of a liquid, for example, sugar in grape juice, in

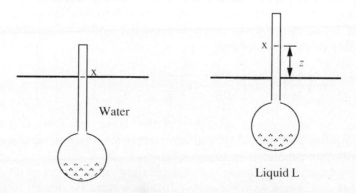

**Figure E2.8**  Pycnometer in water and test liquid.

which case pycnometers are scaled further to provide the component concentration directly.

## 2.2.2 Stability of Static Equilibrium

Equations similar to Eqn. (2.22) are used to deduce stability characteristics of a static equilibrium. The static equilibrium of a fluid is stable if an element of the fluid moved in a direction opposite to gravity, or any other potential field, such as a centrifugal, tends to return to its original position; unstable if the fluid element tends to travel away from its original position; and neutrally stable if the fluid element stays where it has been relocated. Based on Archimedes' principle of buoyancy, for stable, unstable, or neutrally stable static equilibrium, the density of the fluid must decrease, increase, or be constant with distance in the direction opposite to that of gravity (i.e., away from the center of the earth). Thus

$$
\frac{\partial \rho}{\partial z} \begin{cases} < 0, & \text{stable equilibrium} \\ = 0, & \text{neutrally stable equilibrium} \\ > 0, & \text{unstable equilibrium,} \end{cases} \tag{2.34}
$$

where $z$ increases away from the center of the earth. In homogeneous isothermal fluids acted upon only by gravity, the density is constant and therefore their equilibrium is neutrally stable. In homogeneous nonisothermal liquids, as is the case with shallow lakes during freezing conditions, the temperature decreases from nearly 4°C deep under the ice to 0°C at the surface and so does the water density,[6] and therefore any piece of surface ice being forced to sink tends to return to the free surface. Atmospheric air is heated by contact with the soil and cooled by contact with seawater during the daytime, which results in a continuous motion of air from sea to land. The opposite occurs during nighttime, where soil cools due to infrared radiation, whereas water remains warm. Due to the same phenomenon, cool air from higher atmospheric layers of lower temperature tends to descend to earth's surface where it is heated and ascends, being displaced by coming behind cooler layers of air. In fact, Eqn. (2.22) can be used together with other influencing factors (such as humidity and wind conditions) to predict the stability of the atmosphere:

$$
\text{sin}\left[\frac{\partial \rho}{\partial z}\right] = \text{sin}\left[\left(\frac{Mg}{\beta R} - 1\right)\frac{-\beta}{T_0}\left[\frac{T_0 - \alpha z}{T_0}\right]^{(Mg/\beta R)-2}\right] = \text{sin}\left[-\frac{\beta}{T_0}\right]. \tag{2.35}
$$

Thus the atmosphere is stable if the lapse rate, $\beta$, is positive, given the assumption that humidity and air circulation do not significantly alter density distribution predicted by Eqn. (2.22).

Density distribution or stratification may occur with nonhomogeneous liquids and solutions for example, ocean water. To deduce the stability of these liquids, an equation of state that relates density to composition must be used to calculate the sign of $\partial \rho / \partial z$ according to Eqn. (2.35). A nice application of these principles are meterorological and ballast balloons in the atmosphere, as detailed below.

### 2.2.3 Balloons in the Atmosphere

Balloons in the atmosphere are used routinely to conduct meteorological measurements at high elevations. Load balloons are used extensively for entertaining, advertising, and commercial purposes. Balloons are made by elastic materials of known, well-defined properties. The pressure differential across the thin wall of the thickness $H$ of a spherical balloon of radius $R$, and its elastic wall tension $\sigma_E$, defined by

$$\sigma_E = \frac{F_E}{l}, \tag{2.36}$$

where $F_E$ is tangential elastic force over surface length $l$, are related by

$$\Delta p = p_{bal} - p_{atm} = \frac{2\sigma_E}{R}. \tag{2.37}$$

For an ideal elastic wall, the tension $\sigma_E$ is related to the wall thickness $H$ by the Hooke's law of elasticity,[6]

$$\sigma_E = -E(H - H_0), \tag{2.38}$$

where $E$ is the modulus of elasticity of the wall material and $H_0$ is the uninflated wall thickness. Equations (2.36) to (2.38) combine to

$$\Delta p = \frac{2E(H_0 - H)}{R}. \tag{2.39}$$

Equation (2.37) is similar to the Young–Laplace equation for a spherical droplet when the surface tension $\sigma$ is replaced by the tension $\sigma_E$.

The volume of the thin wall must be conserved during inflation or deflation from a reference state to any other state. Thus

$$4\pi R_0^2 H_0 = 4\pi R^2 H, \tag{2.40}$$

$$\frac{H}{H_0} = \frac{R_0^2}{R^2}. \tag{2.41}$$

Equation (2.41) gives the wall thickness $H$ of the inflated balloon in terms of its radius $R$, and its uninflated thickness $H_0$ and radius $R_0$. Equations (2.39) and (2.41) combine to yield

$$\Delta p = \frac{2EH_0}{R}\left(1 - \frac{R_0^2}{R}\right) \simeq \frac{2EH_0}{R}. \tag{2.42}$$

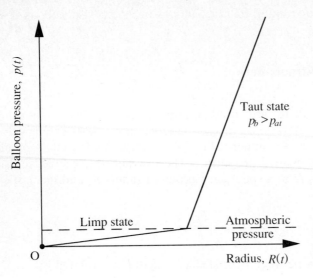

**Figure 2.6** Pressure evolution, or resistance to inflation, when inflating a balloon.

A balloon is in the *limp state* if

$$\Delta p = p_{\text{bal}} - p_{\text{atm}} = 0, \tag{2.43}$$

which is true when the balloon is not fully stretched. A balloon is in the *taut state* if

$$\Delta p = p_{\text{bal}} - p_{\text{atm}} = \frac{2EH_0}{R}, \tag{2.44}$$

which is true of fully stretched balloons. Figure 2.6 shows the pressure evolution when inflating balloons, which, for example, is proportional to the resistance one encounters when inflating balloons by mouth.

Examples 2.9 and 2.10 illustrate the principles governing the vertical motion of balloons in the atmosphere. The scenario when a *closed balloon* is released from earth is initially a series of limp states during which the inside pressure increases slowly to the external atmospheric pressure (which decreases with elevation), which causes the balloon to ascend. Eventually, the balloon reaches its stretching point where it enters the taut state and may equilibrate. This equilibrium condition may be altered by releasing or adding either gas or ballast. An *open balloon* may discharge gas in order to cope with decreasing external pressure as it ascends in the atmosphere. Since the balloon communicates with the atmosphere, discharging ceases when the two pressures across the release valve become equal. Thus the pressure inside an open balloon is identical to (actually slightly above) that of the surrounding atmosphere. In contrast, if the balloon is closed, discharging cannot happen, and the only way to cope with the decreasing external pressure is by volume increase, which generates wall tension. This results in a higher pressure inside the balloon than in the surrounding atmosphere and therefore higher density that tends to descend the balloon resisting the buoyancy force.

**Example 2.9: Closed Balloon Equilibrium**

An uninflated balloon together with its ballast (Fig. E2.9) weighs 200 kg and is to be released in a standard atmosphere with constant lapse rate of 0.00357°F/ft and sea-level conditions of 0°C and 1 atm. When the balloon is inflated with 224 m³ of helium at sea-level conditions, to what altitude must it rise to come to static equilibrium? (From Ref. 7, by permission.)

**Solution**    Newton's law of motion applied to the balloon yields

$$F = V\rho_{air}g - V\rho_{He}g - mg = (m + V\rho_{He})a. \qquad (2.9.1)$$

Here $V$ is the volume of the balloon, $M$ the mass of the load, $g$ the acceleration of gravity, and $a$ the resulting acceleration of the balloon. At sea level Equation (2.9.1) yields

$$F = \left[224 \text{ m}^3\left(\frac{29 \text{ kg}}{22.4 \text{ m}^3} - \frac{4 \text{ kg}}{22.4 \text{ m}^3}\right) - 200 \text{ kg}\right] \times 9.81 \text{ m/s}^2 = 490 \text{ N}$$

where the terms in parentheses are the densities of air and He calculated by the ideal gas law:

$$\rho_i = \frac{M_i p}{RT} = M_i\left[\frac{1 \text{ atm}}{0.0821 \cdot \text{atm/mol} \cdot \text{K} \ (273 \text{ K})}\right] = \frac{M_i}{22.4} \text{ kg/m}^3.$$

This ground force gives initial acceleration,

$$a = \frac{F}{M + V\rho_{He}} = \frac{490 \text{ N}}{(40 + 200) \text{ kg}} = 2 \text{ m/s}^2,$$

and therefore the balloon will start ascending from a limp state.

   Assuming that the balloon remains forever in its limp state, for which

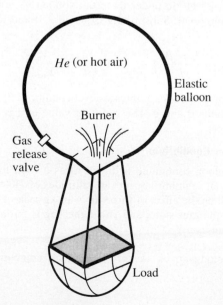

He (or hot air)

Elastic balloon

Burner

Gas release valve

Load

**Figure E2.9**    Balloon with ballast.

$p_{in} = p_{out} = p$, the driving force

$$F = Vg(\rho_{air} - \rho_{He}) - m = \frac{Vgp}{RT}(M_{air} - M_{He}) - m \qquad (2.9.2)$$

is constant independent of $z$ and the balloon would ascend forever while being inflated by the decreasing atmospheric pressure, under essentially constant temperature. However, this is avoided when the balloon reaches its maximum stretching point, where it goes into a taut state. Then the inside pressure becomes slightly higher than atmospheric, and further motion is controlled by releasing He or ballast.

Assume that the balloon comes to equilibrium at elevation $z$, where its pressure is different from the outside atmospheric pressure, which is calculated by

$$\frac{dp}{dz} = -\rho g = -\frac{pM_{air}g}{RT} = -\frac{pM_{air}g}{R(T_0 - \beta z)}, \qquad (2.9.3)$$

which is integrated to

$$p = p_0\left(\frac{T_0 - \beta z}{T_0}\right)^{M_{air}g/R\beta} \qquad (2.9.4)$$

The density of air corresponding to this pressure is

$$\rho_{air} = \rho_0\left(\frac{T_0 - \beta z}{T_0}\right)^{(M_{air}g/R\beta) - 1} \qquad (2.9.5)$$

At the taut state, $p_{He} \neq p_{air}$, and therefore, Eqn (2.9.2) does not apply. The density of He is simply

$$\rho_{He} = \frac{M_{He}}{V_{max}} = \rho_{0,He}\frac{V_0}{V_{max}}, \qquad (2.9.6)$$

where $\rho_{0,He}$ is the density of He under its initial volume $V_0$, and $V_{max}$ its maximum volume at the stretching point. Substituting Eqns. (2.9.5) and (2.9.6) in Eqn. (2.9.1) results in

$$V_{max}\left[\rho_{0,air}\left(\frac{T_0 - \beta z}{T_0}\right)^{(M_{air}g/R\beta a) - 1} - \rho_{0,He}\frac{V_0}{V_{max}}\right] - m = 0, \qquad (2.9.7)$$

which can be solved for $z$ given $V_{max}$, the sea-level conditions $T_0$, $p_0$, $\rho_0$, and $V_0$, and the atmospheric constants $\beta$ and $g$. The resulting value of $z$ is the altitude at which the balloon equilibrates.

### Example 2.10: Open Balloon Equilibrium

An open research balloon containing helium carries a total load of 1000 kg and is currently motionless at equilibrium at an altitude of 3000 m in an adiabatic atmosphere. What will be the effect of throwing a 10-kg sack of sand overboard (a) if the bag is initially in the taut state, and (b) if the bag is initially in the limp state? (From Ref. 7, by permission.)

**Solution**  From thermodynamics, the pressure of an adiabatically stratified atmosphere is given by

$$p = cT^{\gamma/(\gamma-1)}, \qquad \gamma = \frac{c_p}{c_v}. \qquad (2.10.1)$$

Differentiation with respect to altitude $z$ yields

$$\frac{dp}{dz} = c\left[\frac{\gamma}{\gamma - 1}\right] T^{[\gamma/(\gamma-1)]-1} \frac{dT}{dz}, \qquad (2.10.2)$$

which combines with the barometric equation

$$\frac{dp}{dz} = -\rho g = -\frac{pMg}{RT} \qquad (2.10.3)$$

to yield the temperature distribution of the adiabatic atmosphere,

$$\frac{dT}{dz} = -\frac{M(\gamma - 1)g}{R\gamma}. \qquad (2.10.4)$$

Thus in an adiabatically stratified atmosphere there is a constant lapse rate,

$$\beta = -\frac{dT}{dz} = \frac{M(\gamma - 1)g}{R\gamma} = 0.00977 \, \text{K/m}, \qquad (2.10.5)$$

which distributes the temperature according to

$$T \simeq (300 - 0.00977z) \, \text{K}. \qquad (2.10.6)$$

Balloon equilibrium, independent of limp or taut state, occurs when

$$\frac{F}{g} = V(\rho_{\text{air}} - \rho_{\text{He}}) - m = 0, \qquad (2.10.7)$$

where $F$ is the total force on the balloon, $V$ its volume, and $m$ the total mass. Before and after the release of 10 kg, Eqn. (2.10.7) yields

$$V_1(\rho_{\text{air}} - \rho_{\text{He}})_1 = m_1$$

and

$$V_2(\rho_{\text{air}} - \rho_{\text{He}})_2 = m_2.$$

Thus

$$\frac{(\rho_{\text{air}} - \rho_{\text{He}})_2}{(\rho_{\text{air}} - \rho_{\text{He}})_1} = \frac{990}{1000} \frac{V_1}{V_2}. \qquad (2.10.8)$$

*Bag initially in taut state.* For an open balloon in a taut state the conditions are $p_{\text{air}} = p_{\text{He}}$, $V_1 = V_2$, and $T_{\text{air}} = T_{\text{Fe}}$. Thus

$$\frac{p_2 T_1}{T_2 p_1} = \frac{T_1}{T_2}\left(\frac{T_2}{T_1}\right)^{\gamma(\gamma-1)} = 0.99, \qquad (2.10.9)$$

which reduces further to

$$\frac{T_1 - 0.00977\Delta z}{T_1} = 0.99^{0.4}, \qquad (2.10.10)$$

which yields $\Delta z = 112 \, \text{m} > 0$. Thus the balloon will move to a new equilbrium taut position at $\Delta z = 112 \, \text{m}$ higher than its initial position.

*Bag initially in lump state.* By the results of Example 1.9, the balloon will move to an equilibrium position after reaching the taut state. The volumes $V_0$ of state 1 and $V_{\text{max}}$ of state 2 are different. However, the ideal gas law,

$$\frac{p_1 V_0}{RT_1} = \frac{p_2 v_{\text{max}}}{RT_2}, \qquad (2.10.11)$$

applies to both states and therefore,

$$\frac{p_2 T_1}{T_2 p_1} = 0.99 \frac{V_0}{V_{max}}. \qquad (2.10.12)$$

Equation (2.10.8) for this case is then modified to

$$\Delta z = \frac{T_1}{9.776} \left[ 1 - \left( 0.99 \frac{V_0}{V_{max}} \right)^{0.4} \right] \text{km}, \qquad (2.10.13)$$

which gives the new equilibrium position, given $V_0$, $V_{max}$, and $T_1$, by Eqn. (2.10.6).

## 2.3 HYDROSTATICS OF FLUID INTERFACES

When two immiscible fluids of different densities, $\rho_1 > \rho_2$, are in contact and acted upon by gravity, there is an interfacial zone of molecular dimensions made up by both types of molecules, as shown in Fig. 2.7a, such that there is a continuous transition of density from $\rho_1$ to $\rho_2$, as shown in Fig. 2.7b. In the continuous approximation of liquids, this interfacial zone of molecular width is replaced by a mathematical interface, as shown in Fig. 2.7c, across which there is a sharp discontinuous density change from $\rho_1$ to $\rho_2$.

Due to anisotropic forces on molecules at the interface, as explained in Section 1.4 we have a membrane-like tension, called *surface* or *interfacial tension*, for a gas–liquid free surface and a liquid–liquid interface, respectively. This tension is balanced by a pressure jump across the surface according to the Young–Laplace equation[8]

$$\Delta p = p_2(z = h) - p_1(z = h) = 2H\sigma \simeq \frac{\sigma(d^2h/dx^2)}{\sqrt{1 + (dh/dx)^2}}. \qquad (2.45)$$

The last term approximates the curvature of a generalized surface in Cartesian

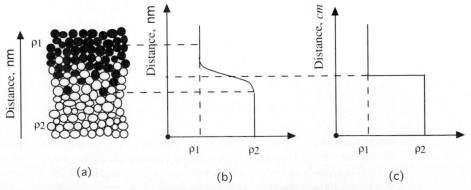

**Figure 2.7** Arrangement of molecules in the interfacial zone (a), molecular representation of density change within the interfacial zone (b), and continuum approximation by an interface line (c).

coordinates. This equation combines with the hydrostatic equations in each fluid,

$$\frac{dp_1}{dz} = -\rho_1 g \tag{2.46}$$

and

$$\frac{dp_2}{dz} = -\rho_2 g, \tag{2.47}$$

to determine pressure distributions and the shape of free surfaces and interfaces, as detailed in Examples 2.11 and 2.12.

### Example 2.11: Capillary Pressure

Figure E2.11 shows the arrangement of three immiscible liquids in a cylindrical container of radius $R = 2\,\text{cm}$ under gravity. Small amounts of the lightest liquid are trapped at the bottom under an almost semispherical interface of radius $s = 0.1\,\text{cm}$.

**(a)** What is the relation among the three densities?
**(b)** What are the interfacial tensions $\sigma_{21}$ and $\sigma_{23}$ between liquid 2 and liquids 1 and 3, respectively?
**(c)** If liquid 3 has density $\rho_3 = 0.8\,\text{g/cm}$ and surface tension $\sigma_3 = 30\,\text{dyn/cm}$, liquid 2 has density $\rho_2 = 0.9\,\text{g/cm}^3$, and liquid 1 has density $\rho_1 = 1\,\text{g/cm}^3$ and interfacial tension $\sigma_{13} = 40\,\text{dym/cm}$, what is the pressure in the trapped liquid 3 at the bottom?
**(d)** How was liquid 3 trapped at the bottom?

### Solution

**(a)** The densities follow $\rho_1 > \rho_2 > \rho_3 > \rho_{\text{air}}$, according to the explanation given in Example 2.1a.
**(b)** The interfacial tensions $\sigma_{21}$ and $\sigma_{23}$ are zero because the corresponding interfaces are planar, according to Eqn. (2.45) and since $dh/dr = 0$.

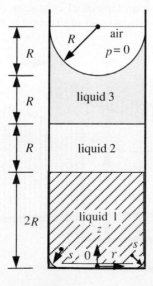

**Figure E2.11** Immiscible liquids at equilibrium under gravity.

**(c)** The capillary pressure inside the oil trapped at the bottom is

$$p_b = p_1(z = 0) - \frac{2\sigma_{13}}{s} = p_1(z = 2R) + 2\rho_1 gR - \frac{2\sigma_{13}}{s}$$

$$= p_2(z = 2R) + 2\rho_1 gR - \frac{2\sigma_{13}}{s}$$

$$= p_2(z = 3R) + \rho_2 gR + 2\rho_1 gR - \frac{2\sigma_{13}}{r}$$

$$= p_3(z = 3R)\rho_2 gR + 2\rho_1 gR - \frac{2\sigma_{13}}{s}$$

$$= p_3(z = 4R) + 2\rho_3 gR + \rho_2 gR + 2\rho_1 gR - \frac{2\sigma_{13}}{r}$$

$$= p_{air} - \frac{2\sigma_3}{R} + (2\rho_3 + \rho_2 + 2\rho_1)g - \frac{2\sigma_{13}}{s}$$

$$= p_{air} + gR(2\rho_3 + \rho_2 + 2\rho_1) - \frac{2\sigma_3}{R} - \frac{2\sigma_{13}}{s}.$$

For the values given,

$$p_b = p_{air}$$

$$= \left[ 981 \ cm/s^2 \times 2 \ cm(1.6 + 0.9 + 2) \ \text{g/cm}^3 - \left( \frac{60}{2} + \frac{80}{0.1} \right) \frac{\text{dyn}}{\text{cm}^2} = 7999 \right] \frac{\text{dyn}}{\text{cm}^3}.$$

**(d)** Liquid 3 was trapped at the bottom after draining of the container filled with liquid 3. Surface tension force along the contact lines of the remaining traces at the bottom corners overcome gravity and hold the traces in place. Such traces could be removed by means of surfactants that decrease surface tension. Heating also decreases surface tension.

### Example 2.12: Liquid Meniscus Shape

Consider an infinitely long horizontal container of cross section as shown in Fig. E2.12, in contact with stationary air of pressure $p = 0$. The surface tension $\sigma$ of the

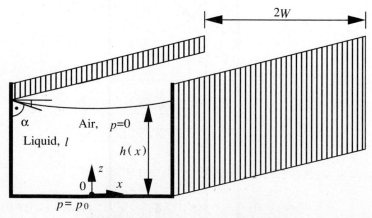

**Figure E2.12** Static meniscus configuration.

liquid is small such that the curvature of the free surface is small and therefore $(dh/dx)^2 \ll 1$, in which case

$$2H = \frac{d^2h/dx^2}{\sqrt{1 + (dh/dx)^2}} \simeq \frac{d^2h}{dx^2}.$$

The slope angle $\alpha$ between the liquid and the side walls is also small, such that $\alpha \sim \sin \alpha \approx \tan \alpha$.

(a) Derive the equation that describes the free-surface shape by combining the Young–Laplace equation across the free surface with the hydrostatic equation within the liquid, given that the pressure at the bottom at $z = 0$ is uniform, $p(z = 0) = p_0$.

(b) Show that in the case of $\sigma = 0$, a planar free surface is obtained.

(c) Find the shape of the free surface when $\sigma \neq 0$.

(d) Show that the solution for (d) yields the result in (c) in the limit of $\sigma \to 0$.

**Solution**

(a) The hydrostatic pressure within the liquid is given by

$$p(x, z) = p_0 + \rho gz, \tag{2.12.1}$$

with respect to the system of coordinates shown. The predicted pressure at the free surface, of elevation $z = h(x)$, is

$$p[x, z = h(x)] = p_0 + \rho gh(x), \tag{2.12.2}$$

where $h(x)$ is the distance of the free surface from the bottom. This pressure is related to the atmospheric pressure by the Young–Laplace equation,

$$p[x, z = h(x)] = p_{\text{atm}} + \sigma \frac{d^2h}{dx^2}. \tag{2.12.3}$$

By combining Eqns. (2.12.2) and (2.12.3), a differential equation that describes the free-surface profile, $h(x)$, results:

$$\sigma \frac{d^2h}{dx^2} - \rho gh(x) - p_0 + p_{\text{atm}} = 0. \tag{2.12.4}$$

(b) For a zero surface tension, Eqn. (2.12.4) predicts that

$$h(x) = \frac{p_0 - p_{\text{atm}}}{\rho g}, \tag{2.12.5}$$

which is a planar interface, independent of $x$.

(c) Equation (2.12.4) is a second-order linear differential equation with general solution

$$h(x) = \left[ c_1 e^{x\sqrt{\rho g/\sigma}} + c_2 e^{-x\sqrt{\rho g/\sigma}} \right] + \frac{p_0 - p_{\text{atm}}}{\rho g}, \tag{2.12.6}$$

where $c_1$ and $c_2$ are arbitrary constants of integration, calculated by applying the necessary boundary conditions of the physical process:

(i) At $x = 0$, $dh/dx = 0$ due to symmetry, and therefore,

$$\left[\frac{dh}{dx}\right]_{x=0} = c_1\sqrt{\frac{\rho g}{\sigma}} - c_2\sqrt{\frac{\rho g}{\sigma}} = 0, \tag{2.12.7}$$

and thus $c_1 = c_2$, and Eqn. (2.12.6) becomes

$$h(x) = c_1\left[e^{x\sqrt{\rho g/\sigma}} + e^{-x\sqrt{\rho g/\sigma}}\right] + \frac{p_0 - p_{atm}}{\rho g}. \tag{2.12.8}$$

(ii) At the solid wall the given angle $\alpha$ is identical to the slope $dh/dx$, and therefore,

$$\left[\frac{dh}{dx}\right]_{x=w} = c_1\sqrt{\frac{\rho g}{\sigma}}\left[e^{w\sqrt{\rho g/\sigma}} - E^{-w\sqrt{\rho g/\sigma}}\right] = \tan\alpha \approx \alpha, \tag{2.12.9}$$

and thus,

$$c_1 = \frac{\alpha\sqrt{\sigma/\rho g}}{e^{w\sqrt{\rho g/\sigma}} - e^{-w\sqrt{\rho g/\sigma}}}, \tag{2.12.10}$$

and Eqn. (2.12.8) becomes

$$h(x) = \alpha\sqrt{\frac{\sigma}{\rho g}}\left[\frac{e^{x\sqrt{\rho g/\sigma}} + e^{-x\sqrt{\rho g/\sigma}}}{e^{w\sqrt{\rho g/\sigma}} - e^{-w\sqrt{\rho g/\sigma}}}\right] + \frac{p_0 - p_{atm}}{\rho g}, \tag{2.12.11}$$

which fully determines the profile of the free surface as a function of $x$.

(d) In the limit of $\sigma = 0$, Eqn. (2.12.11) reduces to

$$h(x) = \frac{p_0 - p_{atm}}{\rho g z}, \tag{2.12.12}$$

which is identical to Eqn. (2.12.5), as it should be.


# PROBLEMS


1. *Pressure measurement.* The device shown in Fig. 2.4 is utilized to measure pressure of flowing fluids in pipes. The U-shaped thin tube contains mercury of density $\rho_m = 13.6\,\text{g/cm}^3$, which is forced to climb the right part of the tube, due to the transmitted pressure $p$ of the flowing fluid.
   (a) Show that the value of the pressure $p$ can be deduced from the density of the measuring liquid in the U-shaped tube and the height difference, $h$.
   (b) What is the value of $p$ for the case shown?
   (c) How will $h$ change if mercury is replaced by color water in order to measure low pressures of flowing gases?
   (d) How will things change for a closed manometer, with $m$ kilograms of air trapped in the higher closed end?

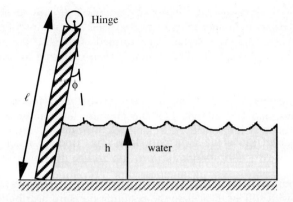

Hinge

$\ell$

$\phi$

h    water

**Figure P2.3**   Flap gates used to prevent flooding.

2. *Dam foundation.* Find the magnitude and direction of the external force at the foundation of the 10,000-ton wall of Example 2.6 required to keep the dam wall in place.

3. *Flap gate.* To prevent flooding, flap gates are used as shown in Figure P2.3. A 3-m$^2$ gate weighing 2 tons, hinged by its upper side, is placed at an angle 5° with the vertical direction. At what height, $h$, will the water rise behind the gate before it will open?

4. *Tank pressure distribution.* Consider the cylindrical tank of Fig. P2.4 with $d_0 = 1$ ft, $d_1 = 0.5$ ft, $d_2 = 1$ ft, $h_w = 1$ ft, $h_0 = 0.5$ ft, $h_a = 1$ ft, and $h_t = 0.5$ ft, open to an atmosphere of 25°C and 1 atm.
   (a) What is the pressure at points $A$, $B$, $C$, $D$, $E$?
   (b) What amount of air is trapped in the air chamber?
   (c) Find the new stratification if 0.5 ft$^3$ if oil is added.

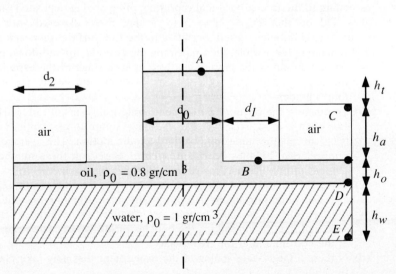

**Figure P2.4**   Fluid stratification in tank.

5. *Fluid interface in solid-body rotation.* Consider two immiscible liquids of negligible surface and interfacial tension in an open cylindrical container under gravity, such that a perfectly planar interface and free surface are formed. Calculate the resulting pressure distribution, $p(r, z)$, throughout. The system is then subjected to rotation at angular speed $\Omega$. Calculate the free surface and interface configurations and the resulting new pressure distribution. Consider oil of density $0.95 \text{ g/cm}^3$ on top of water of density $1 \text{ g/cm}^3$, at heights 10 and 30 cm, respectively, before rotation, filling a container of 50 cm radius, and $\Omega = 10$ rpm. Plot the pressure distribution along the radius of circular cross sections at distances 0, 30, and 40 cm from the bottom, before and during rotation. How would the existence of finite surface and interfacial tensions alter these plots qualitatively?

6. *Free-surface profiles.* Consider a cubic container 2 m on a side that contains $0.5 \text{ m}^3$ of water at rest in contact with still air. Calculate the resulting pressure distributions and free surface profiles, in the absence of surface tension, for the following solid-body-like motions: (a) no motion at all; (b) vertical upward translation at 2 m/s; (c) horizontal translation at the same speed, parallel to itself; (d) horizontal translation at the same speed in the direction of one of its diagonals; (e) rotation at 10 rpm.

7. *Stability of stratified fluids.* A fluid is in stable equilibrium if any volume of it removed from its initial equilibrium position tends to return; otherwise, it is unstable. Show that density-stratified fluid such as atmospheric air are stable. Extend your considerations to mineral water and to salt water, and to temperature-induced stratification in frozen lakes and boiling water heated from below—consider the motion of air bubbles. Do these stability conditions apply to cases of horizontal stratification such as centrifugation and horizontal motion of atmospheric air? How does atmospheric air density stratification, induced by gravity or temperature, influence air circulation?

8. *Gravity vs. buoyancy.* My son Charis, a fourth grader, came home excited one day from his school science class and performed the following experiment for us: He placed four resin samples coated with baking soda in a glass, then added water, then drops of vinegar. This gave rise to the formation of gas bubbles, some of them ascending to the free surface and rupturing, others being deposited on the walls of the glass and on the resins. After 3 to 5 min, resins covered with attached nearly semispherical bubbles started ascending to the free surface, then sinking to the bottom and remaining for a while, then repeating the periodic up-and-down motion.

   (a) Explain in detail the physics involved in each stage of the experiment, and justify the use of resins and soda–vinegar. Simplify and clarify the explanation for a fourth grader.

   (b) To quantify the phenomenon, assume spherical resins of radius 0.8 cm and density $1.05 \text{ g/cm}^3$, in a soda–vinegar foaming solution of depth 8 cm and density $0.9 \text{ g/cm}^3$, at the same time of the periodic motion. The average diameter of the semispherical bubbles deposited is 1 mm. Based on these numbers, calculate the number of bubbles on each resin at the inception of motion.

   (c) Study the periodic motion by assuming that during the ascent, all bubbles remain attached while growing in size, due to the diminishing hydrostatic pressure; that at the free surface as many bubbles are ruptured as required to cause the resin to sink; and that during sinking the bubbles shrink in size. The resin has zero velocity at the bottom and at the free surface.

   (d) Is there a large-scale analog of the experiment that may take place in lakes or in the air?

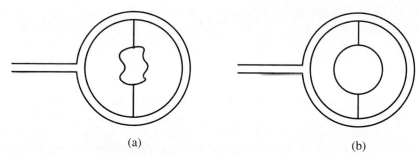

(a)                                                                    (b)

**Figure P2.10**   Surface tension forces in soap films.

9. *Spherical shapes.* Consider droplets of fluid $A$ of density $\rho_A$ and radius $R$ in another fluid $B$ of density $\rho_B$ and interfacial tension $\sigma_{AB}$, under conditions of static equilibrium at rest or at solid-body rotation. Among these conditions and physical properties, what are the most favorable for spherical droplet shape? Consider gravity and/or centrifugal fields or the absence of them, for example in space. Or, consider externally applied pressure or vacuum without droplet evaporation, and combinations of them.

10. *Surface tension action.* A metal wire ring equipped with a concentric cotton ring hung in its interior as shown in Fig. P2.10a is placed in soapy water and then taken out, with a soap film having formed in its interior, where the cotton ring floats loose. Then the film in the interior of the cotton is ruptured by a needle, and the configuration of Fig. P.210b is obtained.

    **(a)** Explain the physics of the two configurations and comment on the action of the surface tension forces for a strictly horizontal arrangement.

    **(b)** What is the resulting pressure within the remaining soap film, considering the equilibrium of the appropriate film parts: Is the film thickness uniform?

    **(c)** Describe qualitatively the evolution of the configuration of Fig. P1.25b as the system shifts from horizontal to vertical.

11. *Surface tension measurement.* Figure P2.11 shows a method for documenting and measuring surface tension based on a $\Pi$-shaped wire equipped with a freely sliding straight wire to which weights can be attached. Upon passage from soapy water, a thin soap film is formed whose surface tension forces are greater than the weight of the sliding part of the wire, $B_1$, and the film is kept somewhat vertical. With the addition of a weight, $B_2$, the film is stretched to a new position at distance $S$ from the original.

    **(a)** By considering the work done against the surface tension forces by the attached

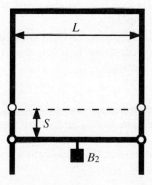

**Figure P2.11**   Work done by surface tension forces.

weight, show that

$$\sigma = \frac{B_2}{2L},$$

where the factor 2 accounts for the two faces of the film.

**(b)** By considering the hydrostatic pressure distribution, show that the thickness of the film cannot be uniform throughout. What is the vertical thickness profile?

**12.** *Rotation in the presence of surface tension.* A high-temperature-resistant cylindrical container of radius $R = 20$ cm contains $V_0 = 0.0628$ m$^3$ of molten steel of density $\rho = 8$ g/cm$^3$ and surface tension $\sigma = 1200$ dyn/cm. In a forming process, it is rotated at 20 rpm. Calculate the resulting pressure distribution and free surface profile in the absence of solidification phenomena. What are the limiting distributions for negligible surface tension? In this process, is the presence of surface tension important? How do the results compare to those with water of density $\rho_w = 1$ g/cm$^3$ and surface tension $\sigma_w = 73$ dyn/cm under the same conditions?

**13.** *Buoyancy at interfaces.* Find the force of buoyancy on a cubic body of volume $V$ submerged at the interface of two immiscible liquids with $0.3V$ in the heavier liquid of density $\rho_A$ and the rest in the other liquid, of density $\rho_B$. Reconsider the situation in the presence of interfacial tension $\sigma_{AB}$, which makes the interface concave toward the lighter liquid such that its contact angle with the body is $\theta$. Do the same results apply to free surfaces?

**14.** *Capillary force.* Consider two solid surfaces parallel to each other placed vertically a small distance $d = 5$ mm apart. The gap is partly filled with a cylindrical water droplet of volume $V = 39.25$ cm$^3$, density 1 g/cm$^3$, and surface tension 73 dyn/cm. The contact angle of the droplet to each of the plates is 45°.

**(a)** Show that the capillary force that keeps the two plates together is given by

$$F = \sigma \Pi \sin \theta + \frac{2\sigma A \cos \theta}{d},$$

where $\Pi$ is the perimeter of contact, called the *contact line,* and $A$ is the enclosed droplet–plate contact area.

**(b)** Sketch the configuration and show that such contact may indeed overcome gravity to keep the droplet in place.

**(c)** Estimate the force $F$. (From Ref. 7, by permission.)

**15.** *Capillary rise.* Express the height $h$ over the level of the pool liquid of density $\rho$ and surface tension $\sigma$ to which the liquid climbs a capillary of diameter $d$, as shown in Fig. P2.15. The *contact angle* $\theta$ here is such that the liquid *wets* and therefore climbs the capillary. Sketch the corresponding arrangement where a liquid *does not wet* and therefore descends in the capillary.

**16.** *Vapor pressure on nonplanar surface.*[9] Kelvin showed that the vapor pressure, $p_{g,r}$, on a free surface of radius of curvature $r$, is less than that on a planar free surface, $p_g$, according to the "Kelvin equation"[10],

$$p_{g,r} = p_g - \frac{2\sigma}{r} \frac{\rho_g}{\rho_l - \rho_g},$$

where $\sigma$ is the surface tension and $\rho_g$ and $\rho_l$ are the densities of saturated vapor and liquid respectively.

**(a)** Derive Kelvin's equation for the equilibrium configurations of Fig. P2.16, taking into account that usually, $\rho_l \gg \rho_g$.

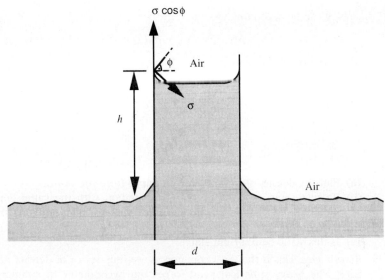

**Figure P2.15**   Capillary rise of wetting liquid.

**(b)** For the two situations, plot qualitatively the climbing or dipping distances on the same plot in capillaries of different diameters, $2r$.

**(c)** What dimensionless number controls climbing or dipping vertical capillaries under equilibrium conditions?

17. *Reservoir pressure.* A vertical cylindrical porous reservoir of oil of radius $R = 1$ km and height $H = 100$ m is trapped undergound by impermeable rock at its two bases and natural gas under high pressure at its cylindrical surface. A pressure gauge registers 2000 psig at the top base. The specific gravity of oil is 0.75 and that of natural gas 0.60, at 15°C compressibility factor 0.80.
   **(a)** What is the natural gas pressure per mole of gas?
   **(b)** What is the value of the same quantity in the presence of 30 dyn/cm surface tension between oil and gas?

18. *Pressure distribution in stratified liquid.* The temperature of a lake may vary nearly linearly from $T_b = 5$°C at depth $h = 100$ m, to $T_s = 20$°C at the surface. Salt concentration varies nearly linearly from $c_b = 0.1$ g/cm$^3$ to $c_s = 0.01$ g/cm$^3$. The water

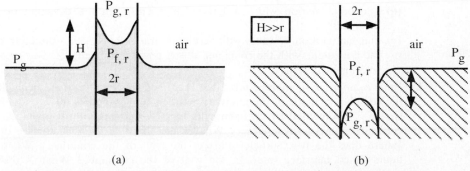

(a)                                                          (b)

**Figure P2.16**   Water wets the glass capillary (a), whereas mercury does not (b), which results in climbing and dipping, respectively.

density can be approximated by

$$\rho(c, T) = p_0 + b_T(T_0 - T) + b_c(c_0 - c),$$

where the coefficients are

$$b_T = \left(\frac{\partial \rho}{\partial T}\right)_c = 0.05 \text{ g/cm}^3 \cdot {}^\circ\text{C}$$

$$b_c = \left(\frac{\partial \rho}{\partial c}\right)_T = -0.1$$

and $\rho_0$ is the density of clear water at $T_0 = 20°\text{C}$ and $c = 0$.
(a) Plot the density stratification with depth from the bottom to the free surface.
(b) Calculate and plot the pressure distribution with the same depth. Also plot the pressure distribution in clear water of $T = 20°\text{C}$ and $c = 0$ throughout, and compare the two.
(c) Is this static equilibrium stable?

19. *Pycnometer.* This is the most common measuring device of density and is shown in Fig. E2.8. The heavy spherical head forces the pycnometer to submerge to depth $X$ in water. The same pycnometer, submerged in a liquid $L$ of unknown density $\rho_x$, reads depth $z - x$.
(a) Find a working equation that estimates the unknown density $\rho_x$ in terms of $\rho_w$ and $z$.
(b) The sugar concentration of a natural grape juice alters its density according to the expression

$$\rho_x = (1 - y)\rho_w + y\rho_s,$$

where $y$ is the mole fraction of the dissolved sugar of density $\rho_s = 1.3 \text{ g/cm}^3$. Derive the working equation of $y$ vs. $z$. (The price at which wine plants buy grapes from producers is determined based on the indication $y$.)

20. *Free surface profile.* Consider a cubic container of 2-m sides that contains $0.8 \text{ m}^3$ of water, at rest open to still air. Calculate the resulting pressure distributions and free surface profiles for the following solid-body motions. In some cases you may not be able to find definite answers. State the reasons why. Then guess qualitative answers.
(a) No motion at all;
(b) vertical upward motion with acceleration $a = 2 \text{ m/s}^2$;
(c) horizontal motion with acceleration $a = 2 \text{ m/s}^2$, parallel to itself;
(d) horizontal motion with acceleration $a = 2 \text{ m/s}^2$, in the direction of one of its diagonals;
(e) diagonal motion upwards with acceleration $a = 2 \text{ m/s}^2$, in a direction of 45° with the horizontal, with the container always parallel to itself;
(f) rotation at 10 rpm.

21. *Static equilibrium of rotating meniscus.* If surface-tension effects are negligible, what is the equilibrium shape of the interface between equal volumes of two liquids, $A$ and $B$ ($\rho_B > \rho_A$), contained by an open cylindrical vessel rotating about its axis, oriented vertically at the earth's surface? What is the shape of the free surface of liquid $A$? Where does the free surface intersect the wall of the container? Where does the liquid–liquid interface intersect the wall of the container? What is the maximum volume of liquid (equal volumes of $A$ and $B$) that can be contained by the vessel rotating at a given angular velocity?

**22.** *Bubble shape.* Consider a gas bubble of pressure $p$, temperature $T$, and volume $V$ trapped in a static liquid of density $\rho$, in depth $H$ much bigger than the average diameter $\bar{d}$ of the bubble. Examine the shape evolution of the bubble as the interfacial tension $\sigma$ increases from nearly zero to finite values and then to infinitely large values, in the following cases of equilibrium: (a) system under gravity alone; (b) system under solid-body translation with constant acceleration $\alpha_x$; (c) system under solid-body rotation with constant angular speed $\Omega$; (d) system under superposition of (a) and (b); (e) system under superposition of (a) and (c); (f) system under superposition of (a), (b), and (c).

**23.** *Droplet shape.* Consider a liquid droplet of volume $V$ and density $\rho$ and average diameter $\bar{d}$ much smaller than its depth $H$ in a second immiscible liquid of the same density.

   **(a)** Examine the shape evolution of the droplet in equilibrium, as the interfacial tension, $\sigma$, decreases from infinitely large to finite values and then to nearly zero.

   **(b)** Repeat when in addition to gravity there is constant linear acceleration of the entire system $a_x$ in the $x$-direction.

   **(c)** Repeat when in addition to gravity there is solid-body rotation with angular speed $\Omega$.

   **(d)** What will happen when (b) and (c) are superimposed?

**24.** *Meniscus shape.* Consider a residual amount of volume $V = 1 \, \text{cm}^3$ of wine of density $\rho = 0.9 \, \text{g/cm}^3$ and surface tension $\sigma = 60 \, \text{dyn/cm}$ in a perfectly spherical cup of radius $R = 3 \, \text{cm}$, made of glass fully wetted by that wine with contact angle between free surface and glass wall of $\theta = 0°$. How would you compute the shape of the wine's free surface?

# REFERENCES

1. K. Wark and C. F. Warner, *Air Pollution: Its Origin and Control,* Harper & Row, New York, 1985.

2. J. H. Perry, *Perry's Chemical Engineer's Handbook,* 6th ed., McGraw-Hill, New York, 1984.

3. Archimedes, *On Floating Bodies,* Books I and II (in T. L. Heath, *The Works of Archimedes,*) Cambridge University Press, Cambridge, 1897.

4. Staff of Research and Education Association, *Handbook of Mathematical Formulas, Tables, Functions, Graphs and Transforms,* REA, New York, 1980.

5. R. C. Reid, J. M. Prausnitz, and T. K. Sherwood, *The Properties of Gases and Liquids,* McGraw-Hill, New York, 1977.

6. R. Hooke, *Lectures de Potentia Restitution,* John Martyn, London, 1678; reprinted by R. T. Gunter, *Early Science in Oxford,* 8, 331 (1931).

7. L. E. Scriven, *Intermediate Fluid Mechanics Lectures,* University of Minnesota, 1980.

8. B. V. Deryagin and S. M. Levi, *Folm Coating Theory,* Focal Press, New York, 1964.

9. N. G. Koumoutsos, *Transport Phenomena,* National Technical University of Athens, Athens, Greece, 1975.

10. J. H. Seinfeld, *Atmospheric Chemistry and Physics of Air Pollution,* Wiley, Sons, New York, 1986.

# 3

# Mass, Energy, and Momentum Balances

## 3.1 SYSTEM AND SURROUNDINGS

The study of processes is facilitated by considering the *system* under investigation and its *surroundings*. *Systema* in Greek may be anything that follows certain rules of order. In fluid mechanics, we define a system to be any subset of a continuum, which may include fluids, or solids, or both that exhibit properties that follow certain rules of *conservation*. Conservation of mass, momentum, and energy means that each may change form; however, none may be created or decay to zero. Mass may change form by a chemical reaction; however, the summation of the mass of all species is constant. Momentum may be converted into force, and vice versa, according to the discussion in Section 1.3, however, the summation of the two is constant. Mechanical energy may be converted into heat by frictional processes, and heat may be generated or consumed by a chemical reaction; however, the summation of different forms of energy, or total energy, remains constant.

Pictorially, a fluid system with its surroundings is as shown in Fig. 3.1. For example, a lake may be the *system* under consideration, the water–soil and water–air contact surfaces its *boundaries,* and everything beyond its boundaries the *surroundings*. The boundaries that enclose the system can be real or imaginary, and are chosen to facilitate the particular investigation. The characteristic properties of interest of each portion $i$ of the boundary are its *area* $\Delta A_i$, and its *unit normal vector* $\mathbf{n}$, which is a vector of length 1 perpendicular to the

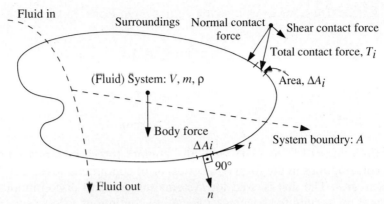

**Figure 3.1**  Fluid system and surroundings.

surface and pointing outward, from the system to the surroundings. A *unit tangent vector* **t** is defined accordingly, perpendicular to **n**. A system may interact with its surroundings in any of the ways described below. These ongoing interactions define the values of the properties of the fluid system (e.g., velocity, temperature, concentration, pressure, etc), as explained in Section 3.2.

### 3.1.1 Contact and Body Forces

The system is in contact with its surroundings according to the continuum approximation, and is therefore subjected to forces from the surroundings through its boundaries. This type of force is the *contact force*, and may vary in magnitude and direction from point to point along the boundary. The total contact force, $F^C$, is given by

$$F^C = \lim_{k \to \infty} \sum_{i=1}^{k} \tau_i \, \Delta A_i = \int_A \tau \, dA, \tag{3.1}$$

where $\tau_i$ is the contact force per area $\Delta A_i$, or *stress*. In the limit $k \to \infty$, and therefore $\Delta A_i \to 0$, it becomes the integral of stress over the entire boundary area. Any contact force may conveniently be decomposed into a *normal component*, in the direction of the unit normal, and a *shear component* perpendicular to the normal component, as shown in Fig. 3.1. The shear component of the force divided by the area of action yields the *shear stress* $\tau_{nt}$, which acts in the tangential **t**-direction on a surface segment of unit normal **n**. The normal component of the contact force divided by the area of action is the *normal stress* $\tau_{nn}$, which acts perpendicularly in the **n**-direction on a surface segment of unit normal **n**. The total, shear, and normal stresses are related by

$$\tau_{nn}^2 + \tau_{nt}^2 = \tau^2. \tag{3.2}$$

When the system is surrounded by fluid in static equilibrium, the only contact force and stress present are the normal ones, identical to the *hydrostatic pressure,* perpendicular and toward any submerged surface, according to Section

2.2.2.2. Thus, for cases of static equilibrium, Eqns. (3.1) and (3.2) take the form

$$F^C = \lim_{k \to \infty} \sum_{i=1}^{k} p_i \, \Delta A_i = \int_A p \, dA, \tag{3.3}$$

$$\tau_{nn} = \tau = p, \tag{3.4}$$

$$\tau_{nt} = 0. \tag{3.5}$$

Equations (3.1) to (3.5) yield the contact forces in terms either of discretized contributions in the form of a series, or of continuous stress distributions under an integral. The first is used for systems subjected to discontinuous or point forces, and the second for systems subjected to continuous forces (e.g., to the pressure of a surrounding fluid). For the system of Fig. 3.1, taken to represent a lake, there is a contact force from air of pressure $p_{air}$ acting on the top free surface of the water of area $A_f$, identical to

$$F_f^C = \int_{A_f} p_{air} \, dA = p_{air} \int_{A_f} dA = p_{air} A_f, \tag{3.6}$$

and from the soil to water boundary, of area $A - A_f$, identical to

$$F_s^C = \int_{A - A_f} p_w \, dA, \tag{3.7}$$

where $p_w$ is the hydrostatic pressure calculated as in Section 2.1.2.1. A case of local contact force or stress arises when considering a standing man as a system. Besides the continuous contact pressure force from the air on its entire body area, $A_b$, given by

$$F_b^C = \int_{A_b} p \, dA, \tag{3.8}$$

there is an additional normal contact force from the ground to his body, through his shoes, of contact area $A_s$, identical to

$$F_s^C = \sum_{i=1}^{2} \frac{B}{2A_s} \Delta A_{s,i} = \frac{B}{2A_s} 2A_s = B, \tag{3.9}$$

where $B$ is the man's weight, and the ratio $B/A_s$ is a normal stress resulting from the reaction of the ground to the weight supported.

Contact forces are due entirely to the contact boundary area between system and surroundings, and their magnitude is in fact proportional to the contact area: Eqns. (3.1) and (3.2), in the absence of contact area, predict zero contact force.

The second important type of force in fluid mechanics are the *body forces*, which arise from other systems not necessarily in contact with the system being considered. The most common and dominant body force is the *gravity force*, directed from the system's center of mass to the center of the earth, which is also identical to the *weight* of the system:

$$F^B = F_g = \lim_{k \to \infty} \sum_{i=1}^{k} g\,\Delta m_i = \int_m g\,dm = \int_V \rho g\,dV = \rho g V = B. \qquad (3.10)$$

Here $g = 9.81\ \text{m/s}^2 = 981\ \text{cm/s}^2 = 32.25\ \text{ft/s}^2$ is the acceleration of gravity, $\rho$ the density of the system, $V$ its volume, $m$ its mass, $B$ its weight, and $F_g$ the gravity force. The gravity force is common for static and flowing fluids, independent of the system's boundary area and proportional to the volume and mass of the system. For systems of regular geometric shape and constant density the discretized series form of Eqn. (3.10) is used; otherwise, the integral form must be employed, in order to calculate the total body force. *Centrifugal forces* arising in rotational motions (e.g., outward force experienced around curves by drivers of cars) and *electromagnetic forces* are also types of body force.

**Example 3.1: Continuous Contact Forces**

Calculate the total pressure force on a vertical thin surface of 1 m width and 2 m height, submerged in two stratified liquids of densities 1 and 0.85 g/cm³ and depths of 0.5 and 1.5 m, as shown in Fig. E3.1.

**Solution**   The normal pressure contact force, according to Eqn. (3.3), is

$$\begin{aligned}
F^C &= \int_A p\,dA = W \int_0^H p\,dz = W \int_0^{h_2} p_2(z)\,dz + W \int_{h_2}^H p_1(z)\,dz \\
&= Wg\left[ \int_0^h \rho_2 z\,dz + \int_{h_2}^H (\rho_2 h_2 + \rho_1 z)\,dz \right] \\
&= Wg\left[ \frac{\rho_2 h_2^2}{2} + \rho_2 h_2 (H - h_2) + \rho_1 \frac{H^2 - h_2^2}{2} \right] \\
&= 9.81\left[ \frac{850 \times 2.25}{2} + 850 \times 1.5 \times 0.5 + 1000 \times \frac{4 - 2.25}{2} \right] \text{N} \\
&= 9.81 \times (956 + 637.5 + 875)\ \text{Nt} = 24{,}216\ \text{N}.
\end{aligned}$$

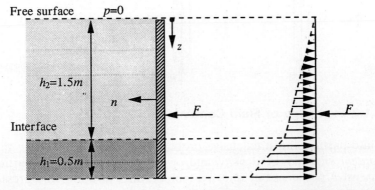

**Figure E3.1**   Continuous-pressure contact force on surface submerged in stratified fluids.

The continuous distribution of the contact force, which is the actual force experienced by the wall, is shown in Fig. E3.1. The summation of these forces is given by the result of the preceding equation. Due to the nature of hydrostatic pressure, the force is toward the wall surface, opposite to its unit normal vector **n**.

The wall in this case will remain in place, or in equilibrium, if a force $F = 24,216\,\text{N}$ and of opposite direction to $F^C$ (i.e., in the direction of the unit normal vector **n**) is applied in such a position that the overall moment of all applied forces on the wall is zero.

### Example 3.2: Discontinuous or Point Contact Forces

Consider a table weighing 15 N that has four legs of cross-sectional area $5\,\text{cm}^2$ each, as shown in Fig. E3.2. Calculate the resulting contact and body forces on the table, in addition to the continuous atmospheric pressure force.

**Solution**  Four contact point forces result on the table, at each of the four areas where a leg contacts the ground. They arise due to the ground reaction to the pressure applied to it by the weight of the table. The total upward contact force is

$$F^C = \sum_{i=1}^{k=4} \tau_i A_i = \sum_{i=1}^{k=4} \left(\frac{B}{4A}\right)A = B = 15\,\text{N}.$$

This force is equal and opposite to the weight $B$, which is a body force. This equation is also an *equilibrium condition*, which states that the summation of forces on the table is zero. It can easily be shown that the summation of moments is also zero. That is why the table stands in place by itself.

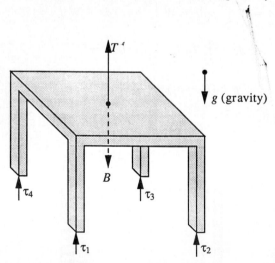

**Figure E3.2**  Contact point forces.

## 3.1.2 Exchange of Fluid Content

Fluid particles have extensive properties of mass and volume that determine the *content* of fluid in density and/or concentration. Fluid particles also possess intensive properties of velocity, temperature, position, and pressure that together with the extensive properties define the fluid content in kinetic, thermal, dynamic, and elastodynamic energy, respectively, and in momentum. These are *possessed and transferred* by the fluid motion or flow in much the same way that a walking or running boy carries and transfers with him his volume, weight, hand, cloth, and other things. Thus, fluid entering or leaving a system through a boundary surface of area $\Delta A_i$ with average velocity $\bar{u}_i$ adds or substracts a *content quantity* $q_i$, at a rate

$$\dot{q}_i = \hat{q}_i \bar{u}_i \Delta A_i, \tag{3.11}$$

where $\hat{q}_i$ is the quantity per unit volume or *quantity density*, $\bar{u}_i$ the *normal average velocity* crossing the area $\Delta A_i$ perpendicularly, and $\dot{q}_i$ the resulting rate of flow of quantity through the $\Delta A_i$ portion of the boundary area. The average velocity is positive for transfer into the system, and negative for transfer out of the system, since it induces positive or negative change of the corresponding content of fluid, respectively. If there are $k$ finite entrances and exits of fluid, the rate of net transfer is

$$\dot{q} = \sum_{i=1}^{k} \bar{u}_i \hat{q}_i \Delta A_i. \tag{3.12}$$

In case of continuous transfer throughout the entire boundary (e.g., across a porous wall boundary), these equations take the form

$$\dot{q} = \lim_{k \to \infty} \sum_{i=1}^{k} \bar{u}_i \hat{q}_i \Delta A_i = \int_A u\hat{q}\, dA, \tag{3.13}$$

where $u$ and $\hat{q}$ are now continuous distributions over the surface area $A$. Table 3.1 lists important quantities convected by fluids.

The mass, energy, heat, and solute contents of fluids are easily understood. *Momentum*, however, although extremely important in fluid mechanics and in mechanics in general, is different from the above. Momentum $J$, possessed by mass $m$ translating with velocity $u$, is a vector quantity in the direction of the

**TABLE 3.1**  IMPORTANT CONVECTED QUANTITIES WITH FLUID FLOW

| Convected quantity, $q$ | Quantity density, $\hat{q}$ | Convection rate, $\dot{q} = \hat{q}\bar{u}A$ |
|---|---|---|
| Mass, $m$ | $\hat{m} = \dfrac{m}{V} = \rho$ | $\dot{m} = \rho \bar{u}A \left[ \dfrac{\text{mass}}{\text{time}} \right]$ |
| Heat, $H$ | $\hat{H} = \dfrac{H}{V}$ | $\dot{H} = \hat{H}\bar{u}A \left[ \dfrac{\text{energy}}{\text{time}} \right]$ |
| Energy, $E$ | $\hat{E} = \dfrac{E}{V}$ | $\dot{E} = \hat{E}\bar{u}A \left[ \dfrac{\text{energy}}{\text{time}} \right]$ |
| Solute, $m_c$ | $\hat{m}_c = \dfrac{m_c}{V} = c$ | $\dot{m}_c = c\bar{u}A \left[ \dfrac{\text{mass of solute}}{\text{time}} \right]$ |
| Momentum, $J = mu$ | $\hat{J} = \dfrac{mu}{V} = u\rho$ | $\dot{J} = \hat{J}\bar{u}A = \rho\bar{u}\bar{u}A \left[ \dfrac{\text{momentum}}{\text{time}} = \text{force} \right]$ |

velocity $u$ defined by

$$J = mu. \tag{3.14}$$

Thus, momentum is a content proportional to the product of mass and velocity.

Newton's law of motion,[1] expressed in the form

$$F = \frac{dJ}{dt} = \frac{d}{dt}(mu) = \dot{m}u + m\alpha, \tag{3.15}$$

where $F$ is force, $\dot{m}$ rate of time change of mass, and $\alpha$ acceleration, relates momentum to force and acceleration caused by the action of the force. According to Eqn. (3.15), momentum represents the ability of any amount of continuous matter or fluid to apply a thrust force $F$, to any obstacle, solid or fluid, that alters its velocity.

Consider a vehicle of mass $m$ driven at velocity $u$ against a wall. It possesses momentum $J = mu$. Upon collision with the wall its velocity vanishes momentarily, resulting in a large decrease in its momentum, given by $dJ/dt = m\, du/dt$. According to Eqn. (3.15), this change of momentum results in a thrust force $F$ on the wall, which may collapse, depending on the relative magnitude of $F$ to the support force of the wall.

Equivalently, momentum is generated to any amount of matter of fluid by the action on it of a force that changes its mass or its velocity at time rates $\dot{m}$ and $\dot{u} = \alpha$, respectively. Consider a sphere of mass $m$ at equilibrium on top of a horizontal plane. Its momentum is zero. A force $F$ is applied and then eliminated momentarily, which sets the sphere in motion with initial velocity $u$ and momentum $J = mu$. Thus, its momentum changed by $dJ/dt = m\, du/dt = F$, due to the action of the contact force $F$. Eventually, the sphere will stop and its momentum vanish, due to the resisting action of a frictional force $R$, coming from the sphere's relative motion and friction with the plane.

Since momentum is a quantity that can be stored, possessed, and transferred by a fluid, an exchange of momentum at the rate

$$\dot{j} = \lim_{k \to \infty} \sum_{i=1}^{k} \hat{J}_i \bar{u}_i \, \Delta A_i = \sum_{i=1}^{k} (\rho \bar{u}_i)\bar{u}_i \, \Delta A_i = \int_A (\rho u)u \, dA \tag{3.16}$$

occurs whenever fluid enters or leaves a system with velocity $u$ through an infinitesimal boundary area $dA$, or with average velocity $\bar{u}_i$ over a finite area $\Delta A_i$.

### Example 3.3: Exchange of Mass and Momentum

Water at temperature $T_0 = 20°C$ of density $\rho = 1\,\text{g/cm}^3$, heat capacity $c_p = 1\,\text{cal/(cm}^3\,\text{C)}$, viscosity $\eta = 1\,\text{cP}$, and salt concentration $c_0 = 0.1\,\text{g/cm}^3$ enters a mixer at average velocity $\bar{u}_1 = 1\,\text{m/s}$ through a pipe of cross-sectional area $A_1 = 10\,\text{cm}^2$. Pure water at the same temperature enters the same mixer at an average velocity $\bar{u}_2 = 0.5\,\text{m/s}$ through a pipe of cross-sectional area $A_2 = 5\,\text{cm}^2$. The

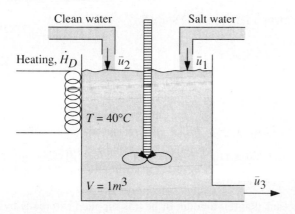

**Figure E3.3** Mass, salt, momentum, and energy transfer to mixer.

mixer of $V = 1\,m^3$ of fluid at any moment is kept at uniform temperature $T = 40°C$ by external heating, with just enough external work to overcome any frictional energy losses. The water leaves the mixer at the temperature and concentration of the mixer, $c$, through an exit pipe of cross-sectional area $A_3 = 7\,cm^2$. Figure E3.3 shows the process.

**(a)** What is the exit average velocity $\bar{u}_3$?

**(b)** What is the required rate of external heat supply $\dot{H}_D$?

**(c)** What is the resulting mixer concentration $c$?

**(d)** What are the rates of convection of mass, momentum, solute, and kinetic and internal energy in and out of the mixer? What are the corresponding net rates of convection?

**(e)** What will happen to the mass, salt concentration, and temperature of the mixer under the previous conditions, without any external heat supply?

Assume that the thermophysical properties of the water remain constant throughout, neglect any volume and other changes due to mixing, and assume well-mixed conditions everywhere in the mixer such that the concentration and temperature are uniform throughout the mixer.

**Solution**

**(a)** It is obvious that if the amount of water is to be kept constant in the mixer, the entering and exiting rates of mass convection must add to zero. Thus,

$$\dot{m} = \sum_{i=1}^{k=3} \rho_i \bar{u}_i A_i = \rho \bar{u}_1 A_1 + \rho \bar{u}_2 A_2 - \rho \bar{u}_3 A_3 = 0,$$

and therefore,

$$\bar{u}_3 = \frac{\bar{u}_1 A_1 + \bar{u}_2 A_2}{A_3} = \frac{100 \times 10 + 50 \times 5}{7}\,cm/s = 179\,cm/s.$$

**(b)** The net heat convection rate to the mixer is

$$\dot{H} = \sum_{i=1}^{k=3} (\hat{H}_i \bar{u}_i) a_i$$

$$= \sum_{i=1}^{k=3} (c_p T_i \bar{u}_i) A_i = c_p T_0 \bar{u}_1 A_1 + c_p T_0 \bar{u}_2 A_2 - c_p T \bar{u}_3 A_3$$

$$= c_p [T_0(\bar{u}_1 A_1 + \bar{u}_2 A_2) - T \bar{u}_3 A_3]$$

$$= 1 \times [20 \times (1000 + 250) - 40 \times 178 \times 7] \, \text{cal/s}$$

$$= -24,840 \, \text{cal/s}.$$

The minus sign means that this amount of heat is supplied externally by heating.

**(c)** The prevailing concentration is found by the solute mass balance,

$$\dot{m}_c = \sum_{i=1}^{k=3} (c_i \bar{u}_i) A_i = c_0 \bar{u}_1 A_1 + 0 - c \bar{u}_3 A_3 = 0,$$

which yields

$$c = \frac{c_0 \bar{u}_1 A_1}{\bar{u}_3 A_3} = \frac{1000}{1250} c_0 = 0.8 \, \text{gr/L}.$$

**(d)** The various in, out, and net convection rates, or *fluxes*, are

$$\dot{m} = \dot{m}_{in} - \dot{m}_{out} = [\rho(\bar{u}_1 A_1 + \bar{u}_2 A_2)] - (\rho \bar{u}_3 A_3)$$

$$= (1250 \, \text{g/s}) - (1250 \, \text{g/s}) = 0$$

$$\dot{J} = \dot{J}_{in} - \dot{J}_{out} = [\rho(\bar{u}_1^2 A_1 + \bar{u}_2^2 A_2)] - (\rho \bar{u}_3^2 A_3)$$

$$= (107.5 \, \text{kg cm/s}^2) - (221.78 \, \text{kg cm/s}^2) = -114.26 \, \text{kg cm/s}^2$$

$$\dot{H}_D = \dot{H}_{in} - \dot{H}_{out} = [c_p T_0(\bar{u}_1 A_1 + \bar{u}_2 A_2)] - (c_p T \bar{u}_3 A_3)$$

$$= (25,000 \, \text{cal/s}) - (50,284 \, \text{cal/s}) = -24.84 \, \text{kcal/s}$$

$$\dot{E} = \dot{E}_{in} - \dot{E}_{out} = [\rho(\bar{u}_1^4 A_1 + \bar{u}_2^3 A_2)] - \rho \bar{u}_3^3$$

$$= (1375 \, \text{kg cm}^2/\text{s}^3) - (39,478 \, \text{kg cm}^2/\text{s}^3)$$

$$= -38,103 \, \text{kg cm}^2/\text{s}^3 = -3.8 \, \text{N m/s} = -3.8 \, \text{W}.$$

Notice that unlike the zero net rate of mass transfer, none of the other net rates is zero. This is due to the fact that there are other nonconvective mechanisms for momentum and energy transfer to a system, for example by heating and stirring in the case at hand, whereas the only mechanism of (total) mass transfer is convection. Solute mass can also be transferred by diffusion, which, however, is absent in the case at hand.

**(e)** With no external heating, the temperatures of the mixer and that of the exiting stream will be identical to those of the entering streams. The exit solute concentrations will not be altered.

Besides *convective transfer of content*, due to relative motion between the system and surroundings and force interactions between the two, there may be

additional supply or removal of content to the system by mechanisms independent of relative motion (i.e., *nonconvective exchange*). Examples of such transfer
are the supply of heat by *conduction*, for example whenever a fluid is in contact
with a boundary of different temperature, and diffusion of solute to a system in
contact with surroundings of different solute concentration, for example absorption of oxygen at a lake's free surface. The net rate of such nonconvective transfer
to a system, generally called *diffusion*, is

$$\dot{q}_D = \sum_{i=1}^{k} \dot{q}_{D,i} = \sum_{i=1}^{k} \mathcal{D}_i \left( -\frac{dc_i}{dx} \right) \Delta A_i \tag{3.17}$$

for discontinuous transfer, and

$$\dot{q}_D = \lim_{k \to \infty} \sum_{i=1}^{k} \mathcal{D} \left( -\frac{dc_i}{dx} \right) A_i = -\int_A \mathcal{D} \frac{dc_i}{dx} dA \tag{3.18}$$

for continuous transfer across the entire boundary area $A$. Here $\mathcal{D}$ is a heat
conduction or solute *diffusion coefficient*, $dc_i/dx$ a temperature or concentration
*gradient* at the system's boundary, and $\Delta A_i$ an elementary boundary area through
which diffusion takes place. The minus signs in Eqns. (3.17) and (3.18) account
for the fact that *conduction or diffusion occurs in the direction of diminishing
concentration and therefore negative gradient or difference* (e.g., conduction of
heat takes place from high to low temperature, and diffusion of solute from high
to low concentration).

Finally, there may occasionally be supply or removal of work and power to
or from the system by pumps and other devices that process or are operated by
the fluid of the system, respectively, for example the stirring mechanism in Fig.
E3.3.

### 3.1.3 Conversion of Fluid Content

Content may be produced from or destroyed to other forms of content. Solute
and heat may be produced or consumed by a *chemical reaction* at rate $r = \pm kc^\alpha$,
where $c$ is solute concentration and $k$ is an appropriate rate constant. Mechanical
energy is continuously being lost to heat by *viscous dissipation* or friction arising
due to layerwise relative motion between adjacent fluid layers. These rates of
*conversion* are quantified by equations of the form

$$\dot{m}_r = \lim_{k \to \infty} \sum_{i=1}^{k} r_i \Delta V_l = \int_V r \, dV = \int_V \pm kc^a \, dV \tag{3.19}$$

for solute, and by

$$\dot{H}_r = \lim_{k \to \infty} \sum_{i=1}^{k} \dot{H}_{r,i} \Delta V_i = \int_V \hat{H}_r \, dV \tag{3.20}$$

for heat, where $\hat{H}_r$ is positive for exothermic and negative for endothermic
reactions. Equation (3.20) may also represent mechanical energy losses by viscous
dissipation, in which case $\hat{H}_r$ is negative, or equivalent internal energy gains, in
which case $\hat{H}_r$ is positive.

**Example 3.4: Solute and Heat Conversion in System**

Consider a continuous well-mixed reactor filled with liquid of volume $1 \, m^3$, where a first-order exothermic chemical reaction at the rate $r = 0.1c$ (mol/L)/s takes place accompanied by the release of 5 cal/mol of heat. For an initial temperature of 25°C and reactant concentration $c_0 = 1 \, mol/L$, calculate the rates of solute consumption and heat generation due to the chemical reaction. What may happen to the heat produced?

**Solution**     According to Eqn. (3.19), the rate of reactant consumption is

$$\dot{m}_r = \int_V kc(t) \, dV = -\int_V 0.1c(t) \, dV = -0.1c(t)V = -100c(t) \, \text{mol/s}, \qquad (3.4.1)$$

because the time-varying concentration is constant throughout a well-mixed reactor. Similarly, heat is produced at the rate of

$$\dot{H}_r = \int_V \hat{H}_r \, dV = \int_V [5 \times 0.1c(t)] \, dV = 500c(t) \, \text{cal/s}. \qquad (3.4.2)$$

If the reactor is closed and insulated, there will be a continuous decrease of reactant according to the balance equation

$$V\frac{dc}{dt} = \dot{m}_r = -100c(t) \, \text{mol/s}, \qquad c(t = 0) = c_0 = 1 \, \text{mol/L} \qquad (3.4.3)$$

which has the solution

$$c(t) = \exp(-0.1t) \, \text{mol/L}. \qquad (3.4.4)$$

Simultaneously, the reacting fluid will be heated according to

$$V\frac{dH}{dt} = V\rho c_p \frac{dT}{dt} = \dot{H}_r = 500c(t) \, \text{cal/s} = 500 \exp(0.1t) \, \text{cal/s} \qquad T(t = 0) = 25°C,$$
$$(3.4.5)$$

where $\rho$ and $c_p$ are heat capacity and density. The solution for the temperature evolution is

$$T(t) = T_0 + \frac{0.5}{\rho c_p}[1 - \exp(-0.1t)]. \qquad (3.4.6)$$

If the reactor is open (i.e., allowed to exchange solute and heat by convection and/or diffusion), the considerations of Example 3.3 must be taken into account in the conservations equations, Eqns. (3.4.3) and (3.4.5), as detailed in Section 3.2.

### 3.1.4 Principles of Fluid Content Conservation

Extensive properties or content of fluids are conserved, which means that although they may change form, they cannot be created from or decay to zero. Mass may change form by chemical reaction, but the summation of all species masses, or total mass, of reactants and products is constant. Potential energy may change into kinetic energy, amounts of which may turn into internal energy by friction; however, the summation of the three is constant, in the absence of an external supply of heat or work. Similarly, momentum may change into force action, and vice versa, but the summation of the two is constant. These facts are represented by the general *conservation statement* for a system under consideration:

$$
\begin{bmatrix}
\text{rate of change of} \\
\text{quantity conserved} \\
\text{in system volume}
\end{bmatrix}
=
\begin{bmatrix}
\text{net transfer of quantity} \\
\text{by convection} \\
\text{through system boundaries}
\end{bmatrix}
$$

$$
+
\begin{bmatrix}
\text{net transfer by} \\
\text{nonconvective mechanisms} \\
\text{through system boundaries} \\
\text{(e.g., contact force, diffusion)}
\end{bmatrix}
+
\begin{bmatrix}
\text{net transfer by} \\
\text{nonconvective mechanisms} \\
\text{through system volume} \\
\text{(e.g., body force)}
\end{bmatrix}
$$

$$
+
\begin{bmatrix}
\text{conversion of} \\
\text{quantity conserved} \\
\text{in system volume}
\end{bmatrix}. \tag{3.21}
$$

The conservation statement of Eqn. (3.21) simply states that the mechanisms in the right-hand-side brackets combine to yield changes within the system, expressed by the left side of the equation. Figure 3.1 illustrates these mechanisms, which are quantified in this section. By identifying the conserved quantity as mass, energy, momentum, heat, or solute, the corresponding conservation equations of these quantities result, as detailed in the rest of this chapter.

## 3.2 CONSERVATION OF MASS

By *mass* hereafter we denote *total mass,* as opposed to a particular species mass, which will be referred to as *solute* (mass). The only mechanism of transfer of mass in a system is by convection, and therefore Eqn. (3.21) takes the form

$$
\frac{dm}{dt} = \sum_{i=1}^{k} \rho(\bar{u}_i A_i)_{\text{in}}, \tag{3.22}
$$

where $\bar{u}_i$ is positive for entry to and negative for exit from the system through a surface of area $A_i$. If, instead, there is a *porous boundary* where the relative velocity is $u$, the resulting equation is

$$
\frac{dm}{dt} = \int_A \rho u \, dA, \tag{3.23}
$$

where $A$ is the area that allows mass exchange. For fluids of constant density,

$$
dm = \rho \, dV,
$$

and therefore Eqns. (3.22) and (3.23) take the form

$$
\frac{dV}{dt} = \sum_{i=1}^{k} (\bar{u}_i A_i) \tag{3.24}
$$

and

$$\frac{dV}{dt} = \int_A u \, dA, \tag{3.25}$$

respectively.

### Example 3.5: Water Level in Tank

Figure E3.5 shows a simple level control device, $C$, that activates a pump to remove $\dot{V}$ ft$^3$/min of water from a cylindrical tank of radius $R = 0.5$ ft in order to maintain a constant height of water $H_0 = 2$ ft, under time-dependent loading through a pipe of diameter $d = 2$ in. with average velocity $\bar{u} = [1 + \sin(\pi t)]$ ft/s. Starting from $t = 0$ where the tank is empty, describe the process of filling and control.

**Solution**  The system here is conveniently chosen within the dashed line shown. Equation (3.24) for water of constant density, $\rho = 1$ g/cm$^3$ takes the form

$$\frac{dV}{dt} = \sum_{i=1}^{k=2} \bar{u}_i A_i = \bar{u} \pi \frac{d^2}{4} - Q. \tag{3.5.1}$$

The volume of the water in the tank changes by

$$\frac{dV}{dt} = \frac{d}{dt}(\pi R^2 H) = \pi R^2 \frac{dH}{dt}, \tag{3.5.2}$$

and Eqn. (3.5.1) becomes

$$\pi R^2 \frac{dH}{dt} = \bar{u} \pi \frac{d^2}{4} - Q, \tag{3.5.3}$$

which predicts how the height $H$ will evolve with time starting from $H(t = 0) = 0$ to $H_0$, where the controller is activated to start removing water at rate $Q$.

*Filling stage.* For $H \leq H_0$, $Q = 0$, and Eqn. (3.5.3) takes the form

$$\frac{dH}{dt} = \frac{d^2}{R^2} \frac{\bar{u}}{4} = \frac{d^2}{4R^2}[1 + \sin(\pi t)], \tag{3.5.4}$$

which is integrated to

$$\int_0^{H(t)} dH = 0.25 \frac{d^2}{R^2} \int_0^t [1 + \sin(\pi t)] \, dt, \tag{3.5.5}$$

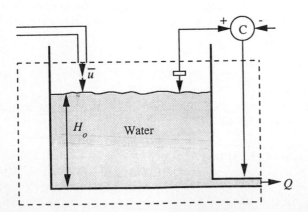

**Figure E3.5**  Water level control.

which yields

$$H(t) = 0.0278\left[t + \frac{1 - \cos(\pi t)}{\pi}\right] \text{ ft.} \tag{3.5.6}$$

Thus the time $\tau$ it takes to reach $H(t = \tau) = H_0 = 2$ ft is given by

$$\tau + 0.318[1 - \cos(\pi\tau)] = 72, \tag{3.5.7}$$

which yields approximately $\tau = 72$ s.

   *Controlled stage.* For this stage, Eqn. (3.5.3) applies with $dH/dt = 0$, since the level must remain constant equal to $H_0 = 2$ ft. Thus the pump must remove water at the rate

$$Q = \pi\bar{u}\frac{d^2}{4} = \frac{3.14}{144} \times [1 + \sin(\pi t)] \text{ ft}^3/\text{s} \tag{3.5.8}$$

in order to maintain the target level.

### Example 3.6: Bifurcation of Flow

For the piping system shown in Fig. E3.6, the following data are given: $R_1 = 2.5$ cm, $\bar{u}_1 = 1$ cm$^3$/s, $R_2 = 2.5$ cm, $\bar{u}_2 = 10$ cm$^3$/s, $R_5 = 1.25$ cm, $\bar{u}_5 = 5$ cm/s, $R_6 = R_4 = 2$ cm. Calculate the average velocities at points $A$, $B$, $C$, and $D$.

**Solution**   The system enclosed by the dashed line is chosen to calculate the average velocity at point $A$. Mass balance under steady-state conditions yields

$$\sum_{i=1}^{k=3} \rho\bar{u}_i A_i = \rho(\bar{u}_1 A_1 + \bar{u}_2 A_2 - \bar{u}_3 A_3) = 0,$$

and therefore,

$$\bar{u}_A = \bar{u}_3 = \frac{\bar{u}_1 A_1 + \bar{u}_2 A_2}{A_3} = \frac{1 \times 6.25 + 10 \times 6.25}{6.25} \text{ cm/s} = 11 \text{ cm/s.}$$

To calculate the average velocity at point $B$, the system enclosed by the dashed-dotted line is considered. By the same mass conservation principles,

$$\bar{u}_B = \bar{u}_4 = \frac{\bar{u}_3 A_3 - \bar{u}_5 A_5}{A_4} = \frac{11 \times 6.25 - 5 \times 1.56}{6.25} \text{ cm/s} = 9.75 \text{ cm/s.}$$

To calculate the common average velocity at points $C$ and $D$, the system enclosed

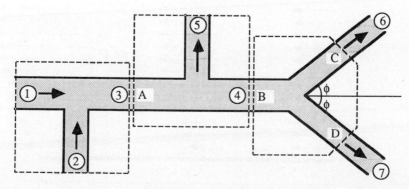

**Figure E3.6**   Flow bifurcation.

by the dotted line is considered. Then

$$\bar{u}_C = \bar{u}_6 = \bar{u}_7 = \bar{u}_D = 0.5 \times \frac{\bar{u}_4 A_4}{A_6} = \frac{0.5 \times 9.75 \times 6.25}{4} \text{ cm/s} = 7.62 \text{ cm/s}.$$

As a check for consistency of the calculations, the entire piping system is now considered:

$$\sum_{i=1}^{k=5} \rho \bar{u}_i A_i = \rho (\bar{u}_1 A_1 + \bar{u}_2 A_2 - \bar{u}_5 A_5 - \bar{u}_6 A_6 - \bar{u}_7 A_7)$$

$$= \rho \frac{\pi}{4} (1 \times 6.25 + 10 \times 6.25 - 5 \times 1.56 - 7.62 \times 4 - 7.62 \times 4)$$

$$= \rho \frac{\pi}{4} (68.75 - 68.75) \equiv 0.$$

This equation shows that mass is conserved, and therefore the calculations performed are correct.

### Example 3.7: Mass Balance with Continuous Velocity

A free stream at velocity $U = 10$ m/s of fluid of viscosity $\eta = 1$ cP and density $\rho = 1$ gr/cm$^3$ overtakes a plate of length $L = 10$ m and width $W = 2$, placed parallel to the stream as shown in Fig. E3.7. A *boundary layer* of thickness

$$\delta(x) = 5 \sqrt{\frac{xv}{U}} \tag{3.7.1}$$

is formed, under which the velocity is nearly linear, approximated by

$$u(x, y) = U \frac{y}{\delta(x)}, \tag{3.7.2}$$

with respect to the system of coordinates shown. Everywhere outside the boundary layer, the velocity is identical to $U$.

**(a)** Plot qualitatively the boundary layer thickness $\delta(x)$ vs. $x$. Then plot velocity

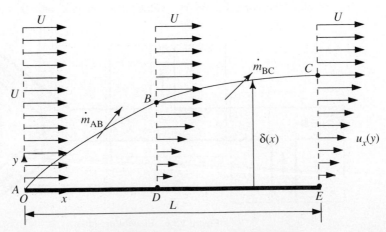

**Figure E3.7**   Boundary layer flow over plate.

profiles qualitatively, at the leading edge of $x = 0$, at a distance $x = 0.4L$ downstream and at the end of the plate, at $x = L$.

**(b)** Find the mass flow rates, $\dot{m}_{AB}$, $\dot{m}_{BC}$, $\dot{m}_{AC}$, that cross the boundaries of projections $AB$, $BC$, and $AC$, respectively.

**Solution**

**(a)** Thickness and velocity profiles are shown in Fig. E3.7, estimated according to Eqns. (3.7.1) and (3.7.2), respectively.

**(b)** Equation (3.23) applied within the $ABD$ system under steady-state conditions for which $dm/dt = 0$ results in

$$W\rho\left[\int_{AB} u_{AB}\,dS - \int_0^{\delta(x=0.4L)} u(x = 0.4L, y)\,dy + \int_0^{x=0.4L} dx\right] = 0, \qquad (3.7.3)$$

and therefore,

$$\dot{m}_{AB} = W\int_{AB} \rho u_{AB}\,dS = W\int_0^{\delta(x=0.4L)} \rho U\frac{y}{\delta(x = 0.4L)}\,dy$$

$$= 0.5W\rho U\delta(x = 0.4L) = 2.25W\rho U\sqrt{\frac{0.4Lv}{U}} \qquad (3.7.4)$$

$$= 28.46\ \text{kg/s}.$$

By the same arguments, for the system projected as $DBCF$,

$$W\rho\left[\int_0^{\delta(x=0.4L)} \frac{Uy}{\delta(x = 0.4L)}\,dy + \int_{BC} u_{BC}\,dS - \int_0^{\delta(x=L)} \frac{Uy}{\delta(x = L)}\right] = 0,$$

$$(3.7.5)$$

and therefore,

$$\dot{m}_{BC} = W\rho\int_{BC} u_{BC}\,dS$$

$$= W\rho\left[-\int_0^{\delta(x=0.4L)} \frac{yU}{\delta(x = 0.4L)}\,dy + \int_0^{\delta(x=L)} \frac{Uy}{\delta(x = L)}\,dy\right] \qquad (3.7.6)$$

$$= -28.46\ \text{kg/s} + 2.5W\rho U\sqrt{\frac{Lv}{U}} = -28.46\ \text{kg/s} + 45\ \text{kg/s} = 16.54\ \text{kg/s}.$$

Similarly, for the system projected as $ACE$, it is

$$W\rho\left[\int_{AC} u_{AB}\,dS - \int_0^{\delta(x=L)} \frac{Uy}{\delta(x = L)}\,dy\right] = 0, \qquad (3.7.7)$$

and therefore,

$$\dot{m}_{AC} = W\rho\int_{AC} u_{AC}\,dS = 0.5W\rho U\delta(x = L) = 2.25W\rho U\sqrt{\frac{Lv}{U}} = 45\ \text{kg/s}.$$

$$(3.7.8)$$

Notice that

$$\dot{m}_{AC} = \dot{m}_{AB} + \dot{m}_{BC}, \qquad (3.7.9)$$

as it should. This is a consistency check that verifies the accuracy of the calculations performed.

## 3.3 CONSERVATION OF ENERGY

The *mechanical energy* of a body, system, or fluid is the summation of the *potential and kinetic energies. Kinetic energy, $E_k$,* is energy possessed by a moving body, system, or fluid, defined by

$$E_k = \tfrac{1}{2}mu^2, \tag{3.26}$$

where $m$ is mass and $u$ is velocity. *Potential energy $E_\phi$* is energy possessed due to a position, which can be turned into kinetic energy once a retarding force is removed. The most characteristic case in fluid mechanics is potential energy associated with elevation beyond the earth's surface, since any body, system, or fluid at any elevation may spontaneously fall under gravity, producing kinetic energy. This form of potential energy is defined by

$$E_\phi = mgz, \tag{3.27}$$

where $m$ is the mass, $g$ the acceleration of gravity, and $z$ the elevation. Thus *mechanical energy* is defined as

$$E_M = E_k + E_\phi = \tfrac{1}{2}mu^2 + mgz. \tag{3.28}$$

To make the *total energy* the *internal energy, U,* associated with thermal content, must be added to the mechanical energy. The internal energy at temperature $T$ is defined by equations of the form

$$U = U_0 + mc_i(T - T_0), \tag{3.29}$$

where $U_0$ is internal energy at reference temperature $T_0$ and $c_i$ heat capacity. Thus, the internal energy difference between two states of different temperatures $T_1$ and $T_2$ is given by

$$\Delta U = mc(T_1 - T_2). \tag{3.30}$$

Thus, the total energy $E_T$ is defined by

$$E_T = E_M + U = E_K + E_\phi + U = \tfrac{1}{2}mu^2 + mgz + U, \tag{3.31}$$

and the *density of energy* is

$$\hat{E}_T = \frac{E_T}{V} = \frac{E_M}{V} + \frac{U}{V} = \frac{E_k}{V} + \frac{E_\phi}{V} + \frac{U}{V} = \frac{1}{2}\rho u^2 + \rho gz + \hat{U} \tag{3.32}$$

Energy, being part of the fluid content, is convected in and out of a system with entering or leaving fluid, according to Eqns. (3.11) to (3.13) and Table 3.1. Also, the existence of any contact forces at a portion of the boundary where the velocity is $u$ results in a *pressure energy* or *work* done to or by the system at the rate

$$E_p = pu \tag{3.33}$$

where $p$ is the local absolute pressure, under the assumption that viscous forces are vanishingly small. The physical significance of Eqn. (3.33) is that it represents the work required to push the unit of mass of fluid into the system by overcoming the local pressure, $p$, or work released by the system in pushing the unit of fluid mass out under pressure $p$. Finally, there may be supply to or production of *shaft work* by the system at rate $\dot{W}_s$, equivalent to the rate of addition or subtraction of energy, and *heat* at rate $\dot{Q}$, also equivalent to the rate of addition or subtraction of energy. The convection used in this book is that $\dot{W}_s$ and $\dot{Q}$ are positive when supplied to the system and negative when removed from the system. The conservation statement of Eqn. (3.21) for total energy $E_T$ takes the form

$$\underbrace{\frac{dE_T}{dt}}_{\text{rate of change of energy}} = \underbrace{\sum_{i=1}^{k} [\hat{E}_T, u_i, A_i]}_{\text{net convection of energy}} + \underbrace{\left[\sum_{i=1}^{m} p_i, u_i, A_i\right]}_{\text{work done by pressure}} +$$

$$\underbrace{+ \quad \dot{W}_s}_{\text{work done to }(+)\text{ or by }(-)\text{ system}} + \underbrace{\dot{Q}}_{\text{heat supply to }(+)\text{ or from }(-)\text{ system,}} \tag{3.34}$$

with

$$E_T = \tfrac{1}{2}mu^2 + mgz + U \tag{3.35}$$

and

$$\hat{E}_{T,i} = \tfrac{1}{2}\rho u_i^2 + \rho gz + \hat{U}_i. \tag{3.36}$$

The summations may take the form of integrals along the lines of Eqn. (3.23), in case of continuous distribution of velocities. The various terms of Eqn. (3.36) are shown pictorially in Fig. 3.2 for a system with a single entry and a single exit, under external heating and production of shaft work.

When the entering the Eqn. (3.34) quantities $u_i$, $p_i$, $U_i$, and $z_i$ are not

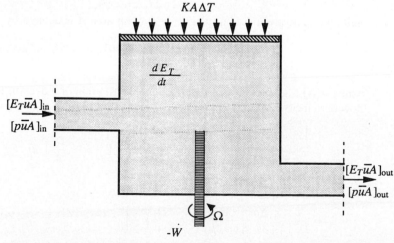

**Figure 3.2**  Total energy conservation for a system with a single entry and a single exit, under heating at rate $\dot{Q}$ and production of shaft work at rate $-\dot{W}$.

uniform over their area $A_i$, their *average values* over $A_i$ are used. The average of a quantity $q$ over an area $A$ is defined by

$$\bar{q} = \frac{\int_A q(A)\,dA}{A}, \tag{3.37}$$

where $q(A)$ is a function that describes the distribution of $q$ over $A$. Averages of powers of these quantities, such as $\bar{u}_i^3$ of the kinetic energy term in Eqn. (3.33), different from $\bar{u}_i^3$, are defined accordingly:

$$\overline{u_i^3} = \frac{1}{A}\int_A u_i^3\,dA = \alpha\bar{u}_i^3 \neq \bar{u}_i^3. \tag{3.38}$$

Here $\alpha$ is a factor of velocity nonuniformity, with $\alpha = 1$ for plug velocity. As demonstrated in Example 3.8, $\alpha = 2$ for laminal flow in a pipe and $\alpha \simeq 1$ for turbulent flow in a pipe. For other kinds of cross-sectional areas and velocity distributions, $\alpha$ is calculated as demonstrated in the same example. By utilizing these averages and the mass flow rate equation

$$\dot{m}_i = \rho\bar{u}_i A_i, \tag{3.39}$$

Eqn. (3.34) takes the form

$$\sum_{i=1}^{k}\left[\left(\frac{\alpha\bar{u}_i^2}{2} + g\bar{z}_i + \frac{\bar{U}_i}{\rho} + \frac{\bar{p}_i}{\rho}\right)\dot{m}_i\right]_{\text{in}} + \dot{W}_s + \dot{Q} = 0. \tag{3.40}$$

Unlike the velocity $u$, in most cases $\hat{U}_i$, $p_i$, and $z_i$ are practically uniform over the area $A_i$ and therefore will be considered identical to their average values.

### Example 3.8: Average Quantities in Fluid Flow

Calculate the average of $u, u^2, u^3, \ldots, u^k$ powers of velocity in flow of a fluid of viscosity $\eta$, driven by pressure gradient in a pipe of diameter $D$. As will be shown in Chapter 3, the axial velocity varies with the radial distance $r$ according to the equation

$$u = \frac{1}{4\eta}\frac{\Delta p}{\Delta L}\left(r^2 - \frac{D^2}{4}\right). \tag{3.8.1}$$

**Solution**   The average of a quantity $q$ over an area $A$ is defined by

$$\bar{q} = \frac{\int_A q(A)\,dA}{A}, \tag{3.8.2}$$

where $q(A)$ is a function that describes the distribution of $q$ over $A$. For laminar pipe flow of velocity profile given by Eqn. (3.8.1), Eqn. (3.8.2) takes the form

$$\begin{aligned}
\overline{u^k} &= \frac{\int_0^{D/2} 2\pi r u^k(r)\,dr}{\pi(D^2/4)} = \frac{8}{D^2}\int_0^{D/2} r\left(\frac{1}{4\eta}\left(\frac{\Delta p}{\Delta L}\right)^k\right)\left(r^2 - \frac{D^2}{4}\right)^k dr \\
&= \frac{8}{D^2}\left(\frac{1}{4\eta}\frac{\Delta p}{\Delta L}\right)^k\int_0^{D/2} r\left(r^2 - \frac{D^2}{4}\right)^k dr \\
&= \frac{8}{D^2}\left[\frac{1}{4\eta}\frac{\Delta p}{\Delta L}\frac{1}{2(1+k)}\left(r^2 - \frac{D^2}{4}\right)^{k+1}\right]_0^{D/2} \\
&= \frac{4}{D^2(k+1)}\left(\frac{1}{4\eta}\frac{\Delta p}{\Delta L}\right)^k\left[0 - \left(-\frac{D^2}{4}\right)^{k+1}\right] = \left(-\frac{1}{16\eta}\frac{\Delta p}{\Delta L}D^2\right)^k\frac{1}{k+1}.
\end{aligned} \tag{3.8.3}$$

The average velocity $\bar{u}$ is obtained from Eqn. (3.8.2) for $k = 1$:

$$\bar{u} = \left(-\frac{1}{16\eta}\frac{\Delta p}{\Delta L}D^2\right)\frac{1}{2} = -\frac{1}{32\eta}\frac{\Delta p}{\Delta L}D^2.$$  (3.8.4)

Thus, any power of the average velocity is

$$\bar{u}^k = \left(-\frac{1}{32\eta}\frac{\Delta p}{\Delta L}D^2\right)^k,$$  (3.8.5)

and its relation to the average of any velocity power, $\overline{u^k}$, is

$$\frac{\overline{u^k}}{\bar{u}^k} = \frac{(-(1/16\eta)(\Delta p/\Delta L)D^2)^k}{(-(1/32\eta)(\Delta p/\Delta L)D^2)^k}\frac{1}{k+1} = \frac{2^k}{k+1}.$$  (3.8.6)

Now, by assigning values to $k$, the following relations are obtained:

$$\overline{u^1} = \bar{u}^1\frac{2}{2} = \bar{u}^1 = \bar{u},$$  (3.8.7)

$$\overline{u^2} = \frac{2^2}{2+n}\bar{u}^2 = \frac{4}{3}\bar{u}^2,$$  (3.8.8)

$$\overline{u^3} = \frac{2^3}{3+1}\bar{u}^3 = 2\bar{u}^3.$$  (3.8.9)

The coefficients $\beta = \frac{4}{3}$ in Eqn. (3.8.8) and $\alpha = 2$ in Eqn. (3.8.9) enter the momentum and energy equations, respectively, for laminar pipe flow. Both are unity in plug flows.

Turbulent flow. A good approximation of the velocity profile in turbulent flow is

$$u = u_{max}\left(1 - \frac{2r}{D}\right)^n.$$  (3.8.10)

By following the same procedure, the results are

$$\bar{u}^k = \frac{8}{D^2}\int_0^{D/2} r\left[\left(1 - \frac{2r}{D}\right)^n u_{max}\right]^k dr$$

$$= \frac{8u_{max}^k}{D^2}\int_0^{D/2} r\left(1 - \frac{2r}{D}\right)^{kn} dr = \frac{8u_{max}^k}{4(1+kn)(2+kn)}.$$  (3.8.11)

From this general expression the following results are obtained:

$$\bar{u} = \frac{2u_{max}}{(1+n)(2+n)},$$  (3.8.12)

$$\bar{u}^k = \left[\frac{2u_{max}}{(1+n)(2+n)}\right]^k,$$  (3.8.13)

$$\frac{\overline{u^k}}{\bar{u}^k} = \frac{2u_{max}^k}{(1+kn)(2+kn)}\frac{(1+n)^k(2+n)^{kk}}{(2u_{max})} = \frac{2^{(1-k)}(1+n)^k(2+n)^k}{(1+kn)(2+kn)}.$$  (3.8.14)

By assigning values to $k$, the following relations are obtained for $n \approx 0.13$, which is a commonly observed value in turbulent pipe flow:

$$\overline{u^1} = \bar{u}^1 \frac{(1+n)(2+n)}{(1+n)(2+n)} = 1 = \bar{u}, \qquad \overline{u^2} = \frac{\bar{u}^2}{2} \frac{(1+n)^2(2+n)^2}{(1+2n)(2+2n)} \approx \bar{u}^2,$$

$$\overline{u^3} = \frac{\bar{u}^3}{4} \frac{(1+n)^3(2+n)^3}{(1+3n)(2+3n)} \approx \bar{u}^3. \tag{3.8.15}$$

Thus in turbulent flow all factors are nearly unity. This is not surprising since the velocity profile predicted by Eqn. (3.8.10) is nearly flat and therefore close to its average value everywhere, except within thin layers near the wall that do not contribute much to the overall average value anyway.

The important conclusion to be kept in mind here is that the average value of a power of a velocity in general is different from the same power of the average velocity. This difference decreases with velocity uniformity and becomes negligible for turbulent flows and zero in plug flows.

### 3.3.1 Bernoulli's Equation

In Eqn. (3.40), the sum

$$\hat{E}_M = \rho\bar{z} + \bar{p} + \rho\alpha\frac{\bar{u}^2}{2} \approx \rho z + p + \rho\alpha\frac{\bar{u}^2}{2} \tag{3.41}$$

represents the *average mechanical energy* per unit volume of flowing fluid. The term

$$\Delta\hat{E}_w = \frac{\rho\dot{W}_s}{\dot{m}} = \frac{\dot{W}_s}{\dot{V}} \tag{3.42}$$

is rate of mechanical energy or shaft power produced or consumed by the unit volume of flowing fluid. The sum

$$\Delta\hat{E}_l = (\hat{U}_{\text{out}} - \hat{U}_{\text{in}}) - \frac{\dot{Q}}{\dot{V}} = \frac{\dot{\mathscr{F}}}{\dot{V}}\mathscr{F} \tag{3.43}$$

represents mechanical energy lost per unit volume of flowing fluid due to *viscous dissipation* at rate $\dot{\mathscr{F}}$, to be quantified in Chapter 4.

For a system with a single entry and a single exit, *as well as for any two points on the same streamline*—the orbit traced by a fluid particle—the mechanical energy conservation equation takes the form

$$\Delta\hat{E}_M = [\hat{E}_M]_{\text{in}} - [\hat{E}_M]_{\text{out}} = \Delta\hat{E}_w + \Delta\hat{E}_l. \tag{3.44}$$

For flowing fluids in the absence of any device supplying work such as a pump, or consuming work such as a turbine, Eqn. (3.44) takes the form

$$\Delta\hat{E}_M + [\hat{E}_M]_{\text{in}} - [\hat{E}_M]_{\text{out}} = \Delta\hat{E}_l, \tag{3.45}$$

and in the absence of heating or cooling by the surroundings or due to viscous dissipation, it reduces further to

$$\Delta \hat{E}_M = [\hat{E}_M]_{in} - [\hat{E}_M]_{out} = 0, \tag{3.46}$$

or equivalently to

$$\left[ g\bar{z} + \frac{\bar{p}}{\rho} + \frac{\alpha \bar{u}^2}{2} \right]_{in} = \left[ g\bar{z} + \frac{\bar{p}}{\rho} + \frac{\alpha \bar{u}^2}{2} \right]_{out}. \tag{3.47}$$

The latter is the celebrated *Bernoulli's equation*,[2] which applies between:

1. Any two points on the same streamline of an ideal frictionless flow, in which case $\alpha = 1$ and $\bar{q} = q$, with $q$ the local velocity, elevation, or pressure.
2. Any two points of a fluid in static equilibrium, in the form of Eqn. (1.47).
3. The unique entry and unique frictionless exit of a static system, in which case the average quantities enter.

### Example 3.9: Bernoulli's Equation: Water Jet under Gravity and Initial Velocity

Figure E3.9 shows a water jet emerging from a hole of diameter $d$ at the base of a tank of water of constant height $H$, positioned at elevation $z_0$.

**(a)** How do the diameter of the jet, $d$, and its velocity, $u$, change with $z$?
**(b)** How would things change under finite surface tension $\sigma$?
**(c)** Can you deduce the shape of the jet?

### Solution

**(a)** A mass balance between the exit and any point along the jet, at a distance $z$ from the reference plane, yields

$$\pi \frac{d_0^2}{4} u_0 = \pi \frac{d^2}{4} u. \tag{3.9.1}$$

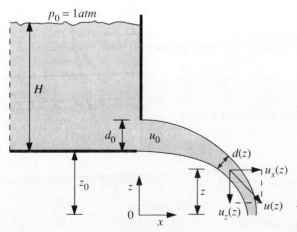

**Figure E3.9**   Water jet under initial velocity and gravity.

Bernoulli's equation between the same points yields

$$\frac{p_0}{\rho} + gz_0 + \frac{u_0^2}{2} = \frac{p}{\rho} + gz + \frac{u_0^2}{2}, \tag{3.9.2}$$

where

$$p_0 = p = p_{atm}. \tag{3.9.3}$$

The exit velocity $u_0$ is estimated by applying Bernoulli's equation between the water free surface, where the pressure is atmospheric and the velocity $u = dH/dt \approx 0$, and the exit, where the velocity is $u_0$ and the pressure is given by Eqn. (3.9.3):

$$g(z_0 + H) + \frac{p_{atm}}{\rho} + 0 = gz_0 + \frac{p_{atm}}{\rho} + \frac{u_0^2}{2}. \tag{3.9.4}$$

Thus

$$u_0 = \sqrt{2gH}. \tag{3.9.5}$$

Equations (3.9.3) and (3.9.5) are substituted in Eqn. (3.9.2) and the result is

$$u(z) = \sqrt{2g(z_0 + H - z)}. \tag{3.9.6}$$

The diameter of the jet at the same position is obtained by Eqn. (3.9.1), given Eqn. (3.9.6):

$$d(z) = d_0\sqrt{\frac{u_0}{u(z)}} = d_0\sqrt[4]{\frac{H}{z_0 + H - z}}. \tag{3.9.7}$$

Notice that Eqns. (3.9.6) and (3.9.7) for $z = z_0$ predict $u(z) = u_0$ and $d(z) = d_0$, as they should, and may serve as a consistency check for the calculations performed.

**(b)** In the presence of surface tension $\sigma$, Eqn. (3.9.1) remains valid. However, the pressure under the jet changes to

$$p(z) = p_{atm} + \frac{2\sigma}{d(z)}, \tag{3.9.8}$$

$$p_0 = p_{atm} + \frac{2\sigma}{d_0}, \tag{3.9.9}$$

where $d(z)/2$ is the curvature of the jet. Then Eqn. (3.9.4) changes accordingly and the predicted velocity discharge velocity is

$$u_0 = \sqrt{2gH - \frac{4\sigma}{\rho d_0}}, \tag{3.9.10}$$

which suggests that the jet will form if

$$\frac{2\sigma}{\rho d_0 gH} < 1, \tag{3.9.11}$$

which of course includes the case of $\sigma = 0$ examined earlier. In fact, for water,

$$\frac{2\sigma}{\rho d_0 gH} = \frac{2 \times 72 \text{ dyn/cm}}{1 \text{ g/cm}^3 \times 981 \text{ cm/s}^2 \times d_0 H} = \frac{0.147}{d_0 H}, \tag{3.9.12}$$

and the jet will always form.

In the case of high surface tension, Eqns. (3.9.8) and (3.9.9) reduce to

$$\frac{u^2(z)}{2} + \frac{2\sigma}{\rho d(z)} = g(z_0 + H - z). \tag{3.9.13}$$

Equation (3.9.1), with $u_0$ given by Eqn. (3.9.10), substituted in Eqn. (3.9.13) eliminates $d(z)$:

$$u^2(z) + \frac{4\sigma}{\rho d_0}\left[\frac{u(z)}{2gH - 4\sigma/\rho d_0}\right]^{1/2} - 2g(z_0 + H - z) = 0. \tag{3.9.14}$$

Equation (3.9.14) can be solved for $u(z)$. For $\sigma = 0$, the results of Eqns. (3.9.6) and (3.9.7) are recovered. For $0 < 2\sigma < (gH/\rho d_0)$ such that inequality (3.9.11) is satisfied, the resulting value of $u(z)$ is less than that predicted by Eqn. (3.9.6) under negligible surface tension, and the corresponding jet diameter predicted by Eqn. (3.9.1) is less than that of Eqn. (3.9.7). Thus, surface tension decelerates the liquid and also reduces the jet diameter. Overall, surface tension reduces the discharge rate.

(c) The shape of the jet cannot be deduced by any of the equations above. A force balance, Newton's law of motion, or momentum balance is required, all of which reduce to

$$m\alpha_x = 0, \tag{3.9.15}$$

where $\alpha_x$ is acceleration in the $x$-direction. The same laws applied in the vertical direction result in

$$m\alpha_z = -B = -mg, \tag{3.9.16}$$

where $B$ is the weight of $m$ mass of flowing liquid. By simple kinematic principles,

$$\alpha_x = \frac{du_x}{dt} = 0, \qquad \alpha_z = \frac{du_z}{dt} = -g, \tag{3.9.17}$$

and therefore,

$$u_x = c = u_0, \qquad u_z = -gt, \tag{3.9.18}$$

where $u_x$ and $u_z$ are the two components of the velocity $u(z)$, calculated by Eqn. (3.9.6), as shown in Fig. E3.9. Distances traveled are given by

$$z = z_0 + \int_0^t u_z\, dt = z_0 - \frac{gt^2}{2}, \qquad x = x_0 = \int_0^t u_x\, dt = 0 + \int_0^t u_0\, dt = t\sqrt{2gH}. \tag{3.9.19}$$

By eliminating the time $t$ from the last two equations, a parabolic equation results,

$$z = z_0 - \frac{x^2}{H}, \tag{3.9.20}$$

which describes the shape of the jet.

The important conclusion to be kept in mind from part (c) is the fact that the complete solution to the jet problem was made possible by utilizing the two momentum equations, Eqns. (3.9.14) and (3.9.15), in addition to the mass and energy conservation equations, Eqns. (3.9.1) and (3.9.2). The momentum conservation equations are examined in Section 3.2.4.

### 3.3.2 Enthalpy Equation

Equations (3.34) to (3.38) can also be written in terms of *enthalpy H,* defined by

$$H = U + pV = U + \frac{mp}{\rho}, \tag{3.48}$$

and therefore,

$$\Delta H = \Delta U + p\,\Delta V + V\,\Delta p = \Delta U + \frac{m\,\Delta p}{\rho} + p\,\Delta V. \tag{3.49}$$

The resulting *enthalpy equation* is

$$\begin{aligned}
\frac{dE_T}{dt} &= \frac{d}{dt}\left(\frac{1}{2}m\bar{u}^2 + mgz + \hat{U}\right) \\
&= \sum_{i=1}^{k}\left[\left(\frac{\alpha\bar{u}_i^2}{2} + gz_i + \frac{\hat{H}_i}{\rho}\right)\dot{m}_i\right] - p\,\Delta\hat{V} + \dot{W}_s + \dot{Q},
\end{aligned} \tag{3.50}$$

where $\hat{V} = 1/\rho$ is the *specific volume.* At steady state,

$$\sum_{i=1}^{k}\left[\left(\frac{\alpha\bar{u}_i^2}{2} + gz_i + \hat{H}_i\right)\dot{m}_i\right] - p\,\Delta\hat{V} + \dot{W}_s + \dot{Q} = 0, \tag{3.51}$$

where the term

$$\dot{W} = -p\,\Delta\hat{V} \tag{3.52}$$

represents *work done by or to the system by expansion $(\Delta\hat{V} > 0)$ or contraction $(\Delta\hat{V} < 0)$ of its volume.* Thus for a system of constant density, $\dot{W} = p\,\Delta\hat{V} = 0$.

#### Example 3.10: Enthalpy Change Equations

Calculate the enthalpy change of an inviscid fluid of density $\rho = 1$ g/cm$^3$ between its entrance in and exit from a vertical insulated pipe of radius $R = 2.5$ cm and length 4 m, flowing under gravity alone, with average velocity $\bar{u} = 1$ m/s, throughout. Show that the opposite flow against gravity, at the same flow rate, occurs only with the addition of shaft work by a pump. What amount of shaft work is required?

**Solution**  Equation (3.51) applies with the following approximations: $\Delta\hat{V} = 0$, because the density is constant; $\dot{Q} = 0$, because the pipe is insulated; $k = 2$ and $u_1 = u_2$, because the average velocity is constant throughout; $\dot{W}_s = 0$, for spontaneous gravity flow.

The resulting form of Eqn. (3.52) is then

$$\dot{m}\left(gz_1 + \frac{\hat{H}_1}{\rho}\right) - \dot{m}\left(gz_2 + \frac{\hat{H}_2}{\rho}\right) = 0, \tag{3.10.1}$$

and therefore

$$\hat{H} = \hat{H}_2 - \hat{H}_1 = \rho g(z_1 - z_2) = 39{,}240 \frac{\text{N m}}{\text{m}^3}, \tag{3.10.2}$$

which suggests that the enthalpy density of the exit, $\hat{H}_2$, is larger than that at the inlet, by the amount of lost potential energy in flowing from higher to lower elevations, in the absence of viscous dissipation. The enthalpy difference $\Delta\hat{H}$, which is per unit volume of flowing fluid, can be expressed in enthalpy per unit mass of flowing fluid by

$$\Delta\tilde{H} = \Delta\frac{\hat{H}}{\rho} = 39.24 \frac{\text{N m}}{\text{kg}}, \tag{3.10.3}$$

and in the form of rate of enthalpy by

$$\Delta\dot{H} = \dot{m}\,\Delta\tilde{H} = \dot{m}\frac{\Delta\hat{H}}{\rho} = \pi R^2 \bar{u}\,\Delta\hat{H} = 77\frac{\text{N m}}{\text{s}} = 77\text{ W}. \tag{3.10.4}$$

For the case of upward flow, a pump is required to provide power:

$$\dot{W}_s = 77\text{ W}. \tag{3.10.5}$$

### Example 3.11: Mass and Energy Conservation

Under the assumption of total frictional energy losses at rate $\mathscr{F} = 23$ W, find the pressure $p_5$ in the flow system of Fig. E3.6, for inlet pressures $p_1 = p_2 = 10$ atm and atmospheric exit pressures, $p_6 = p_4 = p_7 = 1$ atm.

**Solution**   Equation (3.45) does not apply, because the mass flow rate $\dot{m}_i$ changes from point to point. Thus the original equation, Eqn. (3.40), has to be used, with $\dot{W}_s = \dot{Q} = 0$, and

$$\frac{1}{\rho}\sum_{i=1}^{k=5}\hat{U}_i\dot{m}_i = \mathscr{F}, \tag{3.11.1}$$

by Eqn. (3.43). The resulting energy equation is

$$\sum_{i=1}^{k=5}\left[\left(\frac{\alpha\bar{u}_i^2}{2} + gz_i + \frac{p_i}{\rho}\right)\dot{m}_i\right] - \dot{\mathscr{F}} = 0, \tag{3.11.2}$$

which solved for $p_5$ yields

$$\begin{aligned}
p_5 = {} & \frac{\rho\dot{\mathscr{F}}}{\dot{m}_5} - \frac{\alpha\rho}{2\dot{m}_5}(\bar{u}_i^2\dot{m}_1 + \bar{u}_2\dot{m}_2 - \bar{u}_6\dot{m}_6 - \bar{u}_7\dot{m}_7) \\
& - \frac{\rho g}{\dot{m}_5}(z_1 + z_2 - z_5 - z_6 - z_7) - \frac{p_1\dot{m}_1 + p_2\dot{m}_2 - p_6\dot{m}_6 - p_7\dot{m}_7}{\dot{m}_5}.
\end{aligned} \tag{3.11.3}$$

For horizontal arrangement for which $\sum_i z_i = 0$, and laminar flow for which $\alpha = 2$, and given the relations

$$\dot{m}_i = \rho\bar{u}_i\pi R_i^2 = \rho\dot{V}_i, \tag{3.11.4}$$

Eqn. (3.11.3) reduces to

$$p_5 = \frac{1}{\dot{V}_5}[\dot{\mathscr{F}} - \rho(\bar{u}_1^2\dot{V}_1 + \bar{u}_2^2\dot{V}_2 - \bar{u}_6^2\dot{V}_6 - \bar{u}_7^2\dot{V}_7) - (p_1\dot{V}_1 + p_2\dot{V}_2 - p_6\dot{V}_6 - p_7\dot{V}_7)].$$

$$\tag{3.11.5}$$

For the piping system of Fig. 3.6, it is $\dot{V}_1 = \pi R_1^2 \bar{u}_1^2 = 19.6\ \text{cm}^3/\text{s}$, $\dot{V}_2 = \pi R_2^2 \bar{u}_2^2 = 196\ \text{cm}^3/\text{s}$, $\dot{V}_5 = \pi R_5^2 \bar{u}_5^2 = 24.5\ \text{cm}^3/\text{s}$, and $\dot{V}_6 = \dot{V}_7 = 95.6\ \text{cm}^3/\text{s}$. By inserting these values in Eqn. (3.11.5), the resulting pressure is

$$p_5 = [9.36 \times 10^6 - 348 - 8.1 \times 10^6]\ \text{N/m}^2 = 1.24\ \text{atm}.$$

Thus the pressure drops due to frictional losses, which will cause the flow rate to diminish. To maintain the flow rate, external supply of energy is required, for example by a pump, as discussed in Chapter 4.

### 3.3.3 Applications of Bernoulli's Equation

**Liquid discharge under pressure.** Figure 3.3 shows outflow from high to low atmospheric pressure, through a hole of diameter $d$. When the discharge velocity is high, gravity effects in the vicinity of the hole—which would tend to bend the jet downward—are negligible, and the jet contracts to a diameter $Cd$, where $C$ is a *coefficient of contraction*. The governing equations are for mass conservation,

$$\bar{u}_1 D = \bar{u}_2 Cd \quad \text{and} \quad \bar{u}_1 \pi \frac{D^2}{4} = \bar{u}_2 \pi \frac{C^2 d^2}{4}, \tag{3.53}$$

for rectangular and axisymmetric discharge, respectively, and the Bernoulli's equation for the two cases

$$\frac{p_1}{\rho} + \frac{\bar{u}_1^2}{2} = \frac{p_2}{\rho} + \frac{\bar{u}_2^2}{2}. \tag{3.54}$$

Combining Eqns. (3.53) with Eqn. (3.54) yields

$$\bar{u}_2 = \left[ \frac{2(p_1 - p_2)}{\rho} \right]^{1/2} \left[ 1 - C^2 \left( \frac{d}{D} \right)^{2k} \right]^{-1/2} \tag{3.55}$$

and

$$\dot{V} = \left( \frac{\pi d}{4} \right)^{k-1} Cd = \left[ \frac{2(p_1 - p_2)}{\rho} \right]^{1/2} \left[ 1 - C^2 \left( \frac{d}{D} \right)^{2k} \right]^{-1/2}, \tag{3.56}$$

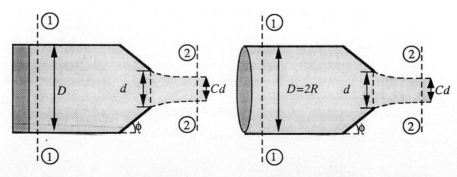

**Figure 3.3** Liquid discharge under pressure from rectangular and cylindrical containers.

**TABLE 3.2**  CONTRACTION COEFFICIENTS
FOR HORIZIONTAL RECTANGULAR
AND AXISYMMETRIC DISCHARGE AT HIGH SPEED
AS SHOWN IN FIG. 3.3 AND FOR VERTICAL
DISCHARGE UNDER GRAVITY (FROM REF. 3,
BY PERMIGGION.)

| $d/D$ | Contraction coefficient, $C$ | | | |
|---|---|---|---|---|
|  | $\gamma = 45°$ | $\gamma = 90°$ | $\phi = 135°$ | $\gamma = 180°$ |
| 0.0 | 0.746 | 0.611 | 0·537 | 0.500 |
| 0.1 | 0.747 | 0.612 | 0.546 | 0.513 |
| 0.2 | 0.747 | 0.616 | 0.555 | 0.528 |
| 0.3 | 0.748 | 0.622 | 0.566 | 0.544 |
| 0.4 | 0.749 | 0.631 | 0.580 | 0.564 |
| 0.5 | 0.752 | 0.644 | 0.599 | 0.586 |
| 0.6 | 0.758 | 0.662 | 0.620 | 0.613 |
| 0.7 | 0.768 | 0.687 | 0.652 | 0.646 |
| 0.8 | 0.789 | 0.722 | 0.698 | 0.691 |
| 0.9 | 0.829 | 0.781 | 0.761 | 0.760 |
| 1.0 | 1.000 | 1.000 | 1.000 | 1.000 |

where $\dot{V}$ is volumetric flow rate, and $k = 1$ for square and $k = 2$ for circular hole. The coefficient $C$ depends on the amount of frictional energy losses in the vicinity of the exit. Helmholtz and Kirchhoff calculated $C$ in the case of $d/D \ll \frac{1}{4}$ as,

$$C\left(\frac{d}{D} \le 1\right) = \frac{\pi}{\pi + 2} = 0.611, \tag{3.57}$$

and following them, von Mises extended the calculations to the situations of Fig. 3.3, as tabulated in Table 3.2. Table 3.2 also applies to vertical discharge under gravity and to discharge of a liquid into another liquid, if in Eqns. (3.55) and (3.56) the pressure difference, $p_1 - p_2$, is corrected to include the contribution of the external liquid to the jet pressure, $p_2$.

**Orifice-plate flow-rate meter.**    This device consists of a circular disk with a central hole of diameter $d$ bolted between the flanges in a pipe of diameter $D$, as shown in Fig. 3.4. By measuring the pressure difference, $\Delta p = p_1 - p_2$, the flow rate is calculated by means of Eqn. (3.56) where the discharge coefficient is found from Fig. 3.5 in terms of the Reynolds number, defined by

$$\text{Re} = \frac{\rho D \bar{u}_2}{\eta}, \tag{3.58}$$

and the ratio of pipe to hole diameter, $d/D$. The orifice-plate meter is among the common devices to measure flow rates in large-scale piping systems.[4]

**Liquid discharge under gravity.**    When the discharge velocity is small, the effect of gravity in Fig. 3.3 cannot be neglected and the flow patterns of Fig. 3.5 are observed.[5] For horizontal discharge under gravity as shown in Fig. 3.5a,

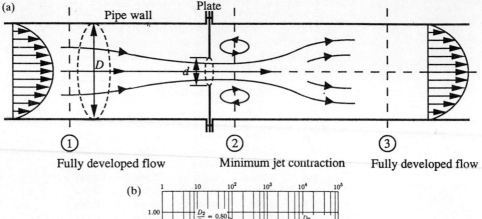

(a)

Fully developed flow     Minimum jet contraction     Fully developed flow

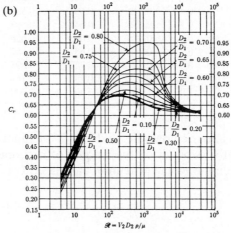

**Figure 3.4** Orifice-plate flow-rate meter and contraction coefficients. (From Ref. 3, by permission from *Instruments and Control Systems*.)

the equations are

$$Du_1 = u_2 Cd \tag{3.59}$$

and

$$\frac{p_1}{\rho} + g\frac{D}{2} + \frac{u_1^2}{2} = gz_2 + \frac{u_2^2}{2} \simeq gH + \frac{u_2^2}{2}, \tag{3.60}$$

where $D/2$ is the average elevation of the streamlines at the inlet where the pressure is $p_1$ and the velocity $u_1$, and $z_2$ is the elevation where the pressure is atmospheric (and the gauge pressure is zero), the velocity is $u_2$, and the cross section is $Cd$. The resulting velocity is

$$u_2 = d\left[1 - C^2\left(\frac{d}{D}\right)^2\right]^{-1/2}\left[2\frac{p_1}{\rho} + g(D - 2H)\right]^{1/2}, \tag{3.61}$$

and the flow rate is

$$\dot{V} = Cd\left[1 - C^2\left(\frac{d}{D}\right)\right]^{-1/2}\left[\frac{2p_0}{\rho} + g(D - 2H)\right]^{1/2}. \tag{3.62}$$

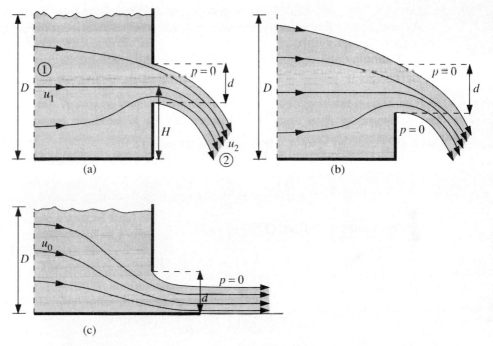

**Figure 3.5** Liquid discharge under gravity (a), free-surface discharge (b), and bottom discharge (c). (From Ref. 5, by permission.)

In the common case of $D \gg d$, Eqns. (3.61) and (3.62) become

$$u_2 = \left[ g(D - 2H) + \frac{2p_1}{\rho} \right]^{1/2} \qquad (3.63)$$

and

$$\dot{V} = Cd \left[ g(D - 2H) + \frac{2p_1}{\rho} \right]^{1/2}. \qquad (3.64)$$

If, in addition, the hole is symmetric (i.e., for $H = D/2$), the contraction coefficient $C$ is found in Table 3.2. In the case of $D \gg d$, a good approximation in general is $C = 0.611$. However, for $D$ and $d$ of the same order of magnitude, $z_2 \neq H$, and knowledge of $C$ in terms of $z$ is required in Eqns. (3.59) and (3.60). The resulting flow rate in this case is

$$\dot{V} = \tfrac{2}{3} C \sqrt{2g} \left[ \left( \frac{u_1^2}{2g} + \frac{d}{2} + D - H \right)^{3/2} - \left( \frac{u_1^2}{2g} + D - H - \frac{d}{2} \right) \right], \qquad (3.65)$$

with the contraction coefficient found from Table 3.2. In this case, however, Eqn. (3.59) needs to be solved numerically since the flow rate $\dot{V}$ depends on $u_1$. Commonly, $u_1^{(1)} = 0$ is assumed initially and the flow rate $\dot{V}(u_1^{(1)} = 0)$ is calculated. A new $u_1^{(2)}$ is calculated by means of Eqn. (3.65) and then $\dot{V}(u_1^{(2)})$ is calculated. The calculations are repeated until two consecutive values of $\dot{V}$ do not differ much. For a nonsymmetric hole, the contraction coefficient must be found by experiment.

The case of Fig. 3.5b, representing *free-surface discharge,* is modeled accordingly by Eqn. (3.65) by setting $d/2 = D - H$, which yields

$$\dot{V} = \tfrac{2}{3}C\sqrt{2g}\left[\left(\frac{u_1^2}{2g} + \frac{d}{2}\right)^{3/2} - \left(\frac{u_1^2}{2g}\right)^{3/2}\right], \tag{3.66}$$

with $C$ calculated from Table 3.2, with $d/D$ replaced by $D/(D + 2H - d)$.

For discharge represented by Fig. 3.5c, Eqn. (3.60) becomes

$$\frac{p_1}{\rho} + g\frac{D}{2} + \frac{u_1^2}{2} = \frac{p_2}{\rho} + g\frac{Cd}{2} + \frac{u_2^2}{2} = gCd + \frac{u_2^2}{2}, \tag{3.67}$$

which combines with Eqn. (3.59) to yield

$$u_2 = \left(1 + C\frac{d}{D}\right)^{-1/2}[2gD]^{1/2} \tag{3.68}$$

and

$$\dot{V} = Cd\left(1 + C\frac{d}{D}\right)^{-1/2}[2gD]^{1/2}. \tag{3.69}$$

In this case, the contraction coefficient is $C = 0.611$ in the case of $d \ll D$ and $C \approx 0.605$ for $d/D < 0.5$.

These approximate expressions for discharge are utilized in hydraulics and large-scale flow applications to calculate magnitudes of discharge from reservoirs and dams.

**Pitot tube.** Its prototype is shown in Fig. 3.6a and it is used to find the speed of traveling vehicles such as boats and airplanes moving through a stationary fluid, or to find the velocity of fluid streams overtaking stationary bodies.[4] This *J*-shaped tube is attached to a body that travels with velocity $u_1$

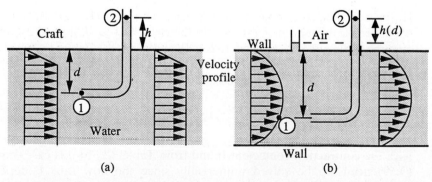

(a)                                             (b)

**Figure 3.6** (a) Pitot tube, and (b) Pitot static tube, for velocity measurement. (From Ref. 5, by permission.)

relative to a fluid of density $\rho$. Bernoulli's equation yields

$$\frac{p_1}{\rho} + \frac{u_1^2}{2} - gd = \frac{p_2}{\rho} + gh, \tag{3.70}$$

with elevations taken with respect to the free surface. The pressure at point 1, away from the moving surface, is hydrostatic, identical to

$$p_1 = p_2 + \rho g d. \tag{3.71}$$

Equations (3.70) and (3.71) combine to yield the unknown velocity

$$u_1 = \sqrt{2gh}. \tag{3.72}$$

The *Pitot static tube* shown in Fig. 3.6b is used to find *velocity distributions* in the interior of fluid streams.[4] By the same principles, based on Bernoulli's equation and the hydrostatic equation, the unknown velocity $u_1(d)$, at location $d$ away from the wall, is

$$u_1 = \sqrt{2gh(d)}, \tag{3.73}$$

where $h(d)$ changes with the location $d$. Thus, by this method, nonuniform velocity fields such as the parabolic profile of Fig. 3.6b can be quantified.


## 3.4 CONSERVATION OF MOMENTUM

The equations of conservation of mass and mechanical energy or Bernoulli's equation were employed in Section 3.3.3 to calculate discharge velocity and flow rate. However, as demonstrated in Example 3.9, the shape of the resulting liquid jet cannot be calculated by these two alone. In fact, in Example 3.9, a simplified momentum balance was performed which allowed approximation of the jet shape by Eqn. (3.9.20).

A body or fluid particle of density $\rho$, traveling with velocity $u$, possesses momentum $\hat{J} = \rho u$, per unit volume, which it may transfer to a system at a rate $j = \rho u u = \dot{m} u$ per unit area of entering or exiting surface. Change of momentum within the system is also induced by gravity force, $F_g = mg = \rho V g$, and by contact force, $F^C = \sum_i \tau_i \Delta A_i$, where $\tau_i$ is stress and $\Delta A_i$ the contact surface area. Thus, Newton's law of motion applied to a fluid system produces the momentum conservation equation, which along the lines of Eqn. (3.21) can be written as

$$\frac{d}{dt}(\rho V u) = \sum_{i=1}^{k} (\rho u_i u_i A_i) + \rho V g + \sum_{j=1}^{m} \tau_j A_j, \tag{3.74}$$

for a system of volume $V$, with $k$ entries and exits and $m$ contact surfaces with its surroundings.

Unlike the mass and energy conservation equations, Eqn. (3.74) is a *vector equation* because the included terms are vector quantities (e.g., force and

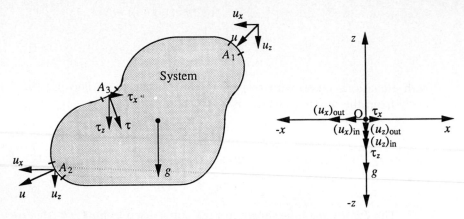

**Figure 3.7**  Momentum balance for a general two-dimensional system.

momentum), and therefore *Newton's law and any equilibrium conditions must be satisfied in three mutually perpendicular directions, x, y, and z,* as shown in Fig. 3.7 for a two-dimensional system. For the system of Fig. 3.7, momentum balances in the $x$- and $z$-directions take the form

$$\frac{d}{dt}(\rho V u_x)$$

$$= [(\rho u_x)u_x + (\rho u_x)u_z]_1 A_1 - [(\rho u_x)u_x + (\rho u_x)u_z]_2 A_2 + [\rho V g_x] + [(\tau_{ix})_3 A_3]$$

(3.75)

and

$$\frac{d}{dt}(\rho V u_z)$$

$$= [(\rho u_z)u_z + (\rho u_z)u_x]_1 A_1 - [(\rho u_z)u_z + (\rho u_z)u_x]_2 A_2 + [\rho V g_z] + [(\tau_{iz})_3 A_3].$$

(3.76)

The terms $(\rho u_i)u_j$, $i = x, z$ and $j = x, z$, represent $\rho u_i$ momentum per unit volume entering or leaving the system with velocity $u_j$. In other words, the product $(\rho u_i)$ is the content and can be transferred by each of the velocity components. Consider fluid moving with velocity $u$ that penetrates a boundary surface of the system at angle $\theta$ such that the velocity can be decomposed into two components, $u_n = u \cos \theta$ perpendicular to the surface and $u_t = u \sin \theta$ parallel to the surface. Although $u_t$ will never make it into the system, $u_n$ will penetrate the system with normal velocity $u_n$. In doing so, this component convects into the system mass $\rho$ at rate $\rho u_n$, energy $\hat{E}$ at rate $\hat{E} u_n$, as well as $n$-momentum $\rho u_n$ at rate $(\rho u_n)u_n$ and $t$-momentum $\rho u_t$ at rate $(\rho u_t)u_n$. The convective terms of Eqns. (3.75) and (3.76) are obtained in the same way by replacing $n$, $t$ with $x$, $z$ and $z$, $x$, respectively.

   In these momentum equations, $u^2 = u_z^2 + u_x^2$, $g^2 = g_z^2 + g_x^2$, $\tau_i^2 = (\tau_{iz})^2 + (\tau_{ix})^2$ where $\tau_{ij}$ is a stress acting on the $i$-surface in the $j$-direction and $\tau_i$ is the

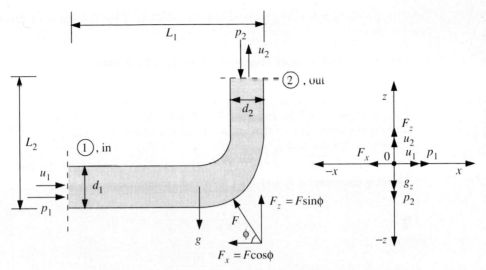

**Figure 3.8**  Momentum balance for two-dimensional regular system.

total stress on the *i*-surface. In macroscopic balances all viscous stresses are relatively small and therefore omitted from the corresponding equations unless otherwise stated or anticipated.

In most cases, one or more of the directions $x$, $y$, and $z$ is conveniently chosen to coincide with one or more of the directions $u_i$, $\tau_i$, or $g_i$, in which case Eqns. (3.75) and (3.76) simplify considerably; as shown for the curved pipe of Fig. 3.8. For the system of Fig. 3.8, $(\bar{u}_x)_1 = \bar{u}_1$, $(\bar{u}_z)_1 = 0$, $(\bar{u}_x)_2 = 0$, $(\bar{u}_z)_2 = \bar{u}_2$, $g_z = g$, $g_x = 0$, $F_x = F \cos \phi$, $F_z = F \sin \phi$, $(\tau_x)_1 = p_1 \pi(d_1^2/4)$, and $(\tau_z)_2 = -p_2 \pi(d_2^2/4)$, and Eqns. (3.75) and (3.76) at steady state simplify to

$$\rho a \bar{u}_1^2 \pi \frac{d_1^2}{4} + p_1 \pi \frac{d_1^2}{4} + F \sin \phi = 0 \qquad (3.77)$$

and

$$-\rho a \bar{u}_2^2 \pi \frac{d_2^2}{4} - \rho g V - p_2 \pi \frac{d_2^2}{4} + F \cos \phi = 0, \qquad (3.78)$$

where $\beta$ is a velocity uniformity coefficient for the average value of the square of the velocity, as calculated in Example 3.8. According to Example 3.8, it takes the value $\beta = 1$ for plug and turbulent flows and $\beta = 4/3$ for laminar flow. Equations (3.77) and (3.78) are the most usable forms and are used hereafter; *however, the reader should always keep in mind the original, general equations, Eqns. (3.79) and (3.80)*, for a system not conforming to convenient choices of coordinate directions.

If we now go back to Example 3.9, we discover that Eqns. (3.9.17) and (3.9.18) are indeed the application of Eqns. (3.77) and (3.78), respectively, to that situation. Of course, for more than one mass inlet and/or outlet and force, the components of the general equation, Eqn. (3.74), apply. Also, for nonuniform velocity profiles, integral forms along the lines of Eqn. (3.13) and Example 3.1 can be used.

### Example 3.12: Application of Momentum Equation: Bent Tube

Calculate the magnitude and direction of the required external force $F$ needed to support the system of Fig. 3.8, given that $\bar{u}_1 = 10$ m/s, $\rho = 1$ g/cm³, $d_1 = 20$ cm, $d_2 = 10$ cm, $L_1 = 1$ m, and $L_2 = 2$ m, that the weight of the thin plastic pipe wall is negligible compared to that of the liquid contained, that the flow is turbulent, and that $p_1 = 10$ atm and $p_2 = 1$ atm.

**Solution**  Equations (3.77) and (3.78) apply with $\beta = 1$ and the unknown force having components $F_x$ and $F_z$ of unknown magnitude:

$$\rho \bar{u}^2 \frac{d_1^2}{4} + p_1 \pi \frac{d_1^2}{4} + F_x = 0, \tag{3.12.1}$$

$$-\rho \bar{u}_2^2 \pi \frac{d_2^2}{4} - \rho g \frac{\pi (d_1^2 + d_2^2)(L_1 + L_2)}{8} - p_2 \pi \frac{d_2^2}{4} + F_z = 0. \tag{3.12.2}$$

The unknown velocity $\bar{u}_2$ is calculated by the mass balance equation,

$$\rho \bar{u}_1 \pi \frac{d_1^2}{4} = \rho \bar{u}_2 \pi \frac{d_2^2}{4}, \tag{3.12.3}$$

which yields

$$\bar{u}_2 = \bar{u}_1 \frac{d_1^2}{d_2^2} = 10 \times \frac{400}{100} \text{ m/s} = 40 \text{ m/s}. \tag{3.12.4}$$

Equation (3.12.1) predicts that

$$F_x = \left( -1 \times 10^6 \times 3.14 \times \frac{400}{4} - 10 \times 1.013 \times 10^6 \times 3.14 \times \frac{400}{4} \right) \text{dyn}$$

$$= -3.8 \times 10^8 \text{ dyn} = -3.8 \times 10^3 \text{ N}. \tag{3.12.5}$$

Similarly, Eqn. (3.12.2) predicts that

$$F_z = \left[ 1 \times 16 \times 10^6 \times 3.14 \times \frac{100}{4} + 1 \times 981 \times \frac{3.14 \times (400 + 100)(200 + 100)}{32} \right.$$

$$\left. + 1.013 \times 10^6 \times 3.14 \times \frac{100}{4} \right] \text{dyn} \tag{3.12.6}$$

$$= (12.56 + 1444 + 0.0078) \times 10^8 \text{ dyn}$$

$$= 145 \times 10^9 \text{ dyn} = 145 \times 10^4 \text{ N}.$$

Thus the magnitude of the force is

$$F = \sqrt{F_x^2 + F_z^2} = 14,500 \text{ N}, \tag{3.12.7}$$

and its angle with the positive $x$-direction is defined by

$$\tan \phi = -\frac{F_z}{F_x} = \frac{145}{0.38} = 383, \tag{3.12.8}$$

which is true for $\phi = 89.85°$, as shown in Fig. 3.8. This nearly vertical force is due to the fact that the dominant force that needs to be overcome is the weight of the liquid, as can be seen from the relative magnitudes of the terms of Eqns. (3.12.5) and (3.12.6).

### Example 3.13: Application of Momentum Equation: Steady Pipe Flow

Calculate the total shear stress and force at the wall of a horizontal pipe of diameter $d$ and length $L$, due to steady flow of water of density $\rho$ and viscosity $\eta$, at mass flow rate $\dot{m}$ under pressure difference $\Delta p = p_2 - p_1 < 0$ (Fig. E3.13).

**Solution**    A convenient system here is the entire flowing liquid alone, excluding the rigid pipe wall. Mass balance yields

$$\bar{u}_1 = \bar{u}_2 = \bar{u} = \frac{4\dot{m}}{\rho \pi d^2}, \tag{3.13.1}$$

and the velocity is constant along the pipe. An energy balance yields

$$\frac{p_1}{\rho} - \frac{p_2}{\rho} - \frac{\mathscr{F}}{\dot{m}} = 0. \tag{3.13.2}$$

For momentum balance, Eqn. (3.77) is a convenient form:

$$\sum_{i=1}^{k-2} \pi \frac{d_i^2}{4} u_i + F + \pi \frac{d^2}{4}(p_1 - p_2) = 0, \tag{3.13.3}$$

where $F$ is shear force by the wall on the fluid. Since $u_1 = u_2$, it is

$$F = \pi \frac{d^2}{4}(p_2 - p_1) = \pi \frac{d^2}{4}\Delta p < 0. \tag{3.13.4}$$

This total shear force is opposite to the shear force by the fluid on the wall. Therefore,

$$\tau_{rz}^w = -\frac{F}{\pi d L} = -\frac{d\,\Delta p}{2\,L} \tag{3.13.5}$$

is the shear stress by the liquid on the wall.

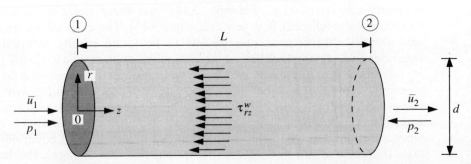

**Figure E3.13**   Momentum balance in steady pipe flow, and resulting shear stress by the liquid on the wall.

## 3.5 SIMULTANEOUS APPLICATION OF CONSERVATION EQUATIONS

For an isothermal system with $k$ inlets and outlets and $m$ contact and external forces, the macroscopic conservation equations at steady state are

*Mass*:

$$\sum_{i=1}^{k} \rho \bar{u}_i A_i = 0 \tag{3.79}$$

*Mechanical energy*:

$$\sum_{i=1}^{k} \left[ \left( \frac{\alpha_i}{2} \rho \bar{u}_i^2 + \rho g z_i + p_i \right) \bar{u}_i A_i \right] + \dot{W}_s + \dot{\mathscr{F}} = 0 \tag{3.80}$$

*Bernoulli's equation (alternative to mechanical energy)*:

$$\rho \alpha_1 \frac{\bar{u}_1^2}{2} + p_1 + \rho g z_1 = \rho \alpha_2 \frac{\bar{u}_2^2}{2} + p_2 + \rho g z_2 \tag{3.81}$$

*Momentum*:

$$\sum_{i=1}^{k} (\beta_i \rho \bar{u}_{x,i}^2 A_i) + \rho V g_x + \sum_{j=1}^{m} [(p_{x,j} + \tau_{zx,j}) A_j \rho + F_x] = 0 \tag{3.82}$$

$$\sum_{i=1}^{k} (\beta_i \rho \bar{u}_{z,i}^2 A_i) + \rho V g_z + \sum_{j=1}^{m} [(p_{z,j} + \tau_{xz,j}) A_j + F_z] = 0 \tag{3.83}$$

Here $\bar{u}_i^2 = \bar{u}_{x,i}^2 + \bar{u}_{z,i}^2$, $g^2 = g_x^2 + g_z^2$, $F_i^2 = F_{x,i}^2 + F_{z,i}^2$, $p_x^2 = p_{x,i}^2 + p_{z,i}^2$, and so on. Also, $\tau_{zx,j}$ and $\tau_{xz,j}$ are primarily shear stresses to the fluid, and $F_x$ and $F_z$ are any external forces. The velocities and corresponding flow rates are positive for entry to and negative for exit from the system. Pressure is always toward the boundary considered (e.g., Example 3.12), and shear stress is opposite to the direction of the adjacent flow (e.g., Example 3.13). The external forces may have any specified direction, or else be left to be found by the solution to these equations (e.g., Example 3.12).

In most flow situations all three conservation equations need to be used, especially when forces other than those due to pressure and gravity, such as viscous and external support forces (e.g., Fig. 3.8) and inertia forces equivalent to acceleration, enter the calculations. Usually, the easier-to-apply mass and mechanical (or Bernoulli's, in the absence of frictional losses) equations are used first. If these two are not sufficient to provide all the answers, the momentum equations in one or more directions are used. Characteristic cases of simultaneous application of all conservation equations follow.

**Orifice plate with sudden expansion.**    The mass conservation equation and Bernoulli's equation applied to the frictionless flow between inlet and complete jet contraction at point 2 of Fig. 3.5 result in

$$\frac{p_1 - p_2}{\rho} = \frac{\alpha(\bar{u}_2^2 - \bar{u}_1^2)}{2} = \frac{\alpha \bar{u}_1^2}{2}\left(\frac{D^4}{C^4 d^4} - 1\right), \tag{3.84}$$

with $\alpha = 1$ for turbulent and $\alpha = 2$ for laminar flow. Bernoulli's equation does not apply downstream from the complete jet contraction, because there are significant frictional losses and $\dot{\mathscr{F}} \neq 0$ in Eqn. (3.80). The mass conservation equation between 2 and 3 yields

$$\pi \frac{d^2}{4}\bar{u}_2 = \pi \frac{D^2}{4}\bar{u}_3 = \pi \frac{D^2}{4}\bar{u}_1, \tag{3.85}$$

and therefore $\bar{u}_3 = \bar{u}_1$. The energy equation, not Bernoulli's equation, between 1 and 3 then yields

$$p_1 - p_3 + \rho\dot{\mathscr{F}} = 0, \tag{3.86}$$

with both $p_3$ and $\dot{\mathscr{F}}$ unknown. Thus a momentum equation, along the flow direction between 2 and 3, close enough together such that shear friction with the wall is negligible, results in

$$\rho\pi \frac{C^2 d^2}{4}\beta\bar{u}_2^2 - \rho\pi \frac{D^2}{4}\beta\bar{u}_3^2 + \pi \frac{D^2}{4}(p_2 - p_3) = 0, \tag{3.87}$$

with $\beta = 1$ for turbulent flow and $\beta = 1.25$ for laminar flow. Eliminating $\bar{u}_3$ between Eqns. (3.85) and (3.87) results in

$$\frac{p_2 - p_3}{\rho} = \beta\bar{u}_1^2\left(1 - \frac{D^2}{C^2 d^2}\right) \leq 0, \tag{3.88}$$

and by adding Eqn. (3.84) to Eqn. (3.88), the overall pressure drop is

$$\frac{p_1 - p_3}{\rho} = \bar{u}_1^2\left[\frac{\alpha}{2}\left(\frac{D^4}{C^4 d^4} - 1\right) + \beta\left(1 - \frac{D^2}{C^2 d^2}\right)\right] \geq 0 \tag{3.89}$$

Thus, since $D \gg Cd$, it is always $p_1 > p_3$, due to frictional energy losses between 1 and 3. In fact, the pressure always decreases in the flow direction of a fully developed horizontal flow in order to push the liquid through. The frictional energy losses are then calculated by Eqns. (3.86) and (3.89), to be

$$\dot{\mathscr{F}} = \frac{p_1 - p_3}{\rho} = \bar{u}_1^2\left[\frac{\alpha}{2}\left(\frac{D^4}{C^4 d^4} - 1\right) + \beta\left(1 - \frac{D^2}{C^2 d^2}\right)\right] \geq 0. \tag{3.90}$$

Thus $\dot{\mathscr{F}}$ in the energy equation is always positive in the direction of a fully developed flow. However, this is not necessarily true for an undeveloped flow: For example, $p_1 > p_2$ for the converging flow between 1 and 2, and $p_2 < p_3$ for the expanding flow between 2 and 3, as can easily be seen by Eqn. (3.88).

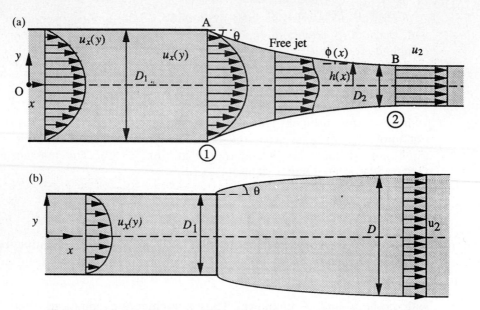

**Figure 3.9** Contraction and swelling of a free jet at high and low Reynolds number 4, respectively.

**Contraction of free jet.**   Planar and axisymmetric jets at high laminar Reynolds number contract to a conduit $D_2$ less that the pipe or channel diameter $D_1$, far enough downstream from the nozzle exit, as shown in Fig. 3.9. At high Reynolds numbers, a parabolic velocity profile, given by $u_x(y) = k\bar{u}_1[1 - (4y^2/D_1^2)]$, where $\bar{u}_1$ is the average velocity with $k = \frac{3}{2}$ for planar and $k = 2$ for round jet, persists down to the exit. In the absence of surface tension, the pressure is uniform everywhere under the free jet and therefore $p_1 = p_2 = p_{\text{atm}}$. At high Reynolds number where viscous forces are negligible compared to inertia or convective forces. Under these approximations, the mass conservation equation between points 1 and 2 yields

$$\bar{u}_1 = D_1^m = \bar{u}_2 D_2^m, \tag{3.91}$$

with $m = 1, 2$ for planar and axisymmetric jets, respectively. The mechanical energy equation yields

$$\alpha_1 \bar{u}_1^2 = \bar{u}_2^2, \tag{3.92}$$

where according to Example 3.8, $\alpha_1 = \frac{5}{6}$ and 2, respectively, for planar and round jets. Equations (3.91) and (3.92) yield

$$D_2 = \left(\frac{1}{\sqrt{\alpha_1}}\right)^{1/m} D_1 \quad = \begin{cases} 0.84 D_1 & \text{round jet} \\ 0.92 D_1 & \text{planar jet.} \end{cases} \tag{3.93}$$

These values agree with experimental observations[6] and numerical predictions.[7]

**Expansion or swelling of free jet.**   At low Reynolds numbers, $\text{Re} < 1$, the laminar parabolic velocity profile does not persist down to the exit. Due to exit effects on the slow flow, the exit velocity profile is approximately[8]

$$u_x(y) = 0.85k\bar{u}_1\left(1 - \frac{4y^2}{D_1^2}\right)^{1/2}, \tag{3.94}$$

where $\bar{u}_1$ is the average velocity of the fully developed channel or pipe flow far upstream from the exit. In this case the jet expands or swells to a conduit of value,

$$D_2 = \begin{cases} 1.13D_1 & \text{planar jet} \\ 1.185D_1 & \text{round jet.} \end{cases} \tag{3.95}$$

These values have been observed experimentally[6] and predicted numerically.[7,8]

Figure 3.9b is typical of the *extrusion* of highly viscous polymeric liquids through dies to form plastic parts. The ratio $D_2/D_1$ in the polymer processing literature is called the *extrudate swell*. The extrudate swell of polymeric liquids is much higher than the values predicted by Eqn. (3.95) for a Newtonian liquid. It may reach up to 3 for highly elastic liquids, where large *viscoelastic normal stresses* developed, as discussed in Chapter 9. Values of the extrudate swell of viscoelastic liquids are approximated by Tanner's equation,[9]

$$\frac{D_2}{D_1} = 0.13 + \left[1 + \frac{1}{2}\left(\frac{\tau_{zz} - \tau_{rr}}{2\tau_{rz}}\right)_w^2\right]^{1/6}, \tag{3.96}$$

where $\tau_{zz}$, $\tau_{rr}$, and $\tau_{rz}$ are normal and shear stresses at the wall far upstream from the exit. For a Newtonian liquid for which $\tau_{zz} = \tau_{rr} = 0$, the equation yields a swelling of 1.13, in agreement with Eqn. (3.95).

## 3.6 ANGULAR MOMENTUM CONSERVATION

A mass $m$ that rotates with angular speed $\Omega$ about a center at distance $r$ possesses *angular momentum*

$$J_r = rJ_\theta, \tag{3.97}$$

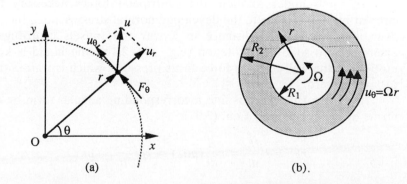

**Figure 3.10**  Angular kinematics for point mass $m$, and (b) for fluid of density $\rho$. A strictly circular motion is obtained for $u_r = 0$.

where

$$J_\theta = mu_\theta = mr\Omega \tag{3.98}$$

is the corresponding linear momentum in the azimuthal direction. Figure 3.10 illustrates such a rotational motion. Thus the *density of angular momentum* of a fluid of azimuthal linear velocity $u_\theta(r) = \Omega r$ is

$$\hat{\jmath}_r = \frac{J_r}{V} = \frac{rJ_\theta}{V} = \rho r^2 \Omega. \tag{3.99}$$

For such purely circular motion with constant angular speed $\Omega$, Newton's law of motion for fluid of constant density is

$$\hat{F}_r = \frac{d\hat{J}_\theta}{dt} = \frac{d}{dt}(\rho u_\theta) = \rho u_\theta \Omega = \rho \alpha_r, \tag{3.100}$$

where $\alpha_r$ is *centrifugal acceleration* in the radial direction. Since the magnitude of $u_\theta$ is constant for constant $r$, the centrifugal acceleration measures the change in the direction of $u_\theta$ along the circular orbit. The arising centifugal force acceleration is in the radial direction, too, and tends to push the body away from its circular motion. Therefore, to maintain such a motion, a *centripetal force*, equal in magnitude and opposite in direction to the centrifugal force, is required.

Two profound examples of circular motion of solid bodies are: (a) For a solid sphere attached to a thread and set in circular motion by a hand holding the other end of the thread. The centripetal force is applied by the hand to the sphere through the thread. If the thread breaks, there is no means to transmit such a force and the centrifugal force expels the sphere from its circular orbit. (b) A vehicle traveling around along a curving road. The centrifugal force that tends to expel the vehicle away. To avoid such a scenario, road terrains are slightly inclined toward the center of rotation. This gives rise to the necessary centripetal force, developed at the wheel–terrain contacts, which is directed toward the center and offsets the centrifugal force.

In fluids under rotation, the centripetal forces necessary to maintain a circular motion are due to the developed normal stresses along the radius that are distributed with radial distance in a way that offsets centrifugal forces. For example, in a solid-body rotation (e.g., Fig. 2.1b), the liquid is kept in circular motion by the wall reaction force to its pressure, which is transmitted from layer to layer inward.

Equation (3.100) turns into a corresponding angular form by multiplying all terms by $r$, as done with Eqn. (3.97):

$$r\hat{F}_r = r\frac{d\hat{J}_\theta}{dt} = r\frac{d}{dt}(\rho u_\theta) = r\rho\frac{du_\theta}{dt} = \rho r^2 \Omega^2 = \rho r \alpha_r. \tag{3.101}$$

Equation (3.101) is cast in *a torque-to-angular momentum relation for rotational motions, similar to the force-to-linear momentum relation,* or Newton's law of motion, for linear motion:

$$\hat{T} = r\hat{F}_r = r\frac{d\hat{J}_\theta}{dt} = \frac{d\hat{J}_r}{dt} = rp\alpha_r = \rho\Omega^2 r^2, \tag{3.102}$$

where $\hat{T}$ is *torque* per unit of volume.

The required power $\hat{P}$ to apply torque $\hat{T}$ (e.g., in centrifugation of liquids), or the power produced by a fluid torque $\hat{T}$ [e.g., in windmills and waterwheels (Example 3.15) and in lawn sprinklers (Example 3.14)] is given by

$$\hat{P} = \hat{F}u_\theta = \hat{F}\Omega r = \Omega\hat{T}. \tag{3.103}$$

The angular momentum conservation equation is easily obtained by multiplying all the terms of the (linear) momentum conservation equation by $r$:

$$r_m \frac{d}{dt}(V\hat{J}_\theta) = \sum_{i=1}^{k}[\alpha_i\rho(r_i u_{\theta,i}^2 A_i] + \rho r_m V g_\theta + \sum_{j=1}^{m} r_j\rho F_{\theta,j}A\dot{\rho}. \tag{3.104}$$

Here $r_m$ is the distance between the center of mass and the center of rotation, and $F_{\theta,j}$ represents any contact forces due to pressure. The resulting equation of conservation of angular momentum is

$$\frac{d}{dt}(V\hat{J}_r) = \sum_{i=1}^{k}\alpha_i\rho\Omega^2 r_i^3 A_i + T_g + \sum_{j=1}^{m} T_j \tag{3.105}$$

for $k$ mass inlets and/or outlets and $m$ contact torques $T_j$, and $T_g$ is torque due to any gravity component in the azimuthal direction.

### Example 3.14: Water Sprinkler

The sprinkler shown in Fig. P3.22 of Problem 3.22 discharges water at the rate of $2 \text{ m}^3/\text{min}$. If the rotating unit overcomes a resisting torque, $T = 53\Omega^2 \text{ N m}$, where $\Omega$ is its angular speed in rad/s, and the water is discharged through nozzles of diameter $d = 4 \text{ cm}$ and $\theta = 0°$, what is the resulting angular speed $\Omega$?

**Solution**  Equation (3.104) applies with $d/dt = 0$ at steady rotation $g_\theta = 0$ for a horizontal sprinkler, and

$$\sum_{j=1}^{m} T_j = T = 53\Omega^2. \tag{3.14.1}$$

For water discharge at zero velocity from the rotating sprinkler, the absolute azimuthal velocity at the nozzle exit, would be

$$[u_{\theta,i}]_{\dot{V}=0} = \Omega L. \tag{3.14.2}$$

For a nonrotating sprinkler discharging at rate $\dot{V}$, the absolute azimuthal velocity at the nozzle exit would be

$$[u_{\theta,i}]_{\Omega=0} = \frac{4\dot{V}}{2\pi d^2}. \tag{3.14.3}$$

For the rotating discharging sprinkler of Fig. P3.22 the relative azimuthal velocity between rotating nozzle and the discharged water is

$$u_{\theta,i} = \frac{4\dot{V}}{2\pi d^2} - \Omega L, \tag{3.14.4}$$

and the corresponding rate of water discharge is

$$\dot{m} = 2\rho \left( \frac{4\dot{V}}{2\pi d^2} - \Omega L \right) \frac{\pi d^2}{4}. \tag{3.14.5}$$

Equation (3.105) then becomes

$$L\frac{4\dot{V}}{2\pi d^2}\dot{m} = 53\Omega^2, \tag{3.14.6}$$

which yields

$$2 \times 1 \times 100 \times 5300 \times \left( \frac{4 \times 2 \times 10^6}{60 \times 2 \times 3.14 \times 4} - 100\Omega \right) \times 3.14 \times \tfrac{16}{4} = 53\Omega^2 \times 10^7, \tag{3.14.7}$$

and finally,

$$\Omega^2 + 2.512\Omega - 13.32 = 0, \tag{3.14.8}$$

with solution

$$\Omega = \frac{-2.512 + \sqrt{6.31 + 53.28}}{2} = 2.6 \, \text{rad/s}.$$

### Example 3.15: Power by Waterwheel

Figure E3.15 shows how a waterwheel works in principle. Water falls under gravity, tangentially on the alternating horizontal wings, and rotation about the center

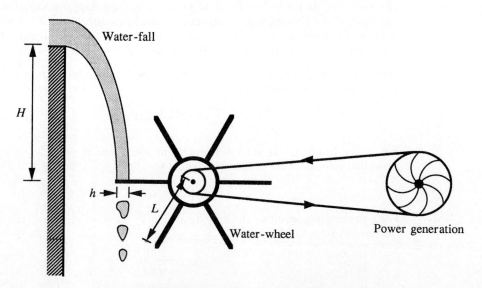

**Figure E3.15**   Water-wheel operation.

accompanied by torque generation and power is induced. For a height $H = 4\,\text{m}$, wing length $L = 2\,\text{m}$ from the center of rotation, wing width $W = 20\,\text{cm}$, impinging depth $h = 10\,\text{cm}$, and negligible wind weight, calculate:

**(a)** the torque generated, $T$;

**(b)** the resulting power $P$ and the angular speed of the power-consuming machine if frictional losses are negligible.

**(c)** What is the significance of the wing spacing?

**Solution**

**(a)** Consider the system containing the waterwheel, for which Eqn. (3.105) takes the form

$$\frac{d}{dt}(V\hat{J}_r) = \rho\Omega^2 L^2 Wh + 0 - T_e, \tag{3.15.1}$$

and at steady state,

$$T_e = \rho\Omega^2 L^3 Wh = \rho u_\theta^2 LWh = \rho(2Hg)LWh$$

$$= [1000\,\text{kg/m}^3 \times (2 \times 4 \times 9.81)\,\text{m}^2/\text{s}^2 \times (2 \times 0.2 \times 0.1)\,\text{m}^3] \tag{3.15.2}$$

$$= 3139\,\text{N m}.$$

**(b)** Under the assumption of negligible frictional torque losses, the power produced is given by Eqn. (3.103):

$$P = \Omega T_e = \frac{\sqrt{2Hg}}{L} T_e$$

$$= \left(\frac{\sqrt{2 \times 4 \times 9.81}}{2} \times 3139\right) \frac{\text{N m}}{\text{s}} \tag{3.15.3}$$

$$= 13{,}904\,\frac{\text{N m}}{\text{s}} = 13{,}904\,\text{J/s} = 13.904\,\text{kW}.$$

**(c)** Wing spacing must be such that the torque generated by the falling water is enough to bring a new wing to the impinging place, such that a constant torque is always applied to maintain a constant rotation. Otherwise, the rotation would alternate between acceleration and deceleration.

## PROBLEMS

1. *Shear stress in boundary layer.* By appropriate momentum balance, given the results of Example 3.7, calculate the resulting shear forces on the plate surfaces of projections *AD*, *DE*, and *AE*. What are the magnitude and direction of the external force required to keep the plate in place?

2. *Average velocity.* By using the considerations of Example 3.8, calculate the average of the $k$th power velocity and its relation to the $k$th power of the average velocity, for the following flow situations, shown in Fig. P3.2.

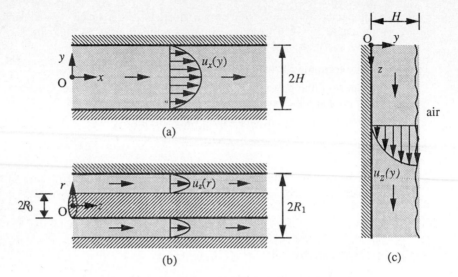

**Figure P3.2** Simple flows: (a) channel, (b) annular, and (c) film.

**(a)** Channel laminar flow of velocity profile,

$$u_x(y) = 1.5\bar{u}\left(1 - \frac{y^2}{H^2}\right)$$

**(b)** Vertical laminar film flow, where

$$u_x(y) = 1.5\bar{u}\left(\frac{2y}{H} - \frac{y^2}{H^2}\right)$$

**(c)** Annular laminar flow, where

$$u_z(r) = \frac{R^2\,\Delta p}{4\eta\,\Delta L}\left[1 - \frac{r^2}{R^2} + \frac{1 - R_1/R_0}{\ln R_0/R_1}\ln\frac{r}{R}\right]$$

In these expressions, $H$, $R_0$, and $R_1$ are geometrical dimensions, as shown in Fig. P3.2.

**(d)** What are the resulting average velocities and flow rates in each case?

3. *Water leak.* An open cylindrical container of radius $R = 20$ cm is brought to depth $H = 30$ cm into a large water reservoir of density $\rho = 1$ g/cm$^3$ open to the atmosphere of pressure $p = 1$ atm.

   **(a)** What is the hydrostatic pressure at the bottom of the container and at its cylindrical wall?

   **(b)** At time $t = 0$, a hole of diameter $d = 2$ cm at the center of the bottom opens and water starts to enter the container. How long does it take to fill the container? What is the final volume of water? Neglect any entrance friction losses.

   **(d)** How will things change if the container of height $H_c = 1$ m is closed initially, filled with air at ambient temperature $T = 20°C$ and pressure $p_0 = 1$ atm?

(e) Develop an expression for the air pressure rate of increases.

(f) If your results appear unrealistic, which approximation(s) made led to the erroneous results?

4. *Fluid content convection.* A Newtonian liquid of viscosity $\eta = 1$ cP, density $\rho = 1$ g/cm³, and heat capacity $c_p = 1$ cal/(g K) approaches with uniform velocity $u = 5$ m/s a surface of area $A = 0.5$ m² placed at an angle of 30° with respect to the stream.

(a) Calculate the total convected mass, momentum, and energy through the surface.

(b) What percentages of the total fluid contents do they represent? What happens to the rest?

5. *Atmospheric and hydrostatic pressure.* Derive the equation that governs the pressure distribution in the atmosphere and in the ocean by means of mass and momentum balance on appropriate synthetic systems.

6. *Flow bifurcation.* Figure P3.6 shows flow in a pipe of diameter $D$ and velocity profile $u = (32\dot{V}/\pi D^4)(D^2/4 - r^2)$, bifurcating into two streams of diameters $d_1$ and $d_2$ and velocities $u_i = a_i(d_i^2/4 - r^2)$, respectively. The inlet pressure of the main flow is $p_0$ and the two streams exit at atmospheric pressure. By assuming that the flow is fully developed at the inlet and the two exits, that gravity is negligible, and that the angle $2\theta$ is small such that viscous stresses at the inlet and the outlets are offset, calculate:

(a) the resulting flow rate throughout each of the exiting streams;

(b) the force required to keep the system in place.

7. *Energy conservation equations.* Characterize and explain the terms of the equations of mechanical, thermal, and total energy change. Then write down the appropriate forms for four simplified cases of transfer of energy (e.g., heat conduction in solids, fluid flow in heated pipe). What is the equation of temperature change of a batch reactor if the thermophysical properties of the reacting mixture remain constant throughout? What are the appropriate conservation equations of steady, isothermal, compressible flow in an insulated pipe?

8. *Drag force.* For the flow situations of Problem 3.2, calculate the resulting drag or shear force per unit length of solid boundary. Justify the direction of the drag. What is the drag at the free surface of the film flow? Why is this so?

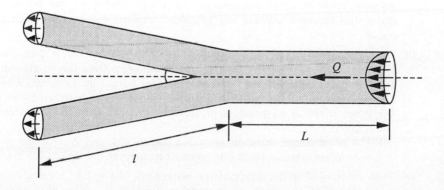

**Figure P3.6**  Bifurcation of flow.

9. *Film thickness.* For the film flow of Problem 3.2, calculate the film thickness of water under ambient conditions, falling down a vertical plate of 1 m width, at a flow rate of $20 \, m^3/s$. What is the resulting film thickness for flow rates of 40, 60, 80, and $100 \, cm^3/s$? Plot the thickness vs. flow rate and deduce their interrelation.

10. *Reservoir dynamics.* A cylindrical reservoir of 2 ft radius and 9 ft height is initially empty. A pump starts delivering water at $20 \, ft^3/min$, and simultaneously water leaves the tank through a circular hole 1 in. in diameter.
    (a) Estimate how the water elevation within the tank evolves with time.
    (b) Will the tank ever be filled? When?
    (c) What are the momentum and mechanical energy of the escaping water? What is its potential energy?
    (d) If the bottom of the reservoir is at 10 ft elevation, what are the momentum and kinetic and potential energy just prior to the impinging of the water jet with the ground?

11. *Gas reservoir.* A rigid spherical reservoir of $1 \, m^3$ volume contains 10 kg of air at 25°C. A small circular gate of 0.5 in. diameter opens and air starts leaking to the atmosphere, at a flow rate proportional to the gauge pressure of the reservoir, i.e., $\dot{V} = k(p_r - p_{atm})$. If the tank pressure reduces to that of the atmosphere within 30 s, what is the value of the constant $k$? How do the internal energy and enthalpy of the air reduce with time if the reservoir is perfectly insulated? What is the momentum of the exiting airstream? What amount of air remains trapped in the tank? How could all residual air be made to escape from the reservoir?

12. *Hydraulic descend.* Figure P3.12 shows a hydraulic arrangement opposite that of the commonly known hydraulic jump. Hydraulic descends are used to increase liquid velocity by decreasing its depth, under constant width $W = 5 \, m$ here, toward the sheet.
    (a) Under the assumption of frictionless flow, calculate $h_2$ and $h_2$. What do they represent? Which of the two will be attained by the water?
    (b) Repeat in the presence of friction, under reasonable assumptions as required.
    (c) Repeat (a) and (b) in the case of a hydraulic jump going from depth $h_2$ to larger depth $h_1$. What is the value of $h_1$?

13. *Laminar and turbulent flow.* Consider flows of different fluids in a horizontal pipe of 10 cm diameter at $600 \, cm^3/s$. Determine if the flow is laminar or turbulent for the following fluids at ambient conditions: oxygen, nitrogen, air, gasoline, water, mercury, glycerin, crude oil. Plot the resulting Reynolds number vs. the kinematic viscosity, distinguish between laminar and turbulent regions, and position each of the fluids above.

14. *Nozzle ejection.* Figure P3.14 shows a nozzle attached to a 3-in.-diameter pipe, ejecting water against a disk held a distance 1 ft apart from the nozzle exit.
    (a) For a flow rate of $100 \, m^3/s$, the emerging jet has a 1-in. diameter. What pressure is measured at point 1, under negligible frictional losses between 1 and 2?
    (b) What is the corresponding pressure in the presence of 73 dyn/cm of surface tension?
    (c) The jet impinges vertically on a 5-in.-radius disk and is deflected parallel to the disk. What external force $F$ is required to support the disk in the two cases?

15. *Tank drainage.* For the axisymmetric water tank of Fig. P3.15, calculate the resulting drain rate $\dot{V}$ at the onset of discharging, through the 3-in. discharge tube, in the absence of frictional losses. What is the significance of the inclined part of the wall?

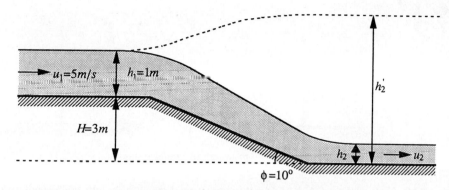

**Figure P3.12**   Hydraulic descend arrangement.

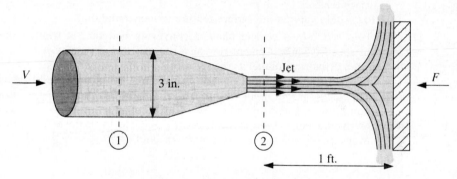

**Figure P3.14**   Nozzle jet and vertical impingement.

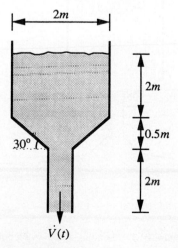

$\dot{V}(t)$

**Figure P3.15**   Tank drainage.

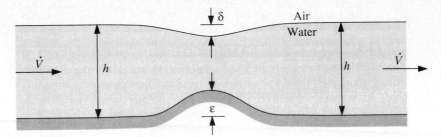

**Figure P3.16** Venturi flume.

16. *Venturi flume.* A cylindrically symmetric obstacle of height $\varepsilon$ is placed on the bed of a slightly inclined open rectangular channel of width $W$ toward the sheet plane shown in Fig. P3.16. The free surface forms a shallow dip of depth $\delta$, as shown.
    (a) Explain the shape of the free surface.
    (b) Calculate the flow rate $\dot{V}$, in terms of $W$, $h$, $\delta$, and $\varepsilon$, under negligible friction and surface tension.
    (c) How would varying the surface tension influence the dip?

17. *Conveying belt.*[10] In a cement plant, a conveying belt of 2 m width is moving at a constant speed 0.5 m/s. Cement raw material of $2\,\text{g/cm}^3$ density is deposited on the belt from a cylindrical funnel as shown in Fig. P3.17.
    (a) Calculate the belt tension required to maintain a constant speed throughout.
    (b) What electric power input is required if a yield coefficient of $n = 0.7$ is assumed to account for all frictional and conversion losses?
    (c) How would you design a rough control scheme to achieve this time-dependent power input, given the results of parts (a) and (b)?

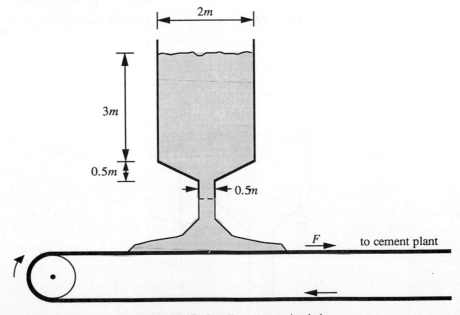

**Figure P3.17** Loading a conveying belt.

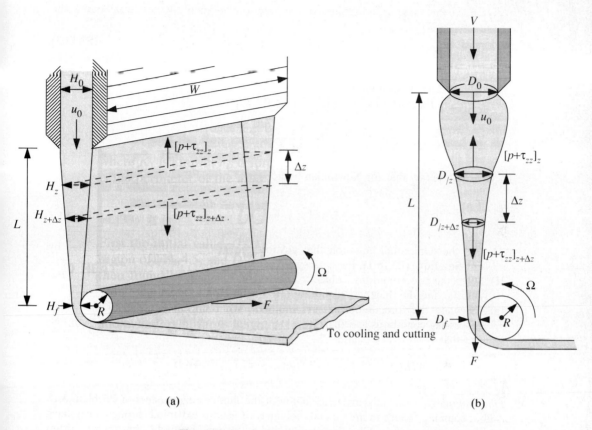

(a)                                                                      (b)

**Figure P3.18**    Processes of (a) film casting and (b) fiber spinning.

18. *Film casting.*[11] Polymer and metal films and sheets are manufactured by extruding polymer melts or molten metals through dies to form liquid curtains taken downstream by tension-applying cylinders, as shown in Fig. P3.18a.

   (a) For a molten metal of flow rate $\dot{V} = 0.1 \text{ m}^3/\text{s}$, die width $W = 1 \text{ m}$, die gap $H = 1 \text{ cm}$, wet sheet length $L = 2 \text{ m}$, and cylinder radius $R = 20 \text{ cm}$ rotating at 80 rpm, find the sheet thickness $H_f$ obtained.

   (b) Show that conservation of mass over the liquid system of width $W$, thickness $H(z)$, and height $\Delta z$ results in

$$H(z)u(z) = \dot{V}, \qquad (P3.18.1)$$

   where $u(z)$ is axial velocity uniform over each cross section of the sheet.

   (c) Similarly, in the absence of any shear stress within the liquid and drag by the air, and assuming negligible gravity, surface tension, and net convection in the system, show that momentum balance yields

$$WH(z)[p(z) + \tau_{zz}(z)] = F, \qquad (P3.18.2)$$

   where $F$ is the tension applied by the cylinder transmitted upstream throughout the sheet, and $\tau_{zz}(z)$ developed viscous normal stress in the direction of casting, also uniform over each cross section of the sheet.

**(d)** Under the same assumptions, shown that the momentum equation across the thickness is

$$p(z) + \tau_{xx}(z) = 0, \tag{P3.18.3}$$

where $\tau_{xx}(z)$ is the normal viscous stress developed across the sheet, also uniform over each cross section.

**(e)** Combine Eqns. (P3.18.1) to (P3.18.3) in the form

$$\frac{\tau_{zz}(z) - \tau_{xx}(z)}{u(z)} = \frac{F}{W\dot{V}}. \tag{P3.18.4}$$

Given the fact that for Newtonian liquid it is

$$\tau_{zz}(z) = -\tau_{xx}(z) = 2\eta \frac{du}{dz}, \tag{P3.18.5}$$

find the differential equation that governs the velocity profile $u(z)$ and then by utilizing Eqn. (P3.18.1), the corresponding equation for the sheet thickness $H(z)$.

**(f)** Solve the two equations, subject to the die-exit boundary conditions, to find the velocity and the thickness distribution in this casting process.

**(g)** Calculate the force required and the final sheet thickness achieved under the processing conditions specified for a polymer melt of 100 P viscosity, by using the fact that

$$F = WH_f(p + \tau_{zz})_{z=L} = WH_f(\tau_{zz} - \tau_{xx})_{z=L} = 4WH_f \frac{du}{dz}\Big|_{z=L}. \tag{P3.18.6}$$

**19.** *Fiber spinning.*[12] The axisymmetric analog of the film casting process of Problem 3.18 is fiber spinning, where molten metal, polymer, or glass is extruded through a capillary spinnerett to form a wet fiber that is solidified and wound-up under tension by a drum at a distance $L$, as shown in Fig. P3.18b.

**(a)** For flow rate $\dot{V} = 100 \text{ cm}^3/\text{s}$, spinnerett diameter $D_0 = 0.5 \text{ cm}$, length $L = 0.5 \text{ m}$, drum radius $R = 5 \text{ cm}$, and angular speed $\Omega = 100 \text{ rpm}$, calculate the diameter $D_f$ of the fiber obtained.

**(b)** Consider the cylindrical system shown. Show that mass balance under constant melt density yields

$$\pi R^2(z)u(z) = \dot{V}, \tag{P3.19.1}$$

and momentum balance in the spinning direction, under negligible shear stresses, air drag, gravity, surface tension, and net and momentum convection, results in

$$\pi R^2(z)[p(z) + \tau_{zz}(z)] = F, \tag{P3.19.2}$$

where $F$ is tension applied by the drum and transmitted unchanged upstream. Similarly, momentum balance in the vertical direction results in

$$p(z) + \tau_{rr}(z) = 0. \tag{P3.19.3}$$

In Eqns. (P3.19.1) to (P3.19.3), $u(z)$, $\tau_{zz}(z)$, and $\tau_{xx}(z)$ are axial velocity, axial stress, and radial stress, all uniform over each cross section of the fiber, and $R(z)$ is the diminishing radius of the fiber.

(c) Combine the three equations in the form

$$\frac{\tau_{zz} - \tau_{rr}}{u(z)} = \frac{F}{\pi \dot{V}},$$   (P3.19.4)

and given the fact that for Newtonian liquid it is

$$\tau_{zz} \approx -2\tau_{rr} \approx 2\eta \frac{du}{dz},$$   (P3.19.5)

find the final equation that governs the velocity profile $u(z)$. Then by utilizing Eqn. (P3.19.1), find the equation for the fiber radius $R(z)$.

(d) Solve the two equations subjected to die-exit conditions for melt of 1000 P viscosity under $F = 10^6$ dyn drawing force, and plot the velocity and the radius along the fiber.

(e) What is the radius of the fiber obtained? How does it depend on melt viscosity, drawing force or tension, and melt flow rate?

20. *Pressure drop in mixing.* Figure P3.20 shows a commonly used approach to disperse fluid $B$ into fluid $A$ to form a solution $AB$.

(a) Calculate the resulting pressure drop for $D = 5$ in., $d = 1$ in., $\rho_A = \rho_B = \rho_{AB} = 1$ g/cm³, $u_1^A = 1.5$ m/s, $u_1^B = 4$ m/s, and $u_u^{AB} = 2$ m/s under negligible frictional losses.

(b) How is this mixing pressure drop compared to that over the same length in the absence of the dispersing tube and liquid $B$ for $u = u_2^{AB} = 2$ m/s and $u = u_1^A = 1.5$ m/s?

(c) Examine how the resulting pressure drop increases with the ratios $\rho_A/\rho_B$, $u_1^A/u_1^B$, and $(\rho_A u_1^A)/(\rho_B u_1^B)$ in the case of unequal densities.

21. *Jet swelling.* Consider slow, creeping flow in the geometry of Fig. 3.9. A planar jet of width $W$ emerges from the slit of opening $H$ much less than $W$. Far downstream, it swells to thickness $1.185H$, where the velocity becomes plug.

(a) If the velocity at the exit of the slit is approximated by

$$u_x(y) = c\bar{u}\left(1 - \frac{4y^2}{H^2}\right)^n,$$

where $\bar{u}$ is average velocity upstream from the exit, what are the values of $c$ and $n$?

(b) Repeat for a creeping round jet that swells to $1.13H$, where $H$ is the diameter of the capillary and $\bar{u}$ its average velocity.

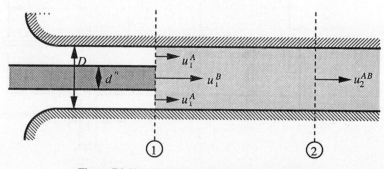

**Figure P3.20**   Dispersion of fluid $A$ into fluid $B$.

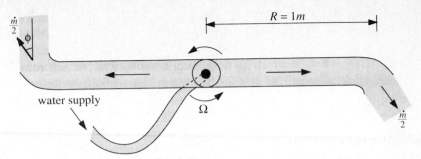

**Figure P3.22**   Lawn sprinkler.

22. *Lawn sprinkler.* The sprinkler of Fig. P3.22 discharges water from two identical
    nozzles of 1 cm diameter and $\phi = 30°$ angle with its arms, at a total flow rate
    $\dot{m} = 0.785$ L/s.
    **(a)** Show that a rotational motion will be induced under the discharge rate. If
    frictional losses are neglected, what is the resulting speed of rotation $\Omega$?
    **(b)** What amount of external torque $T_e$ would keep the sprinkler from rotating?
    **(c)** What mechanism causes rotation of the sprinkler? Can you quantify it?

23. *Tank discharge.*[13] The tank of Fig. P3.23 discharges water as shown, where $D \gg S$ and
    $H \gg h$. Show that the magnitude of the supporting horizontal force is

$$F = 2(\rho g h - p_{\text{atm}})\frac{\pi d^2}{4}.$$

24. *Stream deflection.* A vane deflects water as shown in Fig. P3.24. Find the magnitude
    and direction of the external force $F$ required for flow rate 2000 gpm, deflection angle
    $\phi = 45°$, and stream of thickness $h = 0.1$ m and width $W = 1$ m.

25. *Rocket propulsion.* A rocket of initial gross mass $m_0 = 10$ tons including fuel, is
    propelled vertically from earth by releasing exhaust gases at rate $\dot{m} = 1$ ton/min under
    atmospheric pressure, through a rear discharging gate of area $A = 0.5$ m$^2$.

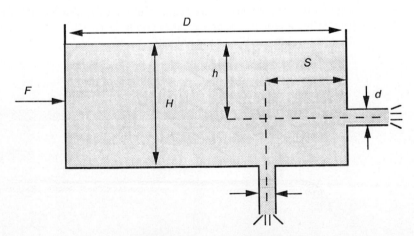

**Figure P3.23**   Tank discharge.

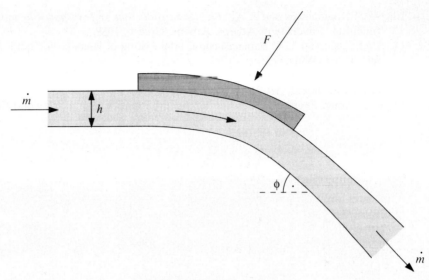

**Figure P3.24**   Deflection vane.

**(a)** Show that the momentum equation for the rocket is

$$m\frac{du}{dt} = \dot{m}v + pA,$$

where $u$ is the rocket speed and $v$ the gas exit velocity with respect to the traveling rocket gate.

**(b)** At what time and speed does the rocket exit the atmosphere?

# REFERENCES

1. I. Newton, *Principia* (by S. Pepys, London, 1886), Vol. 1, *The Motion of Bodies*; translated by F. Cajori, University of California Press, Berkeley, Calif., 1966.
2. T. Carmady and H. Kobus, *Hydrodynamics by Daniel Bernoulli and Hydraulics by Johan Bernoulli,* Dover, New York, 1968.
3. G. L. Tuve and R. E. Sprenkle, Orifice discharge coefficients for viscous liquids, *Instruments,* 6, 201 (1933).
4. J. H. Perry, *Perry's Chemical Engineer's Handbook,* 6th ed., McGraw-Hill, New York, 1984.
5. G. K. Noutsopoulou, *Lectures on Hydraulics,* National Technical University of Athens, Athens, Greece, 1975.
6. D. C. Huang and J. L. White, Extrudate-swell from slit and capillary dies, *Polym. Sci. Engrg. Rep.* 113, University of Tennessee, Knoxville, Tenn., 1977.
7. G. C. Georgiou, T. C. Papanastasiou, and J. O. Wilkes, Laminar Newtonian jets at high Reynolds numbers and high surface tension, *AIChE J.,* 34, 1559 (1988).
8. W. J. Silliman and L. E. Scriven, Separation flow near a static contact line: Slip at wall and shape of free surface, *J. Comput. Phys.,* 34 287 (1980).
9. R. I. Tanner, *Engineering Rheology,* London Press, Oxford, 1985.

10. N. G. Koumoutsos and J. A. Palevos, *Introduction to Transport Phenomena,* National Technical University of Athens, Athens, Greece, 1989.

11. S. M. Alaie and T.-C. Papanastasiou, Film casting of viscoelastic liquid, *Polym. Engrg. Sci.,* 31, 67 (1991).

12. T. C. Papanastasiou, C. W. Macosko, L. E. Scriven, and Z. Chen, Fiber-spinning of viscoelastic liquid, *AIChE J.,* 33, 834 (1987).

13. R. R. Long, *Engineering Science Mechanics,* Prentice Hall, Englewood Cliffs, NJ, 1963.

# 4

# Viscous Flow and Friction: Confined, Open, Free Stream, and Porous Media Flows

## 4.1 FRICTIONAL FORCES AND FRICTIONAL ENERGY LOSSES

In engineering applications, viscous fluids are transferred through pipelines, in which viscous resistance gives rise to *frictional forces* that dissipate mechanical energy into heat and internal energy. This mechanical energy loss, called *viscous dissipation,* is due to the shear and normal viscous stresses. The mechanism is analogous to what had been utilized by the first inhabitants of Earth to produce fire: By shearing a pair of wood pieces, mechanical energy dissipates into heat that ignites the inflammable wood.

In laminar flow, viscous dissipation is due entirely to contact and friction of adjacent fluid layers or particles flowing at different velocities. In turbulent flow, additional dissipation occurs due to in-place oscillation with time of fluid particles, which enhances particle–particle contact and promotes friction. In both cases the underlying mechanism is the same: friction due to contact of fluid particles moving at different velocities and dissipation of mechanical energy to overcome it.

The frictional forces that arise tend to inhibit the relative motion of neighboring fluid parts, as well as the relative motion between fluid streams and bounding or submerged solid bodies. The portion of frictional force due to shear

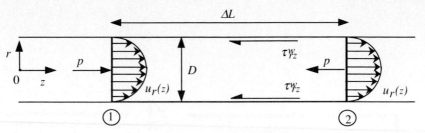

**Figure 4.1** Steady flow in pipe.

stresses is called *friction drag* and that due to normal viscous stresses is called *form drag*. A third form of drag is due to the vertical pressure force (i.e., a *pressure drag*). For a boundary or surface of arbitrary shape the three are added vectorially to provide the total drag.

In confined flows in channels and pipes, such frictional energy losses are overcome primarily either by means of a pump providing power to the fluid, or arranging for the fluid to fall under gravity from a higher to a lower elevation, such that potential energy is reduced. In free-stream flows overtaking submerged bodies (e.g., wind around smokestacks) or in motion of solid bodies in stationary fluids (e.g., traveling vehicles, aircrafts, submarines), of interest are the forces required to keep the body in place and the power required to maintain the motion, respectively.

For the pipe flow shown in Fig. 4.1, macroscopic mass, momentum, and energy balances can be performed, common for both laminar and turbulent flow. The conservation equations of mass and momentum, Eqns. (3.79) and (3.82), applied to the pipe flow under consideration yield

$$\bar{u}_1 \frac{D^2}{4} = \bar{u}_2 \frac{D^2}{4},\qquad(4.1)$$

and therefore $\bar{u}_1 = \bar{u}_2$. Since the profile $u_z(r)$ persists throughout, $\tau_{zz} = \eta\, \partial u_z/\partial z = 0$, and the momentum equation reduces to

$$p_1 \pi \frac{D^2}{4} - p_2 \pi \frac{D^2}{4} - \tau_{rz}^w \pi D\, \Delta L = 0,\qquad(4.2)$$

and therefore,

$$\tau_{rz}^w = -\left(\frac{p_1 - p_2}{\Delta L}\right)\frac{D}{4},\qquad(4.3)$$

where $\tau_{rz}^w$ is the wall shear stress from wall to fluid, with direction as shown in Fig. 4.1, that resists the motion of the fluid. The mechanical energy equation is

$$\frac{p_1}{\rho} = \frac{p_2}{\rho} + \mathscr{F} \qquad \left[\frac{\text{energy}}{\text{mass}}\right],\qquad(4.4)$$

where $\mathscr{F}$ represents mechanical energy losses per unit of flowing mass. Therefore,

$$\hat{\mathscr{F}} = (p_1 - p_2) \geq 0, \tag{4.5}$$

since $p_1 \leq p_2$, with $p_1 = p_2$ being the case of static equilibrium of $\hat{\mathscr{F}} = 0$. $\mathscr{F}$ is related to the total energy losses per unit time, or lost power,

$$\dot{\mathscr{F}} = \mathscr{F}\bar{u}A\rho = \mathscr{F}\bar{u}\frac{D^2}{4}\rho \qquad \left[\frac{\text{energy}}{\text{time}}\right]. \tag{4.6}$$

Equations (1.38), (4.3) and (4.4) combine as

$$\mathscr{F} = kf\left(\frac{\bar{u}^2}{2}\right)\left(\frac{\Delta L}{D}\right) \qquad \left[\frac{\text{energy}}{\text{mass}}\right]. \tag{4.7}$$

The coefficient $f$, which depends on the Reynolds number, is called the *friction factor*. For $k = 4$ the *Fanning friction*[1] *coefficient*, known also as Darcy's coefficient, is obtained. This definition is used primarily in chemical and mechanical engineering applications:

$$f_F = \frac{\tau_{rz}^w}{\frac{1}{2}\rho\bar{u}^2}. \tag{4.8}$$

For $k = 1$, the *Moody friction*[2] *coefficient* is obtained. This definition is used primarily in hydraulics:

$$f_M = \frac{\tau_{rz}^w}{\frac{1}{8}\rho\bar{u}^2}. \tag{4.9}$$

Equation (4.7) takes the forms:

$$\mathscr{F} = \frac{\hat{\mathscr{F}}}{\rho} = f_M\left(\frac{\bar{u}^2}{2}\right)\left(\frac{\Delta L}{D}\right) \qquad \left[\frac{\text{energy}}{\text{mass}}\right] \tag{4.10}$$

and

$$\mathscr{F} = \frac{\hat{\mathscr{F}}}{\rho} = 4f_F\left(\frac{\bar{u}^2}{2}\right)\left(\frac{\Delta L}{D}\right) \qquad \left[\frac{\text{energy}}{\text{mass}}\right] \tag{4.11}$$

when the Moody or the Fanning factor is used, respectively. The Fanning factor $f_F$ and Eqn. (4.11) are used primarily in this book. $\mathscr{F}$ and $\hat{\mathscr{F}}$ represent energy losses per unit of flowing mass or volume respectively. The power loss is found by multiplying each by the mass or the volumetric flow rates, $\dot{m}$ or $\dot{V}$, respectively.

Thus, in steady pipe flow, a pressure difference over distance $\Delta L$,

$$\frac{\Delta p}{\Delta L} = \frac{p_2 - p_1}{\Delta L} < 0, \tag{4.12}$$

is needed to force the fluid through in a parabolic-like velocity profile, $u_z(r)$, which persists downstream from the short entrance length. The motion is accompanied by shear friction along the wall quantified by Eqn. (4.3). The wall shear friction gives rise to energy frictional losses, quantified by Eqn. (4.11).

## 4.2 FRICTION IN LAMINAR FLOW

It is evident that the velocity profile $u_z(r)$ and its average $\bar{u}$, which Eqns. (4.6) to (4.11) depend on, cannot be obtained by the macroscopic balances as applied so far. Since $u_z(r)$ is a velocity distribution in the interior of the fluid, a system in the interior of the fluid must be considered, as shown in Fig. 4.2 The system is a cylinder of length $\Delta z$ and radius $r$. The interior system of Fig. 4.2 is similar to that of Fig. 4.1. $p|_z$ and $\tau_{zz}|_z$ are the pressure and normal viscous stress at point 1, or equivalently, at distance $z$ from the origin at $z = 0$. Similarly, $p|_{z+\Delta z}$ and $\tau_{zz}|_{z+\Delta z}$ are the pressure and normal stress at distance $z + \Delta z$. The shear stress $\tau_{rz}|_r$ at radial distance $r$ from the axis of symmetry is due to the retarding action of the fluid surrounding the system, which, being close to the stationary wall, travels slower than the fluid within the system. Due to this choice of the system, there is continuous flow in, through the surface at distance $z$, and flow out through the other surface at distance $z + \Delta z$, whereas there is no flow at all in the radial direction.

The conservation equations are applied as follows. The mass conservation equation is

$$\rho[u_z \pi r^2]_z = \rho[u_z \pi r^2]_{z+\Delta z}, \tag{4.13}$$

and therefore $u_z|_z = u_z|_{z+\Delta z}$, which means that the velocity profile $u_z$ is not a function of $z$, and therefore,

$$\frac{\partial u_z}{\partial z} = 0. \tag{4.14}$$

The momentum equation is

$$[\rho \beta u_z^2 \pi r^2]_z - [\rho \beta u_z^2 \pi r^2]_{z+\Delta z} + [(p - \tau_{zz})\pi r^2]_z$$

$$- [(p - \tau_{zz})\pi r^2]_{z+\Delta z} + [2\pi r \Delta z \tau_{rz}]_r = 0. \tag{4.15}$$

Since $u_z|_z = u_z|_{z+\Delta z}$, the first two terms cancel each other. Also, $\tau_{zz} = \eta \, \partial u_z/\partial z = 0$ by virtue of Eqn. (4.14). The remaining terms are rearranged to the

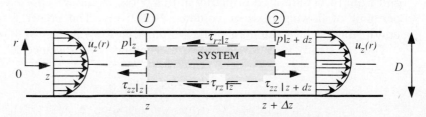

**Figure 4.2**   Interior system for laminar flow in pipe.

form

$$\tau_{rz}\big|_r = -\left[\frac{p\big|_z - p\big|_{z+\Delta z}}{\Delta z}\right]\frac{r}{2} = \left[\frac{\Delta p}{\Delta z}\right]\frac{r}{2} < 0, \tag{4.16}$$

which means that the assigned direction to the shear stress in Fig. 4.2 is the correct one. Since the points 1 and 2 in Fig. 4.2, and thus locations $z$ and $z + \Delta z$, have been chosen arbitrarily, Eqns. (4.15) and (4.16) suggest that

$$\left[\frac{\Delta p}{\Delta z}\right] = \frac{\Delta p}{\Delta L} = \frac{p_L - p_0}{\Delta L} < 0, \tag{4.17}$$

and the pressure gradient $\Delta p/\Delta z$ at any distance $z$ is constantly identical to $\Delta p/\Delta L$. Thus, Eqn. (4.16) becomes

$$\tau_{rz}\big|_r = \frac{\Delta p}{\Delta L}\frac{r}{2}. \tag{4.18}$$

The shear stress $\tau_{rz}$ is related to the velocity by

$$\tau_{rz} = \eta \frac{du_z}{dr}, \tag{4.19}$$

according to the constitutive equation [e.g., Eqn. (1.23) and Appendix F], and therefore Eqn. (4.18) becomes

$$\eta \frac{du_z}{dr} = \frac{\Delta p}{\Delta L}\frac{r}{2}, \tag{4.20}$$

which is integrated twice to

$$u_z(r) = \frac{1}{4\eta}\frac{\Delta p}{\Delta L}r^2 + c_1, \tag{4.21}$$

where $c_1$ is an arbitrary constant of integration at this point. The velocity of the fluid at the solid wall must be zero, that is,

$$u_z\left(r = \frac{D}{2}\right) = \frac{1}{4\eta}\frac{\Delta^2}{\Delta L}\frac{D^2}{4} + c_1 = 0, \tag{4.22}$$

and therefore,

$$c_1 = -\frac{1}{4\eta}\frac{\Delta p}{\Delta L}\frac{D^2}{4}. \tag{4.23}$$

The constant $c_1$ is substituted in Eqn. (4.22) to yield the complete velocity profile,

$$u_z(r) = -\frac{1}{4\eta}\frac{\Delta p}{\Delta L}\left(\frac{D^2}{4} - r^2\right), \tag{4.24}$$

a parabolic function of $r$ which satisfies the boundary conditions $u_z(r = D/2)$ and $du_z/dr\big|_{r=0} = 0$, as shown in Figs. 4.1 and 4.2.

Equation (4.24) suggests that the velocity decreases from its *maximum value* at $r = 0$, which is

$$(u_z)_{max} = u_z(r = 0) = -\frac{1}{4\eta}\frac{\Delta p}{\Delta L}\frac{D^2}{4},$$  (4.25)

to its minimum value,

$$(u_z)_{max} = u_z\left(r = \frac{D}{2}\right) = 0,$$  (4.26)

at the solid wall. Therefore, momentum is continuously transferred from the interior to the exterior fluid layer, which in turn retards the motion of the interior layer through shear stress $\tau_{rz}$, in Fig. 4.2. The *average velocity* is

$$\bar{u} = \frac{\dot{V}}{A} = \frac{\int_0^{D/2} 2\pi r u_z(r)\,dr}{\pi(D^2/4)} = -\frac{1}{32\eta}\frac{\Delta p}{\Delta L}D^2 = \tfrac{1}{2}(u_z)_{max},$$  (4.27)

where $A = \pi D^2/4$ is the cross-sectional area and

$$\dot{V} = \int_0^{D/2} 2\pi r u_z(r)\,dr = -\frac{\pi}{8\eta}\left(\frac{\Delta p}{\Delta L}\right)\left(\frac{D}{2}\right)^4 = \bar{u}A$$  (4.28)

is the *volumetric flow rate*. The mass flow rate is

$$\dot{m} = \rho\dot{V} = -\frac{\pi\rho}{8\eta}\left(\frac{\Delta p}{\Delta L}\right)\left(\frac{D}{2}\right)^4.$$  (4.29)

The rates $\dot{V}$ or $\dot{m}$ are achieved under the imposed pressure gradient $\Delta p/\Delta L$.

For laminar flow, knowledge of the velocity profile by Eqn. (4.24) allows an analytic evaluation of the friction factor by utilizing Eqns. (4.4), (4.10), and (4.27):

$$-\frac{\Delta p}{\rho} = \mathscr{F} = 4f_F\frac{\bar{u}^2}{2}\left(\frac{\Delta L}{D}\right) = -2f_F\bar{u}\left(\frac{1}{32n}\frac{\Delta p}{\Delta L}D^2\right).$$  (4.30)

Thus, for laminar pipe flow the Fanning friction coefficient is

$$f_F = \frac{16}{Re} = \frac{16\eta}{\rho D\bar{u}}, \qquad Re < 2100.$$  (4.31)

### Example 4.1: Flow of Crude Oil in Pipe

Crude oil of viscosity 100 times that of water and density 0.9 g/cm³ is to be transferred through a 1000-m-long horizontal steel pipe of internal diameter 30 cm, from tanker to refinery, at the operating rate of 400 gpm.

(a) What inlet pressure and power input are required for such delivery at atmospheric pressure for water and for crude oil?

(b) Calculate and plot the velocity and shear stress distribution for oil and water, and comment on their similarities and differences.

(c) What are the resulting average velocities, friction factors, and total frictional energy losses for water and crude oil?

**Solution**

(a) Assuming laminar flow, the required pressure gradient is given by Eqn. (4.29):

$$-\frac{\Delta p}{\Delta L} = \frac{128 \eta \dot{V}}{\pi D^2} = \frac{128 \times 1 \times (40 \times 0.00378 \times 10^6)}{3.14 \times 900 \times 60} \frac{\mu\,\text{bar}}{\text{cm}}$$

$$= 114.1 \frac{\mu\,\text{bar}}{\text{m}} = 0.0141 \frac{\text{atm}}{\text{m}}.$$

Thus the inlet pressure for crude oil is

$$p_{\text{in}} = p_{\text{out}} + \frac{\Delta p}{\Delta L}\Delta L = (1 + 0.0141 \times 1000)\,\text{atm} = 15.1\,\text{atm}.$$

The corresponding pressure gradient and required inlet pressure for water are 0.000141 atm/m and 1.14 atm, respectively. The differences in the values for oil and water demonstrate the dominant role of viscosity on the pressure drop and the power required to overcome it, as explained in part (c).

(b) The resulting velocity distributions for oil and water are given by Eqn. (4.24):

$$[u_z(r)]_{\text{oil}} = [u_z(r)]_w = \frac{1}{4\eta}\left(\frac{\Delta p}{\Delta L}\right)\left(r^2 - \frac{D^2}{4}\right) = -28.5(r^2 - 0.0225).$$

The corresponding shear stress distributions are different:

$$[\tau_{rz}]_{\text{oil}} = \eta_{\text{oil}}\frac{\partial u_z}{\partial r} = -57r$$

$$[\tau_{rz}]_w = \eta_w\frac{\partial u_z}{\partial r} = -0.57r.$$

The velocity and stress distributions are plotted as shown in Fig. E4.1.

(c) The resulting average velocities are

$$\bar{u}_{\text{oil}} = \bar{u}_w = \frac{\dot{V}}{A} = \frac{400 \times 0.00378}{3.14 \times 0.0225 \times 60} \frac{\text{m}}{\text{s}} = 0.357 \frac{\text{m}}{\text{s}}.$$

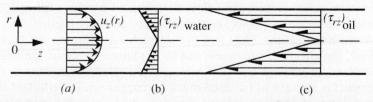

$(a)$ $(b)$ $(c)$

**Figure E4.1**   Velocity and shear stress distribution in pipe flow: (a) velocity of both oil and water at the same volumetric flow rate; (b) shear stress of water; (c) shear stress of oil of viscosity 100 times that of water.

The corresponding Reynolds numbers are

$$(\text{Re})_{\text{oil}} = \frac{\rho_{\text{oil}} \bar{u} D}{\eta_{\text{oil}}} = \frac{0.9 \times 35.7 \times 30}{1} = 964$$

and

$$(\text{Re})_w = \frac{\rho_w \bar{u} D}{\eta_w} = \frac{1 \times 35.7 \times 30}{0.01} = 107{,}100.$$

Thus, under the same average velocity in the same pipe, the flow of oil is laminar, whereas that of water is turbulent. Therefore, Eqn. (4.3.1) applies only to oil:

$$f_F = \frac{16}{\text{Re}} = 0.0166$$

$$\mathscr{F} = 4f_F \frac{\bar{u}^2}{2} \frac{\Delta L}{D} = 4 \times 0.0166 \times \frac{0.0357^2}{2} \times \frac{1000}{0.3} \frac{\text{m}^2}{\text{s}^2} = 0.141 \frac{\text{J}}{\text{kg}}$$

$$\hat{\mathscr{F}} = \mathscr{F}\rho_{\text{oil}} = (0.141 \times 900) \frac{\text{J}}{\text{m}^3} = 127 \frac{\text{J}}{\text{m}^3}$$

$$\dot{\mathscr{F}} = \frac{\hat{\mathscr{F}}}{\dot{V}} = \frac{127 \times 60}{400 \times 0.00378} \frac{\text{J}}{\text{s}} = 5 \frac{\text{kJ}}{\text{s}} = 5 \text{ kW}.$$

The value of $\mathscr{F}$ is energy input per unit mass of oil; that of $\hat{\mathscr{F}}$ is the same energy per unit volume of oil; and that of $\dot{\mathscr{F}}$ is power input. These amounts are provided externally to compensate for the lost mechanical energy due to friction, in order to maintain the steady flow. The corresponding quantities for water flow are determined as detailed in Section 4.3, for turbulent flow.

## 4.3 FRICTION IN TURBULENT FLOW

In turbulent flow, there are no exact analytic expressions for the friction factor, because the steady analytic velocity profile of Eqn. (4.24) no longer applies. In addition, convection proportional to velocity $\bar{u}$ dominates over frictional resistance proportional to viscosity $\eta$, as suggested by a high value of the Reynolds number, which characterizes turbulent flow. The velocity profile flattens out and the friction coefficient is reduced as defined by Eqn. (4.8). Furthermore, the friction factor in turbulent flow is increased by the *wall roughness, $\varepsilon$,* which at high turbulent velocity contributes to further friction as explained below. According to Eqns. (4.8) and (4.9), which are common for laminar and turbulent flow, the effects of roughness can be explained by their significance to the viscous stresses: In laminar flow the arising shear stresses are due to transverse velocity gradients that are not altered much by roughness; in turbulent flow, in addition to this kind of laminar viscous stresses, there are also turbulent stresses, called Reynolds stresses (e.g., Chapter 10), that depend on the velocity fluctuations. These fluctuations to all possible directions are retarded by roughness and so are the total shear stress and the friction coefficient, according to Eqn. (4.8).

Figure 4.3 known as the *Moody friction diagram*,[3] which is based largely on Nikuradse's experiments with smooth and rough pipes,[4] is the most widely used source of friction factors for turbulent flow. Colebrook's equation,[5]

$$\frac{1}{\sqrt{f_F}} = -4\log\left(\frac{\varepsilon}{D} + \frac{4.64}{\text{Re}\sqrt{f_F}}\right) + 2.28, \tag{4.32}$$

is a fairly good approximation of the friction factor in the turbulent region, whereas Blasius's equation,

$$f_F = 0.079\,\text{Re}^{-1/4}, \qquad \text{Re} > 100{,}000, \tag{4.33}$$

applies to turbulent flow beyond Re = 100,000 in smooth pipe. Equation (4.32) is difficult to use, because $f_F$ appears implicitly in a nonlinear equation. Wood's approach,[6] which is easier to use than Eqn. (4.32), predicts friction factors as accurate as those predicted by Eqn. (4.32), within typical experimental errors:

$$f_F = \alpha + \beta\,\text{Re}^{-c}$$

$$\alpha = 0.0235\left(\frac{\varepsilon}{D}\right)^{0.225} + 0.1325\left(\frac{\varepsilon}{D}\right),$$

$$\beta = 22\left(\frac{\varepsilon}{D}\right)^{0.44}, \quad c = 1.62\left(\frac{\varepsilon}{D}\right)^{0.134}.$$

Equations (4.32) to (4.34) are represented graphically in plots of the form of Fig. 4.3, which are the commonly used sources for estimating *friction factors, from the Reynolds number and the relative roughness, $\varepsilon/D$, where $\varepsilon$ is a representative*

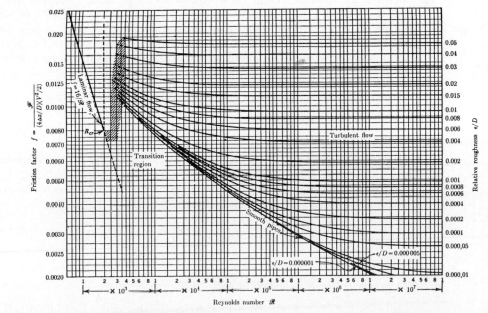

**Figure 4.3**  Fanning friction factors for pipe and close or open channels. [From L. W. Moody, Friction factors for pipe flow, *Trans. ASME*, 66, 672 (1944), by permission of ASME.]

**TABLE 4.1**  ROUGHNESS $\varepsilon$ OF COMMONLY USED
PIPE WALL SURFACES

| Wall surface | $\varepsilon$ (ft) | $\varepsilon$ (mm) |
| --- | --- | --- |
| Concrete | 0.001–0.01 | 0.3–3.0 |
| Wood iron | 0.0006–0.003 | 0.0072–0.036 |
| Cast iron | 0.00085 | 0.25 |
| Galvanized iron | 0.0005 | 0.15 |
| Asphalted cast iron | 0.0004 | 0.0048 |
| Commercial steel | 0.00015 | 0.046 |
| Drawn tubing | | |
| (glass, plastic, lead) | 0.000005 | 0.0015 |
| Riveted steel | 0.003–0.03 | 0.036–0.36 |

length of surface valleys and hills, or roughness, and $D$ is the diameter of the pipe. Values of roughness for commonly used pipes are given in Table 4.1.

The Reynolds number Re and the relative roughness $\varepsilon/D$ define a unique point on the Moody diagram of Fig. 4.3, which in turns defines a unique friction factor. For example, for the turbulent flow of water in Example 4.1 in a smooth pipe at Re = 107,100, the resulting friction factor is $f_F = 0.0045$, which indeed is significantly lower than that of oil flow in the same pipe, $f_F = 0.0166$, corresponding to Re = 964, also found from Fig. 4.3. Example 4.2 illustrates more friction factor estimations in laminar and turbulent flows.

**Example 4.2: Calculation of Friction Factor**

Calculate the frictional energy losses, the resulting pressure drop, and the power input required by a pump to transfer a fluid of $1$ g/cm$^3$ density and $0.995$ cP viscosity through a 100-m-long horizontal pipe of internal diameter 4 cm at flow rates of 0.0001, 0.001, 0.01, 0.1, and 1 m$^3$/s, initially when the wetted pipe surface is smooth and after a year of operation where surface exhibits relative roughness $\varepsilon/D = 0.008$ due to corrosion, at the same flow rates.

**Solution**   The Reynolds number at each flow rate is determined first:

$$\text{Re} = \frac{\rho D \bar{u}}{\eta} = \frac{\rho D}{\eta} \frac{4\dot{V}}{\pi D^2} = \frac{4\rho \dot{V}}{\pi \eta D} = 32\dot{V}.$$

The resulting friction factor $f_F$ is found from Fig. 4.3 given the Reynolds number and the values of relative roughness, 0 and 0.008, for smooth and rusted pipes, respectively. Then the resulting total frictional losses per unit mass of flowing fluid are estimated by Eqn. (4.11), given $\rho$, $D$, $\Delta L$, $f_F$, and $\bar{u} = 4\dot{V}/\pi D^2$:

$$\mathscr{F} = 4f_F\left(\frac{L}{D}\right)\frac{\bar{u}^2}{2} = 2f_F\left(\frac{L}{D}\right)\left(\frac{4\dot{V}}{\pi D^2}\right)^2 = [31.695 f_F \dot{V}^2]\,\text{cm}^2/\text{s}^2$$

$$= [0.0032 f_F \dot{V}^2]\,\text{J/Kg}$$

The pressure drop is estimated by Eqn. (4.5).

$$\Delta p = p_2 - p_1 = -\hat{\mathscr{F}} = -\mathscr{F}\rho = [-31.695 f_F \dot{V}^2 \rho]\,\text{dyn/cm}^2 = -[3.17 f_F \dot{V}^2]\,\text{N/m}^2$$

$$= -[3.12 f \dot{V}_\rho^2]10^{-5}\,\text{atm}.$$

Finally, the power required is estimated by Eqn. (1.41):

$$P = -\tau_{rx}^w[\pi DL)\bar{u} = -\left(\frac{D}{4}\frac{\Delta p}{L}\right)\pi DL\bar{u}$$

$$= [-\Delta p\dot{V}]\,\text{erg/s} = [-10^{-10}\,\Delta p\dot{V}]\,\text{kW}.$$

Table E4.2 gives the results from these calculations. Several important observations and conclusions are in order:

1.  For Re < 2000, for which Eqn. (4.31) applies, rough pipes exhibit friction factors independent of relative roughness and identical to that of smooth pipe at the same Reynold number (Fig. 4.3).

2.  At the transition region, for 2000 < Re < 4000, there is no unique value of friction factor, which may fluctuate between a lower limit predicted by Eqn. (4.31) and an upper limit corresponding to the particular relative roughness $\varepsilon/D$ curve. For design purposes the upper limit is commonly used.

3.  At high Reynolds numbers, such that

$$\left(\frac{\varepsilon}{D}\right)\text{Re} > 2500,$$

    the friction factor becomes independent of the Reynolds number.

4.  The units of $\mathscr{F}$ are those of power per mass flow rate or energy loss per unit of flowing mass,

$$\frac{\text{W}}{\text{kg/s}} = \frac{\text{N(m/s)}}{\text{kg/s}} = \frac{\text{kg(m/s}^2)(\text{m/s})}{\text{kg/s}} = \frac{\text{m}^2}{\text{s}^2},$$

    and the power input is therefore,

$$P = \dot{V}\,\Delta p = \dot{V}\rho\mathscr{F} = \dot{V}\hat{F}.$$

5.  Increases in frictional losses, accompanied by pressure drop and power consumption, may occur with the operation age of a piping system. These factors must be taken into account when designing and purchasing such systems.

6.  For combinations of Reynolds numbers and relative roughness not covered by Moody's diagram of Fig. 4.3 (e.g., Re = $3.2 \times 10^7$ in smooth pipe), Eqn. (4.33) applies. Indeed,

$$f_F = 0.79\,\text{Re}^{-1/4} = 0.079 \times (0.32 \times 10^8)^{-1/4} = 0.00105,$$

    in agreement with that of Table E4.2.

7.  For laminar flow of Re < 2100, the friction factor $f_F$ is calculated quickly by Eqn. (4.31).

8.  The maximum average velocity is $\bar{u} = 796$ m/s, which is below the speed of sound, $C \approx 1500$ m/s, in the flowing liquid of denisty $\rho = 1$ g/cm$^3$ and $\eta = 0.99$ cP. For higher than $c$ average velocities the flow becomes supersonic and the preceding analysis does not apply, as discussed in Section 4.13.

**TABLE E4.2**  FRICTIONAL MECHANICAL ENERGY LOSSES PER KILOGRAM OF FLOWING FLUID $\mathscr{F}$, PRESSURE DROP $\Delta p$, AND POWER $P$ REQUIRED TO MAINTAIN FLOW, IN SMOOTH PIPE AND IN ROUGH PIPE OF RELATIVE ROUGHNESS 0.008 [EXAMPLE 4.2]

| Common value | | | Smooth pipe, $\varepsilon/D = 0$ | | | | Rough pipe, $\varepsilon/D = 0.008$ | | | |
| $\dot{V}$ ($m^3/s$) | $\bar{u}$ (m/s) | Re | $f_F$ | $\mathscr{F}$ [kW/(kg/s)] | $-\Delta p$ (atm) | $P$ (kW) | $f_F$ | $\mathscr{F}$ [kW/(cg/s)] | $-\Delta p$ (atm) | $P$ (kW) |
|---|---|---|---|---|---|---|---|---|---|---|
| $10^{-4}$ | $7.96 \times 10^{-2}$ | $3.2 \times 10^{-3}$ | 0.005 | 0.00016 | 0.00157 | $1.60 \times 10^{-5}$ | 0.0125 | 0.0004 | 0.039 | $4 \times 10^{-4}$ |
| $10^{-3}$ | $7.96 \times 10^{-1}$ | $3.2 \times 10^{-4}$ | 0.00575 | 0.0184 | 0.180 | 0.0184 | 0.0094 | 0.030 | 0.293 | $3 \times 10^{-2}$ |
| $10^{-2}$ | 7.96 | $3.2 \times 10^{5}$ | 0.0036 | 1.152 | 11.37 | 11.52 | 0.0088 | 2.816 | 27.45 | 28.16 |
| $10^{-1}$ | 79.6 | $3.2 \times 10^{6}$ | 0.0024 | 76.8 | 758 | 7680 | 0.0088 | 281.6 | 2,746 | 2,816 |
| 1 | 796 | $3.2 \times 10^{7}$ | 0.0015 | 480 | 47,980 | $4.8 \times 10^{6}$ | 0.0088 | 28,160 | 274,600 | 2,816,000 |

## 4.4 MORE CONSIDERATIONS ON FRICTION

### 4.4.1 Commercial Steel Pipes and Pipe Strength

Commercial steel pipes are often referred to by their nominal size and schedule number, and in fact there are standard sizes of commercial steel pipes, a selection of which with their characteristics is shown in Table 4.2. The *schedule number* is defined by[7]

$$n = 1000\frac{p_{max}}{S}, \tag{4.35}$$

where $p_{max}$ is the maximum allowable pressure from the liquid and $S$ the allowable azimuthal tensile stress within the pipe wall beyond which the circular wall breaks into two semicircular pieces. Thus, for a pipe wall of thickness $\delta$ and

**TABLE 4.2   REPRESENTATIVE SIZES OF STEEL PIPE**

| Nominal size (in.) | Outside diameter (in.) | Schedule number | Wall thickness (in.) | Inside diameter (in.) |
|---|---|---|---|---|
| $\frac{1}{2}$ | 0.840 | 40 | 0.109 | 0.622 |
|  |  | 80 | 0.147 | 0.546 |
| $\frac{3}{4}$ | 1.050 | 40 | 0.113 | 0.824 |
|  |  | 80 | 0.154 | 0.742 |
| 1 | 1.315 | 40 | 0.133 | 1.049 |
|  |  | 80 | 0.179 | 0.957 |
| 2 | 2.375 | 40 | 0.154 | 2.067 |
|  |  | 80 | 0.218 | 1.939 |
| 3 | 3.500 | 40 | 0.216 | 3.068 |
|  |  | 80 | 0.300 | 2.900 |
| 4 | 4.500 | 40 | 0.237 | 5.026 |
|  |  | 80 | 0.337 | 4.407 |
|  |  | 160 | 0.531 | 3.438 |
| 6 | 6.625 | 40 | 0.280 | 6.065 |
|  |  | 80 | 0.432 | 5.761 |
|  |  | 160 | 0.718 | 5.189 |
| 8 | 8.625 | 40 | 0.322 | 7.981 |
|  |  | 80 | 0.500 | 7.625 |
|  |  | 160 | 0.906 | 6.813 |
| 10 | 10.75 | 40 | 0.365 | 10.020 |
|  |  | 80 | 0.593 | 9.564 |
|  |  | 160 | 1.125 | 8.500 |
| 12 | 4.500 | 40 | 0.406 | 11.938 |
|  |  | 80 | 0.687 | 11.376 |
|  |  | 160 | 1.312 | 10.126 |
| 16 | 4.500 | 40 | 0.500 | 15.000 |
|  |  | 80 | 0.843 | 14.314 |
|  |  | 160 | 1.562 | 12.876 |
| 24 | 4.500 | 40 | 0.687 | 22.626 |
|  |  | 80 | 1.218 | 21.564 |
|  |  | 160 | 2.312 | 19.376 |

*Source*: Data from Refs. 7 and 8.

internal diameter $D$, the relation

$$\frac{\delta}{D} = \frac{p_{max}}{2S} \tag{4.36}$$

exists that, together with Eqn (4.36), defines the safety limits of a pipe as demonstrated in Example 4.3. These limits are important design considerations for piping systems and are given as additional specifications.[7,8] Other considerations may be accounted for in Eqns. (4.35) and (4.36) as detailed elsewhere.[7]

### Example 4.3: Characteristics of Steel Pipe

A commercial steel pipe to transfer water at $0.005 \text{ m}^3/\text{s}$ is to be purchased. Pumping facilities along the lengthy horizontal pipe may provide up to 1 kW power per meter of pipe length. By utilizing Table 4.2 give specifications for the appropriate pipe.

**Solution**   The relation between input of power per length and frictional losses is

$$\frac{P}{\Delta L} = \dot{V}\frac{\Delta p}{\Delta L} = \frac{\rho \dot{V}}{\Delta L}\mathscr{F} = 2f_F\rho\dot{V}\bar{u}^2\frac{1}{D} = \frac{32\dot{V}\rho}{\pi^2 D^5}f_F, \tag{4.3.1}$$

and therefore,

$$D = \sqrt[5]{\frac{32\dot{V}^3\rho f_F}{\pi^2(P/\Delta L)}} = (0.0525 f_F^{1/5})\,\text{m} = 2f_F^{1/5}\,\text{in.} \tag{4.3.2}$$

From Table 4.2 a nominal size 2-in. pipe of schedule number 80 is picked, with internal diameter $D = 1.939$ in. For that value, Table 4.1 predicts relative roughness, $\varepsilon/D = 0.0009$. The resulting Reynolds number is

$$\text{Re} = \frac{\rho D\bar{u}}{\eta} = \frac{4\rho\dot{V}}{\pi\eta D} = 130,000. \tag{4.3.3}$$

From these Reynolds numbers and relative roughness values, the friction number from Fig. 4.3 is 0.0053. For that value of $f_F$, Eqn. (4.3.2) predicts that

$$D^{(1)} = 2(0.0053)^{1/5}\,\text{in.} = 0.7\,\text{in.}, \tag{4.3.4}$$

which disagrees with the picked pipe of $D^{(0)} = 2$ in. Obviously, a smaller value of $D$ is needed. Such a value is $D^{(2)} = 0.742$ in. for a schedule 80, nominal size $\frac{3}{4}$, steel pipe of Table 4.2. For that value, $\varepsilon/D = 0.0024$, and the Reynolds number is 340,000, in which case $f_F = 0.00625$, from Fig. 4.3. For these values, $D^{(3)} = 0.724$ in., which is close enough to $D^{(2)} = 0.742$ in. Therefore, the best available steel pipe is a nominal size $\frac{3}{4}$, schedule 80, of internal diameter $D = 0.742$ in.

For this commercial pipe, $\varepsilon/D = 0.00244$, Eqn. (4.3.3) predicts Re = 348,000, and Fig. 4.3 indicates $f_F = 0.0065$. The resulting flow rate is calculated by means of Eqn. (4.3.1),

$$\dot{V} = \left[\frac{\pi^2 D^5(P/\Delta L)}{32\rho f_F}\right]^{1/3} = 0.0049\,\text{m}^3/\text{s}, \tag{4.3.5}$$

which is close enough to the required $\dot{V} = 0.005 \text{ m}^3/\text{s}$.

According to Eqns. (4.35) and (4.36),

$$\frac{\delta}{D} = \frac{p_{max}}{2S} = \frac{n}{2000} = 0.04. \tag{4.3.6}$$

The thickness $\tau$ from Table 4.2 is 0.308 in. Thus

$$\frac{\delta}{D} = \frac{0.154}{0.742} = 0.20 > 0.04, \tag{4.3.7}$$

and the pipe is safe.

### 4.4.2  Friction Due to Fittings and Equivalent Length

The expressions for friction developed so far were for smooth or rough wetted surfaces of regular rectilinear planar or cylindrical walls. Additional friction and therefore pressure drop are induced by geometrical irregularities and obstacles to flow, such as sudden contractions or expansion of pipes and channels (e.g., Fig. 3.3), elbows (e.g., Fig. E3.6), valves and nozzles (e.g. Fig. P3.14) and others. These fittings also induce acceleration (by forcing the fluid to change direction), hence the flow fields are radically different from, and more energy-dissipative than, pipe flows. This additional friction over these options or fittings is quantified by adding an *equivalent length* ($\Delta L_e/D$) to the actual length ($\Delta L/D$) of the pipe or channel in Eqn. (4.10), which results in an increased $\mathscr{F}$ and increased pressure drop by Eqn. (4.4). Values of equivalent lengths for common fittings and options are listed in Table 4.3. More extensive listings are given elsewhere.[7–10]

### 4.4.3 Noncircular-Cross-Section Conduits and Open Channels

Not all commercially available and industrially or otherwise applied pipes are of circular cross-sectional area. For these conduits, Eqn. (4.11) is easily shown to

**TABLE 4.3** EQUIVALENT LENGTHS OF PIPE FITTINGS

| Fitting or option | $\Delta L_e/D$ |
|---|---|
| Angle valve (open) | 160 |
| Gate valve (open) | 6.5 |
| Globe wall | 330 |
| Square | 30 |
| Standard "T" elbow | 70 |
| 45° | 15 |
| Sudden contraction | |
| 4:1 | 15 |
| 2:1 | 11 |
| 4:3 | 6.5 |
| Sudden expansion | |
| 1:4 | 30 |
| 1:2 | 20 |
| 3:4 | 6.5 |

take the form

$$\mathscr{F} = -\frac{\Delta p}{\rho} = -\frac{\tau_{rz}^w \, \Pi}{\rho \, A} \Delta L = \frac{\tau_{rz}^w}{\frac{1}{2}\rho \bar{u}^2}\left(\frac{1}{2}\bar{u}^2\right)\frac{\Delta L}{4A/\Pi} = 4f_F \frac{\bar{u}^2}{2}\frac{\Delta L}{D_e}, \qquad (4.37)$$

where

$$D_e = \frac{4A}{\Pi} \qquad (4.38)$$

is an *equivalent diameter,* defined as four times the ratio of the cross-sectional area $A$ of flow to the perimeter of the cross-sectional area $\Pi$, in contact to the wall bounding the flow. The factor 4 is placed to account correctly for the case of a circular pipe for which Eqn. (4.37) applies, too, since $D_e = \pi D^2/\pi D = D$ for a circular cross-sectional area. Example 4.4 demonstrates the applicability of the friction factor diagram of Fig. 4.3 to an open inclined channel flow, where the equivalent diameter defined by Eqn. (4.38) is used. Comparisons between theoretical and experimental values of friction factor with noncircular ducts further justify the use of the equivalent diameter.[11,12]

### Example 4.4: Friction in Conduits with Open Channel

An open channel for irrigation of 2 m width, made with concrete of 1.4 mm roughness, carries water downhill over a distance 2 km at average ground inclination $\theta = 10°$ with the horizontal. What is the resulting depth of water in the channel at flow rate 14 m³/s?

**Solution**   Assume depth $H^{(0)} = 1$ m. The equivalent diameter is

$$D_e = \frac{4A}{\Pi} = \frac{4WH^{(0)}}{(W + 2H^{(0)})} = \frac{4 \times 2 \times 1}{2 + 2} \, \text{m} = 2 \, \text{m}, \qquad (4.4.1)$$

and the average velocity is

$$\bar{u} = \frac{\dot{V}}{WH^{(0)}} = 7 \text{ m/s}. \qquad (4.4.2)$$

Table 4.1 yields $\varepsilon/D = 0.0007$. The Reynolds number is

$$\text{Re} = \frac{\rho \bar{u} D_e}{\eta} = 14 \times 10^6, \qquad (4.4.3)$$

and the resulting friction factor from Fig. 4.3 is $f_F = 0.0045$. The energy equation for the inclined flow is

$$g \, \Delta z \cos \theta = 4f_F \frac{\bar{u}^2}{2}\frac{\Delta L}{D_e}, \qquad (4.4.4)$$

and the resulting average velocity is

$$\bar{u} = \left[\frac{g \sin \theta D_e}{2f_f}\right]^{1/2} = \left[\frac{9.81 \times 0.174 \times 2}{2 \times 0.0045}\right]^{1/2} \frac{\text{m}}{\text{s}} = 19.48 \frac{\text{m}}{\text{s}}, \qquad (4.4.5)$$

which is very different from that of Eqn. (4.4.2). Thus, a smaller depth, say $H^{(1)} = 0.5$ m, must be tried. Then

$$D_e = \frac{4A}{\Pi} = \frac{4 \times 2 \times 0.5}{2 + 1} = 0.8 \text{ m} = 1.33 \text{ m},$$

$$\bar{u} = \frac{\dot{V}}{WH^{(1)}} = 14 \text{ m/s},$$

$$\text{Re} = \frac{\rho \bar{u} D_e}{\eta} = \frac{1000 \times 14 \times 1.33}{0.001} = 1.86 \times 10^7,$$

$$\frac{\varepsilon}{D} = 0.00105,$$

and from Fig. 4.3,

$$f_F = 0.005.$$

Equation (4.5) now predicts that

$$\bar{u} = \left( \frac{9.81 \times 0.174 \times 0.8}{2 \times 0.00525} \right)^{1/2} \frac{\text{m}}{\text{s}} = 11.4 \frac{\text{m}}{\text{s}},$$

which in this case is smaller than that assumed. Thus a new value $H^{(2)} = 0.45$ m between the two previous values assumed for $H$ is considered. The resulting new values of the entering variables are $D_e = 1.24$ m, $\bar{u} = 14.55$ m/s, Re $= 1.8 \times 10^7$, $\varepsilon/D = 0.0011$, $f_F = 0.0051$. With these values the predicted average velocity is

$$\bar{u} = \left( \frac{9.81 \times 0.174 \times 1.24}{2 \times 0.0051} \right)^{1/2} \frac{\text{m}}{\text{s}} = 14.41 \frac{\text{m}}{\text{s}},$$

which is close to the earlier one of 14.55 m/s.

Thus the depth of the water in the channel is about

$$H = 0.45 \text{ m}.$$

## 4.5 CHARACTERISTIC PROBLEMS OF FRICTION IN PIPES

There may be three main categories of such problems, I, II, and III, summarized below in increasing degree of difficulty:

**I.** *Direct problems in specified geometry and flow conditions.* The data known are the volumetric flow rate $\dot{V}$, the length of the flow conduit including any equivalent length due to fittings $\Delta L$, the hydraulic diameter $D$ or $D_e$, the roughness $\varepsilon$, and the viscosity $\eta$ and density $\rho$ of the fluid. The primary variable to be calculated is the pressure drop $\Delta p$ through the energy friction loss per time

$\hat{\mathscr{F}}$, or per volume $\hat{\mathscr{F}} = \mathscr{F}/\dot{V}$, or per mass $\mathscr{F} = \hat{\mathscr{F}}/\rho$. First the Reynolds number,

$$\text{Re} = \frac{\rho \bar{u} D_e}{\eta} = \frac{\rho \bar{u}}{\eta} \frac{4A}{\Pi}, \tag{4.39}$$

and the relative roughness $\varepsilon/D$ are computed. Then the friction factor $f_F$ is found either by means of Eqns. (4.32) to (4.34) or, most commonly, from the diagram of Fig. 4.3. Then the calculation of $\mathscr{F}$ is done by means of Eqn. (4.11). Once $\mathscr{F}$ is calculated, other quantities, such as $\hat{\mathscr{F}}$ and $\hat{\mathscr{F}}$, the pressure drop, $\Delta p$, by Eqn. (4.4) or (4.37), and the required power, $P = \dot{V} \Delta p$, to move the fluid are calculated.

**II.** *Indirect problems in specified geometry.* The known data in this more difficult case are $\mathscr{F}$ or $\Delta p$, $\Delta L$, $D$, $\rho$, $\eta$, and $\varepsilon$. The primary variable to be calculated is the average velocity $\bar{u}$ or the volumetric flow rate, $\dot{V} = A\bar{u}D^2/4$. Unlike *problem I*, the friction factor $f_F$ cannot be found directly, because although the relative roughness $\varepsilon/D$ is known, the Reynolds number is unknown. The solution proceeds *iteratively* as follows: A value of the friction factor $f_F$ is assumed in the curve of known $\varepsilon/D$ of Fig. 4.3, preferably at a high Reynolds number, where the curve is almost flat and $f_F$ independent of Re, unless it somehow is known or strongly anticipated that the Reynolds number must be low. This guessed value $f_F^{(0)}$ is used in Eqn. (4.11) and an average velocity $\bar{u}^{(0)}$ is calculated. Then a Reynolds number based on $\bar{u}^{(0)}$ is computed. This number and the known $\varepsilon/D$ determine a new value $f_F^{(1)}$ for the friction factor. If $f_F^{(1)} = f_F^{(0)}$, calculations terminate and the value of the mean velocity sought is $\bar{u} = \bar{u}^{(0)}$. If $f_F^{(1)} \neq f_F^{(0)}$, the cycle is repeated by using $f_F^{(1)}$ in Eqn. (4.11), computing a new value $\bar{u}^{(1)}$ for the average velocity and a new Reynolds number which together with the fixed $\varepsilon/D$ determined a new $f_F^{(2)}$ in Fig. 4.3. The cycle is repeated until $f_F^{(n+1)} \simeq f_F^{(n)}$, where $n$ is number of cycles or iterations. The velocity sought is $\bar{u} = \bar{u}^{(n)}$, the volumetric flow rate is $\dot{V} = A\bar{u}$, the required power is $P = \dot{V} \Delta p = \dot{V}\rho\mathscr{F} = \dot{V}\hat{\mathscr{F}}$, and so on.

**III.** *Indirect problems under specified flow conditions.* In this case, the known data are $\mathscr{F}$ or $\Delta p$, $\dot{V}$, $\Delta L$, $\rho$, $\eta$, $\varepsilon$, and the primary variable sought is the diameter $D$, or equivalent diameter $D_e$. This is the most difficult type of friction problem, because neither the relative roughness nor the Reynolds number can be computed directly. However, the volumetric flow rate expression, $\dot{V} = \pi\bar{u}D^2/4$, is replaced in Eqn. (4.11) to yield

$$\mathscr{F} = 4f_F \left(\frac{\Delta L}{D}\right) \frac{\dot{V}^2}{2(\pi D^2/4)^2}, \tag{4.40}$$

and therefore,

$$D^5 = \frac{32(\Delta L)\dot{V}^2}{\pi^2 \mathscr{F}} d_F = c_1 f_F, \tag{4.41}$$

where $c_1$ is known. Also,

$$\text{Re} = \frac{\rho \bar{u} D}{\eta} = \frac{4\dot{V}\rho}{\pi\eta D} = \frac{c_2}{D}, \tag{4.42}$$

with $c_2$ also known. Given Eqns. (4.41) and (4.42), the solution proceeds as follows: An initial value $f_F^{(0)}$ for the friction coefficient, preferably independent of Reynolds number, is guessed. A diameter $D^{(0)}$ is calculated by means of Eqn. (4.41) and then a $\text{Re}^{(0)}$ by Eqn. (4.42) and a relative roughness $\varepsilon/D^{(0)}$. From these values a new $f_F^{(1)}$ is computed from Fig. 4.3. If $f_F^{(1)} = f_F^{(0)}$, calculations terminate, the diameter value sought is $D = D^{(0)}$, and other secondary variables are calculated accordingly. If $f_F^{(1)} \neq f_F^{(0)}$, a new $D^{(1)}$ is computed by Eqn. (4.41), then a new $\text{Re}^{(1)}$ by Eqn. (4.42) and a new roughness $\varepsilon/D^{(1)}$, and then a new $f_F^{(2)}$ from Fig. 4.3. The cycle or iteration is repeated until $f_F^{(n+1)} \simeq f_F^{(n)}$, where $n$ is the iteration number, in which case the diameter value sought is $D = D^{(n)}$, and other secondary variables are calculated accordingly based on $D^{(n)}$.

Example 4.5 demonstrates the numerical procedure for the three types of friction problems in confined flows.

### Example 4.5: Three Distinct Types of Problems on Friction

*I. Direct problem.* Calculate the frictional losses per unit length of a smooth pipe of 5 cm diameter, transferring oil of density 0.98 g/cm³, and viscosity 10 cP, at a flow rate 0.01 m³/s.

*II. Indirect problem—fixed geometry.* The pressure drop per unit length of a horizontal pipe of 5 cm diameter transferring oil of 0.98 g/cm³ density and 10 cP viscosity is 0.19 atm/m. What flow rate is achieved?

*III. Indirect problem—fixed flow rate.* A commercial pipe is required to transfer oil of density 0.98 g/cm³ and viscosity 10 cP at a flow rate of 0.01 m³/s over a distance of 5 km, horizontally. If the maximum available pumping capacity is 1000 kJ per flowing kg, what pipe diameter is required?

**Solution**    *Problem I:*

$$\frac{\mathscr{F}}{\Delta L} = 4f_F \frac{\bar{u}^2}{2D} = \frac{32 f_F \dot{V}^2}{\pi D^5}. \tag{4.5.1}$$

The Reynolds number is

$$\text{Re} = \frac{\rho \bar{u} D}{\eta} = \frac{4\dot{V}\rho}{\eta \pi D} = \frac{4 \times 980 \times 0.01}{0.01 \times 3.14 \times 0.05} = 2.50 \times 10^4, \tag{4.5.2}$$

and the corresponding friction factor from Fig. 4.3 is $f_F = 0.006$. Thus

$$\frac{\mathscr{F}}{\Delta L} = \frac{32 \times 0.006 \times 10^{-4}}{3.14 \times 0.05^5} \frac{\text{m}}{\text{s}^2} = 19.56 \frac{\text{m}^2/\text{s}^2}{\text{m}} = 19.56 \frac{(\text{J/m})}{\text{kg}}$$

(i.e., there is 19.56 J energy loss per meter of pipe per kg of flowing oil.) The power loss per meter of pipe length is,

$$\frac{\dot{\mathscr{F}}}{\Delta L} = \frac{\mathscr{F}\dot{m}}{\Delta L} = (19.56 \times 980 \times 0.01) \frac{\text{W}}{\text{m}} = 0.19 \text{ kW/m}.$$

*Problem II:*

$$\frac{\mathscr{F}}{\Delta L} = \frac{1}{p} \frac{\Delta p}{\Delta L} = \frac{0.19 \times 1.013 \times 10^5}{980} \frac{\text{N/m}^2}{\text{kg/s}} = 19.63 \text{ m/s}^2. \tag{4.5.3}$$

A friction factor value $f_F^{(0)} = 0.0045$ along the curve for smooth pipe in Fig. 4.3 is guessed where the Reynolds number is $\mathrm{Re}^{(0)} = 10^5$. The resulting average velocity is given by Eqn. (4.11) as

$$\bar{u}^{(0)} = \left[\left(\frac{\mathscr{F}}{\Delta L}\right)\frac{D}{2f_F}\right]^{1/2} = \frac{19.63 \times 0.05}{0.009}\,\frac{\mathrm{m}}{\mathrm{s}} = 10.44\ \mathrm{m/s}. \tag{4.5.4}$$

The resulting Reynolds value is

$$\mathrm{Re}^{(1)} = \frac{\rho \bar{u}^{(0)} D}{\eta} = \frac{980 \times 10.44 \times 0.05}{0.01} = 51{,}156\ \mathrm{Re}^{(0)}.$$

The new friction factor value corresponding to $\mathrm{Re}^{(1)}$ is, from Fig. 4.3, $f_F = 0.00525$, and the resulting velocity is $\bar{u}^{(1)} = 9.66\ \mathrm{m/s}$, and

$$\mathrm{Re}^{(2)} = \frac{980 \times 9.66 \times 0.005}{0.01} = 47{,}361 \neq \mathrm{Re}^{(1)},$$

for which $f_F^{(2)} = 0.0055$ and $\bar{u}^{(2)} = 9.43\ \mathrm{m/s}$. The cycle continues with new values $\mathrm{Re}^{(3)} = 46{,}233$, $f_F^{(3)} = 0.0055$, and $\bar{u}^{(3)} = 9.4\ \mathrm{m/s}$. Since $\bar{u}^{(3)} = \bar{u}^{(2)}$, the cycle terminates with average velocity $\bar{u} = \bar{u}^{(3)} = 9.4\ \mathrm{m/s}$, for which the flow rate is

$$\dot{V} = \frac{\pi D^2 \bar{u}}{4} = \frac{3.14 \times 0.005^2 \times 9.4}{4}\,\frac{\mathrm{m}^3}{\mathrm{s}} = 0.018\ \mathrm{m^3/s}. \tag{4.5.5}$$

*Problem III*: The maximum power loss that can be overcome by the pumping facilities is $\mathscr{F} = 1000\ \mathrm{kW}$. Equations (4.41) and (4.42) for this case become

$$D = 0.277 f_F^{1/5} \tag{4.5.6}$$

and

$$\mathrm{Re} = \frac{1274}{D}. \tag{4.5.7}$$

Assume initially that $f_F^{(0)} = 0.006$, in which case Eqn. (4.5.6) yields $D^{(0)} = 0.099\ \mathrm{m}$ and Eqn. (4.5.7) yields $\mathrm{Re}^{(0)} = 12{,}795$. The new friction factor value is $f_F^{(1)} = 0.00725$, which results in $D^{(1)} = 0.103\ \mathrm{m}$ and $\mathrm{Re}^{(1)} = 12{,}368$. The new value of friction factor is $f_F^{(2)} = 0.0073$, which results in $D^{(2)} = 0.103\ \mathrm{m}$. Since $D^{(2)} = D^{(1)}$, the diameter required is $D = 10.3\ \mathrm{cm}$.

## 4.6 FRICTION AND RESISTANCE IN FREE-STREAM FLOWS

Confined flows in pipes and channel flows under gravity that have been considered so far belong to one of two fundamental forms of flow. The other extreme are free-stream flows overtaking submerged flows (e.g., wind flow around buildings), or equivalently, fluid flow induced by bodies traveling in stationary fluids (e.g., vehicles traveling in air and ships and submarines in the ocean). Other applications of interest to process engineering include small-scale flows around solid spheres, liquid droplets and gas bubbles, which are important in air pollution and its control, and in heterogeneous chemical reactions and mass and heat transfer processes. One of the most important variables in these situations is the resisting force that the body poses to flow, or the *hydrodynamic drag force* that

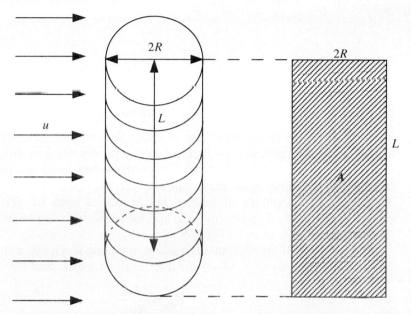

**Figure 4.4**  Free stream of approaching velocity $u$ overtaking a cylinder of radius $R$ and height $L$ that results in projected area $A = 2RL$.

the fluid exerts on the body. The prototype flow is shown in Fig. 4.4, which may represent a stationary cylinder overtaken by a fluid stream approaching with velocity $u = \bar{u}$, or else a cylinder traveling parallel to itself with velocity $u$ in a stationary fluid. Indeed, the two situations are fully equivalent, with respect to an observer that rides the cylinder in both cases. In the former case an external force opposite to the drag force is needed to hold the cylinder in place, whereas in the latter an external force is needed to maintain the steady translation.

In accordance with Eqn. (4.8), the resulting drag force along the direction of flow, $x$, is

$$F_D = F_x = \int_S \tau_{rx}\big|_{r=R} \cos\theta \, dS = \int_S \tau_{rx}\big|_{r=R} \, dA = \int_A f\rho \frac{u^2}{2} \, dA = C_D \rho \frac{u^2}{2} A,$$

$$(4.43)$$

where $F_D$ is the *drag force* opposing the relative motion of the body with respect to the fluid, $C_D$ a *drag coefficient* that in general depends on the Reynolds number, $u$ the *approach velocity* of the free stream, and $A$ the *projected area* of the body to the flow direction, as shown in Fig. 4.4. For these cases, since the approaching stream has a uniform, plug-like velocity, the average velocity, $\bar{u}$, is identical to the velocity $u$. Note also that the use of $C_D$ removes the need to solve for the complicated flow around the object in order to evaluate the integrals.

Values of the drag coefficient for a submerged sphere are given in Fig. 4.5, in terms of the Reynolds number, $\text{Re} = \rho D u / \eta$, where $\rho$ and $\eta$ are the density and viscosity of the fluid, $u$ the approach velocity, and $D$ the diameter of the sphere. Figure 4.5 shows that for small values of Reynolds number, $\text{Re} < 1$, the drag coefficient can be expressed in terms of the Reynolds number as

$$C_D = \frac{24}{\text{Re}}, \tag{4.44}$$

which, replaced in Eqn. (4.43), results in

$$F_D = 6\pi R \eta u, \tag{4.45}$$

which is known as *Stokes' law*[13] for *creeping flow* around submerged sphere. By analyzing the flow around the sphere, it can be shown that $\frac{1}{3}$ of this drag force is a *form drag* coming from normal stresses on the sphere, and the remaining $\frac{2}{3}$ is a *friction drag* coming from shear stresses on the sphere. Values of the drag coefficient for a cylinder of infinite height placed with its axis of symmetry perpendicular to the direction of flow, and for a disk with its planar base opposing the stream, are also given in Fig. 4.5.

In all cases of Fig. 3.5, there is a region of low Reynolds numbers, $\text{Re} < 1$, where $C_D$ is given by Eqn. (4.44) for a sphere and a disk, and by

$$C_D \simeq \frac{7}{\text{Re}} \tag{4.46}$$

for a cylinder. Then there is a region, $1 < \text{Re} < 10^3$, where the drag coefficient decreases sublinearly with the Reynolds number, and in the region $10^3 < \text{Re} < 1 \times 10^5$, corresponding to high turbulence, the drag coefficient becomes practically independent of the Reynolds number. At about $\text{Re} = 2 \times 10^5$ the drag coefficient decreases abruptly due to a *boundary layer separation* in the form of vortices carried by the fluid downstream, removing a significant amount of shear stress from the back of the body. These qualitative features are helpful in providing a good initial guess for the drag coefficient to indirect problems requiring trial and error, as detailed in Section 4.8. The diagrams of Fig. 4.5 hold for velocities less than about half the local speed of sound in the same fluid. Information for higher velocities is given elsewhere.[14] Other limitations and suggested ways to overcome them are discussed by Churchill.[15]

## 4.7 ADDITIONAL CONSIDERATIONS ON FRICTION IN FREE-STREAM FLOWS

### 4.7.1 Liquid Droplets and Gas Bubbles

Drag around liquid droplets of viscosity $\eta_2$ submerged in a fluid of viscosity $\eta_1$ is reduced, compared to those on solid bodies, and the resulting force for creeping flow of $\text{Re} = \rho u D / \eta_1 < 1$ is given by[16]

$$F_D = 4\pi \eta_1 R u \frac{\eta_1 + 1.5\eta_2}{\eta_1 + \eta_2}. \tag{4.47}$$

In the limit of $\eta_1/\eta_2 \to 0$, Eqn. (4.47) reduces to Eqn. (4.45), as it should, since

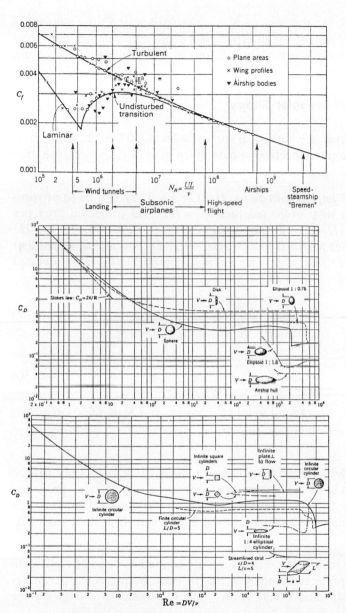

**Figure 4.5** Drag coefficient $C_f$ for plates and $C_D$ for spheres, cylinders, and disks overtaken by free streams of fluids. [Adapted from *NACA Tech. Mem. 1218*, 117 (1949); L. Prandtl, *Ergebnisse de aerodynamischen Versuchsanstalt zu Cöttingen*, R. Oldenbourg, Munich and Berlin, 1923; F. Eisner, *Das Widerstands-problem*, *Proc. 3rd Internat. Congr. Appl. Mech.*, 1930; *NACA Rept. 619*, 1938, and *Rept. 137*, 1972.]

the viscosity of solids is infinite; and in the limit of $\eta_1/\eta_2 \rightarrow \infty$, the same equation reduces to

$$F_D = 4\pi\eta_1 Ru, \tag{4.48}$$

the drag force on a gas bubble of vanishing small viscosity, $\eta_2 \rightarrow 0$, submerged in a fluid of finite viscosity, $\eta_1 \gg \eta_2$. The difference in drag force, depicted by Eqns. (4.45), (4.47), and (4.48), are due to different boundary conditions that apply at the interface of contact between the spherical particle, droplet, of bubble and its fluid surroundings. At the surface of a solid particle the fluid velocity is identical to that of the particle, at the interface of a droplet of different liquid the inner and outer fluid velocities are identical, and at the free surface of a gas bubble the shear stress is zero.

### 4.7.2. Drag on Bodies of Irregular Shape

Not all bodies or particles are spherical or cylindrical. When a problem is encountered with no obvious connection to data given here or elsewhere, the drag coefficient and resulting drag force can be approximated as follows:

**I.** *Slender bodies of large aspect ratio, $H/D > 10$.* They can be approximated as long cylinders of total projected area $A = HD$ and projected area per diameter height $\hat{A} = A/(H/D) = D^2$, of total perimeter $\Pi = 2(H + D) \simeq 2H$ and perimeter per diameter height $\hat{\Pi} = 4D$, $2(\hat{H} + D)/D = 4$, and of equivalent diameter, $D_e = 4\hat{A}/\hat{\Pi} = D^2/4D = D$. The Reynolds number is therefore defined by

$$\text{Re} = \frac{\rho u D_e}{\eta}, \tag{4.49}$$

and the drag force by

$$F_D = \sum_{i=1}^{H/D} (C_D)_i \frac{\rho u_i^2}{2} A, \tag{4.50}$$

where $(C_D)_i$ is found from Fig. 4.5 for cylinders. Notice that Eqn. (4.50) may also apply to cases of different plug-like velocity $u_i$ over different portions $A_i$ of projected area.

**II.** *Nonslender bodies of aspect ratio nearly unity, $H/D \simeq 1$.* They can be approximated as spheroids of equivalent diameter $D_e = \sqrt{D_1 D_2}$ and projected area $A = \pi D_e^2/4$, where $D_1$ is a minimum and $D_2$ a maximum diameter, and perimeter $\Pi = \pi D_e$. Equations (4.43) and (4.49) apply, with $C_D$ found from Fig. 4.5 for spheres.

**III.** *Disks of aspect ratio $H/D \ll 1$.* These are treated as disks of equivalent diameter $D_e = \sqrt{D_1 D_2}$, projected area $A = \pi D_e^2/4$, and the projected perimeter $\Pi = \pi D_e$. Equations (4.43) and (4.49) apply, with $C_D$ found from Fig. 4.5 for disks.

A more efficient way to calculate drag with irregular shapes is by means of shape factors that determine the deviation of a shape from that of a sphere, as detailed elsewhere.[17]

### Example 4.6: Drag Force on Needle-, Sphere-, and Disk-like Particles

Calculate the required horizontal force in order to keep in place the following objects, with their axis of symmetry placed perpendicularly to overtaking a stream of water at approach velocity 1 m/s: (a) sphere of 0.20 m diameter; (b) cylinder of aspect ratio $L/D = 10$ and of surface area identical to that of the sphere; (c) cylinder of aspect ratio $L/D = 0.1$, of surface area identical to that of the sphere, placed with its axis of symmetry parallel to the direction of the stream. (d) What external force would be required to translate the same objects steadily in stationary water?

**Solution**

**(a)** *Sphere or spheroid.* The Reynolds number is

$$\text{Re} = \frac{\rho u D}{\eta} = \frac{1200 \times 1 \times 0.2}{0.001} = 2 \times 10^5.$$

The resulting drag coefficient from Fig. 4.5 for spheres is $C_D = 0.4$, and therefore the drag force is

$$F_D = C_D \frac{\rho u^2}{2} \frac{\pi D^2}{4} = (0.4 \times 1000 \times 0.5 \times 3.14 \times 0.01) \, \text{N} = 6.28 \, \text{N}.$$

**(b)** *Long cylinder or needle.* According to the problem statement, the diameter of the cylinder, $D_c$, is given by

$$\frac{2\pi D_c^2}{4} + \pi D_c(10 D_c) = \pi D^2,$$

which yields

$$D_c = \frac{D}{\sqrt{10.5}} = 0.062 \, \text{m}$$

and

$$L = 10 D_c = 0.62 \, \text{m}.$$

The Reynolds number is

$$\text{Re} = \frac{\rho u D_c}{\eta} = \frac{1000 \times 1 \times 0.062}{0.001} = 6.2 \times 10^4,$$

and the drag coefficient is, from Fig. 4.5 for cylinders, $C_D = 1.2$. The force required is

$$F_D = C_D \frac{\rho u^2}{2} L D_c = (1.2 \times 500 \times 10 \times 0.062^2) \, \text{N} = 23.1 \, \text{N}.$$

**(c)** *Short cylinder or disk.* According to the problem statement, it turns out that $D_c = 0.258 \, \text{m}$ and $L = 0.0258 \, \text{m}$. The Reynolds number is

$$\text{Re} = \frac{\rho D_c u}{\eta} = \frac{1000 \times 0.258 \times 1}{0.001} = 2.58 \times 10^5.$$

The resulting drag coefficient from Fig. 4.5 for disks is $C_D = 1.17$. Thus the required force in this case is

$$F_D = C_D \frac{\rho u^2}{2} \frac{\pi D_c^2}{4} = \left(1.17 \times 500 \times 3.14 \times \frac{0.258^2}{4}\right) \text{N} = 30.61 \, \text{N}.$$

The same forces are required in order to translate the same objects steadily in stationary water.

The overall conclusion here is that friction and drag force on bodies moving with respect to fluids are drastically reduced by surfaces that allow smooth deviation of streamlines upon overtaking the object to all possible directions. Such surfaces are called *aerodynamic* and are taken into severe consideration in designing automobiles, aircrafts, rockets, submarines, and others.

## 4.8 PROBLEMS ON DRAG IN FREE-STREAM FLOWS

As in the case of friction in confined geometries, there are three distinct types of problems on drag in free-stream flow, as summarized below.

**I.** *Direct problem with specified geometry and flow conditions.* The known data are the approach fluid velocity $u$ or the traveling velocity of the body in a stationary fluid, the shape of the body that defines its equivalent diameter $D_e$ and projected area $A$ to the flow direction, and the viscosity and density of the fluid. The drag coefficient is determined from Fig. 4.5 and from tables of the type of Table 3.6, given the Reynolds number $\text{Re} = \rho u D_e / \eta$. Then the drag force on the body is given by Eqn. (4.45), which can be used in conjunction with Newton's law of motion to study how bodies and particles travel in fluids, as demonstrated in Examples 4.7 and 4.8. Problems of this type are important in estimating wind forces on structures,[18] and in particle mechanics, for example to control the motion of pollutants in air.[19]

**II.** *Indirect problems with fixed-shape bodies.* The known data are the resulting drag force $F_D$ (or description of body motion that can be used to induce $F_D$), the equivalent diameter $D$ and shape of the body, and the density and viscosity of the fluid. The primary unknown variable is the approach velocity, $u$. First a value of the drag coefficient $C_D^{(0)}$ is guessed, preferably in the high Reynolds number region of Fig. 4.5, where $C_D$ is independent of the Reynolds number, and a velocity $u^{(0)}$ is estimated by Eqn. (4.45). The value of the corresponding $\text{Re}^{(0)}$ is then calculated, which is used to find a new value, $C_D^{(1)}$, by means of Fig. 4.5. If $C_D^{(1)} = C_D^{(0)}$, calculations terminate and the value of the velocity sought is $u = u^{(0)}$. Otherwise, a new value, $u^{(1)}$, is calculated by Eqn. (4.45) with the new $C_D^{(1)}$ and a new $\text{Re}^{(1)}$ is computed, and then a new value $C_D^{(2)}$ is found from Fig. 4.5. The cycle or iteration continues until $C_D^{(n+1)} = C_D^{(n)}$, in which case the velocity sought is $u = u^{(n)}$. Problems of this kind are important when dealing with devices for cleaning particles and droplets from industrial gases, to determine the velocity of the carrier gas stream to achieve separation and collection of particles and droplets of certain size and shape.[20,21]

**III.** *Indirect problems with unknown body shape.* The known data are the drag force $F_D$, the fluid properties $\rho$ and $\eta$, and the approach velocity $u$. The primary variable to be determined is the shape of the body such that the force induced is $F_D$ under fluid velocity $u$. Problems of this kind are important in aerodynamics,[22] where optimum vehicle designs are sought to minimize drag that resists motion and affects fuel efficiency. Equation (4.43) is first rearranged in the form

$$\hat{F}_D = \frac{F_D}{D} = C_D \frac{\rho u^2}{2} \frac{A}{D} = C_D \frac{\rho u^2}{2} \frac{\Pi D}{4D} = C_D \frac{\rho u^2}{2} \frac{\Pi}{4}, \qquad (4.51)$$

and therefore,

$$\Pi = \frac{\hat{F}_D}{\rho u^2} \frac{2}{C_D} = \frac{K_1}{C_D}, \qquad (4.52)$$

where $K_1$ is known, $\hat{F}_D$ is force per height equal to the equivalent diameter, and $\Pi$ is the perimeter of the area $A/D$. Also,

$$\text{Re} = \frac{\rho u D}{\eta}. \qquad (4.53)$$

First the initial value, $C_D^{(0)}$, is estimated, preferably in the high Reynolds number region of Fig. 4.5, and $D^{(0)}$ is estimated by Eqn. (4.52). The corresponding $\text{Re}^{(0)}$ is then found by Eqn. (4.53), and for this Reynolds number, a new value, $C_D^{(1)}$, is found from Fig. 4.5. If $C_D^{(1)} = C_D^{(0)}$, calculations terminate; otherwise, the iteration proceeds as detailed for problem II. These calculations are accurate for spheres and cylinders of large height $H = nD$ with $n > 10$, where $D$ is its diameter, and approximate for other nearly spherical or cylindrical bodies of nonslender cross-sectional areas.

A class I problem was examined in Example 4.7, a class II problem is studied in Example 4.7 and a class III problem is highlighted in Example 4.8. More examples on drag in free-stream flows include Examples 4.9 and 4.10.

### Example 4.7: Translational Velocity of Sphere in Water

A buoyancy neutral sphere of density $1 \text{ g/cm}^3$ and diameter 30 cm travels steadily in still water, pulled by an external horizontal force of 1 N. What is the horizontal velocity of the sphere?

**Solution**   Since the sphere is moving steadily, the driving external force must be identical to the drag force. Thus

$$F_D = C_D \frac{\rho u^2}{2} \frac{\pi D^2}{4},$$

and therefore,

$$u = \left( \frac{8 F_D}{\pi D^2 \rho C_D} \right)^{1/2} = \left( \frac{8 \times 1}{3.14 \times 0.09 \times 1000} \right)^{1/2} C_D^{-1/2} = \frac{0.168}{\sqrt{C_D}}.$$

The corresponding Reynolds number is

$$\text{Re} = \frac{\rho u D}{\eta} = \frac{1000 \times 0.3 \times u}{0.001} = 3 \times 10^5 u.$$

Assume that $C_D^{(0)} = 0.4$; then $u^{(0)} = 0.168/\sqrt{0.4} = 0.27 \text{ m/s}$ and $\text{Re}^{(0)} = 8.0 \times 10^4$, and for this value Fig. 4.5 yields $C_D^{(1)} = 0.5$. Then by the same order, $u^{(1)} = 0.237 \text{ m/s}$ and $\text{Re}^{(1)} = 7.11 \times 10^4$, and from Fig. 4.5, $C_D^{(2)} = 0.5$. Since $C_D^{(2)} = C_D^{(1)}$, the translational velocity sought is $u = 0.237 \text{ m/s}$.

**Example 4.8: Terminal Velocity**

In an air-cleaning device, solid particles of density 1.5 g/cm³ settle steadily and are required to be collected within 5 s from the moment they enter the top of the 5-m-tall device. What is the average diameter of particles that can be collected if the air density is 1 kg/m³?

**Solution**  For steady vertical settling, the gravity force which is identical to the weight of the particle must be identical to the opposing drag and buoyancy forces:

$$\frac{4}{3}\pi\frac{D^3}{8}\rho_p g = C_D \frac{\rho_{air}u^2}{2}\frac{\pi D^2}{4} + \frac{4}{3}\pi\frac{D^3}{8}\rho_{air}g. \tag{4.8.1}$$

Thus, the steady settling velocity, also called the *terminal velocity*, is

$$u = \left[\frac{4\pi Dg}{3C_d}\left(\frac{\rho_p}{\rho_{air}} - 1\right)\right]^{1/2}, \tag{4.8.2}$$

and therefore

$$D = \frac{3u^2 C_D}{4\pi g}\left(\frac{\rho_{air}}{\rho_p - \rho_{air}}\right) = \frac{3 \times (5/15)^2 \times [1/(1.5 - 1)]}{4 \times 3.14 \times 9.81}C_D = 0.0054C_D. \tag{4.8.3}$$

The Reynolds number is

$$\text{Re} = \frac{\rho Du}{\eta} = 10^4 D. \tag{4.8.4}$$

Assume that $C_D^{(0)} = 1$; then Eqn. (4.8.3) yields $D^{(0)} = 0.049$ m and Eqn. (4.8.4) yields $\text{Re}^{(0)} = 4410$. For this value, Fig. 4.5 indicates that $C_D^{(1)} \approx 0.6$. Following the same order, new values are obtained as $D^{(1)} = 0.030$ m, $\text{Re}^{(1)} = 2700 C_D^{(2)} \approx 0.6$, and again, $D^{(2)} = 0.007$ m, $\text{Re}^{(2)} = 70$, $c_D^{(3)} = 1.4$; then, repeating the cycle, $D^{(3)} = 0.0076$ m, $\text{Re}^{(3)} = 76$, for which Fig. 4.5 indicates $C_D^{(4)} - 1.4$. Thus $C_D = 0.6$, and the particle diameter is approximately

$$D = 0.0054C_D = 0.0076 \text{ m} = 7.6 \text{ mm}.$$

Obviously, judging by the form of Eqn. (4.8.2), all particles of diameter bigger than 7.6 mm will be collected at the prevailing Reynolds number.

## 4.9 ELEMENTARY PARTICLE MECHANICS

The motion of small particles in fluids[23] is governed by Newton's law of motion as stated in Chapter 3:

$$\rho\frac{4}{3}\pi R^3\frac{du_z}{dt} = (\rho_a - \rho_b)g\frac{4}{3}\pi R^3 - F_{drag} + F_{ext} \tag{4.54}$$

$$\rho\frac{4}{3}\pi R^3\frac{du_x}{dt} = F_{drag} + F_{ext}. \tag{4.55}$$

The left-hand side is the acceleration of sphere-like particles of equivalent radius $R$, and velocities $u_x$ and $u_z$, in the horizontal and vertical directions, respectively. The right-hand side includes the driving forces, which are the buoyancy and the

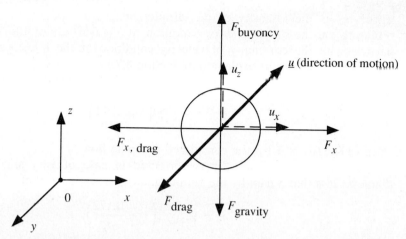

**Figure 4.6**  Composite motion of a particle in the atmosphere under the action of vertical and horizontal forces.

gravity in the vertical direction and the drag force according to Stokes' law in both directions. Under the combined action of these forces the particle will move diagonally as shown by Fig. 4.6. In Eqns. (4.54) and (4.55): $\rho_p$ and $\rho_a$ are the densities of the particle and the air, respectively; $\eta$ is the viscosity of the air; $A_i$ is the projected area on the direction of motion; and $C_d$ is the friction coefficient, which depends on the Reynolds number as shown by Fig. 4.5. Equation (4.54) alone holds for strictly vertical motions and Eqn. (4.55) alone holds for strictly horizontal motions. Horizontal motion may be induced by an initial horizontal *release velocity,* by horizontally blowing wind, and by a horizontal *electrostatic field.*

The left-hand sides of Eqns. (4.54) and (4.55) persist for short times. Soon the motion reaches a steady terminal velocity which is obtained by setting $du_x/dt = du_z/dt = 0$ in these equations. The *terminal velocities* predicted by Eqs. (4.54) and (4.55) are

$$u_z = \left[ \frac{\frac{8}{3}\pi R^3 (\rho_p - \rho_a)g}{A_z C_d \rho_a} + \frac{2F_z}{A_z C_d \rho_a} \right]^{1/2} \qquad (4.56)$$

and

$$u_x = \left( \frac{2F_x}{A_x C_d \rho_a} \right)^{1/2}, \qquad (4.57)$$

where $A_z$ and $A_x$ are the projected surfaces of the spheroid in the $z$- and $x$-directions, respectively, $C_d$ is the friction coefficient, and $F_z$ and $F_x$ are any applied external forces (e.g., *electrostatic* or *magnetic forces*).

Equations (4.54) to (4.55) describe the motion of spherical droplets of liquids in the atmosphere by replacing the viscosity $\eta$ in the drag force by

$$n = \frac{6(\eta + 1.5\eta_d)}{4(\eta + \eta_d)} \qquad (4.58)$$

where $\eta_d$ is the viscosity of the droplet. The correction is introduced when replacing the no-slip boundary condition at the surface of the solid spherical particle with the continuity of velocity condition at the interface of the liquid droplet with the air, as explained in Section 4.7.1.

Stokes' drag coefficient is improved slightly by Oseen's expression,

$$C_D^{(Os)} = C_D^{(St)}\left(1 + \frac{3Re}{16}\right), \tag{4.59}$$

since $(3\,Re/16) \ll 1$ for the considered creeping flow.

Stokes' terminal velocity is corrected in case of very small particles, of diameter less that 5 $\mu$m, by the Cunningham's factor, such that

$$u_t = V_{t,\text{Stokes}}\left(1 + \frac{0.172}{d_p}\right), \tag{4.60}$$

where $d_p$ is diameter in micrometers, to account for the partial slip of the solid molecule-like particle past the air molecules.[19]

### Example 4.9: Terminal Velocity and Stopping Distance

*Particle acceleration and deceleration.* A spherical particle of diameter $d_p$ and density $\rho_p$ starts traveling in air of density $\rho_a$ and viscosity $\eta$ under the influence of an external constant force $F$: gravity, buoyancy, centrifugal, or electrostatic. Describe the motion from $t = 0$ to $t > 0$; then to $t = t_\infty$, where the particle stops if its initial velocity at $t = 0$ was $u_0$.

**Solution**   The governing equation of motion is

$$\rho_p \frac{4}{3}\pi \frac{d_p^3}{8}\frac{du}{dt} = F - 6\pi\eta u \frac{d_p}{2}, \tag{4.9.1}$$

which is integrated to

$$u = \frac{F}{3\pi\eta d_p}\left[1 - \exp\left(-\frac{\rho_p d_p^2}{18\eta}t\right) + u_0\exp\left(-\frac{\rho_p d_p^2}{18\eta}t\right)\right]. \tag{4.9.2}$$

Thus the motion is due to an initial velocity $u_0$, which decays exponentially with time due to the resisting air drag that opposes the driving external force which conserves the motion: If $F = 0$, then $u = 0$ after some time, no matter what $u_0$ is; if $u_0 = 0$, the liquid accelerates and approaches a steady velocity after some time. This behavior is shown by Fig. E4.9a.

To transform time $t$ to traveling distance $x$, the simple kinematic argument

$$dx = u\,dt \tag{4.9.3}$$

is utilized. Then Eqn. (4.9.1) transforms to

$$\rho_p \frac{4}{3}\pi \frac{d_p^3}{8}u\frac{du}{dx} = F - 6\pi\eta u \frac{dp}{2}. \tag{4.9.4}$$

The solution of Eqn. (4.9.5) is given by

$$\left[u + \frac{F}{3\pi\eta d_p}\ln(F - 3\pi\eta d_p u)\right]_{u_0}^{u} = -\frac{18\eta}{\rho_p d_p^2}x. \tag{4.9.5}$$

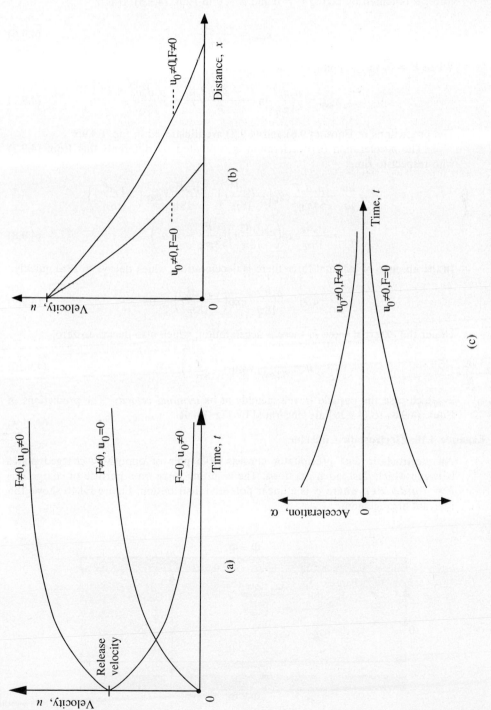

**Figure E4.9** Motion of spherical particle in air released at initial velocity $u_0$, accelerated by constant external force $F$, and decelerated by viscous drag force: (a) velocity vs. time; (b) velocity vs. distance; (c) acceleration vs. time.

The *stopping distance* of a particle released with initial velocity $u_0$, under no external force, is obtained by taking $F = 0$ and $u = 0$ in Eqn. (4.9.5). It is

$$x_{\text{stop}} = \frac{u_0 \rho_p d_p^2}{18\eta}.$$

(4.9.6)

When $F \neq 0$, the stopping distance is given by

$$x_{\text{stop}} = \frac{\rho_p d_p F}{54\pi\eta^2} \ln \frac{F - 3\pi\eta d_p u_0}{F} + \frac{\rho_p d_p^2 u_0}{18\eta}.$$

(4.9.7)

The predictions of Eqns. (4.9.6) and (4.9.7) are illustrated in Fig. E4.9b.

The acceleration or deceleration is estimated by differentiating Eqn. (4.9.2) with respect to time:

$$a = \frac{du}{dt} = \frac{\rho_p d_p F}{54\pi\eta^2} \exp\left(-\frac{\rho_p d_p^2}{18\eta} t\right) - \frac{\rho_p d_p^2 u_0}{18\eta} \exp\left(-\frac{\rho_p d_p^2}{18\eta} t\right)$$

$$= \frac{\rho_p d_p}{18\eta} \exp\left(-\frac{\rho_p d_p^2}{18\eta} t\right)\left(\frac{F}{3\pi\eta} - d_p u_0\right).$$

(4.9.8)

In the absence of external force there is deceleration, which decays to zero quickly:

$$a = -\frac{\rho_p d_p^2 u_0}{18\eta} \exp\left(-\frac{\rho_p d_p^2}{18\eta} t\right) < 0.$$

(4.9.9)

Under the external force $F$, there is acceleration, which also decays to zero,

$$a = \frac{\rho_p d_p F}{54\pi\eta^2} \exp\left(-\frac{\rho_p d_p^2}{18\eta} t\right) > 0,$$

(4.9.10)

in which case the particle travels steadily at its *terminal velocity*. The predictions of Eqns. (4.9.8) to (4.9.10) are illustrated by Fig. E4.9c.

### Example 4.10: Electrostatic Collector

An electrostatic dust precipitator consists of a pair of oppositely charged plates between which dust-laden gas flows. The resulting force on a particle of charge $q$ is $F = -q(d\phi/dx)$, where $\phi$ is a linear potential distribution. Figure E4.10 shows the relevant arrangement.

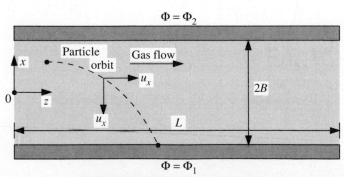

**Figure E4.10**  Electrostatic precipitator.

(a) If the particle is small enough and accelerates to a terminal settling velocity rapidly enough for Stokes' law to be valid, the force of gravity is negligible compared to the electric force, and the gas is not flowing, how long would it take the particle to settle all the way across the gap?

(b) Suppose that the gas moves in plane Poiseuille flow under pressure gradient $\Delta p/\Delta L$, and the streamwise component of the particle's velocity is essentially the gas velocity. Estimate how long the channel should be to ensure that all particles whose radius exceeds $R$ reach a plate rather than passing through the collector. (From Ref. 63, by permission.)

**Solution**

(a) Newton's and Stokes' laws apply here at steady state:

$$F = -q\frac{d\phi}{dx} = 6\pi\eta uR. \qquad (4.10.1)$$

The vertical velocity is

$$u_x = -\frac{q}{6\pi\eta R}\frac{\phi_2 - \phi_1}{2B}. \qquad (4.10.2)$$

The time needed for the particle to settle all the way across the gap is

$$t = \frac{2B}{u_x} = \frac{24\pi\eta RB^2}{q(\phi_2 - \phi_1)}. \qquad (4.10.3)$$

(b) For plane Poiseuille flow, the velocity along the plate is computed as demonstrated in Section 4.2 for axisymmetric pipe flow. It is

$$u_z = \frac{dz}{dt} = -\frac{\Delta p}{\Delta L}\frac{B^2}{2\eta}\left(1 - \frac{x^2}{B^2}\right). \qquad (4.10.4)$$

Vertical and horizontal distances traveled within the same time interval $dt$ are related by

$$dt = \frac{dz}{u_z} = \frac{dx}{u_x}, \qquad (4.10.5)$$

which for the $u_x$ and $u_z$ computed becomes

$$\frac{q(\phi_2 - \phi_1)}{6\pi B^3 R}dz = \left[\frac{\Delta p}{\Delta L}\left(1 - \frac{x^2}{B^2}\right)\right]dx \qquad (4.10.6)$$

and is integrated,

$$\frac{q(\phi_2 - \phi_1)}{6\pi B^3 R}\int_0^z dz = \frac{\Delta p}{\Delta L}\int_{-B}^B\left(1 - \frac{x^2}{B^2}\right)dx, \qquad (4.10.7)$$

to yield

$$z = -8\frac{\Delta p}{\Delta L}\frac{\pi RB^4}{q(\phi_2 - \phi_1)}. \qquad (4.10.8)$$

The requirement therefore is

$$L \geq -\frac{\Delta p}{\Delta L}\frac{8\pi RB^4}{q(\phi_2 - \phi_1)} > 0. \qquad (4.10.9)$$

## 4.10 LIQUID ATOMIZATION AND SPRAYS

Atomization is the process of liquid disintegration into small droplets in a gaseous atmosphere, producing a *spray*. Sprays in the atmosphere exist in mist, cloud, and rain formations. In agriculture they arise with herbicide and insectivide applications. In internal combustion engines they make up the fuel–air mixture for ignition. Sprays also enter in a diversity of industrial applications for painting, printing, and coating the surfaces of vehicles, home appliances, walls, and others.[24] *Atomizers or sprayers* are the devices used to accelerate liquids to form thin films or jets that break into filaments and droplets to produce the spray. The breakup process is often enhanced by air jets impinging on the liquid film or jet, and in selected applications by electrostatic fields, as in electrostatic paint bells, that may also serve to direct the resulting charged droplets. Widely applied atomizers are shown in in Fig. 4.7 and highlighted elsewhere.[25]

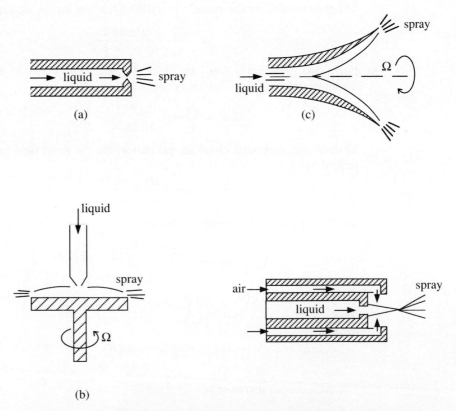

**Figure 4.7**  Liquid atomizers: (a) plain orifice pressure atomizer; (b) disk rotary atomizer; (c) bell atomizer; (d) air-assisted atomization.

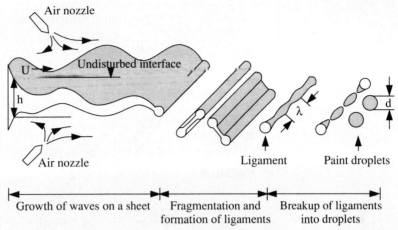

**Figure 4.8**  Atomization of a liquid paint sheet emerging from an orifice. (From Ref. 26, by permission.)

The stages of the breakup of the issuing liquid sheet, curtain, or jet are shown in Fig. 4.8. Initially, *surface waves or instabilities* are induced at the free surface due either to *hydrodynamic instability,* or to *aerodynamic instability, forms surface perforations* expanding toward each other and coalescing to form *threads and ligaments.* Impinging air jets and electrostatic fields further augment these perforations, and in fact in many cases of plane liquid sheets, acceptable disintegration is obtained only by these external factors. The threads and ligaments formed are inherently unstable and break further into *large droplets and smaller satellites* that mix with air to form the spray.

A spray consists of droplets of different size characterized by the statistical means and deviations. To this time, no rigorous theory exists to predict droplet size distribution accurately in sprays. Any predictive capability is based either on empirical correlations or on idealized simplified cases, along the lines of Rayleigh's (1878) classic analysis of inviscid jet breakup at low velocity into droplets of diameter nearly twice that of the liquid jet.[27] Weber extended Rayleigh's analysis to viscous liquids and found that,[28]

$$\left(\frac{D}{d}\right)^3 = \frac{3\pi}{\sqrt{2}}\left(1 + \frac{3\eta_L}{\sqrt{\rho_L \sigma d}}\right)^{1/2}, \qquad (4.61)$$

where $D$ and $d$ are the droplet and jet diameter, respectively, $\eta_L$ and $\rho_L$ the liquid viscosity and density, and $\sigma$ the surface tension. The most commonly used criteria for real sprays were proposed by Ohnesorge,[29] based on experimental data and dimensional analysis. He found the process to depend on two dimensionless numbers: the *Reynolds number* of the liquid,

$$\mathrm{Re} = \frac{\rho_L u_L d}{\eta_L}, \qquad (4.62)$$

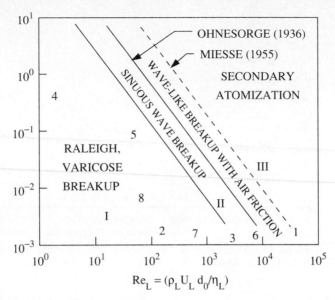

**Figure 4.9** Classification of modes of disintegration proposed by Ohnesorge[29] and modified by Reitz.[30] Numbering refers to Fig. E4.11. (From Ref. 25, by permission.)

and the *Ohnesorge number,*

$$\text{Oh} = \frac{\eta_L}{\sqrt{\rho_L \sigma d}} = \left( \frac{\eta_L u_L}{\sigma} \right)^{1/2} \left( \frac{\eta_L}{\rho_L u_L d} \right)^{1/2} = \sqrt{\frac{\text{Ca}}{\text{Re}}}, \qquad (4.63)$$

where Ca is the capillary number, as illustrated by Fig. 4.9. The figure shows that low Reynolds numbers a jet disintegrates into large droplets as predicted by Rayleigh's and Weber's analyses. At intermediate Reynolds numbers, breakup is induced by surface waves that grow with air resistance, and a wide range of droplet sizes is produced. At high Reynolds numbers, mechanisms for atomization are unclear, but it is completed within a short distance from the discharge orifice.

The equations and figures given here, appropriate for long jets or sheets, by no means cover liquid disintegration modeling, which may in addition be influenced and even controlled by nozzle design and initial velocity profile, air impinging jets, electrostatic fields, non-Newtonian rheology, and other kinds of break-inducing instabilities. Summaries of recent developments, still mostly based on empirical and simplified analyses, can be found elsewhere.[26] These sources are supplemented by alternative approaches, based on computer-aided analyses by finite elements,[31] that predict jet lengths before breakup, dominant size of droplet and satellite, and how viscosity, surface tension, elasticity, yield stress, and other factors influence atomization. In many of these analyses, attention is focused on factors surpressing or enhancing formation of satellite drops: In printing and painting applications, satellites are undesirable; in fuel atomization, satellites of small size lead to more effective combustion.

### Example 4.11: Liquid Sheet Breakup and Droplet Size

Calculate the dominant droplet obtained by atomization of round jets of diameter $d = 0.4$ cm of (a) water at 20°C; (b) water at 90°C; (c) steel at 1810°C, common with jet welding of metals; (d) mercury at 20°C; (e) ink at 20°C. Compare your findings to those of an inviscid liquid jet and to the chart of Fig. 4.9.

**Solution**    First the physical properties required in Eqn. (4.61) and Fig. 4.9 are found and listed in Table E4.11. Then the droplet diameter is calculated by means of Eqn. (4.61). The table illustrates that at low velocities, most common liquids behave as inviscid, unless their capillary number, given by

$$\text{Ca} = \frac{\eta u}{\sigma}, \tag{4.11.1}$$

is significant, as shown in Table E4.11. The table also shows that the diameter of the droplet increases with the capillary and Ohnesorge numbers, and is practically independent of the Reynolds number.

The positioning of each case is shown in Fig. 4.9 by numbering the liquids $1, 2, \ldots, 8$, from inviscid to lead. It is shown that all cases, being at capillary numbers less than unity, adhere to the Rayleigh break-up mechanism.

**Droplet trajectory.**    After disintegration, the resulting droplets and satellites travel—occasionally within electrostatic fields (e.g., Example 4.10)—and are deposited on target substrates (e.g., vehicle panels and house walls). The motion of individual droplets is defined by Eqns. (4.54) to (4.60). If the drag coefficient is defined by

$$C_D = \frac{F}{(\pi d_p^2/4)\rho\,(u^2/2)}, \tag{4.64}$$

where

$$F = 6\pi d_p u \frac{6(\eta + 1.5\eta_p)}{4(\eta + \eta_p)}, \tag{4.65}$$

with $\eta$ and $\rho$ the viscosity and density of air, and $d_p$ and $\eta_p$ the diameter and viscosity of droplet, the following correlations were found to perform satisfactorily in droplet trajectory descriptions:

$$C_D = \frac{24}{\text{Re}}\left[1 + \left(\frac{\text{Re}}{6}\right)^{2/3}\right] \tag{4.66}$$

at low temperature and Re < 1000; and

$$C_D = \frac{24}{Re}\left[1 + \left(\frac{\text{Re}}{6}\right)^{2/3}\right](1 - Y_e) \tag{4.67}$$

**TABLE E4.11.** DOMINANT DROPLET DIAMETER OF ATOMIZATION OF ROUND JET OF DIAMETER 0.4 cm AND VELOCITY 10 cm/s [EXAMPLE 4.11]

| Property | Inviscid | Water | | Fuel oil | | Mercury | Steel | Lead |
|---|---|---|---|---|---|---|---|---|
| | | 20°C | 100°C | 40°C | 105°C | 20°C | 1810°C | 550°C |
| Density (g/cm³) | Any | 0.988 | 0.958 | 0.936 | 0.897 | 13.55 | 7.8 | 11.3 |
| Viscosity (P) | 0 | 0.01 | 0.0028 | 2.15 | 0.13 | 0.015 | 0.05 | 1.7 |
| Surface tension (dyn/cm) | Any | 72 | 57 | 23 | 20 | 480 | 1200 | 1300 |
| Reynolds | ∞ | 340 | 1710 | 1.74 | 28 | 3600 | 750 | 33 |
| Ohnesorge | 0 | 0.0017 | 0.0005 | 0.73 | 0.048 | 0.0003 | 0.001 | 0.02 |
| Capillary | 0 | 0.001 | 0.0005 | 0.93 | 0.065 | 0.0003 | 0.0005 | 0.014 |
| Diameter (cm) (by Eqn. (4.87)) | 1.88*d* | 1.89*d* | 1.88*d* | 2.25*d* | 1.92*d* | 1.88*d* | 1.88*d* | 1.9*d* |

at high temperature with significant ongoing evaporation. In Eqn. (4.67), $Y_e$ is the liquid vapor fraction in the vapor–air mixture under equilibrium conditions.

## 4.11 FLOW IN POROUS MEDIA

A porous medium is a continuous solid phase with intervening void or gas pockets. *Natural porous media* include soil, sand, mineral salts, sponge, wood, and others. *Synthetic porous media* include paper, cloth, filters, chemical reaction catalysts, membranes, and many, many others. *Packed beds* for heat and mass exchange, often accompanied by chemical reaction, are another kind of synthetic porous medium, employed extensively in the chemical industry. *Fluidized beds* provide a special kind of limiting synthetic porous medium in which solid particles float suspended in a continuous fluid phase. A packed bed evolves into a fluidized bed by increasing the flow rate of a penetrating liquid or gas stream.

The *porosity* or void fraction, $\varepsilon$, of a porous medium is defined by

$$\varepsilon = \frac{V - V_s}{V} = 1 - \frac{V_s}{V}, \tag{4.68}$$

where $V_s$ is the solid volume included in the total volume $V$. For a fluid penetrating a porous surface of total cross-sectional area $A$, two different average velocities can be defined: a *superficial* one, $u_s$, defined by

$$u_s = \frac{\dot{V}}{A}, \tag{4.69}$$

where $\dot{V}$ is the volumetric flow rate; and an *interstitial* one, $u_I$, defined by

$$u_I = \frac{\dot{V}}{A - A_s} = \frac{\dot{V}}{\varepsilon A}, \tag{4.70}$$

where $A_s$ is the fraction of the area $A$ occupied by solids. The interstitial velocity is the one that determines the rate of transport of mass, energy, and momentum. However, the superficial velocity can easily be measured and is widely used inasmuch as each can be computed from the other by

$$u_s = \varepsilon u_I, \tag{4.71}$$

given the porosity $\varepsilon$. The hydraulic radius of a porous medium, $\bar{R}$, is defined accordingly by

$$\bar{R} = \frac{V - V_s}{A_w} = \frac{V\varepsilon}{N_P A_P} = \frac{V_P \varepsilon}{(1 - \varepsilon)A_P} = \frac{R_P \varepsilon}{3(1 - \varepsilon)}, \tag{4.72}$$

where $A_w$ is the total contact or wetted area between the fluid and solid, and $N_P$ is the equivalent number of spheroid particles each of surface $A_P$, volume $V_P$, and radius $R_P$, included in the total volume $V$. An effective Reynolds number of a flow through a porous medium is defined by

$$\text{Re} = \frac{\rho u_I (2\bar{R})}{\eta} = \frac{2\rho u_I R_P \varepsilon}{3\eta(1-\varepsilon)} = \frac{2\rho u_s R_P}{3\eta(1-\varepsilon)}, \tag{4.73}$$

Due to the restrictive nature of a porous medium to a penetrating flow through its narrow capillary pores, an overall increased pressure drop occurs, given by

$$\Delta\left(\frac{u_1^2}{z} + \frac{p}{\rho} + gz\right) = -\frac{\hat{\mathscr{F}}}{\rho}, \tag{4.74}$$

for flow of fluid of density $\rho$, between two cross-sectional areas of different velocities, pressures, and elevations, of a continuous porous medium. Equations (4.73) and (4.74) yield a first estimate of the frictional losses, $\mathscr{F}$, of flow through porous media,

$$\mathscr{F} = 3f\left(\frac{u_I^2}{2}\right)\left(\frac{\Delta L}{R_P}\right)\left(\frac{1-\varepsilon}{\varepsilon}\right) = 6f\left(\frac{u_s^2}{2}\right)\left(\frac{\Delta L}{R_P}\right)\left(\frac{1-\varepsilon}{\varepsilon^3}\right). \tag{4.75}$$

This equation shows that frictional losses, and therefore pressure drop and pumping power consumption, decrease with increasing porosity and particle size. The friction coefficient, $f$, in Eqn. (4.75) for flow through porous media is quantified by Ergun's empirical relation,[32]

$$\mathscr{F} = 1.75u_s^2\left(\frac{\Delta L}{d_P}\right)\left(\frac{1-\varepsilon}{\varepsilon^3}\right) + 150u_s\left(\frac{\Delta L}{d_P^2}\right)\left(\frac{1-\varepsilon}{\varepsilon^3}\right)\left(\frac{\eta}{\rho}\right), \tag{4.76}$$

which is plotted in Fig. 4.10. The lower Reynolds limit, $\text{Re} < 10$, gives the

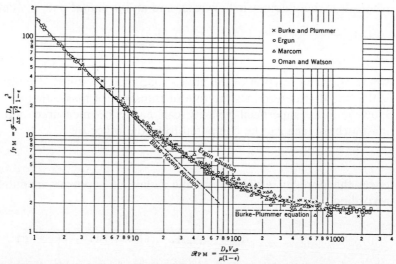

**Figure 4.10** Frictional loss coefficients $f$ for flow of fluid of density $\rho$ and viscosity $\eta$ at superficial velocity $u_s$, through a porous medium of porosity $\varepsilon$ and particle diameter $D_P$. (From Ref. 32, by permission.)

Blake–Kozeny equation[33] for laminar flow,

$$\mathscr{F} = 150u_s\left(\frac{\Delta L}{d_P^2}\right)\left(\frac{1-\varepsilon}{\varepsilon^3}\right)\left(\frac{\eta}{\rho}\right), \tag{4.77}$$

and the upper Reynolds limit, Re > 1000, gives the Burke–Plumber equation[34] for turbulent flow,

$$\mathscr{F} = 175u_s^2\left(\frac{\Delta L}{d_P^2}\right)\left(\frac{1-\varepsilon}{\varepsilon^3}\right). \tag{4.78}$$

Large-scale flows, such as flow of water or oil in underground porous reservoirs, are laminar and Eqn. (4.77) applies. In engineering applications, this equation is customarily cast in a form known as Darcy's equation,[35]

$$\frac{\mathscr{F}}{\Delta L} = \frac{v}{k}u_s, \tag{4.79}$$

where $v = \eta/\rho$ is the *kinematic viscosity* and $k$ is the *permeability* of the porous medium, with units of darcy, defined as

$$1 \text{ darcy} = \frac{(\text{cm/s})\text{cP}}{\text{atm/cm}} = 10^{-8}\text{ cm}^2 = 1.06 \times 10^{-11}\text{ ft}^2.$$

The permeability defines how easily a porous medium is penetrated by viscous fluids. Equation (4.74) combines with Eqn. (4.78) to yield

$$u_x = -\frac{k}{\eta}\frac{d}{dx}(p + \rho g z), \tag{4.80}$$

where $u_x$ is the superficial velocity in the $x$-direction. This equation states that flow through porous media is proportional to the permeability of the medium, to the inverse of the viscosity of the fluid, and occurs from regions of higher to lower pressure, $p + \rho g z$, including the hydrostatic head.

### Example 4.12: Water Penetration in Soil[9]

For irrigation of a land at a slight inclination, $\phi = 5°$, which is infinitely long and of 500 m width, water at a flow rate of $\dot{V}_0 = 0.1 \text{ m}^3/\text{s}$ per meter of width is available (as shown in Fig. P5.23, for example).

As the water advances downhill driven by gravity, evaporation takes place at the rate $\dot{V}_P = 10^{-10} \text{ (m}^3/\text{s})/\text{m}^2$, and penetration into the soil of permeability $k = 10$ darcy. The viscosity of water is $\eta = 1$ cP, and its density $\rho = 1 \text{ g/cm}^3$.

**(a)** What is the maximum length of land downstream from the source that can be irrigated?

**(b)** What is the penetration depth of water in the soil?

**Solution**

(a) Consider a fluid system of length $dx$ and width $W$ (as shown in Fig. P5.23), at steady state. A mass balance over its volume results in

$$\dot{m}(x) - \dot{m}(x + dx) - \rho \dot{V}_p W\, dx - \rho u W\, dx = 0, \qquad (3.12.1)$$

where

$$u = -\frac{k}{\eta} \frac{d}{dy}(p + \rho gz) = -\frac{j}{\eta}\frac{d}{dy}(\rho gh + \rho gz) \approx \frac{k}{\eta} g\rho \cos\phi \qquad (4.12.2)$$

is the velocity of penetration in the soil. Equations (4.12.1) and (4.12.2) combine as

$$\frac{\dot{m}(x) - \dot{m}(x + dx)}{dx} = -\rho\left(r + \frac{k}{\eta}\rho g \cos\phi\right)W. \qquad (4.12.3)$$

In the limit of $dx \to 0$, the left-hand side of this equation becomes the derivative

$$\frac{d\dot{m}}{dx} = -\rho\left(r + \frac{k}{\eta}\cos\phi\right)W. \qquad (4.12.4)$$

Integrating the latter equation and replacing $\dot{m} = \rho \dot{V}$ results in

$$\frac{\dot{V}}{W} = \frac{\dot{V}_0}{W} - \left(r + \frac{k}{\eta}\rho g \cos\phi\right)x, \qquad (4.12.5)$$

which predicts how the flow rate per width diminishes with distance due to evaporation and soil penetration. The flow ceases at length $x = L$, where $\dot{V} = 0$:

$$L = \frac{\dot{V}_0}{r + (k/\eta)\rho g \cos\phi}$$

$$= \frac{10^{-1} \times 10^4}{10^{-10} \times 10^6 + (10^{-7} \times 1 \times 981 \times 0.98)/0.01}\, \text{cm} = 100\, \text{m}. \qquad (4.12.6)$$

(b) The penetration depth $y(t)$ with time is determined by

$$\frac{dy}{dt} = u = \frac{k}{\eta}\rho g \cos\phi, \qquad (4.12.7)$$

which yields

$$y(t) = \left(\frac{k}{\eta}\rho g \cos\phi\right)r, \qquad (4.12.8)$$

and therefore,

$$z(t) = \frac{y(t)}{\cos\phi} = \frac{k}{\eta}\rho g t = (981 \times 10^{-5}t)\, \text{cm}. \qquad (4.12.9)$$

Thus within a time period of 1 h, the penetration depth is about

$$z = 981 \times 10^{-5} \times 3600\, \text{cm} = 35.3\, \text{cm}, \qquad (4.12.10)$$

which is at about the depth of the roots of seasonal cultivation.

**Example 4.13: Permeability of Porous Medium**

Consider a synthetic unidirectional porous medium, made up of parallel pipes of radius $r$ and length $\Delta L$ cemented together such that there are $m$ pipes per $\pi R^2$ cross-sectional area of porous medium. Determine the permeability of this medium by comparing the pressure Poiseuille flow expressions in each pipe to Eqn. (4.79), for a strictly horizontal flow.

**Solution**    According to Eqn. (4.28) for flow in a horizontal pipe, the total flow rate is

$$\dot{V} = m \frac{\pi}{8\eta} \frac{\Delta p}{\Delta L} r^4, \tag{4.13.1}$$

and the superficial velocity is

$$u_s = \frac{\dot{V}}{\pi R^2} = m \frac{1}{8\eta} \frac{\Delta p}{\Delta L} \frac{r^4}{R^2}. \tag{4.13.2}$$

The porosity of the medium is

$$\varepsilon = 1 - \frac{V_s}{V} = 1 - \frac{m\pi r^2 \Delta L}{\pi R^2 \Delta L} = 1 - m \frac{r^2}{R^2}, \tag{4.13.3}$$

and that of the cross-sectional area is

$$\varepsilon_s = \frac{A - A_s}{A} = 1 - \frac{A_s}{A} = 1 - \frac{mr^2}{R^2} = \varepsilon. \tag{4.13.4}$$

Thus, the interstitial velocity is

$$u_I = \frac{\dot{V}}{\varepsilon A} = \frac{\dot{V}}{\pi(R^2 - mr^2)} = \frac{m}{8\eta} \frac{r^4}{(R^2 - mr^2)} \frac{\Delta p}{\Delta L}. \tag{4.13.5}$$

Now, according to Eqn. (4.80), for horizontal flow in porous media the velocity is

$$u_I = -\frac{k}{\eta} \frac{\Delta p}{\Delta L}. \tag{4.13.6}$$

By comparing Eqns. (4.13.5) and (4.13.6), the permeability of the synthetic porous medium is,

$$k = \frac{m}{8} \frac{r^4}{R^2 - mr^2} = \frac{r^2}{8} \frac{1}{(R^2/mr^2) - 1} = \frac{\varepsilon r^2}{8(1 - \varepsilon)} \tag{4.13.7}$$

with limiting values,

$$k = \begin{cases} \infty, & R^2/\eta r^2 = 1, \quad \varepsilon = 1 \text{ (pipe)} \\ 0, & r = 0 \quad \text{impermeable medium.} \end{cases} \tag{4.13.8}$$

Therefore, the permeability is proportional to the fraction of the cross-sectional area that allows flow-through.

## 4.11.1 Packed and Fluidized Beds

Packed beds are utilized in industry to induce large exchange surfaces between different immiscible phases, to achieve increased exchange rates of solute and/or heat. The most common cases involve a liquid falling under gravity or a gas ascending under pressure through particles of catalyst or absorbent that strip one

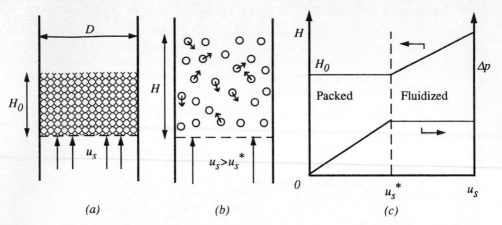

**Figure 4.11** Packed bed when the superficial velocity, $u_s$, is less than the fluidization velocity, $u_s^*$, that evolves into a fluidized bed when $u_s > u_s^*$.

or more of the solutes from the carrier stream. Later stages of such operations, often called regenerations, include flashing of the enriched bed to recover the target solute in a desirable solvent or even to clean the absorbing or reacting solid surface for a new cycle of operation.

Figure 4.11 shows upward flow of a fluid through a packed bed of initial height $H_0$ with solid particles of mean diameter $d_p$. As the superficial velocity $u_s$ increases, the pressure drop $\Delta p$ increases linearly while the height remains constant, as shown in Fig. 4.11c. For particles of a given diameter and therefore beds of given porosity, there is an *incipient velocity of fluidization, $u_s^*$*, beyond which the drag force on the particle overcomes its weight, such that all solid particles start circulating around as if they were fluid particles themselves.[36] Then the height of the bed starts to expand, which allows the pressure drop to remain nearly constant with any further increase in the superficial velocity. There is a second characteristic *entrainment velocity, $\hat{u}_s$*, beyond which solid particles are carried away by the fluid, resulting in destruction of the bed.

Just before fluidization, the energy equation for the packed bed is

$$gH_0 + \frac{p_2 - p_1}{\rho} + \mathscr{F} = 0. \tag{4.81}$$

A force balance on the entire bed yields

$$(p_1 - p_2)\pi \frac{D^2}{4} = gH_0\left[(1 - \varepsilon)\rho_s + \varepsilon\,\rho\pi\,\frac{D^2}{4}\right], \tag{4.82}$$

where the right-hand terms represent the composite weight of the bed, balanced by the pressure force of the left-hand side. Equation (4.81), with the frictional term $\mathscr{F}$ given by Eqn. (4.75), combines with Eqn. (4.82) to yield

$$(u_s^*)^2 + \left[\frac{86(1 - \varepsilon)\eta}{d_p\,\rho}\right]u_s^* - 0.57g\left(\frac{\rho_s}{\rho} - 1\right)d_p\varepsilon^3 = 0, \tag{4.83}$$

which defines the velocity of fluidization of that bed–fluid system as

$$u_s^* = -\frac{43(1-\varepsilon)\eta}{\rho\, d_p} + \left[\frac{1850(1-\varepsilon)^2\eta^2}{\rho^2\, d_p^2} + 0.57g\left(\frac{\rho_s}{\rho}-1\right)\varepsilon^3 d_p\right]^{1/2}. \qquad (4.84)$$

**Example 4.14: Velocity of Fluidization**

Calculate the velocity of fluidization, $u_s^*$, of a catalytic bed of diameter $D = 2$ ft and initial height $H_0 = 3$ ft. The bed is packed with solid particles of density $\rho_p = 2\,\text{g/cm}^3$ and mean diameter $d_p = 0.5$ cm, under continuous flow of fluid of viscosity $\eta = 1$ cP and density $\rho = 1\,\text{g/cm}^3$. What power input is required at fluidization?

**Solution**    The porosity of a catalytic bed in a cubic arrangement is

$$\varepsilon = \frac{V - V_s}{V} = 1 - \frac{V_s}{V} = 1 - \frac{(4/3)\pi(d^3/8)}{d^3} = 1 - \frac{\pi}{6} = 0.48. \qquad (4.14.1)$$

The velocity at inception of fluidization is given by Eqn. (4.84):

$$u_s^* = \left\{-\frac{43(1-0.48)0.01}{0.5} + \left[\left(\frac{1850(1-0.48)^2 0.01^2}{1\times 0.5^2}\right) + 0.57\times 981\left(\frac{2}{1}-1\right)\right.\right.$$

$$\left.\left. \times\, 0.48^3\times 0.5\right]^{1/2}\right\}\frac{\text{cm}}{\text{s}} = -0.45 + (0.19 + 30.9)^{1/2}\ \frac{\text{cm}}{\text{s}} \approx 5.12\ \frac{\text{cm}}{\text{s}}. \qquad (4.14.2)$$

The frictional loss, predicted by Eqn. (4.76), is

$$\mathcal{F} = \left(1.75\times 5.12^2\times\frac{91}{0.5}\times\frac{1-0.48}{0.48^3} + 150\times 5.12\times\frac{91}{0.5^2}\times\frac{1-0.48}{0.48^3}\times 0.011\right)\frac{\text{cm}^2}{\text{s}^2}$$

$$= (3.95\times 10^4 + 1.24\times 10^4)\ \frac{\text{cm}^2}{\text{s}^2} = 5.2\times 10^4\ \frac{\text{cm}^2}{\text{s}^2} = 5.2\ \frac{\text{m}^2}{\text{s}^2}. \qquad (4.14.3)$$

The pressure drop is given by Eqn. (4.81) as

$$\Delta p = \rho g H_0 + \rho\mathcal{F} = (10^3\times 9.81\times 0.91 + 10^3\times 5.6)\ \text{N/m}^2$$

$$= (8.9 + 5.6)\times 10^3\,\text{N/m}^2 = 14{,}500\,\text{Pa} = 0.148\,\text{atm}. \qquad (4.14.4)$$

$$\tag{4.14.4}$$

The pumping power required is

$$P = \dot{V}\,\Delta p = \left(5.12\times 3.14\times\frac{0.61^2}{4}\times 14{,}500\right)\text{W} = 21{,}685\,\text{W} = 16\,\text{hp}. \qquad (4.14.5)$$

## 4.11.2 Filtration

Filters are porous media that are employed to clean slurries of suspended particles in fluids by allowing the fluid through while preventing and holding up particles of dimensions larger than the pores of the filter.[37] For slurries of high particle concentration, a *cake of* captured particles forms on top of the filter as

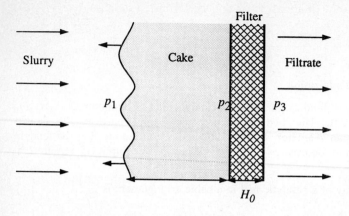

**Figure 4.12** Filtration of a slurry of particles held by the filter in the form of accumulating cake, letting the cleaned liquid (filtrate) through.

fifiltration proceeds and serves as an additional filter layer. Soil, sand, and wood are natural filters. Air and fuel filters in vehicles, and filters for purifying water and cleaning industrial gases before releasing to the atmosphere, are examples of synthetic filters. Membranes employed extensively in separations and biosepara- tions are also examples of thin filters. A simplified representation of filtration is shown in Fig. 4.12.

Darcy's law applied across the cake yields

$$u_s = \frac{\dot{V}}{A} = \frac{1}{a}\frac{dV}{dt} = \frac{k_c}{\eta}\frac{p_1 - p_2}{H(t)} = \frac{k_f}{\eta}\frac{p_1 - p_2}{H_0}, \tag{4.85}$$

where $u_s$ is the filtrate superficial velocity, $\dot{V} = dv/dt$ the filtration rate of the filtrate, $A$ the cross-sectional area, $k_c$ and $k_f$ the permeabilities of cake and filter, respectively, $\eta$ the viscosity of the filtrate, and $H(t)$, $H_0$, $p_1$, $p_2$, and $p_3$ are as shown in Fig. 4.12. By eliminating the intermediate pressure $p_2$, Eqn. (4.84) can be written as

$$u_s = \frac{p_1(t) - p_3}{\eta[H(t)/k_c + H_0/k_f]}, \tag{4.86}$$

where the denominator represents composite *resistance* to filtrate flow.

The increasing thickness of the cake is given by

$$H(t) = \frac{V_c}{A} = \frac{V\hat{V}_s}{(1 - \varepsilon)A}, \tag{4.87}$$

where $V_c$ is the porous volume of the captured particles that form the deposited cake, $\hat{V}_s$ the total volume of solids per volume of filtrate, and $\varepsilon$ the porosity of the cake. Equations (4.85) to (4.87) combine to form

$$\frac{1}{A}\frac{dV}{dt} = \frac{p_1(t) - p_3}{\eta\left[\dfrac{V\hat{V}_s}{(1 - \varepsilon)AK_c} + \dfrac{H_0}{K_f}\right]}, \tag{4.88}$$

which describes a general filtration process.

In processes such as air-cleaning[20] and liquid-purification[38] the resistance of the cake is usually negligible, the driving pressure constant, so the governing equation (4.41) becomes

$$\frac{dV}{dt} = \frac{Ak_f}{\eta H_0}(p_1 - p_3),$$
(4.89)

which is integrated to

$$V(t) = \frac{Ak_f}{\eta H_0}(p_1 - p_3)t,$$
(4.90)

and particle removal of

$$V_s(t) = \hat{V}_s V(t).$$
(4.91)

For filtration of slurries of high particle concentration,[39] the resistance of the thin filter becomes unimportant, and the governing equation reduces to

$$V\frac{dV}{dt} = \frac{A^2(1 - \varepsilon)k_c}{\eta \hat{V}_s}[p_1(t) - p_3].$$
(4.92)

For filtration under constant pressure, Eqn. (4.92) becomes

$$V^2 = \left[\frac{2A^2(1 - \varepsilon)k_c}{\eta \hat{V}_s}(p_1 - p_3)\right]t,$$
(4.93)

while for filtration at a constant flow rate, $dV/dt = \dot{V}_0$, Eqn. (4.92) becomes

$$p_1(t) = p_2 + \frac{\eta \hat{V}_s \dot{V}_0^2}{A^2(1 - \varepsilon)k_c}t.$$
(4.94)

Both the porosity $\varepsilon$ and permeability $k_c$ may, in general, change with time as filtration proceeds. In the usual range of operating conditions a helpful correlation is

$$\frac{1}{(1 - \varepsilon)k_c} = cp^s,$$
(4.95)

where $c$ is a constant determined largely by the size of the particles that form the cake, and $s$ is the cake compressibility, varying from zero for rigid cakes such as sand to unity for highly compressible cakes such as paper pulp. The resistance of the filter itself may also be a function of the pressure applied.[7] In reality, this kind of filtration at high consistency slurries proceeds under mixed conditions of increasing pressure and decreasing flow rate.[40]

### Example 4.15: Filtration with Changing Cake Resistance

Paper pulp of concentration $\hat{V}_s = 10\%$ is to be filtered in a filter press of area $A = 1$ m$^2$ at the rate of $\dot{V} = 1$ m$^3$/s. The paper forms a cake of density 13 lb of dry solids per cubic foot of wet cake deposited on a filter cloth of negligible resistance compared to that of the cake. In addition, the resistance of the cake varies with pressure according to Eqn. (4.95), with $k_c = 10^{-2}$ darcy $-$ 1 atm $-$ 0.5 and $\varepsilon = 0.5$. What pressure of filtration is required?

**Solution**   Equations (4.88) and (4.95) apply, with $H_0/k_f = 0$, $p_3 \approx 0$, $p_1(t) = p(t)$, and $V = \dot{V}t$:

$$\frac{\dot{V}}{A} = \frac{p(t)A}{\eta \dot{V}t\hat{V}_s a} = \frac{p(t)^{0.5}A}{\eta \dot{V}\hat{V}_s k_c} \frac{1}{t} \tag{4.15.1}$$

and

$$p(t) = \left[ \eta \hat{V}_s k_c \left(\frac{\dot{V}}{A}\right)^2 t^2 \right]. \tag{4.15.2}$$

*Application:*

$$p(t) = \left\{ \left[ 1 \times 0.1 \times 10^{-4} \times \left(\frac{10^6}{10^4}\right)^2 \right]^2 t^2(s) \right\} \text{ atm} = [0.01t^2] \text{ atm,}  \quad \textit{with t in sec.} \tag{4.15.3}$$

### 4.11.3 Oil Wells and Oil Recovery

Crude oil (and natural gas) is trapped under pressure in underground porous natural reservoirs, confined between impermeable rock layers. When a well is drilled through the upper impermeable rock formation and penetrates the oil reservoir, a pressure differential is induced between ground level and the trapped oil that may overcome the potential energy difference and drive the oil to the surface, where it is recovered, stored, and processed.[41] Once this process begins, oil migrates in a radial fashion to the well bore and from there to the surface through the pipe. A simplified equipment arrangement is shown in Fig. 4.13.

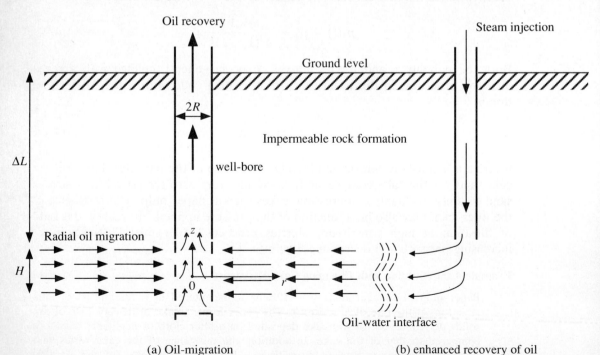

(a) Oil-migration                                      (b) enhanced recovery of oil

**Figure 4.13**   Oil recovery: (a) natural oil migration, and (b) enhanced oil recovery.

The radial migration within the reservoir is governed by the mass conservation equation,

$$u(r, t) = \left[ -\frac{\dot{V}(t)}{2\pi H} \right] \frac{1}{r}, \tag{4.96}$$

where $u$ is the radial velocity at distance $r$ toward the pipe, $\dot{V}(t)$ the total volumetric flow rate of oil, and $H$ the penetrating distance of the pipe within the reservoir, as shown in Fig. 4.13a.

Equation (4.96) combines with Darcy's law, Eqn. (4.80), to yield the radial distribution of pressure,

$$p(r, t) = p_\infty + \frac{\eta \dot{V}(t)}{2\pi Hk} \ln \frac{r}{r_\infty}, \tag{4.97}$$

where $p_\infty$ is pressure at a distance $r_\infty$, far away from the well, known as the static reservoir pressure since the velocity is zero there. The flow toward the pipe is therefore driven by a radial pressure gradient,

$$\frac{dp}{dr} = \frac{\eta \dot{V}(t)}{2\pi Hk} \frac{1}{r}. \tag{4.98}$$

The pressure at the entrance of the pipe is approximately

$$p_0(t) = p(r = R, t) = p_\infty + \frac{\eta \dot{V}(t)}{2\pi Hk} \ln \frac{R}{r_\infty} \tag{4.99}$$

The pressure differential between the entrance and the exit of the pipe, at surface level, is therefore,

$$\Delta p = p_0(t) - p_{\text{atm}} = (p_\infty - p_{\text{atm}}) + \frac{\eta \dot{V}(t)}{2\pi Hk} \ln \frac{r}{r_\infty}. \tag{4.100}$$

To achieve some finite recovery rate $\dot{V}(t)$, this gradient must overcome the hydrostatic pipe head,

$$\Delta p_H = \rho g \, \Delta L, \tag{4.101}$$

where $\rho$ is the oil density and $\Delta L$ the length of the pipe. This is possible if $\Delta p > (\Delta p)_H$, or equivalently, when

$$[p_\infty - p_{\text{atm}} - \rho g \, \Delta L] > \frac{\eta \dot{V}(t)}{2\pi Hk} \ln \frac{R}{r_\infty}. \tag{4.102}$$

In reality, Eqn. (4.102) is altered by the fact that $p_\infty$ is reduced with time, due to the depletion of the reservoir as oil recovery proceeds. Also, $r_\infty$ may be reduced for a reservoir of finite extent, such that recovery will cease when Eqn. (4.101) does not any more hold. Then significant amounts of oil will remain trapped in the capillary pores of the reservoir.

To recover additional amounts of oil, steam may be injected from neighboring wells surrounding the oil well, which, in advancing, converges with and displaces the trapped oil toward the oil well, as shown in Fig. 4.13b. The injection of steam raises the diminished driving pressure $p_\infty$ and also decreases the viscosity significantly, both of which mobilize the trapped oil for further recovery. This stage of recovery is called *enhanced oil recovery.*

Due to the contact of two immiscible liquids in a capillary under conditions of enhanced recovery, an *interface* is formed, across which a capillary pressure differential of approximately[42]

$$\Delta p_c = \frac{4\sigma}{d_p} \tag{4.103}$$

is induced, where $\sigma$ is the *oil–water interfacial tension* and $d_p$ is the diameter of the pore. Enhanced recovery proceeds as long as the driving pressure differential predicted by Eqn. (4.103) is bigger than that of Eqn. (4.98). This is true when

$$\frac{d_p}{2}\left(\frac{dp}{dr}\right)_{r=R} = \frac{\eta \dot{V}(t)d_p}{4\pi HkR} = \frac{\eta \bar{u}d_p}{2k} > \frac{4\sigma}{d_p}, \tag{4.104}$$

where $\bar{u}$ is the average velocity, or equivalently when

$$Ca = \frac{\eta \bar{u}}{\sigma} > \frac{8k}{d_p^2} \approx \frac{\varepsilon}{4(1-\varepsilon)} \tag{4.105}$$

given Eqn. (4.13.8). Here Ca is the dimensionless *capillary number,* expressing the ratio of displacing viscous to resisting capillary forces. Thus Eqn. (4.105) is a necessary condition for liquid–liquid displacement and initiation of enhanced recovery. Otherwise, oil remains trapped in the capillary pores, in the form of droplets of resisting capillary pressure given by Eqn. (4.103). Once recovery is initiated, Eqn. (4.105) suggests that the speed of recovery $\bar{u}$ is proportional to the porosity $\varepsilon$ and inversely proportional to the viscosity $\eta$. Further aspects of oil migration are detailed in the relevant literature.[40,41]

**Example 4.16: Composite Impregnation**

In a typical polymer processing processing operation[43] to manufacture composites, i.e., polymers reinforced with fibers, a polymeric melt of viscosity $\eta = 10^5\,P$ is applied at flow rate $\dot{V} = 20\,cm^3/s$ at a point and allowed to advance radially through a cloth of permeability $k = 1$ darcy, as shown in Fig. E4.16. Calculate how fast the impregnation front advances with time, for $h = 2\,in.$ and $R = 1\,ft.$

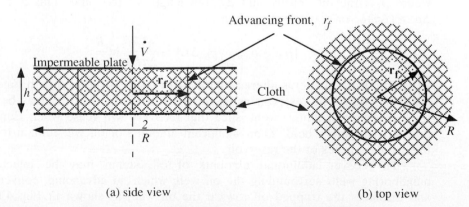

(a) side view                                                          (b) top view

**Figure E4.16**   Composite impregnation: (a) side view; (b) top view.

**Solution**    The radial velocity profile within the melt is obtained by a mass balance within a disk of radius $r$ and height $h$ that contains the entrance

$$\dot{V} = 2\pi r h u_r, \tag{4.16.1}$$

and thus

$$u_r = \frac{\dot{V}}{(2\pi h)r}. \tag{4.16.2}$$

The radial pressure distribution is obtained by Darcy's law, given Eqn. (4.16.2), as

$$u_r = \frac{\dot{V}}{(2\pi h)r} = -\frac{k}{\eta}\frac{dp}{dr}, \tag{4.16.3}$$

and therefore,

$$p(r) = -\frac{\dot{V}\eta}{2\pi h k}\ln r + c, \tag{4.16.4}$$

where $c$ is an arbitrary constant of integration to be evaluated by invoking a boundary condition. At the advancing front the melt pressure is identical to the ambient atmospheric, in the absence of surface tension. Thus

$$p(r = r_f) = -\frac{\dot{V}\eta}{2\pi h k}\ln r_f + c = p_{atm}, \tag{4.16.5}$$

and therefore,

$$c = p_{atm} + \frac{\dot{V}\eta}{2\pi h k}\ln r_f. \tag{4.16.6}$$

Equation (4.16.6) is substituted in Eqn. (4.16.4), and the pressure profile is obtained:

$$p(r) = p_{atm} + \frac{\dot{V}\eta}{2\pi h k}\ln\frac{r_f}{r}. \tag{4.16.7}$$

The position of the advancing front, $r_f$, is obtained by mass balance on the impregnated volume,

$$\dot{V} = 2\pi r_f h u_r(r = r_f) = 2\pi r_f h \frac{dr_f}{dt}, \tag{4.16.8}$$

which yields

$$r_f = \sqrt{\frac{\dot{V}t}{\pi h}}. \tag{4.16.9}$$

The time it takes to impregnate the entire cloth is

$$t_f = \frac{\pi h R^2}{\dot{V}}. \tag{4.16.10}$$

*Application*:

$$r_f = \sqrt{\frac{20}{3.14 \times 5.08}} t^{1/2}$$

$$t_f = \frac{3.14 \times 5.08(30.48)^2}{20} = 370 \text{ s}.$$

The process of impregnation can be carried out under constant flow rate or constant inlet pressure (e.g., Problem 4.39).

## 4.12 FRICTION IN TWO-PHASE FLOW

*Two-phase flow* refers to liquid–gas or liquid–vapor mixtures flowing in pipes and other geometries. A continuous displacement of one phase by the other under the local driving pressure gradient given the compressibility or incompressibility constraints causes an irregular, evolving interface enclosing bubbles or slugs of gas or droplets and films of liquids.[44] Details of flow patterns and related phenomena of momentum, heat and mass transfer exist in several specialized publications.[45]

### 4.12.1 Friction in Two-Phase-Flow

A frictional factor $C_f$ for two-phase flow can be defined by

$$C_f = \frac{\tau_w}{\frac{1}{2}\rho\bar{u}^2}, \tag{4.106}$$

and accordingly, the pressure drop is

$$\frac{dp}{dz} = \frac{4\tau_w}{D} = \frac{2C_f\rho\bar{u}^2}{D}, \tag{4.107}$$

where $D$ is the pipe diameter, $\tau_w$ the wall shear stress, and $\bar{u}$ the average total velocity. A rough estimate of pressure drop can be obtained by using $C_f = 0.005$ for high pressure (e.g., boiler piping), and $C_f = 0.003$ for a flashing flow of water at low pressure. For very low or very high quality two-phase flow, the friction factor is approximated well by that of the corresponding single-phase flow of liquid or gas, respectively. The turbulent friction factor for single-phase flow [e.g., Fig. 4.3 and Eqns. (4.32) to (4.34)] can be used for two-phase turbulent flow, with a Reynolds number based on the equivalent viscosity, $\bar{\eta}$, and the total mass velocity of the actual two-phase flow, i.e.,

$$C_f = f(\text{Re}), \qquad \text{Re} = \frac{(\rho_L u_L + \rho_G u_G)D}{\bar{\eta}}. \tag{4.108}$$

The *equivalent viscosity* can be defined, for example, by

$$\bar{\eta} = \beta\eta_G + (1 - \beta)\eta_L, \tag{4.109}$$

where $\beta$ is defined by

$$\beta = \frac{\dot{V}_G}{\dot{V}_G + \dot{V}_L}, \tag{4.110}$$

where $\dot{V}_G$ and $\dot{V}_L$ are volumetric flow rates of the gas and liquid phases, respectively. With these average expressions, a satisfactory equation for the friction factor is[46]

$$C_f = 0.0014 + 0.125\,\text{Re}^{-0.32}. \tag{4.111}$$

More accurate and general relations for frictional pressure drop in two-phase flow were derived by Baroczy[47] and improved by several other publications.[48]

## 4.12.2 Annular Two-Phase Flow

Among all two-phase-flow patterns, the annular one shown in Fig. 4.14 is the most amenable to analytic approaches.[45] In an annular, two-phase flow in a pipe of radius $R$, wetted by liquid film of thickness $\delta$ and driven by a pressure gradient $dp/dz + \rho_L g$, the velocity profile within the film (i.e., for $R - \delta \leq r \leq R$) is obtained by momentum balance over an annulus within the flowing liquid, along the lines of the analysis in Eqns. (4.13) to (4.29). (Example 6.9 provides details.)

The resulting velocity profile equation is

$$u_L = \frac{1}{\eta_E}\left[\tau_i r_i + \left(\frac{\rho_L g}{2} + \frac{1}{2}\frac{dp}{dz}\right)r_i^2\right]\ln\frac{R}{r} + \frac{1}{4\eta_E}\left[\left(\rho_L g + \frac{dp}{dz}\right)(R^2 - r^2)\right], \tag{4.112}$$

where $\eta_E$ is the effective viscosity defined below, and $r_i = R - \delta$ and the

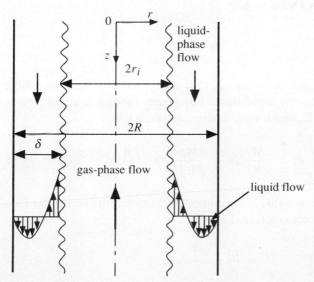

**Figure 4.14.** Two-phase flow in pipe gravity and opposing pressure gradient.

interfacial stress $\tau_i$ at $r = r_i$ is given by

$$\tau_i = -\frac{R - \delta}{2}\left(\frac{dp}{dz} + \rho_G g\right),  \tag{4.113}$$

For laminar flow, $\eta_E = \eta_L$, the ordinary viscosity. For turbulent flow,

$$\eta_E = \eta_L + \rho_L \nu_\varepsilon,  \tag{4.114}$$

where $\nu_\varepsilon$ is the *eddy kinematic viscosity*, calculated by

$$\nu_\varepsilon = \begin{cases} \eta^2(ru)\left[1 - \exp\left(-\frac{\rho_L}{\eta_L}n^2 ru\right)\right], & r > R - \frac{20\eta_L}{(\rho_L \tau_w)^{1/2}} \\ \frac{k(du/dr)^3}{(d^2u/dr^2)^2}, & R - \frac{20\eta_L}{(\rho_L \tau_w)^{1/2}} > r > r_i, \end{cases}  \tag{4.115}$$

where $\tau_w$ is the wall shear stress, and $n$ and $k$ are parameters of turbulent flow estimated as discussed in Chapter 10.

The film liquid flow rate is calculated by

$$\dot{m}_L = \rho_L \int_{R-\delta}^{R} 2\pi ru \, dr.  \tag{4.116}$$

When the thickness $\delta$, or the flow rate $\dot{m}_L$, are known, the flow rate or the film thickness, respectively, are calculated. Equations (4.112) to (4.116) provide a fairly good representation to two-phase annular flows.[49] These expressions can be used to derive analytic equations for frictional losses, according to the procedure followed for laminar flow (Section 4.2).

## 4.13 COMPRESSIBLE FLOW

The preceding material applies to liquids that are practically incompressible and to gases under incompressible conditions. For gases, these are conditions where the density $\rho = Mp/ZRT$, which may change according to

$$dp = \left(\frac{\partial \rho}{\partial p}\right)_T dp + \left(\frac{\partial \rho}{\partial T}\right)_p dt = \frac{M}{ZRT}dp - \frac{Mp}{ZRT^2}dT = \frac{\rho}{p}dp - \frac{\rho}{T}dT,  \tag{4.117}$$

where $M$ is the molecular weight, $Z$ the compressibility factor, and $R$ the ideal gas constant, remains nearly constant, that is, when,

$$\frac{d\rho}{\rho} = \frac{dp}{p} - \frac{dT}{T} \simeq 0.  \tag{4.118}$$

This is true under nearly isothermal and isobaric conditions such that

$$dp \simeq 0 \qquad dT \simeq 0; \tag{4.119}$$

and under conditions such that

$$\frac{dp}{p} = \frac{dT}{T}, \tag{4.120}$$

which prevail under isochoric conditions.

In the absence of external cooling or heating, nearly isothermal and isobaric conditions prevail under small pressure gradients that cannot induce high gas acceleration, keeping the gas velocity below roughly $u = 60$ m/s. Otherwise, high increasing acceleration is induced, associated necessarily with significant gas expansion, which according to thermodynamics turns internal energy into kinetic energy. Thus, the flowing expanding gas cools and accelerates by this self-induced mechanism, often to reach velocities beyond the speed of sound. Such velocities are termed *supersonic*.

The speed of sound, $c$, in a medium of density $\rho$, under pressure $p$ and temperature $T$, is,[50]

$$c = \left(\frac{\partial p}{\partial \rho}\right)_T^{1/2} \equiv \left(\frac{\partial p}{\partial \rho}\right)_S^{1/2} = \begin{cases} \left(\dfrac{k}{\rho}\right)^{1/2} & \text{solids and liquids} \\[2ex] \dfrac{c_p}{c_v \rho} & \text{perfect gas,} \end{cases} \tag{4.121}$$

where the subscripts $T$ and $S$ denote isothermal and isentropic process, respectively, and $k$ is isothermal compressibility. Equation (4.121) shows that at $T = 20°C$ and $p = 1$ atm, the speed of sound in steel of $k = 2.6 \times 10^7$ lb$_f$/in$^2$ and $\rho \simeq 60$ lb$_m$/ft$^3$ is $c_s \simeq 5000$ ft/s, and in air considered an ideal gas of $M = 29$ it is $c_{air} = 1120$ ft/s. Thus, as widely known, sound travels faster in solids than in liquids and slowest in gases, in a fashion opposite to their compressibility as defined by Eqn. (1.10). (The speed of light in air is much, much faster, $c_\rho = 9.83 \times 10^8$ ft/s.)

The speed of sound is practically independent of any flow conditions and is a characteristic material property. Gas flows of velocity $u$ are commonly compared to the speed of sound $c$ by means of their dimensionless *Mach number*, defined by

$$\mathcal{M} = \frac{u}{c} \begin{cases} < 1 & \text{subsonic} \\ = 0 & \text{sonic} \\ > 1 & \text{supersonic.} \end{cases} \tag{4.122}$$

A flow is subsonic, sonic, or supersonic if its maximum velocity is less, equal to, or larger than its characteristic speed of sound in the same medium.

The origin of sound consists of traveling pressure disturbances or waves at speed $c$ given by Eqn. (4.121) caused by a pounding source that compresses a continuous medium: gas, liquid, or solid. They propagate through the medium in all directions in the form of sinusoidal pressure waves of amplitude less than $10^{-3}$ lb$_f$/in$^2$. The speed of propagation of such pressure disturbances, or *speed of sound, c,* defined by Eqn. (4.121), can be determined by mass and momentum balances between an undisturbed point of pressure $p$, density $\rho$, and disturbance velocity $u$ not reached by the sound wave, and an adjacent one of disturbed pressure $p + dp$, density $\rho + d\rho$, and velocity $u + du$ to where the sound wave penetrated.

A sound wave equivalent to a traveling pressure disturbance will penetrate upstream an ongoing flow, remain stationary, or be swept downstream, in a subsonic, sonic, or supersonic flow, respectively. The practical implications of this fluid flow behavior are the following: Consider a subsonic liquid flow in a pipe under constant inlet pressure. By decreasing the outlet pressure, the flow rate will keep increasing according to Eqn. (4.29), until cavitation breaks the continuity of the liquid phase. This is due to the fact that pressure signals of decreasing outlet pressure propagage upstream and accelerate the fluid, given the fact that the speed of sound or pressure signals is higher than the velocity of the liquid in a subsonic flow. Consider now a compressible gas flow in the same pipe. At low velocity where the flow is subsonic, things proceed as explained in the liquid case: gas velocity and mass flow rate keep increasing with decreasing outlet pressure. However, when the gas velocity exceeds the speed of sound and the flow becomes supersonic, any further pressure decrease signal at the outlet will not propagate upstream, the gas will not accelerate any further, and the mass flow rate will cease to increase beyond a maximum value at sonic conditions of about $\mathcal{M} = 1$. The pressure distribution for subsonic and supersonic flow in a pipe is shown in Fig. 4.15. For subsonic liquid and gas flows the pressure drops smoothly and continuously from inlet to outlet. For supersonic flow, this continuous behavior holds only upstream from the outlet, where the local pressure is $p^*$, corresponding to a local $\mathcal{M} = 1$ and sonic or just supersonic gas velocity. The difference $p_{in} - p^*$ drives the maximum mass flow rate achieved, and the pressure drops abruptly from $p^*$ to $p_{out}$, forming a shock wave in the case $p_{out} \ll p^*$. In the case of $p_{out} > p^*$ the flow is, or course, subsonic throughout; at about $p_{out} = p^*$ it attains its maximum mass flow rate and any further reduction of $p_{out}$ such that $p_{out} \ll p^*$ will yield the shock behavior of Fig. 4.15c.

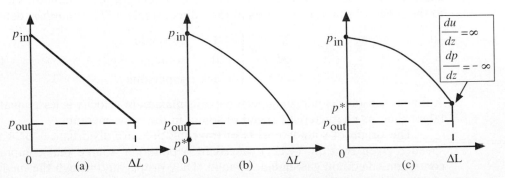

**Figure 4.15** Pipe flows in pipe of length $\Delta L$, inlet pressure $p_{in}$, and outlet pressure $p_{out}$: (a) subsonic liquid flow, (b) subsonic gas flow, and (c) supersonic gas flow.

Consider now the conservation equations for horizontal, compressible ideal gas flow between the inlet of the pipe, where the velocity is $u_0$, the pressure $p_0$, the temperature $T_0$, the density $\rho_0 = MP_0/RT_0$, and the enthalpy $h_0$, and any point downstream where the corresponding variables are $p$, $T$, $\rho$, $u$, and $h$. A mass balance yields

$$\rho_0 \bar{u}_0 = \rho \bar{u} = \frac{\dot{m}}{A}, \tag{4.123}$$

where $\dot{m}$ is the mass flow rate and $A = \pi D^2/4$ the cross-sectional area of the pipe. The total energy conservation equation for steady flow is

$$\frac{h_0}{\rho_0} + \frac{\bar{u}_0^2}{2} = \frac{h}{\rho} + \frac{\bar{u}^2}{2}, \tag{4.124}$$

and equivalently,

$$\frac{h_0}{\rho_0} - \frac{h}{\rho} = c_p(T_0 - T) = \frac{\bar{u}^2}{2} - \frac{\bar{u}_0^2}{2}. \tag{4.125}$$

Equation (4.125) is rearranged to the form

$$\bar{u}^2\left(1 - \frac{\bar{u}_0^2}{\bar{u}^2}\right)\frac{M}{RkT} = \frac{2}{k-1}\left(\frac{T_0}{T} - 1\right), \tag{4.126}$$

where $k = c_p/c_v$. Since $c^2 = RkT/M$ for an ideal gas and $\mathcal{M}_0 = \bar{u}_0/c$, Eqn. (3.76) can also be written as

$$\frac{k-1}{2}\mathcal{M}_0^2\left(\frac{\bar{u}^2}{\bar{u}_0^2} - 1\right) = \frac{T_0}{T} - 1, \tag{4.127}$$

and simplified to

$$\frac{\bar{u}^2(k-1)}{2c^2} = \frac{T_0}{T} - 1, \tag{4.128}$$

under the assumption $\bar{u}_0/\bar{u} \ll 1$. Equations (4.125) to (4.128) indeed suggest that there is cooling of the gas accompanied by decreasing temperature and therefore internal energy that is converted to kinetic energy, giving rise to increased velocity $\bar{u}$. This self-induced acceleration mechanism is absent from incompressible liquid flows due to the fact that unlike compressible gases, liquids cannot expand and cool under the pressure changes from inlet to outlet.

### 4.13.1 High-Speed Gas Flow with Friction

The flow of compressible gases in pipes is practically adiabatic, particularly with heat-insulating walls. However, the flow is never isentropic in the presence of friction, which turns mechanical energy irreversibly into heat. This heat, not being able to escape through the wall, serves to increase internal energy and therefore entropy. This means that the fundamental isentropic relation,

$$\frac{p}{\rho^k} = \frac{p_0}{\rho_0^k},$$ (4.129)

cannot be used unless the flow is assumed frictionless.

To analyze gas flow in the presence of friction, the momentum equation,

$$\dot{m}(\bar{u}_0 - \bar{u}) = \frac{\pi D^2}{4}(p - p_0) + \pi D \tau_{rz}^w \, \Delta z,$$ (4.130)

with

$$\tau_{rz}^w = f_F\left(\rho\frac{\bar{u}^2}{2}\right),$$ (4.131)

must be solved simultaneously with Eqns. (4.123) and (4.125) or (4.128), which hold for any process or flow of ideal gas. The resulting solution is

$$\frac{p}{p_0} = \frac{\bar{u}_0}{\bar{u}}\left\{1 + \left[\frac{(k-1)\mathcal{M}_0^2}{2}\right]\left[1 - \left(\frac{\bar{u}}{\bar{u}_0}\right)^2\right]\right\}$$ (4.132)

and

$$\frac{1}{2k}\left[\frac{2 - (k-1)\mathcal{M}_0^2}{\mathcal{M}_0^2}\right]\left[1 - \left(\frac{\bar{u}_0}{\bar{u}}\right)^2\right] - \left(1 + \frac{1}{k}\right)\ln\left(\frac{\bar{u}}{\bar{u}_0}\right) = \frac{4f_F\,\Delta z}{D}.$$ (4.133)

Equation (4.133) is plotted in Fig. 4.16a and Eqn. (4.132) in Fig. 4.16b.

For a given inlet Mach number $\mathcal{M}_i$ and inlet pressure $p_0$, the velocity of a gas of given $k = c_p/c_v$ will increase and its pressure will decrease with downstream distance $z$, to characteristic values $u^*$ and $p^*$ as shown in Fig. 4.16. For $\mathcal{M}_i \ll 1$, it is approximately $u^* \simeq c$ and $p^* \simeq u_0/c$, where $c$ is the speed of sound in the gas at location $z^*$, where the velocity increases abruptly from $p^*$ to

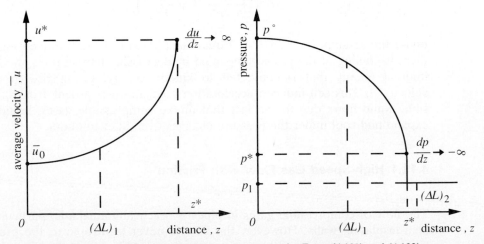

**Figure 4.16**  Predictions of velocity and pressure by Eqns. (4.132) and (4.133).

that applied at the outlet, $p_L$. Thus, if the total length of the pipe $\Delta L$ is smaller than the characteristic $z^*$, the gas flow will proceed smoothly throughout the pipe [case $(\Delta L)_1 < z^*$]. However, if the length of the pipe is larger than $z^*$, the smooth flow persists only down to distance $z^*$ inside the pipe. At $z^*$ the pressure drops abruptly to the externally applied $p_L$ at the outlet [case $(\Delta L)_2 > z^*$]. This means that for a given gas, pipe geometry, and inlet pressure $p_0$, there is an optimum outlet pressure, $p_L = p^*$, where a maximum flow rate $\dot{m}_{max}$ is achieved. Any attempts to increase the flow rate by decreasing the outlet pressure below $p^*$ will fail and the flow rate will continue to be $\dot{m}_{max}$, imposed by the differential $p_L - p^*$. This will proceed unchanged downstream of $z^*$ to the exit, driven by exceedingly high gas expansion. In case of large difference $p_L - p^*$, it may lead to a violent expansion or shock wave.

The maximum flow rate is calculated by means of Eqns. (4.123), (4.132), and (4.133). It is

$$\dot{m}_{max} = A\rho^* u^* = A\frac{Mp^*}{RT^*}u^*$$

$$= \frac{AM}{R}\frac{p_0 u_0^2}{2cT_0}\frac{[2 + (k-1)(\mathcal{M}_0^2 - 1)][2 + \mathcal{M}_0^2(k-1)]^{1/2}[2k + (k-1)\mathcal{M}_0^2]}{\mathcal{M}_0^2(k+1)(k+1)}.$$

$$(4.134)$$

In the common case of $\mathcal{M}_0 = \bar{u}_0/c \ll 1$, it is

$$\dot{m}_{max} = \frac{AM}{R}\frac{p_0 u_0^2}{2cT_0}\frac{(3-k)(2)^{1/2}(2k)}{\mathcal{M}_0^2(k+1)^2}.$$

$$(4.135)$$

### Example 4.17: Isothermal Gas Flow in Pipe

In flow of natural gas through long-distance pipelines, buried in the ground, the flow is kept nearly isothermal.[51] Analyze such isothermal flow of an ideal gas along the lines of the earlier analyzed adiabatic process.

**Solution**   For long pipes the convective term of Eqn. (4.130) is negligible and therefore, for a differential length $dz$, it is

$$\frac{\pi D^2}{4}dp + \pi f_F \frac{\rho \bar{u}^2}{2}dzD = 0.$$

$$(4.17.1)$$

For a perfect gas,

$$\rho = \frac{Mp}{RT},$$

$$(4.17.2)$$

Eqn. (4.17.1) takes the form

$$p\,dp = -2f_F\frac{RT}{DM}\left(\frac{\dot{m}^2}{A^2}\right)dz,$$

$$(4.17.3)$$

where

$$\dot{m} = \rho \bar{u} A. \tag{4.17.4}$$

Integrating Eqn. (4.17.3) yields

$$\dot{m}^2 = \frac{\pi^2 (p_0^2 - p_L^2) D^5 M}{64 f_F (\Delta L) RT} = \frac{\pi^2 (p_0 - p_L^2)(1 + p_L/p_0) \rho_0 D^5}{64 f_F (\Delta L)}, \tag{4.17.5}$$

which is to be compared with the expression for incompressible flow under the conditions given by Eqns (4.27) and (4.29)

$$\dot{m}^2 = \frac{\pi^2 (p_0 - p_L) \rho_0 D^5}{32 f_F (\Delta L)}. \tag{4.17.6}$$

The maximum flow rate predicted by both equations occurs with $p_L = 0$, for which Eqn. (4.17.5) yields $d\dot{m}/dp_L = 0$ and $d^2\dot{m}/dp_L^2 < 0$. Equations (4.17.5) and (4.17.6) combine to give

$$\frac{\dot{m}_{compr.}^2}{\dot{m}_{incom.}^2} = \frac{1}{2}\left(1 + \frac{p_L}{p_0}\right) = 1 + \frac{\Delta p}{2p_0}, \tag{4.17.7}$$

which suggests that isothermal compressible flow is enhanced over the corresponding incompressible flow under the same pressure gradient, by self-induced expansion and acceleration, proportional to $p_L/p_0$.

## 4.14 FLUID TRANSFERRING DEVICES

Pumps are used widely in processing fluids, primarily to overcome frictional losses, to move fluids at desired rates through specified conduits, and deliver them at specified pressure. From a thermodynamic point of view, centrifugal and positive displacement pumps for liquids, and fans and compressors for gases, are machines designed to fulfill these requirements. They consume externally supplied electric or thermal energy in order to increase the kinetic and/or the pressure energy of the transferred fluid. Mixing, regarding temperature and concentration of solute, is also achieved simultaneously by gear pumps. Centrifugal pumps are devices used primarily to deliver fluids at constant pressure, whereas gear and positive displacement pumps are used primarily to deliver fluids at a constant flow rate.

Characteristics of these devices of primary importance include (1) the total required head, defined according to Eqns. (3.41) to (3.47), which represents the difference in mechanical energy between the inlet, or intake (point 1) upstream from the pump, and the outlet or delivery (point 2) downstream from the pump,

$$H = \frac{p_2 - p_1}{\rho g} + \frac{\bar{u}_2^2 - \bar{u}_1^2}{2g} + (z_2 - z_1) + \frac{\mathcal{F}}{g}; \tag{4.136}$$

(2) the power output produced by the pump to achieve such a head,

$$P_{\text{out}} = \rho g H \dot{V} = g H \dot{m} = H \dot{B}. \qquad (4.137)$$

If $\Gamma_{\text{in}}$ is the input to the pump shaft by its motor, the efficiency of the pump is defined by

$$k = \frac{P_{\text{out}}}{P_{\text{in}}} < 1. \qquad (4.138)$$

Thus, Eqns. (4.136) and (4.137) give the required input of electrical or thermal power $P_{\text{in}}$ to intake a fluid of constant density $\rho$ from elevation $z_1$, where its average velocity is $\bar{u}_1$ and pressure $p_1$, move it through a piping system inducing frictional losses $\hat{\mathscr{F}}$ per unit of flowing volume, and finally deliver it at elevation $z_2$ with average velocity $\bar{u}_2$ and pressure $p_2$. A flow diagram for such an arrangement is shown in Fig. 4.17.

The head a specific pump can achieve is determined by means of the energy

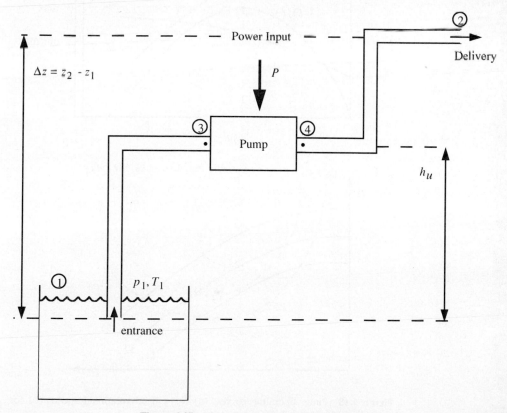

**Figure 4.17**   Flow diagram for pump operation.

equation across the pump, given by

$$H_p = \frac{p_4 - p_3}{\rho g} + \frac{\bar{u}_4^2 - \bar{u}_3^2}{2g} + (z_4 - z_3) + \frac{\mathscr{F}_{3,4}}{g}. \qquad (4.139)$$

A pump operating at rotational speed $\Omega_1$ defines unique curves of achievable head $H_p(\Omega_1)$, of efficiency $k(\Omega_1)$ and of power $P(\Omega_1)$, in terms of achievable delivery rates $\dot{V}$, as shown in Fig. 4.18a and calculated as demonstrated in Example 4.18 (for a centrifugal pump). Called *operating curves*, these are characteristic of and supplied with a pump upon purchase. The optimum operation of a pump at a fixed operating rotational speed occurs at a flow rate $\dot{V}_{opt}$ and total head $H_{opt}$ such that the efficiency is maximum. For rotational speed $\Omega$ expressed in rpm (rounds per minute), a corresponding frequency $N$ in rad/s

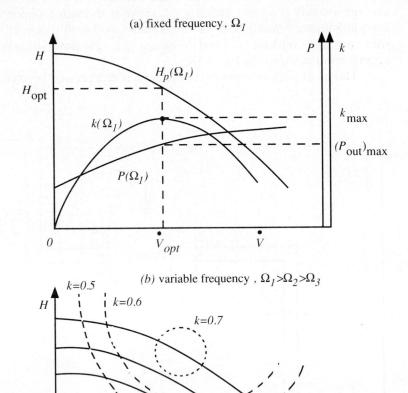

**Figure 4.18**  Pump operating curves: (a) for a fixed rotational speed; (b) for several rotational speeds $\Omega_i$.

(radians per second), or Hz (hertz), is defined by

$$N = \frac{2\pi\Omega}{60} \, \text{Hz}. \tag{4.140}$$

When the rotational speed increases from $\Omega_1$ to $\Omega_2$, the curve for the achievable pump head preserves its shape while being translated upward and to the right, as shown in Fig. 4.18b, where *isoyield* curves are also shown. Figure 4.18b suggests that *optimum operation* is obtained between frequencies $\Omega_2$ and $\Omega_3$. For a given pump, the *operating point* is determined by the intersection of the pump head operating curve of Fig. 4.18a, with the overall required head given by Eqn. (4.136), as shown in Fig. 4.19a.

For the pump represented by Fig. 4.19a the optimum operation is for rotational speeds between $\Omega_2$ and $\Omega_3$ for which the efficiency is maximum. For operation at $\Omega_2$ frequency, the flow rate achieved is $\dot{V}_2$. For a fixed rotational speed the operating point is determined accordingly, as shown in Fig. 4.19b. Due to the frequency constraint, the operation point in Fig. 4.19b is away from the optimum point, which could be achieved by increasing the frequency that would translate upward the curve predicted by Eqn. (4.139).

For liquid pumping, the velocity may change from a small value at the intake point to high values just upstream from the pump, with a simulaneous drop in pressure, according to the mechanical energy equation. Especially when the intake height $h_u$ is large, the pressure may drop to the vapor pressure at the prevailing temperature, in which case liquid starts to turn into vapor. This phenomenon is called *cavitation,* and care must be taken to avoid it. The appearance of vapor reduces the effective cross-sectional area of flow and therefore the flow rate, as discussed in Section 4.12. In addition, the vapor formed travels with the liquid to regions of high pressure within the pump, where it condenses, often with nearly supersonic speed, inducing excessive erosion on sensitive parts of the pump reducing the lifetime of the pump.

Cavitation is inhibited by pump design to include erosion resistant internal parts, and by operation such that the uptake height $h_u$ is not large enough to allow cavitation. The maximum height allowed is

$$(h_u)_{\text{max}} = \frac{p_0 - p^*(T_0)}{\rho g} - \left( \frac{\bar{u}^2}{2g} + \frac{\mathscr{F}}{g} \right), \tag{4.141}$$

where $p_0$ and $T_0$ are the environmental pressure and temperature, $p^*(T_0)$ the vapor pressure of the liquid at $T_0$, $\bar{u}$ the pipe average velocity, and $\mathscr{F}$ the frictional losses between the uptake point and the pump entrance. The phenomenon of cavitation and its prevention continue to be investigated.[53]

In applications where high liquid speeds are unavoidable, for example in rocket fuel feed systems, where a minimum weight-to-volume ratio is vital, cavitation is avoided by means of *pump inducers.*[54] These are axial flow pumps (e.g., screw-type as in Fig. E6.6) with high solidity blades, used in front of the main pump. Such inducers pressurize the liquid in their long helicoidal channels

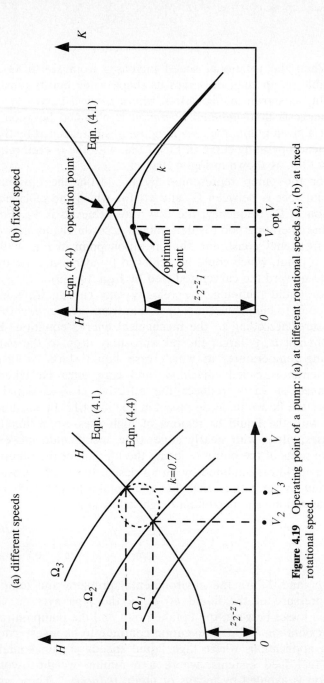

**Figure 4.19** Operating point of a pump: (a) at different rotational speeds $\Omega_i$; (b) at fixed rotational speed.

to a high pressure at the inlet of the main pump. This allows the main pump to operate at high speeds without the risk of cavitation.

### 4.14.1. Centrifugal Pumps

The basic principle of operation of a centrifugal pump is acceleration of the fluid intaken by impinging it at the center of a rotating disk or cap by a shaft that consumes external power, as shown in Fig. 4.20. Initially, the disk is filled with the fluid. Once rotation starts, the fluid is accelerated from the center to the periphery of the disk and a low-pressure region is induced at its center, where the fluid enters the pump. Fluid then flows continuously from the uptake point to the low-pressure center region of the disk and spins outward where it is collected at the periphery by an appropriate housing.

Equation (4.136) applied between points 1 and 2 yields

$$\frac{\bar{u}_2^2 - \bar{u}_1^2}{2} \approx \frac{\bar{u}_2^2}{2} = \frac{W}{m}, \tag{4.142}$$

which estimates the required work $W$ to accelerate the velocity of a fluid mass $m$ from $\bar{u}_1$ to $\bar{u}_2$, under practically constant pressure, $p_2 \simeq p_1$. The velocity is constant between the inlet and outlet of the delivery pipe, and therefore,

$$\frac{\Delta p}{\rho} = \frac{p_3 - p_1}{\rho} = \frac{p_3 - p_2}{\rho} + \frac{p_2 - p_1}{\rho} = \mathcal{F} - \frac{\bar{u}_2^2}{2}, \tag{4.143}$$

with the velocity $\bar{u}_2 \simeq u_3$ given by

$$\bar{u}_2 = \varepsilon \Omega r, \tag{4.144}$$

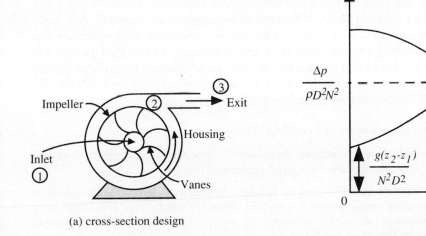

(a) cross-section design

(b) dimensionless curves

**Figure 4.20**  Centrifugal pump: (a) cross-section design; (b) dimensionless characteristic curves.

where $\Omega$ is the rotational speed of the shaft, and $\varepsilon$ is a propeller design parameter. Since $\bar{u}_2$ is prortional to $\dot{V}^n$ in general, as discussed in Section 1.6, Eqn. (4.143) is generalized to the form

$$\frac{\Delta p}{\rho g} = \alpha - \beta \dot{V}^m. \tag{4.145}$$

The pressure head achieved is given by

$$\Delta p = \rho \frac{\bar{u}_2^2}{2} = \rho \frac{\Omega^2 D^2}{8} = \frac{\rho \pi^2 D^2 N^2}{2}, \tag{4.146}$$

where $D$ is the impeller diameter, $\Omega$ its angular speed, and $N$ its frequency of rotation. The flow rate is

$$\dot{V} = (\pi D)(K_1 D)\bar{u}_2 = K D^3 N, \tag{4.147}$$

where $\pi D$ times $K_D$ is the cross-sectional area of the radial flow at point 2. Equations (4.144) to (4.147) define the dimensionless operation curve of the centrifugal pump as shown in Fig. 4.20.

According to the principles of dimensional analysis, as discussed in Section 1.7.1, the two dimensionless numbers shown in Fig. 4.20b fully define the operation of any centrifugal pump of similar design. The figure also shows that the maximum pressure head is achieved under no-flow conditions ($\dot{V} = 0$), whereas for maximum flow rate the head resulting is practically zero. For a given pump design, the required flow rate and/or pressure head are controlled by the rotational speed $\Omega$, which is in turn set by the power $P_{in}$ through Eqns. (4.137) and (4.138), demonstrated in Examples 4.18 and 4.19.

Multistage centrifugal pumps are obtained by combining a number of disks fixed on the same rotating shaft, contained in an appropriate housing that leads the liquid from one disk to another. The overall operation characteristics are obtained by superimposing the individual characteristics. Centrifugal pumps are used to deliver liquids at target pressure in general. Multistage centrifugal pumps are employed where a large pressure head is required, practically over 1000 m, for example to feed industrial boilers and to pump water or oil out of deep wells.

### Example 4.18: Operation Curves of Centrifugal Pump

Discuss methodologies to measure frequency $N$, flow rate $\dot{V}$, and power $P$ and to construct the operation curve of a given centrifugal pump.

**Solution**   The flow rate $\dot{V}$ is easy to measure by means of a volumetric container and by a Venturi meter placed somewhere along the intake line. For larger-scale pumping, the flow rate is also measured by means of drainage weirs[7] for example those shown in Fig. P4.22, for which

$$\dot{V} = \begin{cases} 0.32 \tan \phi h \sqrt{2gh} & \text{(triangular)} \\ 0.66 L h \sqrt{2gh} & \text{(rectangular)}. \end{cases} \tag{4.18.1}$$

The total pump head $H$ is estimated using Eqn. (4.136), by means of two manometers in the intake and delivery pipes, at elevation $z_2 - z_1$ apart, that measure $p_1$ and $p_2$. The absorption and delivery velocities, $u_1$ and $u_2$, are measured from the flow rate $\dot{V}$ and the diameter of the two pipes.

The power input to the pump, $P_{in}$, is measured by means of the electric power consumed. The power produced by the pump, $P_{out}$, is estimated by Eqn. (4.137) and the yield coefficient by Eqn. (4.138) as

$$k = \frac{\dot{V}H\rho}{P_{in}}. \tag{4.18.2}$$

After these measurements are done, curves of the form $k = f(\dot{V})$, $H = f(\dot{V})$, $P_{in} = f(\dot{V})$ are constructed and plotted as shown in Figs. 4.18 and 4.20.

### Example 4.19: Centrifugal Pump Design

Specify a centrifugal pump to deliver $\dot{V} = 1\,\text{m}^3/\text{min}$ of water, raising it from a 10-m-deep well, through a horizontal delivery pipe of $d = 10\,\text{cm}$ internal diameter, to a distance $L = 500\,\text{m}$ away from the well.

**Solution**   For $\dot{V} = 1\,\text{m}^3/\text{min}$ the common velocity of the intake and delivery pipes is

$$u = \frac{4\dot{V}}{\pi d^2} = \frac{4 \times 10^6}{3.14 \times 100 \times 60}\,\frac{\text{cm}}{\text{s}} = 212\,\frac{\text{cm}}{\text{s}}, \tag{4.19.1}$$

and the Reynolds number is

$$\text{Re} = \frac{4\rho\dot{V}}{\eta d} = \frac{4\dot{V}}{\nu d} = \frac{4 \times 10^6}{10 \times 0.01 \times 60} = 6.4 \times 10^6, \tag{4.19.2}$$

which represents a turbulent flow that has friction coefficient $f = 0.008$ from Fig. 4.3. Therefore, the total frictional losses are

$$\mathscr{F} = 4f_F\,\frac{u^2}{2}\,\frac{L}{d} = 0.032 \times \frac{2.12^2}{2} \times \frac{500\,\text{m}^2}{0.1\,\text{s}} = 360\,\frac{\text{m}^2}{\text{s}^2}. \tag{4.19.3}$$

The total required pump head, with $p_2 = p_1$ and $u_2 = u_1$, is at least

$$H = (z_2 - z_1) + \frac{\mathscr{F}}{g} = \left(10 + \frac{360}{9.81}\right)\text{m} = 47\,\text{m}, \tag{4.19.4}$$

corresponding to output power

$$P_{out} = \rho g H \dot{V} = \left(10^3 \times 9.81 \times 47 \times \frac{1}{60}\right)\frac{\text{kg}\,\text{m}^2}{\text{s}^3} = 7.68 \times 10^3\,\text{W} \tag{4.19.5}$$

$$= 7.68\,\text{kW} = 10.3\,\text{hp}. \quad (4.19.5)$$

For $\dot{V} > 0.1\,\text{m}^3/\text{min}$ the yield coefficient is roughly $k = 0.8$, and therefore the input of power required is

$$P_{in} = \frac{P_{out}}{k} = 9.6\,\text{kW} = 12.68\,\text{hp}. \tag{4.19.6}$$

The design of the impeller is based on Eqn. (4.10),

$$\dot{V} = (\pi D)(K_2 D)u_2 = kD^3\Omega = KD^3\,\frac{2\pi N}{60}, \tag{4.19.7}$$

which shows that the specifications can be achieved by different impeller designs rotating with different frequencies $N$.

The possibility of cavitation for cool water is avoided by choosing

$$h_u < \frac{p_0 - p^*(T_0)}{\rho g} \simeq \frac{p_0}{\rho g} \simeq 9 \, \text{m}. \qquad (4.19.8)$$

**Centrifugal fans.**  Based on the same principles of operation as those of centrifugal pumps, centrifugal fans consist primarily of a rotating disk equipped with fins that rotate by air (or gas). Both the pressure and the velocity of the intake air are increased at the delivery side due to the power added by rotation. Often, the air velocity is reduced downstream, accompanied by further pressure increase. The governing equations remain Eqns. (4.136) and (4.137), with appropiate corrections for density variations under conditions of pressure and/or temperature changes across the fan. These devices are commonly used for air ventilation and circulation in confined spaces, such as houses, vehicles, and mines. A different type of application of these devices is to propel boats and helicopters. In automobiles, fans are used for engine cooling and for interior ventilation.

*Fans* in general operate at delivery-to-absorption pressure ratios close to unity, and the input shaft power is used entirely to increase the kinetic energy and therefore the velocity of the airstream. For delivery-to-absorption pressure ratios up to 3, the corresponding devices are known as *blowers,* and beyond that value as *compressors*. In blowers the input energy serves to increase both the velocity and the pressure of the airstream, whereas with compressors most of power is consumed to increase pressure.

### 4.14.2 Piston (or Positive Displacement) Pumps

To deliver liquids at an accurate flow rate, *positive displacement pumps* are used. The basic piping upstream and downstream from the pump remains identical to that shown in Fig. 4.17. However, the principles of operation are different from those of a centrifugal pump. Figure 4.21a shows the fundamental steps of operation. A piston moves back and forth, in and out of a cylindrical chamber that is equipped with two valves, one on the intake side and the other on the delivery side. When the piston moves to the right, suction is induced within the container, the absorption valve opens, and liquid at low pressure fills the container, while the delivery valve remains fixed. Then the piston moves to the left, the liquid in the container is compressed, the intake valve closes, the delivery valve opens, and fluid is delivered. This absorption–compression–delivery cycle continues by the continuous back-and-forth motion of the piston. The piston is driven by a motor that supplies power to the pump in order to overcome frictional loses and to maintain the oscillatory motion of the piston.

Figure 4.21b shows the operating diagram for a displacement pump. The abrupt change of pressure at the two far-end positions of the piston is a consequence of the liquid's incompressibility, whereas the horizontal lines reflect

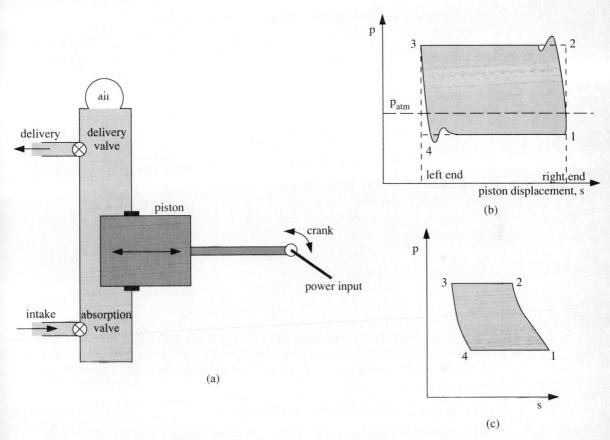

**Figure 4.21** Piston pump: (a) design; (b) operating diagram for liquid; (c) operating diagram for gas.

volume changes, proportional to $\Delta V = (\pi D^2/4)\,\Delta S$, due to incoming or discharged liquid, during the absorbtion and delivery cycles, respectively. The deviations between the theoretical (dashed) and real (solid) curves are due primarily to the rise time that it takes for a valve to open or close.

The difference between atmospheric pressure and absolute vacuum is 1.033 bar, which is equivalent to a 10.33 m column of water. Thus, the theoretical maximum intake height for a liquid of density $\rho$, under atmospheric conditions, to avoid cavitation, is

$$(z_u)_{max} = 10.33\,\frac{\rho_\omega}{\rho}. \tag{4.148}$$

On the other hand, the delivery height can theoretically be infinitely large, inasmuch as the motor inputs the appropriate power and the pump design can maintain a stable operation at high piston speeds and frequencies.

The total head achieved is given by Eqn. (4.136). When the uptake and delivery are done under atmospheric conditions (i.e., $p_1 = p_2$), and the uptake and delivery pipes are of the same diameter (i.e., $\bar{u}_1 = \bar{u}_2$), Eqn. (4.136) simplifies to the form

$$H = (z_2 - z_1) + \frac{\hat{\mathcal{F}}}{\rho},$$
(4.149)

and the corresponding power is given by Eqn. (4.2). The flow rate achieved is

$$\dot{V} = \frac{\pi D^2}{4} \frac{N}{60},$$
(4.150)

where $D$ is the piston diameter and $N$ the frequency in hertz (cycles per second). In general, $50 < N < 500$, to avoid pump material fatigue due to the piston acceleration back and forth.

The output or useful power of a displacement pump is given by Eqn. (4.137) which can be cast in the form

$$P = \dot{V} \, \Delta p.$$
(4.151)

The input or consumed power is

$$P_{\text{in}} = \frac{P_{\text{out}}}{k},$$
(4.152)

where $k$ is a yield coefficient that has values between 0.7 and 0.9 in the presence of a crank, and between 0.85 and 0.95 when the piston is connected directly to the motor. Displacement pumps are commonly applied to feed devices such as boilers, chemical reactors, and absorption beds.

### Example 4.20: Characteristics of Displacement Pumps

What are the specifications of a positive displacement pumped used to feed crude oil of viscosity 0.1 P and density 0.98 g/cm³ to a pilot-scale trickle-bed reactor operating at pressure $p = 50$ atm, through a short horizontal pipe of internal diameter $d = 2$ cm at a volumetric flow rate 1 cm³/s?

**Solution**  The pressure drop due to frictional losses and barometric head is negligible for a short horizontal pipe. The power input required is determined by Eqns. (4.151) and (4.152):

$$P_{\text{in}} = \frac{P_{\text{out}}}{k} = \frac{\dot{V} \, \Delta p}{k} = \frac{1 \text{ cm}^3/\text{s} \times (50 - 1) \text{ atm}}{0.8}$$

$$= \frac{1 \times 49 \times 9.8 \times 10^5}{0.8} \frac{\text{erg}}{\text{s}} = 6 \text{ W} = 0.0006 \text{ hp}.$$

The pump design equation is Eqn. (4.150). For a piston of diameter $D = 5$ cm, the cycle frequency is

$$N = \frac{240 \dot{V}}{\pi D^2} = \frac{240 \times 1}{3.14 \times 36} \frac{\text{cycles}}{\text{s}} \approx 2 \text{ Hz}.$$

Compressors are large piston pumps used to compress and deliver gas at high pressure to pneumatic devices or processes. Due to the compressibility of gas, the operation diagram changes to that of Fig. 4.21c, where the enclosed area represents the useful work given by the compressor. Its value depends on the type of the compression stage 1–2 that is used: e.g., isothermal, isentropic, or polytropic, as detailed in thermodynamics textbooks. The volumetric delivery rate of gas is given by

$$\dot{V} = V_P \frac{N}{60},$$
(4.153)

where $V_P$ is the gas volume per piston cycle at the intake conditions and $N$ is the cycle frequency per minute. The power achieved is estimated by

$$P = \frac{p_1(V_1 - V_4)N}{m}\left[1 - \left(\frac{p_2}{p_1}\right)^m\right],$$
(4.154)

where $m = (n - 1)/n$, $n$ being the exponent of the polytropic process, which follows the ideal gas law,

$$\frac{V_2}{V_1} = \left(\frac{P_1}{P_2}\right)^{1/n},$$
(4.155)

and the pressures and volumes are as shown in Fig. 4.21c. The power supply required is given by Eqns. (4.151) and (4.152).

### 4.14.3 Gear Pumps

The basic design of a gear pump is shown in Fig. 4.22. The liquid enters and fills compartments between teeth that rotate in contact with the housing, such that the liquid is transferred to the discharge position at the opposite end. Each compartment is in contact with the housing such that there is no transfer of liquid from one compartment to another. Thus, the liquid recirculates and mixes in each compartment while being transferred from the intake to the discharge position by the rotating gear.

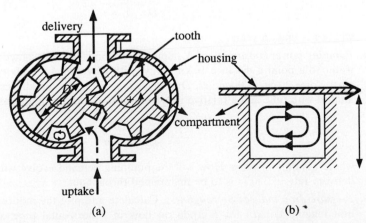

(a)           (b)

**Figure 4.22**  Gear pump and individual tooth.

The flow rate achieved by a gear pump is estimated by

$$\dot{V} = k\pi DhlN, \tag{4.156}$$

where $D$ is the bottom diameter of the tooth, $h$ its height, $l$ its length (toward the paper here), $n$ the frequency of shaft rotation, and $k$ efficiency, commonly $0.9 < k < 0.95$. Gear pumps are used primarily for lubrication of machinery and vehicles and in hydraulic devices, where pressures up to 100 atm may be achieved. They are also used for high-accuracy metering of fluids.

## PROBLEMS

1. *Laminar and turbulent flows.* Calculate the Reynolds number of horizontal flow in a pipe of diameter 3 in. at a flow rate $\dot{V} = 500\ \text{in}^3/\text{s}$ of incompressible air, water, crude oil of viscosity $\eta = 100\ \text{P}$ and specific gravity 0.85, and polycarbonate melt of viscosity $\eta = 10^5\ \text{P}$ and specific gravity 0.90. Classify the flows as laminar or turbulent. For each flow, what is the required input of power to move the fluid if the applied pressure difference is $\Delta p = 60\ \text{psi}$? What is the friction force overcome in each case? (Useful formulas may include $P = \dot{V}\,\Delta p = \mathscr{F}\bar{u}$.)

2. *Commercial steel pipe selection.* Steel pipe of total length 2000 m is required to transfer $0.01\ \text{m}^3/\text{s}$ water to an elevation of 1000 m. The total available power input is 6000 kW at most. What is the appropriate pipe size among those listed in Table 3.3? What additional power will be required after a year of operation which will induce 0.002 relative roughness?

3. *Transfer rates in pipe.* Viscous crude oil of 1 P viscosity and $0.9\ \text{g/cm}^3$ density is transferred through a horizontal pipe over a 10-km distance. If the consumed pumping power is 2000 kW, calculate the transfer rates for a smooth pipe of diameter 5, 10, or 20 cm. How are these rates affected with time if corrosion gives rise to relative roughness according to the empirical equation

$$\frac{\varepsilon}{D} = \frac{0.01t}{1 + t},$$

where $t$ is time in years?

4. *Pumping power consumption.* A pumping station is to be designed to deliver irrigation water to a point a distance 10 km away, with elevation 1 km above the inlet, through plastic pipe 20 cm in diameter. Due to algae growth, the initially smooth wall surface is altered with time and a relative roughness is induced, given approximately by

$$\frac{\varepsilon}{D} = \frac{0.01t}{1 + t},$$

where $t$ is time in years. How will the pumping demand evolve with time if a constant delivery rate of $3\ \text{m}^3/\text{s}$ is to be maintained throughout?

5. *Pressure drop induced by roughness.* Calculate and plot the induced pressure drop per unit length, $\Delta p/\Delta L$, for a crude oil flow in a horizontal pipe of internal diameter $D = 2$ in. The viscosity $\eta = 5\ \text{P}$ and density $\rho = 0.85\ \text{g/cm}^3$. Consider a smooth

surface, and roughness $\varepsilon/D = 10^{-4}$, $10^{-3}$, $10^{-2}$, 0.05. Distinguish and plot viscosity and roughness contributions to the total pressure drop.

6. *Pipe size.* A smooth pipe is required to provide water at 180 gpm to an elevation of 100 ft above the inlet through a 1500-ft-long pipe.
   **(a)** What is the required minimum pipe diameter?
   **(b)** What would be the side length of a closed channel instead of a pipe?
   **(c)** What are the corresponding requirements for horizontal transfer and for transfer to a point 100 ft below the inlet under the same power input?

7. *Delivery rate.* A nominal size 3 in., schedule 80, 1-km-long steel pipe is to be used to transfer water at elevation 100 m above the inlet.
   **(a)** What is the resulting delivery rate?
   **(b)** What would be the corresponding delivery rates horizontally, and to a point 100 m below, under the same power input?

8. *Frictional energy losses.*
   **(a)** Calculate the flow rate achieved initially for the system of Fig. P4.8, for a smooth pipe of 5 cm diameter, transferring water from the upper to the lower reservoir.
   **(b)** What would be the power input required by a pump to achieve the opposite flow, from a lower to an upper reservoir?
   **(c)** Repeat (a) and (b) for two different relative roughness values, 0.001 and 0.01.

9. *Corossion-induced friction.* A commercial pipe is designed to transfer $\dot{V} = 10 \, \text{m}^3/\text{min}$ of corrosive fluid by means of a pump that inputs $P = 10 \, \text{W}$ to the system. After a time, the roughness of the pipe's internal area increases such that the friction coefficient, $f_F$, doubles in value. What new power input and resulting pressure drop are required to transfer the same amount of fluid?

10. *Roughness reduction.* The power required to move a certain amount of a corrosive viscous fluid of viscosity $\eta = 1 \, \text{P}$ and density $0.95 \, \text{g/cm}^3$ at a flow rate $\dot{m} = 10 \, \text{kg/s}$ through a pipe of internal diameter 2 in. over a length of 1 km is reduced by 15% after the pipe is flushed by a surface cleaning agent. To what degree was the surface roughness reduced by such an operation? (Pipeline engineers call it "pigging" the pipe.)

11. *Friction in open channel.* An open square concrete channel of side length $d = 1 \, \text{m}$, at an angle of $10°$ with the horizontal, transfers irrigation water over a 1-km horizontal distance.

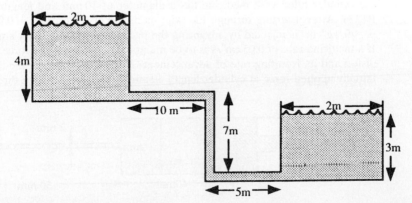

**Figure P4.8**   Reservoir connection.

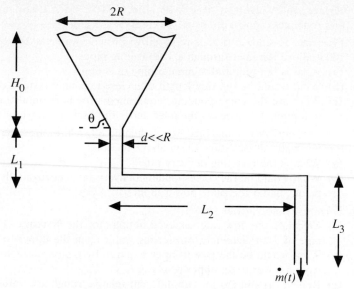

**Figure P4.12**    Drainage of conical tank.

(a) What are the resulting transfer rate and corresponding friction factor?

(b) The friction doubles in value due to the increase in roughness. What is the new flow rate?

(c) What power would be required to move the same liquid at the same flow rates in a closed channel of the same dimensions?

12. *Drainage of conical tanks with piping.*[55] For the conical tank of Fig. P4.12, estimate the time required for complete drainage, as follows.

(a) Find the expression that yields the height of water $H(t)$ at time $t$, where $H(t = 0) = H_0$. Then find the time $t^*$ such that $H(t^*) \approx 0$.

(b) Estimate $t^*$ in the case of $H_0 = 2$ m, $R = 0.5$ m, $d = 2$ cm, $L_1 = 0.5$ m, $L_2 = 2$ m, and $L_3 = 0.5$ m.

13. *Hypodermic needle.* The needle has an inside diameter 2 mm and length of 50 mm, and the cylinder filled with medicine has a diameter of 10 mm and length of 40 mm (Fig. P4.13). After inserting through the skin, medicine of viscosity $\eta = 0.9$ cP and density $\rho = 0.9$ g/cm$^3$ is injected by advancing the piston against the artery's pressure of 7 psi. If a medicine rate of 0.05 cm$^2$/s is to be maintained, what are the force required on the piston and its resulting rate of advancement? What would be the medicine rates if the initially applied force at cylinder length 40 mm were kept constant throughout?

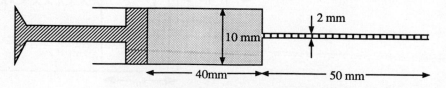

**Figure P4.13**    Hypodermic needle.

**14.** *Friction for flow over flat plate.* When an approaching free stream at velocity $U = 1$ m/s of a fluid of kinematic viscosity $v = 1$ St flows over a flat plate of negligible thickness and length $L = 2$ m placed parallel to the stream, the resulting disturbed velocity over the plate, $u(x, y)$, as shown in Fig. E3.7, is given approximately by

$$\frac{u(x, y)}{U} = \frac{1.5y}{\delta(x)},$$

where $x$ is the distance downstream from the plate leading edge, $y$ the distance over the plate, and $\delta(x)$ the thickness of the resulting boundary layer given approximately by

$$\delta(x) = 4.64\sqrt{\frac{vx}{U}}.$$

**(a)** Plot the thickness $\delta(x)$ vs. downstream distance $x$.
**(b)** Plot the resulting velocity profile at downstream distances of $x = 0.2L$, $x = 0.5L$, and $x = 0.8L$.
**(c)** Show that under laminar flow conditions, the friction factor $f_F$, defined by

$$\frac{F_D}{L} = 2f_F\rho\frac{U^2}{2},$$

where $F_D$ is the total drag force on both sides of the plate, is

$$f_F = \frac{1.328}{\sqrt{Re}} \qquad Re = \frac{UL}{v}.$$

**(d)** Repeat the considerations for turbulent flow with velocity profile

$$\frac{u(x, y)}{U} = \left[\frac{y}{\delta(x)}\right]^{1/7}.$$

**15.** *Friction in shear-thinning liquids.*[56] Show that the relation

$$\tau_{rz}^w = \frac{R}{2}\frac{\Delta p}{\Delta L},$$

for pipe flow, holds for Newtonian as well as for non-Newtonian liquids. For power-law, shear-thinning liquids the same wall stress is also given by the constitutive equation,

$$\tau_{rz}^w = k\left(\left|\frac{du}{dr}\right|\right)^{n-1}\frac{du}{dr} \qquad n < 1,$$

which yields Newtonian behavior in the limit of $k = \eta$ and $n = 0$. The resulting velocity profile is

$$u_z(r) = \left(\frac{1}{4k}\frac{\Delta p}{\Delta L}\right)^{1/n}\left(\frac{n2^{1/n}}{n+1}\right)(R^{1+1/n} - r^{1+1/n}).$$

(a) Show that this profile reduces to a Newtonian profile in the limit of $n = 1$ and $k = \eta$, as it should.

(b) Calculate the mean velocity, defined by

$$\bar{u} = \frac{\dot{V}}{\pi R^2} = \frac{1}{\pi R^2} \int_0^R 2\pi r u_z(r)\, dr.$$

(c) By using the equation

$$\frac{\mathscr{F}}{\Delta L} = \frac{\Delta}{\Delta L} = 4g_F \frac{\bar{u}^2}{2}\left(\frac{\Delta L}{2R}\right),$$

show that the resulting friction factor for laminar shear-thinning flow is

$$f_F = 16\left[8\left(\frac{n}{6n+2}\right)^n \frac{2\rho r^n \bar{u}^{2-n}}{k}\right]^{-1} = \frac{16}{\text{Re}_n}.$$

Does this expression reduce to the Newtonian fluid expression in the appropriate limit?

(d) Compare and comment on the differences between Newtonian and shear-thinning frictional losses under similar flow conditions.

(e) For $n > 1$, a shear-thickening liquid is modeled. Compare the resulting frictional energy losses for this liquid to those for Newtonian and shear-thinning liquids.

(f) Explain your comparisons by observing that according to the constitutive equation, the resulting shear stresses depend on the exponent $n$.

16. *Frictionless gas flow in pipe.* Calculate the temperature, velocity, density, and mass flow rate of an ideal gas flowing in an insulated pipe starting from a large reservoir of temperature $T_0$, pressure $p_0$, and vanishingly small velocity. Why is the flow isentropic? To what physical situation does it correspond?

17. *Inviscid, reversible gas flow.*[57] Show that the governing momentum equation is

$$\frac{du}{u}(1 - \mathscr{M}^2) = -\frac{dA}{A},$$

where $u$ is average gas velocity, $A$ is cross-sectional area of flow,

$$\mathscr{M} = \frac{u}{c} = \frac{u}{dp/d\rho}$$

is the Mach number, and $c$ is the sonic velocity.

(a) Simplify the equation and solve for subsonic, sonic, and supersonic flow.

(b) For flow in horizontal pipe show that the mass flow rate is

$$\dot{m} = \frac{2(\rho_2 A_2)^2 \int_{p_1}^{p_2} (dp/\rho)}{1 - (p_2 A_2/p_1 A_1)^2},$$

between any two points 1 and 2 along the pipe.

(c) Solve the preceding equation for the cases of isothermal and incompressible gas flow, and comment on your answers.

18. *Drag reduction fluids.*[58] The friction factor in pipes is reduced with the addition of polymeric additives to the flowing liquid. An empirical expression in this case is[59]

$$\frac{1}{\sqrt{f_F}} = (4 + \alpha)\log(\mathrm{Re}\sqrt{f_F}) - 0.4 - \alpha \log(\sqrt{2}\,\alpha\beta),$$

where $\alpha$ and $\beta$ are empirical constants depending on the type and concentration of polymer.

(a) What is the value of $\alpha$ in the limit of zero polymer concentration? [Hint: Compare with friction factor equation(s) for a pure liquid.]

(b) What is a realistic range for $\alpha$ and $\beta$ values?

(c) Construct a qualitative plot of $f_F$ vs. $\alpha$, given the limit calculated in (a) and the maximum drag reduction asymptote,

$$\frac{1}{\sqrt{f_F}} = 19\log(\mathrm{Re}\sqrt{f_F}) - 32.4.$$

19. *Energy conservation in open-channel flow.*[60] A typical open-channel flow at high Reynolds number is represented by a liquid layer of depth $h$ and average velocity $\bar{u}$, flowing down an inclined plane at angle $\phi$ with the horizontal. A pressure head–like energy along the channel for such flows is defined by

$$H = h + \frac{u^2}{2g} = h + \frac{1}{2g}\left(\frac{\dot{V}}{Wh}\right)^2,$$

where $\dot{V}$ is the constant volumetric flow rate and $W$ is the channel width. Plot $H$ vs. $h$ and comment on the features of the resulting parabolic curve in relation to the two axes and to the line $H = h$.

20. *Momentum conservation in open channel.*[60] A prototype flow over an obstruction is shown in Fig. P4.20. Show that the momentum change over the obstruction is

$$\Delta J = J_1 - J_2 = \left(\frac{\rho g h_1^2}{2} + \frac{\rho \dot{V}^2}{h_1 W^2}\right) - \left(\frac{\rho g h_2^2}{2} + \frac{\rho \dot{V}^2}{h_2 W^2}\right),$$

where $\dot{V}$ is the constant volumetric flow rate and $W$ is the width of the channel. Plot $J$ vs. $h$ and comment on the qualitative features of the resulting parabolic curve.

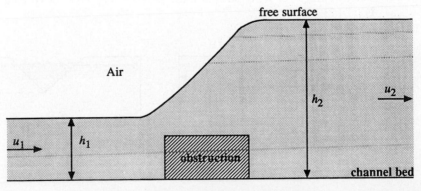

**Figure P4.20**   Open flow over obstruction.

21. *Economical cross section.* A typical aim in designing flow channels is to minimize the conduit's cross-sectional area $A$, for a fixed flow rate $\dot{V}$ or for a given cross-sectional area, to maximize flow rate. Often, the Chezy equation,

$$\frac{\dot{V}}{A} = c\sqrt{\frac{A\cos\phi}{\pi}},$$

is used, where $\pi$ is the wetted perimeter, $\phi$ the inclination, and $c$ is the Chezy coefficient, approximated by

$$c = \frac{R^{1/6}}{n},$$

where $R$ is hydraulic radius. Show that the most economical cross section is the one for which the width is twice its depth. Repeat your considerations for an elliptical cross section of maximum and minimum axes $a$ and $b$, respectively.

22. *Open-channel measurements.*[61] A weir is some type of obstruction, built across an open channel, over which the liquid flows and measured, as shown in Fig. P4.22, for example.

  **(a)** Under zero approach velocity from inside, the liquid layer falls over the weir as an ideal free-falling body. Show that its mean velocity is

$$u = \sqrt{2gh}.$$

  Then show that the resulting ideal flow rate for a rectangular weir is

$$\dot{V} = \tfrac{2}{3}L\sqrt{2gh^3}.$$

  **(b)** In practice, the approach velocity is not zero, in which case the Francis formula,

$$\dot{V} = 3.33\left(L - \frac{nh}{10}\right)\left[\left(h - \frac{U_a^2}{2g}\right) - \left(\frac{U_a^2}{2g}\right)^{3/2}\right],$$

  applies, with $U_a$ the approach velocity, and $n$ an empirical constant, depending on the liquid nature and the velocity range.

  **(c)** What would be the corresponding expressions for a triangular weir?

23. *Hydraulic jumps.* These flows are most commonly used to dissipate the energy of water flowing over weirs, such as those shown in Fig. P4.22, according to the sketch of Fig. P4.23.

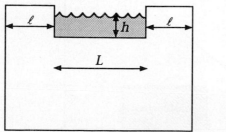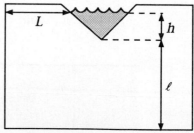

**Figure P4.22**   Open-channel measurement, by rectangular and triangular weirs.

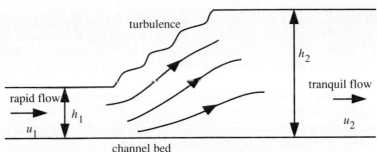

turbulence

$h_2$

tranquil flow

rapid flow

$h_1$

$u_1$

$u_2$

channel bed

**Figure P4.23**   Hydraulic jump.

(a) Find the relation between the depths upstream and downstream from the jump.

(b) Show that the energy is indeed reduced by such an arrangement. What is the resulting energy loss?

24. *Baseball orbit.* It can be shown that the trajectory of a baseball can be represented by the equations

$$m \frac{d^2x}{dt^2} = -k \frac{dx}{dt} \sqrt{\left(\frac{dx}{dt}\right)^2 + \left(\frac{dy}{dt}\right)^2} \tag{P4.24.1}$$

and

$$m \frac{d^2y}{dt^2} = -k \frac{dy}{dt} \sqrt{\left(\frac{dx}{dt}\right)^2 + \left(\frac{dy}{dt}\right)^2} \tag{P4.24.2}$$

with $k$ defined by

$$k = \frac{C_D A \rho}{2},$$

where $\rho = 1 \text{ kg/m}^3$ is the air density, $m = 0.66 \text{ kg}$ is the mass of the ball, $A = 40 \text{ cm}^2$ is the ball's projected area, and $C_D = 0.44$ is the drag coefficient. Also,

$$u_x = \frac{dx}{dt} \qquad u_y = \frac{dy}{dt} \qquad u = \sqrt{u_x^2 + u_y^2} \tag{P4.24.3}$$

are the horizontal, vertical, and total velocities, respectively.

(a) Show how Eqns. (P4.24.1) and (P4.24.2) can be derived, by considering the composite motion in the $(x, y)$ plane.

(b) Describe the orbit of the baseball after being hit in the absence of gravity, under initial strike velocity $u(t = 0) = 100 \text{ m/s}$ at an angle $\theta = 30°$ with the horizontal, i.e., $u_x(t = 0) = 50\sqrt{3} \text{ m/s}$ and $u_y(t = 0) = 50 \text{ m/s}$ and initial hit point at elevation $y_0 = 1.5 \text{ m}$.

(c) In the presence of gravity, how does the variation of air density with altitude influence pitching and striking distances?

(d) Cast the equations in the form

$$\frac{du_y}{du_x} = \frac{u_y}{u_x} + \frac{mg}{ku_x \sqrt{u_x^2 + y_y^2}}, \tag{P4.24.4}$$

and find $u_y$ in terms of $u_x$, analytically or numerically, by advancing from the known values at $t = 0$.

(e) Use Eqn. (P4.24.3) to find and plot the orbit $y(x)$.

25. *Terminal velocity and stopping distance.* A spherical fluid droplet of density $\rho_1 = 1 \, \text{g/cm}^3$ and viscosity $\eta_1 = 1 \, \text{cP}$ is injected in a second immiscible liquid of density $\rho_2 = 1 \, \text{g/cm}^3$ and viscosity $\eta_2 = 0.25 \, \text{cP}$, at initial velocity $u_0 = 1 \, \text{m/s}$. Describe the motion of the droplet for the cases: (a) horizontal release; (b) vertical release upward; (c) vertical release downward; (d) diagonal release at $45°$ upward. Assume that the droplet maintains a spherical profile with radius $R = 0.5 \, \text{cm}$ in the presence of large interfacial tension.

26. *Drag coefficient corrections.*[19] The following analytic equations have been proposed for the drag coefficient, $C_D$, for spherical particles:

$$C_D = \frac{24}{\text{Re}}, \qquad \text{Re} < 0.1$$

$$C_D = \frac{24}{\text{Re}}\left[1 + \frac{3\,\text{Re}}{16} + \frac{9\,\text{Re}^2}{160}\ln(2\,\text{Re})\right] \qquad 0.1 < \text{Re} < 2$$

$$C_D = \frac{24}{\text{Re}}(1 + 0.15\,\text{Re}^{0.687}) \qquad 2 < \text{Re} < 500$$

$$C_D = 0.44 \qquad 500 < \text{Re} < 200{,}000.$$

  (a) Plot these predictions on top of the curve of Fig. 4.5 and compare the two curves.
  (b) Where is the biggest disagreement? Why?
  (c) What would be corresponding expressions for cylindrical particles?
  (d) What are the advantages and disadvantages of Fig. 4.5, compared to these analytical expressions?
     Typical aerosol particles of water-like liquids in air range in diameter from 10 to 1000 $\mu$m. Calculate the resulting Reynolds number, the drag coefficient, the terminal velocity, and their difference from solid particles of such droplets of diameter 10, 50, 100, 200, 400, and 1000 $\mu$m.

27. *Drag on liquid droplet.* Consider the droplets obtained by the breakup of the jets of Example 4. For each case, calculate the terminal velocity and the resulting drag force. Include the case of evaporating droplet, if appropriate, by utilizing Eqn. (4.67) to describe the drag coefficient.

28. *Gravity settling.*[62] Consider gravity settling of particulates from mixed room air onto a tabletop.
  (a) Assuming spherical particles of a single size $d_p$ settling according to Stokes' law and with a particulate loading of $Y_p$, derive an expression for the flux of particle image area onto the tabletop, say in units of (cm$^2$ of particle image area)/(s)(cm$^2$ of table surface).
  (b) If air density is taken as $1.2 \times 10^{-3} \, \text{g/cm}^3$ and air viscosity is taken as $1.8 \times 10^{-4} \, \text{g/cm s}$, if the shiny surface of a table is completely obscured in 20 days and the particulates are 3 $\mu$m in diameter, does the room quality satisfy standards for particulates?
  (c) Assuming that the air does meet air quality standards with regard to particulate loading and that the particulate size is 0.3 $\mu$m instead of 3 $\mu$m, after how many days would the tabletop (or "lung surface") be completely obscured because of

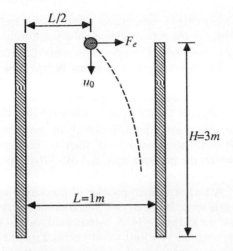

**Figure P4.29** Vertical electrostatic collector.

gravity settling from a mixed air flow over the surface? (From Ref. 62, by permission.)

29. *Electrostatic collector.* A spherical droplet of diameter $d_p = 1$ mm of a liquid of viscosity $\eta_p = 0.2$ cP and density $\rho_p = 1$ g/cm$^3$ enters a horizontal electrostatic field at vertical speed $u_0 = 1$ m/s. (Fig. P4.29). The electrostatic field extends to distance $H = 3$ m and results in a homogeneous horizontal force on a particle of magnitude $F_e = 30$ dyn. The properties of the electrified air at these conditions are $\eta_a = 0.05$ cP and $\rho_a = 0.01$ g/cm$^3$. Will the spherical liquid droplet be collected by the device? If not, what design changes would you make to collect such particles? You can make any reasonable engineering assumptions and approximations to facilitate the solution, but you need to justify these assumptions a priori and evaluate a posteriori. (From Ref. 63, by permission.)

30. *Porosity.* Compute the porosity of synthetic porous media for (a) flow perpendicular to parallel pipes of radius $r$, with centers located at the corners of squares of side length $l$; (b) flow parallel to the same arrangement; (c) flow perpendicular to a similar arrangement with pipe centers located at the corners of isosceles triangles of side length $l$; (d) flow parallel to the latter arrangement; (e) flow through spheres of radius $r$ placed with centers located at the corners of a cube of side length $l$, and then at the corners of an isosceles pyramid of side length $l$. For all cases examine the limits of $l = 2r$ and $r/l \to 0$, and comment on their physical significance.

31. *Filtration under constant pressure.* The following data were obtained with a pilot-scale filter of area 400 cm$^2$ and thickness $H_f = 10$ cm, operating under constant pressure 0.2 atm: At times 0, 9, 25, 36, 49, and 64 min, the amounts of filtrate were 0, 3.1, 5.2, 6.3, 7.4, and 8.5 L. The slurry entering was a water-based particle suspension, and the filtrate is essentially water at 10% less volumetric flow rate.
    (a) Is the assumption of negligible filter resistance valid?
    (b) Calculate the permeability of the accumulated cake.
    (c) In a typical filtration operation, the filter is shut down after time $t_{max}$, the cake is removed, the filtercloth cleaned, and the filter reassembled within a total

regeneration time $t_{off} = 20$ min. What value of $t_{max}$ maximizes the amount of the cake recovered? What are the corresponding values of filtrate and cake produced?

32. *Filtration under constant flow rate.* Consider the same filter of Problem 4.31 of permeability $k_f = 0.1$ darcy, operating at a constant filtrate flow rate $\dot{V} = 1$ m$^3$/min and forming a cake of constant permeability $k_c = 0.1$ darcy and porosity $\varepsilon = 0.65$. Calculate the evolution of pressure with time. When does this filtration become practically impossible? How may filtration proceed beyond that critical point?

33. *Air-cleaning filters.* In a cement plant, particle-rich air of concentration 1-cm$^3$ particles per cubic meter of air is to be cleaned by a filter of permeability 0.2 darcy and thickness 1 cm, across which the pressure drop is 1 psi. What rate of air cleaning and amount of particles collected are achieved in 24 h of operation?

34. *Chromatographic bed.*[38] A bed of sorbent powder of porosity $\varepsilon$ and particle diameter $d_p$ is formed in the annular space between two thin concentric cylindrical porous walls of radii $R_1$ and $R_2$, as shown in Fig. P4.34. A carrier fluid, rich in the compound to be recovered, is pumped through the interior tube at pressure $p_1$ and left to penetrate the bed radially, where the compound is absorbed and later recovered by regenerating the bed.

    **(a)** What is the resulting gas flow rate if absorption of the compound does not alter the porosity or the local pressure?

    **(b)** Under this constant flow rate $\dot{V}$, how does the pressure $p_1$ change with time if absorption results in a porosity decrease given by $\varepsilon(t) = \varepsilon_0 \exp(-kt) \ll 1$ for a carrier gas of concentration of $c$ volume of compound per volume of carrier gas?

35. *Water purification.*[38] A water purification device consists of a hollow vertical cylinder of 2 m height and 0.15 m radius packed with ion-exchange resin particles. For water flow under gravity alone, calculate the resulting pressure drop for different rates of purification, say from 0 to 100 gph. The permeability of the bed is 0.04 darcy and the particle diameter 0.5 cm. If flow is reversed, what pumping capacity is required to maintain the same flow rates? What if the bed were horizontal? Comment on the differences and advantages among the three arrangements.

36. *Fluidization.* A catalytic bed of height 30 cm and radius 10 cm is supported by a metal screen of negligible resistance to flow. The catalytic particles have average diameter 0.5 cm and density 1.5 g/cm$^3$.

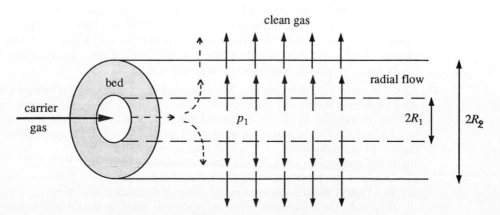

**Figure P4.34**   Annular chromatographic bed for gas cleaning.

(a) Calculate the most likely porosity and permeability of the bed for free-falling water at a rate of 1 ft³/min.
(b) What pumping power is required for upward flow at the same flow rate?
(c) Calculate the fluidization velocity for a flow of crude oil of viscosity 10 cP, water of viscosity 1 cP, and air of viscosity 0.1 cP.
(d) What fluidization power is required in the three cases?

**37.** *Oil recovery.* Consider a well producing oil at the volumetric rate $\dot{V} = 100$ gpm in the arrangement of Fig. 4.13. The oil in the bed flows radially inward, while its pressure is monitored at the porous bore wall of the radial distance 5 in. to be 4 atm and at the radial distance 400 in. to be 8 atm. What would be the pressure difference between the two points if the flow rate increased to 200 gpm? In a later enhanced recovery procedure, what are the required rates of water to displace oil of 10 cP viscosity and 50 dyn/cm interfacial tension through a porous bed of permeability 0.01 darcy and average pore diameter 2 mm? Plot the admissible rates in terms of the ratio $k\sigma/\eta r_p^2$. What do you observe?

**38.** *Water-well operation.* Figure 4.13 may show the horizontal cross section of a well of radius $r_1$ in a bed of fine sand that produces water at a volumetric flow rate $\dot{V}$ per unit depth at pressure $p_1$. The water flows radially inward from the outlying region, with symmetry about the axis of the well. A pressure transducer enables pressure $p_2$ to be monitored at a radial distance $r_2$.
(a) Show that the superficial radial velocity $u$ varies with radial position $r$ according to

$$u = \frac{\dot{V}}{2\pi h r}.$$

(b) For 100 gpm the recorded pressures at radial positions $r_1 = 3$ in. and $r_2 = 300$ in. were 10 and 30 atm, respectively, and for 200 gpm were 20 and 60 atm. What would be the corresponding pressures for 300 gpm?
(c) What is the effective permeability of that soil?

**39.** *Composite impregnation.*[43] A Newtonian polymeric melt of 100 P viscosity is applied at constant pressure of 10 atm at the center of a circular cloth specimen of radius 1 ft, height 10 cm, and permeability 0.1 darcy, and left to advance slowly radially to impregnate the entire cloth, as shown in Fig. E4.6.
(a) What is the resulting radial velocity?
(b) How does the impregnation front, at distance $r_f$, advance with time, and how long does it take to impregnate the entire specimen?
(c) How does the melt flow rate vary with time under constant inlet pressure?

**40.** *Pultrusion of thermoplastic composites.*[64] A typical pultrusion die is shown in Fig. P4.40. Long continuous fibers surrounded by polymer melt are pulled and coextruded through the converging die to form a solid composite part, after cooling downstream from the die.
(a) Show that fiber mass conservation demands that

$$\frac{v(x)}{v_L} = \frac{h}{h(x)}, \tag{P4.40.1}$$

where $v$ is the fiber volume fraction and $h$ the die gap.
(b) Calculate $v(x)$ for the linear die profile shown, of length $L$, inclination $\theta$, and end gap $2h$, if a composite of $v_L$ fiber volume per composite volume is to be obtained.

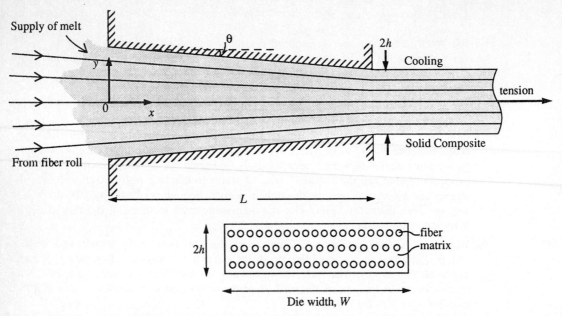

**Figure P4.40**  Pultrusion process and cross section of resulting composite.

(c) Show that the melt average velocity varies along the die, according to

$$\bar{u}(x) = \frac{1 - v_L}{v_L} U v(x) W, \qquad (P4.40.2)$$

where $U$ is the pulling speed.

(d) By considering the pultrusion process as flow of a melt of viscosity $\eta$ at the speed given by Eqn. (P4.40.2) through a regular porous bed of cross-sectional area amenable to flow,

$$A = 2Wh(x)[1 - v(x)], \qquad (P4.40.3)$$

with $v(x)$ given by Eqn. (P4.40.1), derive the form of Darcy's equation for the pressure gradient along the die.

(e) Solve Darcy's equation to obtain an expression for the overall pressure drop, $\Delta p/L$.

**41.** *Foam viscosity.* A good expression for a foam viscosity is[65]

$$\eta_f = \eta_0[1 - (2.3\Gamma)^{0.33}]^{-1},$$

where $\eta_0$ is the viscosity of the base liquid and $\Gamma$ the foam quality, defined by

$$\Gamma = \frac{V_G}{V_G + V_L},$$

where $V_G$ and $V_L$ are the gas and liquid volumes at the prevailing pressure and temperature conditions. Plot the viscosity vs. quality of soap- and water-based foams at atmospheric pressure and temperatures of 20°, 50°, and 90°C. Comment on the form of the resulting curves.

**42.** *Vertical two-phase flow.* For a bubbly flow pattern, show that the liquid and gas superficial velocities, $u_{ls}$ and $u_{gs}$, are related by

$$u_{ls} = u_{gs}\frac{1 - \alpha}{\alpha} - (1 - \alpha)u_r,$$

where $\alpha$ is the gas volume fraction and $u_r$ is the gas–liquid relative velocity, known as the *bubble rise velocity*, by[66]

$$u_r = 1.53\left[\frac{g(\rho_l - \rho_g)\sigma}{\rho_l^2}\right]^{0.25}.$$

Combine the two equations and plot $u_{ls}$ vs. $u_{gs}$ and comment on the resulting transition curve for a water–air system at 25°C and pressure 10 N/cm².

43. *Transition from stratified to annular upward flow.*[67] For high gas flow rates, a stratified two-phase flow evolves into annular two-phase flow, where a liquid film flows upward adjacent to the wall, and gas flows in the center carrying entrained liquid droplets. It has been observed that annular flow cannot exist unless the gas velocity in the gas core is sufficient to lift the entrained droplets.

(a) Show that the minimum velocity for annular flow is derived from the equation

$$0.5C_d\left(\frac{\pi d^2}{4}\right)\rho_G u_G^2 - \left(\frac{\pi d^3}{6}\right)g(\rho_L - \rho_G) = 0,$$

where $d$ is the maximum stable droplet diameter determined by

$$d = \frac{30\sigma}{\rho_G u_G^2},$$

with $\sigma$ the interfacial tension and $C_d = 0.44$ the drag coefficient.

(b) Show that the resulting critical velocity is

$$u_G = \left(\frac{4k}{3C_d}\right)^{0.25}\left[\frac{\sigma g(\rho_L - \rho_G)}{\rho_G^2}\right]^{0.25}.$$

(c) Under the observation that the resulting liquid film is very small, even for relatively high liquid velocities, such that the gas velocity $u_G$ is approximately identical to the superficial velocity $u_{gs}$, show that the criterion for transition to annular flow is

$$\frac{u_{gs}\rho_G^2}{\sigma g(\rho_L - \rho_G)} = 91.$$

44. *Annular two-phase flow.* Calculate the resulting film thickness of a vertical two-phase laminar flow of a water–steam system at 100°C flowing under gravity and opposing gas motion at flow rate $\dot{m}_G = 2$ g/s in a vertical pipe of 10 cm diameter. Then plot the resulting velocity distribution within the thin film for gas flow rates up to 10 g/s and comment on the resulting profiles.

45. *Entrainment velocity in fluidized bed.* Derive an expression that yields the entrainment velocity $\hat{u}_s$ beyond which fluidization is destroyed, by applying Newton's law of motion to a single spherical particle under the combined action of gravity, buoyancy, and drag forces. Consider first conditions of laminar flow for which $C_d = 24/\text{Re}$, then turbulent conditions where $C_d = 0.44$, and finally, a transition region where $C_d = 18.5/\text{Re}^{0.6}$.

46. *Pump alternatives.* A fluid of density 1 g/cm³ and viscosity 10 cP is to be delivered from tank A to tank B. The total equivalent piping length is 10 m and the pipe diameter is

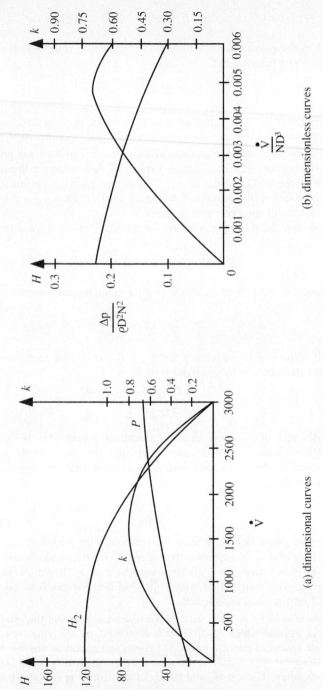

**Figure P4.48** Pump characteristic curves.

(a) dimensional curves

(b) dimensionless curves

2 in. The two free surfaces are at an elevation difference of 3 m apart, and 10 m³ of fluid is to be transferred within 2 min. Give the specifications for a centrifugal, a positive displacement, and a gear pump to deliver the fluid (e.g., power, frequency).

**47.** *Pump efficiency.* The diameter of the discharge pipe of a positive displacement pump is 3 in,, and that of the intake 4 in. The gauge pressure at intake and discharge is 0.5 and 2 atm, respectively. Find the efficiency of the pump at flow rates of 0, 1, 2, 3, 4, and 5 ft/s.

**48.** *Pump operation curves.* The pump of characteristics shown in Fig. P4.48 and operating at 1200 rpm is attached to an 8-in.-diameter pipe of 400 m length, lifting water from one reservoir to another, the free surfaces of which are at elevation differences fluctuating between 20 and 80 ft.

(a) Plot the curve of delivery volumetric flow rate vs. water surface elevation difference and the efficiencies.

(b) A small laboratory centrifugal pump has been evaluated as shown in Fig. P4.48b. Based on these measurements, design a large-scale pump to deliver 1 m³/s of the same fluid at total head 500 m; plot the impeller diameter vs. frequency of rotation, and the pump efficiency. What is the optimum point of operation?

(c) Repeat the considerations of part (b) from laboratory to large-scale transfer of fluid of twice the density and viscosity of the lab fluid. Neglect any variations in the Reynolds number.

**49.** *Pump operating point.* A pump having characteristics shown in Fig. P4.48b is connected to a 300-m-long horizontal pipe of 2 in. diameter. What are the resulting flow rate and the efficiency achieved? Assume a constant friction factor $f_F = 0.005$ and identical entry and exit pressures.

**50.** *Displacement pump.* A small laboratory piston pump having a piston diameter 1.5 in. delivers crude oil of density 0.95 g/cm³ and viscosity 10 cP from a reservoir under 3 atm pressure to a trickle bed reactor operating at 10 atm, through a 1-m-long pipe of 0.5 in. diameter at a flow rate of 1 cm³/s. If the efficiency of the pump is 80%, what power input and cycle frequency are required?

**51.** *Gear pump.* What is the delivery rate of a gear pump of design shown in Fig. 4.4 with $D = 6$ in., $l = 2$ in., and $h = 1.5$ in., rotating at speed 100 rpm and efficiency 90%? What power input is required for a pressure increase of 10 atm?

**52.** *Fountain systems.* The system shown in Fig. P4.52 is used in industry for heat and/or mass exchange between air and water, and in parks and city squares for air-conditioning and aesthetic purposes. Design a centrifugal pump to eject 0.05 m³/min of water at maximum elevation 4 m through a 5-cm-diameter pipe in a pool of 10 m diameter and 1 m water depth.

**53.** *Speed of sound.* A fluid is penetrated by spherical sound waves traveling at speed $c$ that disturb its density and pressure. Consider two points a distance $dx$ apart, where the velocity, pressure, and density have values $c$, $p$, $\rho$ and $c + dc$, $p + dp$, $\rho + d\rho$, respectively.

(a) By mass and momentum balances between the two points at distance $dx$ apart, show that the speed of sound in the fluid is

$$c = \left(\frac{dp}{d\rho}\right)^{1/2}.$$

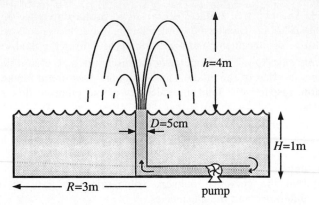

**Figure P4.52** Fountain hydraulics.

**(b)** What are the speeds of sound in ideal and real gases, in liquids and in solids, based on appropriate equations of state?

# REFERENCES

1. J. T. Fanning, *A Practical Treatise on Hydraulics and Water Supply Engineering*, Van Nostrand, New York, 1893.
2. H. Clemens, On the origin of the Chezy formula, *J. Assoc. Engrg.*, 18, 363 (1887).
3. L. W. Moody, Friction factors for pipe flow, *Trans. ASME*, 66, 672 (1944).
4. J. Nikuradse, Stroemungsgesetze in rauhen Rohren (Flow Laws in Rough Tubes), *VDI Forschungsh.*, 361 (1933).
5. C. F. Colbrook, Turbulent flow in pipes with particular reference to the transition region between smooth and rough pipe laws, *J. Inst. Civil Engrg.*, 11, 133 (1938).
6. D. J. Wood, An explicit friction factor relationship, *Civil Engrg.*, 36, 60 (1966).
7. R. H. Perry, *Perry's Chemical Engineer's Handbook*, 6th ed., McGraw-Hill, New York, 1984.
8. L. S. Marks, *Standard Handbook for Mechanical Engineers*, McGraw-Hill, New York, 1958.
9. Crane Co., Flow of fluids through valves, fittings and pipe, *Tech. Paper 410*, 1969.
10. *Pipe Friction Manual*, 3rd ed., Hydraulic Institute, New York, 1961.
11. J. E. Walker, G. A. Wham, and R. R. Rothfns, Fluid friction in noncircular ducts, *AIChE J.*, 3, 484 (1957).
12. F. M. Henderson, *Open Channel Flows*, Macmillan, New York, 1966.
13. H. Lamb, *Hydrodynamics*, Dover, New York, 1932.
14. A. C. Charters and T. N. Thomas, The aerodynamic performance of small spheres from subsonic to high supersonic velocities, *J. Aeronaut. Sci.*, 12, 468 (1945).
15. S. W. Churchill, *Viscous Flow: The Practical Use of Theory*, Butterworth, Boston, 1988.
16. G. K. Batchelor, *An Introduction to Fluid Dynamics*, Cambridge University Press, Cambridge, 1979.

17. J. Happel and H. Brenner, *Low Reynolds Hydrodynamics*, Martinus Nijhoff, The Hague, 1983.

18. D. B. Steinman, Suspension bridges: The aerodynamic stability problem and its solution, *Amer. Sci.,* 42, 397 (1954).

19. J. H. Seinfeld, *Atmospheric Chemistry and Physics of Air Pollution*, Wiley, New York, 1985.

20. K. Wark and C. F. Warner, *Air Pollution: Its Origin and Control*, Harper & Row, New York, 1976.

21. R. R. Hughes and E. R. Gilliland, The mechanics of drops, *Chem. Engrg. Prog.,* 48, 497 (1952).

22. R. Von Mises, *Theory of Flight*, Dover, New York, 1945.

23. C. E. Lapple and C. B. Shepherd, Calculation of particle trajectories, *Ind. Engrg. Chem.,* 32, 605 (1940).

24. T. C. Papanastasiou, A. N. Alexandrou, and W. P. Graebel, Rotating thin films in bell-sprayers and spin-coating, *J. Rheol.,* 32, 485 (1988).

25. A. H. Lefebvre, *Atomization and Sprays*, Hemisphere, New York, 1989.

26. R. A. Brodkey, *The Phenomena of Fluid Motions*, Addison-Wesley, Reading, Mass., 1967.

27. Lord Rayleigh, On the instability of jets, *Proc. London Math. Soc.,* 10, 4 (1878).

28. C. Weber, Disintegration of liquid jets, *Z. Angew. Math. Mech.,* 11, 136 (1931).

29. W. Ohnesorge, Formation of drops by nozzles and the breakup of liquid jets, *Z. Angew. Math. Mech.,* 16, 355 (1936).

30. R. D. Reitz, Atomization and break-up regimes of a liquid jet, Ph.D. thesis, Princeton University, 1978.

31. K. Ellwood, G. Georgiou, T. C. Papanastasiou, and J. O. Wilkes, Laminar jets of Bingham-plastic liquids, *J. Rheol.,* 34, 787 (1990).

32. S. Ergun, Fluid flow through packed columns, *Chem. Engrg. Prog.,* 48, 89 (1952).

33. F. E. Blake, The resistance of packing in fluid flow, *Trans. AIChE,* 14, 415 (1921).

34. S. P. Burke and W. B. Plummer, Gas flow through packed columns, *Ind. Engrg. Chem.,* 20, 1196 (1928).

35. H. P. G. Darcy, *Les Fontaines publique de la Ville de Dijon*, Victor Dalmont (Ed.), Paris, 1856.

36. O. Faltsi-Saravelou and I. A. Vesalos, Simulation of dry fluidized bed process for $SO_2$ removal from flue gases, *Ind. Engrg. Chem. Res.,* 29, 251 (1990); Dynamic modeling of an adiabatic reacting system of small gas fluidized particles, *Comput. Chem. Engrg.,* 15, 639 (1991).

37. G. D. Dickey and C. L. Bryden, *Theory and Practice of Filtration*, Reinhold, New York, 1946.

38. R. G. Rice and B. K. Heft, Separations via radial flow chromatography in compacted particle beds, *AIChE J.,* 37, 629 (1991).

39. F. M. Tiller, How to select solid–liquid separation equipment, *Chem. Engrg.,* Apr. 29, 116 (1974).

40. N. H. Chen, Liquid–solid filtration: Generalized design and optimization calculations, *Chem. Engrg.,* 85, 97 (1978).

41. M. Muskat; *Physical Principles of Oil Production*, McGraw-Hill, New York, 1949.

42. A. E. Scheidegger, *The Physics of Flow Through Porous Media*, Macmillan, New York, 1946.

43. Y. R. Kim, S. P. McCarthy, S. C. Nolet, and C. Koppernaes, Resin flow through fiber reinforcements during composite processing, *SAMPE Quart,* 22, 16 (1991).

44. S. V. Paras and A. J. Kerabelas, Measurements of local velocities inside thin liquid films in horizontal two-phase flow, *Exp. Fluids,* 13, 190 (1992).

45. A. E. Bergles, J. G. Collier, J. M. Delhaye, G. F. Hewitt, and F. Mayinger (Eds.), *Two-Phase Flow and Heat Transfer in the Power and Process Industries,* Hemisphere, New York, 1981.

46. A. E. Dukler, M. Wicks, and R. G. Cleveland, Frictional pressure drop in two-phase flow: B. An approach through similarity analysis, *AIChE J.,* 44 (1964).

47. C. J. Baroczy, A systematic correlation for two-phase pressure drop, *Chem. Engrg. Prog.,* 62, 232 (1966).

48. M. M. Delhaye, Frictional pressure drops, in *Two-Phase Flow and Heat Transfer in the Power and Process Industries* (ed. A. E. Bergles et al.), Hemisphere, New York, 1981.

49. S. V. Paras and A. J. Karabelas, Properties of the liquid layer in horizontal annular flow, *Internat. J. Multiphase Flow,* 17, 455 (1991).

50. A. H. Shapiro, *The Dynamics and Thermodynamics of Compressible Fluid Flow,* Vol. 1, Ronald Press, New York, 1953.

51. D. L. Katz, *Handbook of Natural Gas Engineering,* McGraw-Hill, New York, 1959.

52. G. F. Hewitt, Annular flow, in *Two-Phase Flow and Heat Transfer in the Power and Process Industries* (ed. A. E. Bergles et al.), Hemisphere, New York, 1981.

53. J. W. Hall and G. M. Wood, Eds., *Cavitation Research Facilities and Techniques,* ASME, New York, 1964.

54. B. Lakshminarayana, Three-dimensional flow in rocket pump-inducers, *ASME J. Basic Engrg.,* 95, 567 (1973).

55. J. T. Sommerfeld, Drainage of conical tanks with piping, *Chem. Engrg. Educ.,* 24, 145 (1990).

56. R. Darby, Laminar and turbulent pipe flows of non-Newtonian fluids in *Encyclopedia of Fluid Mechanics* (ed. N. P. Cheremisinoff), Vol. 7, Gulf Publishing, Houston, Tex., 1988.

57. N. P. Cheremisinoff, Properties of gases and overview of compressible flow in *Encyclopedia of Fluid Mechanics* (ed. N. P. Cheremisinoff), Vol. 8, *Aerodynamics and Compressible Flows,* Gulf Publishing, Houston, Tex., 1989.

58. J. L. Lumley, Drag reduction by additives, *Annu. Rev. Fluid Mech.,* 1, 367 (1969).

59. P. S. Virk, Drag reduction fundamentals, *AIChE J.,* 21, 625 (1975).

60. W. R. C. Meyers and J. F. Lyness, Flow in rectangular channels in *Encyclopedia of Fluid Mechanics* (ed. N. P. Cheremisinoff), Vol. 1, *Flow Phenomena and Measurement,* Gulf Publishing, Houston, Tex., 1986.

61. N. P. Cheremisinoff, Industrial flow measuring devices in *Encyclopedia of Fluid Mechanics* (ed. N. P. Cheremisinoff), Vol. 1, *Flow Phenomena and Measurement,* Gulf Publishing, Houston, Tex., 1986.

62. W. Ranz, *Air Pollution Lecture Notes,* University of Minnesota, 1980.

63. L. E. Scriven, *Intermediate Fluid Mechanics,* Lecture Notes, University of Minnesota, 1980.

64. B. T. Astron and R. B. Pipes, Modeling of a thermoplastic pultrusion process, *SAMPE Quart.,* 22, 55 (1991).

65. J. O. Sibree, The viscosity of froth, *Faraday Soc. Trans.,* 30, 735 (1934).

66. T. Z. Harmathy, Velocity of large drops and bubbles in media of infinite or restricted extent, *AIChE J.,* 6, 281 (1960).

67. D. Barnea and Y. Taitel, Flow pattern transition in two-phase gas–liquid flows in *Encyclopedia of Fluid Mechanics,* Vol. 3, *Gas–Liquid Flows* (ed. N. P. Cheremisinoff), Gulf Publishing, Houston, Tex., 1986.

# 5

# *Introduction to Differential Fluid Mechanics*

## 5.1 MACROSCOPIC BALANCES

In Chapters 1 to 4, *fluid systems* have been analyzed primarily from the *macroscopic point of view*: Balances and resulting information were based on *average properties,* without much consideration as to how a given property—pressure, velocity, stress—*is distributed in a flow field.* These macroscopic balances and results are useful and sufficient in many practical applications that require calculation of macroscopic quantities, such as:

1. Flow rate: $\dot{V} = A\bar{u}$.

2. Friction and pumping power: $\mathscr{F} = 2f_F\bar{u}^2 L/D$, $P = \dot{V}\,\Delta p$.

3. Jet thrust: $J = \dot{m}\bar{u}$.

4. Settling velocity of particle: $\bar{u}^2 = 3C_D\rho_f/[4gD(\rho_s - \rho_f)]$.

This information is obtained by applying the conservation principles within the *whole system* up to its *boundaries,* as shown in Fig. 5.1. The conservation statement

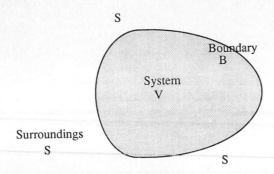

**Figure 5.1** System ($V$), boundary ($B$), and surroundings ($S$).

for the macroscopic system[1] of Fig. 5.1 is

$$\underbrace{\frac{d}{dt}\left(\begin{array}{c}\text{conserved}\\\text{property}\end{array}\right)}_{\text{rate of increase}} = \underbrace{\left(\begin{array}{c}\text{rate of convection of property}\\\text{through boundaries}\end{array}\right)}_{\text{rate of transfer by convection}}$$

$$\underbrace{+\left(\begin{array}{c}\text{rate of transfer other than by}\\\text{convection through boundaries}\end{array}\right) + \left(\begin{array}{c}\text{rate of transfer other than by}\\\text{convection through volume}\end{array}\right)}_{\text{rate of transfer by diffusion}}$$

$$\underbrace{\pm\left(\begin{array}{c}\text{rate of creation or}\\\text{destruction in volume}\end{array}\right)}_{\text{rate of conversion}}.$$

$$(5.1)$$

As explained in Chapter 2, the left-hand term represents changing of a content or property within a system or volume, due to the competing mechanisms represented by the terms on the right-hand side: Its first term may represent convection of content, such as mass, momentum, and energy by fluid that moves relative to the adjacent volume boundary, and is zero in the absence of such motion; the second term may represent diffusion of solute, conduction of heat, or contact force in a solute mass, energy, or momentum balance equation, respectively; the third term may represent body force in a momentum balance equation; and the fourth term may represent conversion of solute by a chemical reaction in a solute-mass balance equation, as well as production or consumption of heat by chemical reaction, due to change of phase or due to viscous dissipation, in energy balance equations. One or more of these mechanisms may be present in a particular flow or process situation, in which case the corresponding conservation or balance equation includes the appropriate term(s). For example, in mass balances the only nonzero term is the convective one because (total) mass is transferred only by convection; in momentum balances, in addition to convection, there is a contribution from body and contact forces; and in mechanical energy balances, in addition to convection, there is conversion to heat by viscous dissipation. The application of macroscopic balances in common flow situations was the topic of Chapter 3, where numerous examples were presented.

## 5.2 MICROSCOPIC DIFFERENTIAL BALANCES

In some processing applications, the average quantities obtained by macroscopic balances are not sufficient to analyze a process fully. In most of these cases, a detailed *distribution* or *profile* of velocity, stress, and pressure within the flowing fluid is required. In a class of *free-surface flows* in the form of thin films which has many applications in materials processing, the most important feature is the *film thickness*. The following situations are representative of the necessity of computing velocity and pressure profiles and film thickness.

**Velocity distribution.**   In a tubular reactor, shown in Fig. 5.2, the residence time of reactant in a liquid phase entering at $z = 0$ is given by[2]

$$t_D(r) = \int_0^{\Delta L} \frac{dz}{u(r)} = \int_0^{\Delta L} \left[ \frac{1}{4\eta} \frac{\Delta p}{\Delta L} (r^2 - R^2) \right]^{-1} dz = \frac{4\eta\, \Delta L}{(\Delta p/\Delta L)(r^2 - R^2)}, \qquad (5.2)$$

where $u(r)$ is the velocity profile over the cross-sectional area with $\eta$ the viscosity of the fluid and $\Delta p$ the pressure differential driving the flow. Thus, the residence-time distribution depends on the radial distance $r$ in a way that is known if and only if the velocity profile $u(r)$ is known. According to this equation, the residence time of the reactant entering at $r = 0$ is much less than the residence time of reactant entering near the solid wall, at $r \approx R$. Thus the rate of conversion is expected to be different across the pipe. Macroscopic balances are based on average quantities and therefore are not applicable to compute velocity or other quantities that change in space. Thus, there is a need for a

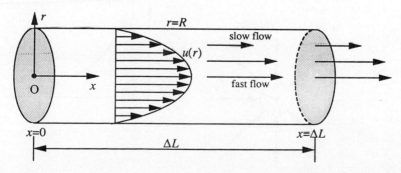

**Figure 5.2**   Residence time in a tubular reactor.

different kind of balance based on point quantities, allowing variation from point to point within and at the boundaries of the system considered. This kind of balance is *microscopic or differential,* being based on quantity microvariations over *infinitesimal or differential distances.*

It is worthwhile noting here that knowledge of the velocity profile by means of differential balances allows determination of the macroscopic or average properties as well: For example, the average velocity is given by

$$\bar{u} = \frac{1}{2\pi R} \int_0^R 2\pi r u(r)\, dr. \tag{5.3}$$

The average concentration is computed by

$$\bar{c} = \frac{1}{2\pi R} \int_0^R 2\pi r c[u(r), r, \mathscr{D}]\, dr, \tag{5.4}$$

where $c[u(r), r, \mathscr{D}]$ is the concentration profile, which in general depends on radial distance $r$, velocity profile $u(r)$, and diffusion coefficient $\mathscr{D}$. Indeed, *all macroscopic quantities can be calculated as a result of microscopic or differential analysis,* starting from the profiles of velocity and pressure that have been computed.

**Pressure distribution.**    In a sink flow, where liquid flows radially and disappears in a sink, as shown in Fig. 5.3, the pressure along a streamline is of the form shown. Theoretically, somewhere along the streamline, cavitation of liquid may occur[3] if the local pressure becomes identical to the liquid's vapor pressure. To study such phenomena, the detailed pressure distribution $p(r)$ must be evaluated along the radial streamline. Similar situations arise around gas bubbles growing or collapsing in liquid baths, for example during boiling, where detailed velocity and pressure distributions around the bubble are required to quantify the evolution of such processes[4] as liquid boiling, mass absorption by bubbles, void evolution in composite impregnation,[5] and cavitation.[6]

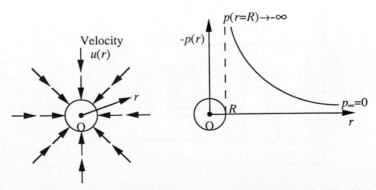

**Figure 5.3**  Sink flow and pressure distribution along radial streamlines.

**Film thickness.**    Thin films are frequently utilized in heat and mass exchange processes,[7] due to the induced large interfacial area of exchange and small penetration distances under the film thicknesses (Fig. 5.4a). Equally important applications of thin films include coating[8] of metals, plastics, and paper for performance (e.g., photographic and magnetic films), protection (e.g., painting of metals used in corrosive environments), or decoration (e.g., plastic films) purposes. These films are deposited onto the corresponding substrate in a liquid state, as shown in Fig. 5.4b, and left to solidify in subsequent stages, forming permanent solid layers. Both the thickness of the film and its uniformity, which are the primary variables of interest in these processes, can be evaluated by microscopic analyses.

Chapters 5 to 10 cover *microscopic or differential fluid mechanics,* which provides complete information about the detailed velocity and pressure distribution in a flowing fluid (i.e., *the exact values of pressure and velocity at any specific*

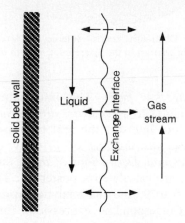

(a) Interfacial exchange

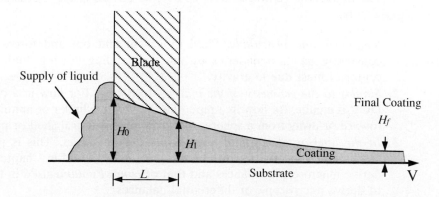

(b) Application of thin film, or coating

**Figure 5.4**   Thin films in heat and mass exchange processes (a), and in coating (b), where a liquid is deposited onto a fast-moving substrate to form a final thin coating.

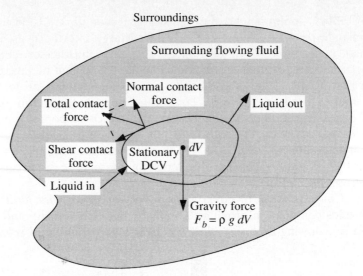

**Figure 5.5** Differential control volume (DCV) and interactions with its surroundings, in contrast to the macroscopic system of Fig. 3.1.

*location within the flow field).* This information is obtained by applying the conservation principles to a *synthetic, differential control volume within the flow field,* which does not intersect the real boundaries of the system, as explained below. Once these *distributions of primary variables* have been computed, *secondary variables and distributions of them,* such as stress, flow rate, frictional force, and others, are calculated a posteriori. The remainder of this chapter, as well as Chapters 6 to 9, and the analytic expressions derived apply to laminar flow conditions. Corresponding expressions for turbulent flow are derived in Chapter 10.

The *differential control volume (DCV),* as shown in Fig. 5.5, is conveniently chosen to be:

1. *Stationary and penetrable.* Fluid moves in and out and forces are acting constantly on its boundaries by the surrounding flowing fluid, and on its center of mass due to gravity.
2. *Similar to the geometry of the macroscopic flow.* For flow in a channel, the DCV is a cube; for flow in a pipe, the DCV is a cylinder or annulus; for flow toward or away from a sphere, the DCV is a spherical shell or part of it.
3. *In the interior of the fluid away from its boundaries.* This is perhaps the biggest difference between the concept of *system* used in Chapters 1 to 4 to derive macroscopic balances and that of *control volume* used in this chapter to derive microscopic or differential balances.
4. *Of infinitesimal dimensions.* As shown in Table 5.1, when distances shrink to zero average properties over them are forced to become point variables.

The last two properties are the fundamental differences between the

**TABLE 5.1**  APPROPRIATE DIFFERENTIAL CONTROL VOLUMES AND SURFACES

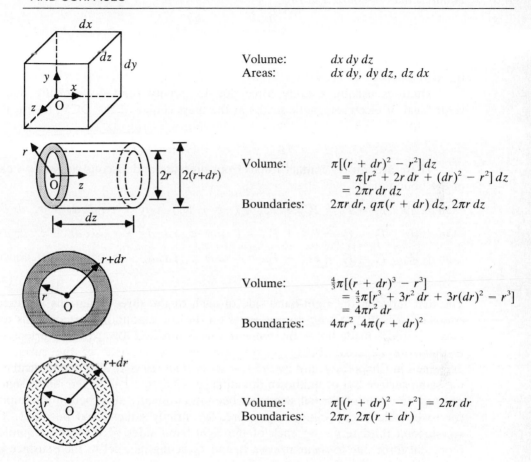

Volume:   $dx\,dy\,dz$
Areas:    $dx\,dy,\ dy\,dz,\ dz\,dx$

Volume:   $\pi[(r + dr)^2 - r^2]\,dz$
          $= \pi[r^2 + 2r\,dr + (dr)^2 - r^2]\,dz$
          $= 2\pi r\,dr\,dz$
Boundaries:   $2\pi r\,dr,\ q\pi(r + dr)\,dz,\ 2\pi r\,dz$

Volume:   $\frac{4}{3}\pi[(r + dr)^3 - r^3]$
          $= \frac{4}{3}\pi[r^3 + 3r^2\,dr + 3r(dr)^2 - r^3]$
          $= 4\pi r^2\,dr$
Boundaries:   $4\pi r^2,\ 4\pi(r + dr)^2$

Volume:   $\pi[(r + dr)^2 - r^2] = 2\pi r\,dr$
Boundaries:   $2\pi r,\ 2\pi(r + dr)$

concept of *system* used in Chapters 1 to 4 to derive macroscopic balances and that of *control volume* used in this chapter to derive microscopic or differential balances.

**Interactions between DCV and its surroundings.**  The interactions between the control volume and its surroundings conform to the general conservation statement, by Eqn. (5.1): An extensive quantity, $dq$, within a cubic DCV (without loss of generality) changes according to

$$\frac{dq}{dt} = \frac{d}{dt}(\hat{q}\,dV) = \frac{d}{dt}(dx\,dy\,dz\hat{q}) = dx\,dy\,dz\,\frac{d\hat{q}}{dt}, \qquad (5.5)$$

where $\hat{q}$ is the density of the quantity (i.e., amount of quantity per unit volume).

Quantity $q$ is convected in an out of the stationary DCV through its infinitesimal boundaries, at rate

$$\underbrace{dx\,dy}_{\substack{\text{surface of} \\ \text{convection}}} \times \underbrace{u\hat{q}}_{\substack{\text{convected} \\ \text{quantity}}} \tag{5.6}$$

through a surface of area $dA = dx\,dy$.

There is usually a body force due to gravity (and occasionally due to centrifugal or electromagnetic fields) at the mass center of the DCV, identical to its weight,

$$F_g = B = \rho g\,dx\,dy\,dz. \tag{5.7}$$

There are always contact forces from pressure and viscous stresses on each of the boundaries:

On $dx\,dy$: $\quad F_T = F_P + P_{VN} + F_{VS} = (-p + \tau_{zz})\,dx\,dy + \tau_{zx}\,dx\,dy + \tau_{zy}\,dx\,dy$

On $dx\,dz$ $\quad F_T = F_P + F_{VN} + F_{VS} = (-p + \tau_{yy})\,dx\,dz + \tau_{yx}\,dx\,dz + \tau_{yz}\,dz\,dy$

On $dy\,dz$ $\quad F_T = F_P + F_{VN} + F_{VS} = (-p + \tau_{xx})\,dy\,dz + \tau_{xy}\,dx\,dz + \tau_{xz}\,dz\,dy.$

$$\tag{5.8}$$

The first term on the right-hand side of each of the three equations represents contact forces due to *total normal stress* on the corresponding surface. This total normal force is made up by the *pressure* that is directed toward the $i$-surface and is therefore negative—being opposite to the unit normal of the surface as explained in Chapter 2—and by the *viscous normal stress* $\tau_{ii}$ also perpendicular to the same surface but of unknown direction. In Chapters 1 to 4, the minus sign of pressure in the total normal stress has been intentionally absorbed in the implied (or to be calculated) value of pressure, for strictly educational purposes. The second and third terms on each of the right-hand sides of the three equations represent force due to *shear stresses* $\tau_{ij}$ and $\tau_{ik}$ acting parallel to the $i$-surface but in an unknown direction along the $jk$ plane.

Thus, the total normal stress and the shear stress on a given surface are added vectorially, as explained in Appendix A, to yield the total stress on the surface of an unknown arbitrary direction. Figure 5.6 illustrates the velocity and the (total) stress decomposition at a $z$-surface, into normal components $u_z$ and $-p + \tau_{zz}$, respectively, and shear components $u_x$, $u_y$ and $\tau_{zx}$, $\tau_{zy}$, respectively.

Among the three velocity components, only the $u_z$ component convects a

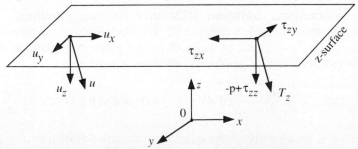

**Figure 5.6**  Velocity and stress decomposition at a $z$-surface.

conserved quantity across the $z$-surface, according to Eqn. (5.4). Thus, mass is convected at rate $\rho u_z \, dA$, $x$-momentum at rate $(\rho u_x)u_z \, dA$, $y$-momentum at rate $(\rho u_y)u_z \, dA$, $z$-momentum at rate $(\rho u_z)u_z \, dA$, energy at rate $\hat{E}u_z \, dA$, and so on. For the same reasons, the normal stress component $-p + \tau_{zz}$ contributes to the $z$-momentum, the shear stress component $\tau_{zx}$ to the $x$-momentum, and the shear stress component $\tau_{zy}$ to the $y$-momentum.

As discussed in Chapter 3, convective mass, momentum, and energy are positive for entry into the DCV. Because the DCV is infinitesimally small, a positive entry $\dot{q}(x)$ through an $x$-surface at distance $x$ from the origin crosses the DCV with minor differential change $[\dot{q}(x)/dx]\,dx = d\dot{q} \ll \dot{q}(x)$ [e.g., Eqn. (A.11) of Appendix A], and appears at the opposite $x$-surface at distance $x + dx$ as a negative exit. The same is true for the stresses, for example: *Assume* that a shear stress, say $\tau_{zx|z+dz}$ *from the surroundings on the z-surface of the DCV*, located a distance $z + dz$ from the origin, is positive. This shear stress is transmitted from layer to layer through the control volume and appears as positive shear stress *from the DCV to the surroundings*, at the opposite $y$-surface located at distance $y$ from the origin. Thus, its opposite shear stress *from the surroundings to the DCV* at the same spot is negative. According to the same mechanism, if a total normal stress on the DCV is assumed positive at one of its surfaces, the corresponding total normal stress at the opposite surface must be negative. It is easy to verify that this rule applies to cases of static equilibrium where the only stress present is the isotropic pressure.

The considerations discussed above are extremely important to digest before entering differential mass and momentum balances. In summary:

1. Velocities are taken to be positive when entering a DCV through one of its surfaces and negative at the opposite surface.

2. Shear stresses are *assumed by convection* to be positive at the distant surfaces from the origin (e.g., at $x + dx$, $y + dy$, and $x + dz$ for a cubic DCV). This assumption being made, the corresponding stresses at opposite surfaces closer to the origin (i.e., at $x$, $y$, and $z$ for a cubic DCV) must be negative.

3. The same rule also applies to the total normal stresses. Because the direction of the pressure contribution to a normal stress is known a priori, total normal stresses are conveniently represented by the sums $-p + \tau_{ii}$.

4. In any balance, convective and force contributions must be sought at all surfaces of the DCV. Special attention must be paid to momentum balances, where, for example, convection and stresses at all surfaces may contribute to each and all of the three momentum equations, in general.

Example 5.1 shows all velocity components upon entry into or exit from a cubic DCV and all stress components acting on the same DCV, according to the principles just laid down. Following examples deal with other types of DCVs.

### Example 5.1: Velocity and Stress Components on Cubic DCV

Consider a cubic differential control volume (DCV) with respect to a given Cartesian system of coordinates, of sides $dx = dy = dz$, as shown in Fig. E5.1. Draw the

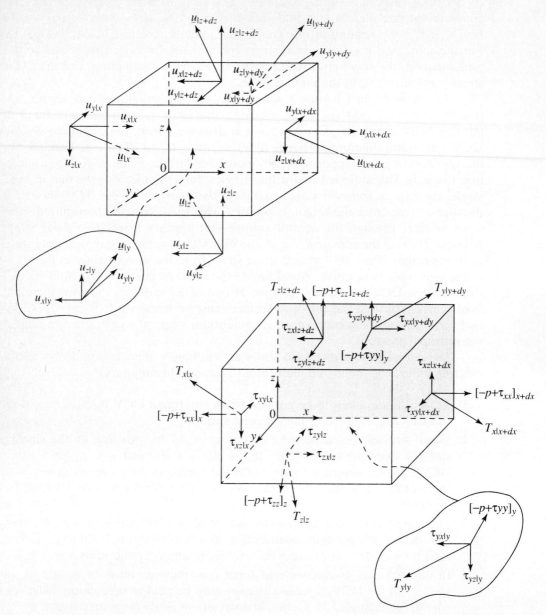

**Figure E5.1** Cubic differential control volume with its velocity and normal and shear stress components.

vectors of all velocity components at its six square surfaces. Draw the vectors of the total normal stress $-p + \tau_{ii}$ and of shear stresses $\tau_{ij}$ on the same surfaces.

**Solution**   The velocity components are shown in Fig. E5.1a. The following observations are in order:

1.  A normal velocity $u_i$, $i = x, y, z$, to an $i$-surface is positive for entry taken a distance $i$ from the origin and necessarily negative at the opposite surface at a distance $i + di$.

2.  Scalar mass and energy may be convected across an $i$-surface by the corresponding normal velocity component $u_i$ alone.

3.  Momentum is a vector and is conveniently decomposed into three scalar momentum components following the three velocity components at each surface. When the $u_i$ normal velocity component crosses the $i$-surface, it convects within the DCV all three momentum components $(\rho u_j)$, $j = x, y, z$, at rates $(\rho u_j)u_i \, dA$, respectively. The same mechanism of momentum transfer is applied to macroscopic balances, as discussed in Section 3.4.

The stress components are shown in Fig. E5.1b. The following observations are in order:

1.  Normal stress components $-p + \tau_{ii}$, $i = x, y, z$, are drawn positive at all surfaces at distances $i + di$, $i = x, y, z$, and negative at all surfaces at distances $i = x, y, z$. It is always a good practice to draw total normal stresses and not pressure and viscous normal stresses separately.

2.  Shear stress $\tau_{ij}$, $i = x, y, z$, $j = x, y, z$, are similarly drawn positive at distances $i + di$ and negative at distances $i$.

3.  A momentum balance in the $i$-direction in general includes (a) the pair of normal stresses $-p + \tau_{ii}$ at distances $i$ and $i + di$, respectively; (b) the two pairs of shear stresses $\tau_{ji}$, $j \neq i$ at distances $j$ and $j + dj$. In many flow situations, most of these components are zero, due to symmetry considerations.

The microscopic or differential conservation equations are obtained by substituting the conserved quantity $q$ in Eqn. (5.1) and quantifying the remaining terms by the present mechanisms of interactions between the control volume and its surrounding fluid. The conservation equations of primary interest in the remaining chapters are:

1.  The scalar *continuity equation*, which expresses conservation of mass and is obtained by identifying mass $m$ or its density $\rho$ in Eqn. (5.1).

2.  The three components of the vector *momentum equation* that express conservation of linear momentum and are obtained by identifying the momentum components $mu_x$, $mu_y$, and $mu_z$ or their densities $\rho u_x$, $\rho u_y$, and $\rho u_z$ in Eqn. (5.1).

Detailed derivation and analysis of differential heat and solute conservation equations are outside the scope of this book. They are treated in transport phenomena[4] and reaction engineering[2] courses and textbooks.

## 5.3 CONSERVATION OF MASS

These are the simplest equations to apply, because the only mechanism of transfer of fluid mass is by convection, and therefore only the first two terms of Eqn. (5.1) need to be quantified. Furthermore, for incompressible liquids that is the topic of the remaining chapters, the left-hand term is zero, and the surviving convection terms yield the ordinary velocity derivatives, as demonstrated in the following two examples. It is emphasized, that hereafter, by *mass* we mean the *total mass* and not the mass of any particular disolved species, which is termed *solute mass.* Balances on solute mass become essential in reacting flows, where solute mass may change form one to another species mass due to an ongoing chemical reaction. Flows examined hereafter are not reacting, and therefore only (total) mass balances are considered.

### Example 5.2: Mass Balance in Spherical Radial Flow

Find the velocity distribution of the purely radial liquid flow induced around a nearly spherical bubble[6] of radius $R = (D + h_b)/4$ that grows steadily in a liquid with rate $dR/dt = \dot{R}$ in terms of $\dot{R}$ and $R$, as shown in Fig. E5.2a. The solution can be found by means of a liquid spherical shell DCV with respect to a spherical system of coordinates, with its origin at the center of the bubble. This DCV has radii $r > h_b$ and $r + dr$ and thickness $dr$, as shown in Fig. E5.2b. By appplying Eqn. (5.1) for liquid mass conservation in this DCV, the resulting equation is

$$\frac{d}{dt}(\rho 4\pi r^2\, dr) = (4\pi r^2 u_r \rho)_r - (4\pi r^2 u_r \rho)_{(r+dr)}. \qquad (5.2.1)$$

The first term represents the change of mass with time within the control volume due to convection of fluid in and out of the control volume, represented by the second and third terms, respectively, forced by the expanding bubble. For an incompressible liquid of constant density, and given the fact that the control volume is constant, the

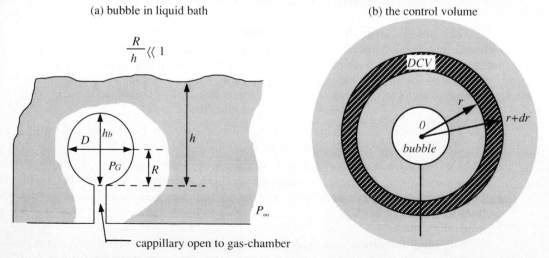

(a) bubble in liquid bath          (b) the control volume

**Figure E5.2**  (a) Radial flow induced by bubble growth or collapse; (b) corresponding differential control volume.

first term is zero. Then, by dividing the remaining terms by $dr$ and taking the limit $dr \to 0$, which gives the derivative (e.g., Section A.3 of Appendix A), the resulting governing equation is

$$\frac{d}{dr}(r^2 u_r) = 0. \tag{5.2.2}$$

Integration of the differential equation gives the general solution,

$$u_r = \frac{c}{r^2}, \tag{5.2.3}$$

where $c$ is an arbitrary constant of integration calculated by the obvious boundary condidtion at the gas–liquid interface that the radial liquid velocity there is identical to the growth rate of the bubble. At $r = R$, $u_r(r = R) = dR/dt = \dot{R}$, so that

$$\dot{R} = \frac{c}{R^2}. \tag{5.2.4}$$

Equation (5.2.4) substituted in the general solution expression, Eqn. (5.2.3), yields the particular solution

$$u_r = \dot{R}\frac{R^2}{r^2}, \tag{5.2.5}$$

which fully determines the *velocity profile* around the growing bubble (i.e., given the bubble growth rate $\dot{R}$ and its instantaneous radius $R$, the velocity at any radial distance $r$ can be computed). The velocity diminishes from its maximum value $u_r(r = R) = \dot{R}R^2/R^2 = \dot{R}$ to $u_r(r \to \infty) = 0$ far away from the bubble, where the liquid is undisturbed by the small expanding bubble.

Other useful quantities can be computed given the velocity profile. For example, the radial viscous normal stress component is

$$\tau_{rr} = 2\eta\frac{du_r}{dr} = -3\eta\dot{R}\frac{R^2}{r^3}. \tag{5.2.6}$$

The corresponding net viscous stress in the radial direction,

$$T_r = \tau_{rr} + (\tau_{\phi\phi} + \tau_{\theta\theta})_r \tag{5.2.7}$$

where $(\tau_{\phi\phi} + \tau_{\theta\theta})_r$ is radial component of $(\tau_{\phi\phi} + \tau_{\theta\theta})$ contributing to the radial motion is constant and therefore its radial gradient zero. Thus, the flow is driven by pressure forces alone. Such flows are called *irrotational* and are examined in more detail in Chapter 7. Due to the absence of viscous forces, there is no loss of mechanical energy by friction, and Bernoulli's equation applies along the radial streamlines. This allows determination of the *pressure distribution*,

$$\frac{d}{dr}\left(\frac{p}{\rho} + \frac{u_r^2}{2} + gz\right) = 0. \tag{5.2.8}$$

It turns out that all types of radial flows of straight streamlines, toward or away from a point (plane sink or source flow), from a line (cylindrical sink or source flow), and to or from a sphere (radial spherical flow), are irrotational. For these particular flows, the velocity is determined by mass balance alone and the pressure distribution follows from Bernoulli's equation. For nonradial flows, however, the velocity distribution cannot be determined by mass balance alone nor the pressure profile by Bernoulli's equation. Example 5.3 demonstrates these features.

### Example 5.3: Mass Balance in Pipe Flows

Apply a mass balance to find the velocity and pressure profiles of steady laminar flow in a horizontal pipe.

**Solution**  The appropriate control volume here is an annulus of length $dz$ along the flow direction and radii $r$ and $r + dr$, as shown in Fig. E5.3, with its axis of symmetry along the centerline of the pipe. The continuity equation for mass conservation in the shown DCV is

$$\frac{d}{dt}(\rho 2\pi r\, dr\, dz) = (\rho 2\pi r\, dr\, u_z)_z - (\rho 2\pi r\, dr\, u_z)_{z+dz}. \tag{5.3.1}$$

At steady state, at the limit $dr \to 0$, Eqn. (5.3.1) yields

$$2\pi \frac{du_z}{dz} = 0; \quad \text{therefore, } u_z = f(r). \tag{5.3.2}$$

Thus, the only information obtained by the applied mass balance is that $u_z$ is constant, independent of $z$, along the pipe and may be a function of radius $r$. The exact form of the function $f(r)$ is determined by a momentum balance on the same control volume, as explained in Section 5.4, given the limited information obtained by mass balance.

Thus, a mass balance alone determines the functional behavior of the velocity profile in radial flows. The pressure profile of these flows, and both the velocity and pressure profiles in other classes of flow (e.g., pipe flow), are determined by momentum balances. Because radial flows (e.g., sink flow) are irrotational, Bernoulli's equation can also be utilized to find the pressure given the velocity profile from a mass balance alone. Actually, it can be shown that the radial momentum equation for radial flows degenerates to Bernoulli's equation.

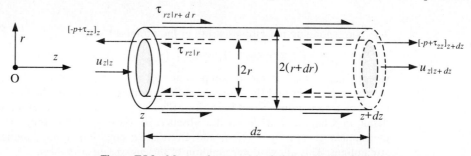

**Figure E5.3**  Mass and momentum balances in pipe flow.

## 5.4 CONSERVATION OF MOMENTUM

The mass balance in Example 5.3 showed that the velocity is constant in the streamwise direction but may change with radial distance similarly over any cross-sectional area:

$$u_z = f(r). \tag{5.9}$$

To determine the exact form of $f(r)$, a momentum balance in the direction of flow ($z$, here) is performed over the same control volume. The statement of Eqn. (5.1) takes the form

$$\frac{d}{dt}(\text{momentum}) - \text{rate of momentum in} - \text{rate of momentum out} \tag{5.10}$$
$$+ \text{ body force} + \text{contact force},$$

where the terms are quantified as follows:

*Accumulation:*

$$\frac{d}{dt}(\rho u_z \, dV) = \frac{d}{dt}(\rho u_z 2\pi r \, dr \, dz) = 2\pi r \, dr \, dz \frac{d}{dt}(\rho u_z), \tag{5.11}$$

which is zero for incompressible ($\rho$ = constant), steady flow ($d/dt = 0$).

*Momentum in:*

$$[(\rho u_z)u_z]_z = 2\pi r \, dr \tag{5.12}$$

*Momentum out:*

$$-[(\rho u_z)u_z]_{z+dz} 2\pi r \, dr \tag{5.13}$$

*Body force (gravity):*

$$(2\pi r \, dr \, dz)\rho g z = 0 \text{ (since } g_z = 0) \tag{5.14}$$

*Viscous normal force: pressure and viscous normal forces:*

$$-[2\pi r(-p + \tau_{zz}) \, dr]_z + [2\pi r(-p + \tau_{zz}) \, dr]_{z+dz} \tag{5.15}$$

*Shear force:*

$$-[2\pi r \tau_{rz}]_r \, dz + [2\pi r \tau_{rz}]_{r+dr} \, dz \tag{5.16}$$

By substituting these terms in Eqn. (5.8), the momentum equation at steady state results:

$$0 = 2\pi r \, dr[(u_z u_z)_z - (u_z u_z)_{z+dz}]\rho - 2\pi r \, dr[(p|_{z+dz} - p|_z)]$$
$$+ 2\pi r \, dr[\tau_{zz}|_z - \tau_{zz}|_{z+dz}] + 2\pi \, dz[(r\tau_{rz})_{r+dr} - (r\tau_{rz})_r]. \tag{5.17}$$

By dividing by $2\pi \, dr \, dz$, this equation becomes

$$\rho r \frac{u_z^2|_z - u_z^2|_{z+dz}}{dz} - r \frac{p|_{z+dz} - p|_z}{dz} + r \frac{\tau_{zz}|_{z+dz} - \tau_{zz}|_z}{dz} + \frac{(r\tau_{rz})_{r+dr} - (r\tau_{rz})_r}{dr} = 0. \tag{5.18}$$

In the limit of $dz \to 0$ and $dr \to 0$, the terms reduce to the corresponding derivatives, so that the difference Eqn. (5.18) reduces to the differential equation

$$\rho \frac{\partial u_z^2}{\partial z} + \frac{\partial p}{\partial z} - \frac{\partial \tau_{zz}}{\partial z} - \frac{1}{r}\frac{\partial}{\partial r}(r\tau_{rz}) = 0. \tag{5.19}$$

The first term is zero by virtue of the result of the mass balance (Example 5.2), $\partial u_z/\partial z = 0$, and therefore

$$r\left(\frac{\partial p}{\partial z} - \frac{\partial \tau_{zz}}{\partial z}\right) - \frac{\partial}{\partial r}(r\tau_{rz}) = 0. \tag{5.20}$$

Equation (5.20) is the momentum equation for flow of any incompressible liquid in a cylindrical horizontal geometry (pipe and annulus) similar to that of Fig. E5.2.

The pressure gradient $\partial p/\partial z$, which drives flows in rectilinear pipes, channels, and annuli, is usually known and identical to the pressure difference between the entry of the liquid (at pressure $p_0$ maintained by a pump, say) and the exit to the ambient (at gauge pressure zero), divided by the length of the flow, i.e.,

$$\frac{\partial p}{\partial z} = \frac{\Delta p}{\Delta L} < 0. \tag{5.21}$$

The pressure distribution for these flows is linear, by integrating Eqn. (5.21):

$$p(z) = p_0 + \frac{\Delta p}{\Delta L}z = p_0 + \frac{0 - p_0}{\Delta L}z = p_0\left(1 - \frac{z}{\Delta L}\right). \tag{5.22}$$

The final form of Eqn. (5.20) then becomes

$$\frac{\Delta p}{\Delta L} = \frac{\partial \tau_{zz}}{\partial z} + \frac{1}{r}\frac{\partial}{\partial r}(r\tau_{rz}). \tag{5.23}$$

Equation (5.23) is still not solvable, unless the two involved unknown stresses, $\tau_{zz}$ and $\tau_{rz}$, can be expressed in terms of a single common unknown, which in this case is the unique velocity component along the pipe, $u_z$. This relation is provided by the *constitutive equation,* as already explained in Chapter 1 and detailed below.

## 5.5 CONSTITUTIVE EQUATION

The constitutive equation is an equation of state under flow and deformation that differentiates the behavior of rheologically different fluids, even when subjected to identical flow conditions. This is consistent with the physics of the liquids involved; not all fluids will behave similarly under the same flow conditions (geometry and pressure gradient here). Some fluids, for example water, will flow easily even under mild pressure and stress gradients. But other liquids, for example ketchup, will require large such a gradient to flow (e.g., pouring ketchup out of its bottle). The equation that specifies how a given fluid flows or deforms when subjected to force gradients (sush as pressure difference, gravity, or shear difference) is called the *constitutive equation.*[9] The constitutive equation relates all the stress components to the velocity and therefore permits solution of Eqn. (5.23). The constitutive equation is *not* a conservation equation. It is a relation between the stress and the velocity and its derivatives and is characteristic of the fluid alone [i.e., it is given (often derived from molecular theories) for a given fluid without changing its form with the flow geometry]. *Constitutive theories* are discussed further in Chapter 9 in connection with non-Newtonian fluids.

A fluid is *Newtonian* if it follows the Newtonian constitutive equation

$$\tau_{ij} = \eta \left( \frac{\partial u_i}{\partial x_j} + \frac{\partial u_j}{\partial x_i} \right), \tag{5.24}$$

where $\tau_{ij}$ is the stress on the *i*-surface (perpendicular to the *i*-axis of the coordinate system) acting in the *j*-direction, and $u_i$ and $x_i$ are the velocity and coordinate, respectively. The derivation and sign convention of these equations follows the approach laid down in Section 1.3. The components of Eqn. (5.24) with respect to the commonly used system of coordinates are tabulated in Appendix F. These are essential for a complete analysis (i.e., calculation of velocity, pressure, and stress) of any flow of Newtonian liquid. Example 5.4 demonstrates the use of Appendix F to solve Eqn. (5.23), representing momentum conservation in pipe flow.

In Eqn. (5.24) and Appendix F, $\eta$ is the *viscosity,* measuring resistance to flow, characteristic of the fluid. The viscosity of gases can be estimated by the kinetic theory of molecules as summarized in Section 1.3, whereas that of liquids is still based largely on empirical or semiempirical correlations. Viscosities of pure compounds and common application liquids are measured by means of *viscometers* and tabulated as in Table 1.4 and Figs. 1.5 and 1.6.

A fluid is *non-Newtonian* if it does not follow Eqn. (5.24). The most characteristic class of non-Newtonian liquids are polymeric liquids of long macromolecules, which can reduce resistance to flow by aligning themselves along the streamlines of flow. Several constitutive equations have been proposed to describe non-Newtonian liquids.[10] The simplest is the *power-law model,*

$$\tau_{ij} = k \left| \frac{\partial u_i}{\partial x_j} \right|^n \left( \frac{\partial u_i}{\partial x_j} + \frac{\partial u_j}{\partial x_i} \right), \tag{5.25}$$

in which the viscosity is not constant, but rather, depends on the absolute value of the shear rate $\partial u_i / \partial x_j$. The parameters $k$ and $n$ are *material parameters,* characteristic of non-Newtonian liquids. The parameter $k$ is a *consistency factor* and $n$ is the *power-law exponent.* The Newtonian liquid is recovered from Eqn. (5.25) in the limit $n = 0$ and $k = \eta$.

Non-Newtonian liquids such as polymer melts and solutions, suspensions, emulsions, paints, sauces (ketchup, mayonnaise, butter), fiber suspensions in composite processing, and others are as important to engineers as Newtonian fluids. Chapter 9 highlights some of the most important differences between Newtonian and non-Newtonian fluids, and the resulting macroscopic behavior of non-Newtonain fluids, under common processing conditions.

### Example 5.4: Use of the Constitutive Equation to Solve the Momentum Equation

Solve the momentum equation, Eqn. (5.23), for flow of Newtonian liquid in a horizontal pipe.

**Solution**   Since the liquid is Newtonian, the relations of Appendix F apply, which allow the elimination of stresses in the momentum equation,

$$\frac{\Delta p}{\Delta L} = \frac{\partial \tau_{zz}}{\partial z} + \frac{1}{r}\frac{\partial}{\partial r}(r\tau_{rz}), \tag{5.4.1}$$

with the pressure gradient $\Delta p/\Delta L$ known. From Appendix F and the mass balance result [Eqn. (5.2.2)], the viscous normal stress is

$$\tau_{zz} = 2\eta\frac{\partial u_z}{\partial z} = 0.$$

The shear stress, from Appendix F, is

$$\tau_{rz} = \eta\frac{\partial u_z}{\partial r}, \tag{5.4.2}$$

and Eqn. (5.4.1) becomes

$$\frac{d}{dr}\left(r\frac{du_z}{dr}\right) = \frac{r}{\eta}\frac{\Delta p}{\Delta L}, \tag{5.4.3}$$

which is integrated to

$$r\frac{du_z}{dr} = \frac{r^2}{2\eta}\frac{\Delta p}{\Delta L} + c_1, \tag{5.4.4}$$

and finally to

$$u_z = \frac{r^2}{4\eta}\frac{\Delta p}{\Delta L} + c_1\ln r + c_2. \tag{5.4.5}$$

Equation (5.4.5) is the *general solution* to the momentum equation (5.24) and predicts the velocity $u_z$, in terms of the radius $r$, of all laminar flows of Newtonian liquid in horizontal cylindrical fields, such as long circular pipes and annuli. The *particular solution* to the flow under consideration is obtained only after invoking the appropriate *boundary conditions,* the most common of which are summarized in Table 5.2 and explained in Section 5.6.

## 5.6 BOUNDARY CONDITIONS

To find the particular solution to a differential equation, appropriate boundary conditions must be supplemented. These are relations that describe how the fluid system interacts with the rest of the natural world along its boundaries. The most common boundary conditions in fluid flows are:

1. *No-slip boundary conditions.* Liquids adhere to the motion of adjacent liquid or solid boundaries.
2. *Slip boundary conditions.* Gases slip along solid boundaries, and this results in zero shear stress there.
3. *Symmetry boundary conditions.* The normal velocity and shear stress along planes or lines of symmetry are zero.
4. *Free-surface boundary conditions.* The shear stress and pressure at a planar free surface are zero, unless otherwise specified.

5. *Interface boundary conditions.* The velocity, pressure, and shear stress are continuous across a two-liquid planar interface unless otherwise specified.

6. *Linear azimuthal velocity $u_\theta$ of a surface rotating with angular velocity $\Omega$ about its axis.* At distance $R$, this value is $u_\theta = R\Omega$.

7. *Sink and source boundary conditions.* The liquid is at rest far away from them.

8. *Flow-rate boundary conditions.* For example:

$$\text{Flow in pipe of radius } R: \dot{V} = \int_0^R 2\pi r u_z \, dr. \tag{5.26}$$

$$\text{Planar radial flow of height } H: \dot{V} = 2\pi R H u_r|_{r=R}. \tag{5.27}$$

$$\text{Spherical radial flow}: \dot{V} = 4\pi R^2 u_r|_{r=R}. \tag{5.28}$$

Table 5.2 lists common flow fields with the required boundary conditions.

### Example 5.5: From the General to the Particular Solution

Find the velocity profile for the horizontal pipe flow shown in Fig. E5.3, starting from the general solution obtained in Example 5.4.

**Solution**    The velocity profile given by Eqn. (5.4.5) is not fully defined until $c_1$ and $c_2$ are assinged specific values. The general solution,

$$u_z(r) = \frac{r^2}{4\eta} \frac{\Delta p}{\Delta L} + c_1 \ln r + c_2, \tag{5.5.1}$$

involves two arbitrary constants, $c_1$ and $c_2$, and therefore two boundary conditions are required. The appropriate boundary conditions are:

$$\text{At } r = 0, \quad \frac{du_z}{dr} = 0 \text{ (symmetry: zero } \tau_{rz})$$

$$\text{At } r = R, u_z = 0 \text{ (no slip)}$$

These boundary conditions are substituted in the general solution, Eqn. (5.5.1), which results in:

$$\text{At } r = 0, \quad \frac{du_z}{dr} = 0 + \frac{c_1}{0} + 0 = 0.$$

$$\text{At } r = R, \quad u_z = \frac{R^2}{4\eta} \frac{\Delta p}{\Delta L} + c_1 \ln R + c_2 = 0.$$

The first equation forces $c_1 = 0$. Then the second equation yields $c_2 = -(R^2/4\eta)(\Delta p/\Delta L)$. The constants $c_1$ and $c_2$ are then substituted into the general solution to produce the particular solution,

$$u_z(r) = \frac{1}{4\eta} \frac{\Delta p}{\Delta L} (r^2 - R^2). \tag{5.5.2}$$

**TABLE 5.2**  BOUNDARY CONDITIONS IN COMMON FLUID FLOWS

| Flow | Flow geometry | Boundary conditions |
|---|---|---|
| Pipe flow | | At $r = 0$, $\partial u_z/\partial r = 0$ <br> At $r = R$, $u_z = 0$ |
| Film flow under gravity | | At $y = 0$, $u_x = 0$ <br> At $y = H$, $\tau_{yx} = 0$ |
| Two-layer channel flow | | At $y = 0$ and $y = H_A + H_B$, <br> $u_x^A = u_x^B = 0$ <br> At $y = H_A$, <br> $u_x^A = u_x^B$ and $\tau_{yx}^A = \tau_{yx}^B$ |
| Torsional flow | | At $r = R$, $u_\theta = R\Omega$ <br> At $r = \infty$, $u_\theta \approx 0$ <br> (also, often at $r = 0$, $u_\theta$ <br> is finite, when the fluid <br> extends to $r = 0$) |
| Sink (source) flow | <br> (a) cylindrical     (b) spherical | At $r = \infty$, $u_r = 0$, $p = p_{\text{hydrost.}}$ <br> $\dot{V} = 2\pi R H u_r|_{r=R}$ (planar) <br> $\dot{V} = 4\pi R^2 u_r|_{r=R}$ (spherical) <br> Bernoulli's equation applies, <br> too: $\Delta(p/\rho + u^2/2 + gz) = 0$ |

As a check for consistency, Eqn. (5.5.2) indeed yields $u_z = 0$ at $r = R$, and $du_z/dr = 0$ at $r = 0$. Unlike the general solution, the particular solution is fully defined and restricted to the specific flow geometry shown in Fig. E5.3.

The step-by-step development of the particular solution to a specific flow, through Examples 5.2 to 5.5, is summarized in Section 5.7. The procedure described is common for most flow situations amenable to analytic solutions.


## 5.7 HOW TO SOLVE PROBLEMS OF LAMINAR FLUID FLOW


There are two fundamental approaches to solve flow problems amenable to analytic solutions. The first is based on control volume principles and is summarized below. The second is based on tabulated universal conservation equations, also derived by generalized control volume principles, and is described in Chapter 7. The strategy adopted in this book is first a systematic control volume-based approach, accompanied by examples and then generalization of these principles to show how the tabulated universal equations are obtained and then how they are utilized to apply to specific flow situations. The differential control volume (DCV) approach follows.

**Step 1.**    Understand the flow situation: What is the geometry? What symmetries exist? What is the most convenient system of coordinates (i.e., where is its origin and its directions)? Which velocity components are zero? Which coordinate(s) do the nonzero velocity component(s) depend on?

*Spherical radial flow*:

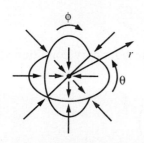

Flow spherically symmetric, (i.e.,
$\partial/\partial\theta = \partial/\partial\phi = 0$).
In addition, flow is radial (i.e., $u_\theta = u_\phi = 0$, $u_r \neq 0$).

*Pipe flow*:

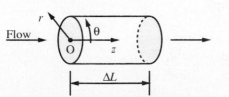

Flow cylindrically symmetric
     (i.e., $\partial/\partial\theta = 0$).
Flow unidirectional (i.e., $u_r = u_\theta = 0$).
Fully developed flow [i.e., $u_z = f(r)$].

*Channel flow*:

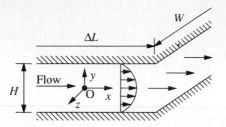

Flow in rectangular geometry.
Flow unidirectional (i.e., $u_y = u_z = 0$).
Fully developed flow [i.e., $u_x = f(y, z)$].
If $W/H \gg 1$, then $u_x = f(y)$.

**Step 2.** If the flow is radially symmetric, use the appropriate mass balance equation to find the velocity profile. Then use the appropriate momentum balance in the direction of motion to find the pressure profile. A quick way to find the pressure from the velocity profile of an inviscid flow (of zero viscosity, e.g., any ideal gas flow), of laminar flow at Re $\to \infty$ away from solid boundaries (e.g., river flow), and of any radial flow, is by Bernoulli's equation (Section 3.3.1). For radial flows in porous media, Darcy's law (Section 4.11) must be used instead, to calculate the pressure, as demonstrated in Example 5.6. If the flow is not radial (e.g., pipe, channel, film, and torsional flows) use the appropriate momentum equation *in the direction of fluid motion* to find the velocity profile. The pressure profile is usually given in most of these flows or else calculated by means of the remaining momentum equations.

**Step 3.** Derive the appropriate differential equation by mass or momentum balances on appropriate control volumes, for example:

*Spherical flow*:

$$\frac{\partial}{\partial r}(r^2 u_r) = 0 \qquad u_r = \frac{c_1}{r^2}. \tag{5.29}$$

This differential equation is of first order, so one boundary condition is needed. At $r = R$, $\dot{V}$ or $u_r$ may be given.

*Pipe flow*:

$$\frac{\partial p}{\partial z} = \eta \frac{1}{r} \frac{\partial}{\partial r}\left(r \frac{\partial u_z}{\partial r}\right). \tag{5.30}$$

Usually, $\partial p/\partial z = \Delta p/\Delta L$ is constant, so that Eqn. (5.30) is a second-order differential equation, needing two boundary conditions: at $r = 0$, $\partial u_z/\partial r = 0$ (symmetry); at $r = R$, $u_z = 0$ (no slip).

*Channel flow*:

$$\frac{\partial p}{\partial x} = \eta \frac{\partial^2 u_x}{\partial y^2}. \tag{5.31}$$

Usually, $\partial p/\partial x = \Delta p/\Delta L$ is constant, so that Eqn. (5.31) is a second-order differential equation, needing two boundary conditions: at $y = 0$, $\partial u_x / \partial y = 0$ (symmetry); at $y = H/2$, $u_x = 0$ (no slip).

**Step 4.**    Solve the differential equation and boundary conditions to find the velocity distribution.

**Step 5.**    Calculate macroscopic quantities as detailed in Section 5.8.

# 5.8 CALCULATION AND MEASUREMENT OF USEFUL FLOW QUANTITIES

*Flow rate*:

$$\dot{V} = \int_{\text{section of flow}} \underbrace{u}_{\substack{\text{normal} \\ \text{velocity}}} \times \underbrace{ds}_{\substack{\text{elementary} \\ \text{area}}} \tag{5.32}$$

*Stress*:

$$\tau_{xy} = \eta \left( \frac{\partial u_x}{\partial y} + \frac{\partial u_y}{\partial x} \right). \tag{5.33}$$

*External force on boundary*:

$$F = - \int_{\text{boundary}} (\text{stress}) \, ds. \tag{5.34}$$

*Residence time*:

$$t = \int_0^x \frac{dx}{u_x}, \; = \int_0^z \frac{dy}{u_y}, \; = \int_0^\theta \frac{r \, d\theta}{u_\theta}, \; = \int_0^r \frac{dr}{u_r}, \; = \int_0^s \frac{ds}{u} \tag{5.35}$$

where $(u_x, u_y)$ and $(u_\theta, u_r)$ are components in the $(x, y)$ direction of the Cartesian or the $(\theta, r)$ direction of the polar system of coordinates, of the *streamline velocity u* in the direction of flow *s*. Thus

$$ds = [(dx)^2 + (dy)^2]^{1/2} = r \, d\theta \tag{5.36}$$

and

$$u = [u_x^2 + u_y^2]^{1/2} = [u_r^2 + u_\theta^2]^{1/2}. \tag{5.37}$$

For unidirectional flows, the simplified velocity relations of Example 5.8 apply.

*Streamlines.* There are imaginary lines along which fluid particles travel with velocity tangential to them. If a fluid particle is marked and pictured as it travels, the recorded line represents a streamline. If $u_x$ and $u_y$ are velocity components, the streamlines are defined by the fact that time $dt$ is required to cover distance

$ds$ with velocity $u$, or distance $dx$ with velocity $u_x$ or distance $dy$ with velocity $u_y$, according to Eqn. (5.35). Thus

$$dt = \frac{dx}{u_x} = \frac{dy}{u_y}, \tag{5.38}$$

which is also the condition that the velocity $(u = u_x + u_y)^{1/2}$ is tangential to the streamline. The streamlines of rectilinear pipe and channel flows are straight lines parallel to the wall, those of radial flows are identical to the radii, and those of torsional flows are concentric circles.

*Friction between two points on streamline*:

$$\frac{\Delta u^2}{2} + \frac{\Delta p}{\rho} + g\,\Delta z + \mathscr{F} = 0 \tag{5.39}$$

The last requirement allows evaluation of the pressure. For rectilinear pipe, channel, and annulus flows, the pressure is calculated directly from the known pressure gradient by means of equations similar to Eqn. (5.22). For radial flows, $\mathscr{F} = 0$ and the pressure is calculated from Bernoulli's equation itself, which in this case is identical to the radial momentum equation. For the remaining cases, the pressure is calculated by means of momentum balances in the remaining directions, usually perpendicular to the flow direction (e.g., torsional flows).

*Vorticity*: This quantity represents the ability of fluid particles to rotate with respect to each other, with angular speed $\Omega$, as they are convected by different velocities. In fact, the vorticity is twice the angular velocity, and is defined by

$$w_i = \frac{\partial u_j}{\partial u_k} - \frac{\partial u_k}{\partial u_j} = 2\,\Omega. \tag{5.40}$$

Thus, a buoyancy neutral spherical particle will rotate about one of its axes at angular speed $\Omega = w/2$, if left in a fluid flow where the vorticity is $w$. Tornados and other swirling flows are characterized by large vorticity, whereas radial flows are of zero vorticity. Vorticity is not a directly measurable quantity; it is rather a property that gives rise to measurable rotation. Thus, vorticity is responsible for many diverse phenomena, such as efficient mixing of highly viscous liquids (e.g., Section 10.4), damages inflicted by tornados (e.g., Problem 7.11), enzyme production induced by cells in vorticity-rich liquid environments, and fiber orientation in composite suspensions.[11] The physics and dynamics of vorticity are detailed in Chapter 7, where *potential irrotational flows of zero vorticity and boundary layer flows dominated by vorticity are examined.*

**Rheometers.**     Flow rates are measured by devices called rheometers. The *Venturi-pipe rheometer* is used to measure *average velocity* of liquids. The *orifice plate* (Fig. 3.4) is used to measure *average velocity* of gases and liquids, at high Reynolds numbers. The *pitot tube* (Fig. 3.6) is used to measure *local velocities* over cross sections of pipes, and therefore to determine *velocity profiles.* *Rotameters* (Fig. P8.2) measure upward flow rates in vertical pipes, based on the equilibrium of a submerged cone under the forces of gravity, buoyancy, and flow pressure drop. *Weirs* (Fig. P4.22) are used to measure flow rates in open channels. Devices based on light deflection by seeded particles or impurities moving with the fluid velocity, such as *hot-wire anemometers* and *laser doppler velocimeters,* measure *local velocity components* with the highest accuracy.[12]

*Wall stresses and pressure* are measured by means of sensitive elastic membranes over holes drilled along the transfering pipe or channels.[13] *Residence times and streamlines* are measured by observing and picturing marked particles or impurities as they travel with the fluid. Local traveling times are measured by the measured velocity of seeded particles to cross measuring optical volumes, as in the laser doppler velocimeter. Seeding particles used to measure velocities and traveling times must be big enough such that Brownian motion is impossible and small enough such that gravity and buoyancy effects are negligible.

*Vorticity* can be measured by recording angular velocity of rotation of slender, buoyancy-neutral fibers traveling with the fluid, or the evolution of relative position of pairs of seed particles as they travel with the fluid.[14]

## 5.9  EXAMPLES OF DIFFERENTIAL FLUID MECHANICS

The best way to overcome difficulties in solving fluid mechanics problems is by solving as many of them as possible. The remainder of this chapter is spent on problem-solving training. Chapter 6 continues along the same philosophy; however, flows are further classified into groups of problems, each group amenable to a common solution approach.

### Example 5.6: Flow Through a Porous Cylindrical Wall

Consider purely radial flow of liquid through the porous wall of an infinitely long cylinder of outer radius $R = 15$ cm (Fig. E5.6). The liquid escapes at flow rate $\dot{V}/L = 0.13$ m³/s per meter length.

**(a)** Calculate the radial velocity $u_r$ at $r = 15$ cm, and as $r \to \infty$. Justify your answer on physical arguments.

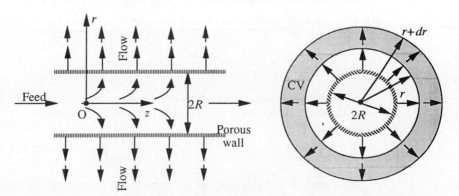

**Figure E5.6**  Flow through porous wall and cross section of annular control volume.

**(b)** What is the resulting pressure distribution outside the cylinder?

**(c)** If the pressure inside the cylinder is 1.2 atm and the thickness of the wall $H = 10$ cm, what are the permeability of the porous wall and the pressure distribution?

**Solution**

**(a)** Mass conservation on an annular control volume of base shown in Fig. E5.6b and height $dz$ yields

$$\frac{d}{dt}(\rho 2\pi r\, dr\, dz)_r = (2\pi r\rho u_r\, dz) - (2\pi r\rho u_r\, dz)_{r+dr}. \qquad (5.6.1)$$

Dividing by $(2\pi r\, dr\, dz)$ and invoking the definition of derivative results in

$$\frac{d\rho}{dt} = \lim_{dr\to 0}\frac{1}{r}\frac{(ru_r\rho)_r - (ru_r\rho)_{r+dr}}{dr} = \frac{1}{r}\frac{d(ru_r\rho)}{dr}. \qquad (5.6.2)$$

For an imcompressible liquid of constant density $\rho$, Eqn. (5.6.2) reduces further to

$$\frac{d}{dr}(ru_r) = 0 \quad \text{and} \quad u_r = \frac{c}{r}. \qquad (5.6.3)$$

To determine the constant $c$, the fact that the total flow rate is $\dot{V}/L$ is used: At $r = R$, $\dot{V}/L = 2\pi Rc/R$ and therefore $c = (\dot{V}/L)(1/2\pi)$. The resulting velocity profile, both through the porous wall and beyond it, is

$$u_r(r) = \frac{\dot{V}}{2\pi Lr}. \qquad (5.6.4)$$

The velocity at $r = 20$ cm is $u_r|_{r=20\,\text{cm}} = 5/20\pi = 0.080$ cm/s. The velocity at $r \to \infty$ is $u_r|_{r=\infty} = 0$, which means that the liquid away from the cylinder is virtually at rest and increases approaching the porous wall. This is due to the fact that the velocity has to decrease with radial distance, in order to maintain a constant flow rate $\dot{V}$ and therefore conservation of mass, since the area $2\pi r$ increases with radial distance.

**(b)** The pressure distribution *outside the cylinder is given by Bernoulli's equation* expressed as

$$p(r) + \frac{u_r^2(r)}{2} = p_\infty + \frac{u_r^2(r \to \infty)}{2}. \qquad (5.6.5)$$

Given that the velocity and pressure far away from the cylinder are zero and hydrostatic, respectively, the resulting pressure distribution is

$$p(r) = p_{\text{hydr.}} - \frac{u_r^2(r)}{2} = p_{\text{hydr.}} - \left(\frac{\dot{V}}{2\pi L}\right)^2\frac{1}{2r^2}, \qquad R \le r \le \infty. \qquad (5.6.6)$$

This expression holds only on the outside nonporous part of flow. Within the porous wall, the pressure is calculated as follows.

**(c)** The pressure distribution *through porous media is obtained by Darcy's law,*

*Eqn.* (4.80):

$$u_r(r) = -\frac{k}{\eta}\frac{dp}{dr},\tag{5.6.7}$$

given the velocity $u_r(r)$ by Eqn. (5.6.4), which is common throughout the flow. It is

$$p(r) = -\frac{\dot{V}\eta}{2\pi Lk}\ln r + c,\tag{5.6.8}$$

where the constant $c$ can be determined by a pressure boundary condition. The common pressure at the outside cylindrical surface given by Eqn. (5.6.5) is

$$p(r = R) = p_{\text{hydr.}} - \left(\frac{\dot{V}}{2\pi L}\right)^2\frac{1}{2R^2} = 1\text{ atm} - 0.55\text{ atm} = 0.45\text{ atm},\tag{5.6.9}$$

which substituted in Eqn. (5.6.8) determines the constant,

$$c = p(r = R) + \frac{\dot{V}\eta}{2\pi Lk}\ln R = 0.45\text{ atm} + \frac{\dot{V}\eta}{2\pi Lk}\ln R,\tag{5.6.10}$$

and the pressure distribution within the porous wall becomes

$$p(r) = 0.45\text{ atm} + \frac{\dot{V}\eta}{2\pi Lk}\ln\frac{R}{r},\qquad R - H \le r \le R.\tag{5.6.11}$$

To calculate the permeability $k$, the known pressure at the inside cylindrical surface, $p(r = 5\text{ cm}) = p_I = 1.2\text{ atm}$, is substituted in Eqn. (5.6.11). Then

$$k = \frac{\dot{V}\eta}{2\pi L}\frac{\ln(R/R_I)}{(p_I - 0.45)} = 3.1 \times 10^{-6}\text{ darcy}.\tag{5.6.12}$$

Similar techniques, based on measuring pressure differences of flow in porous media, are used to measure permeabilities.

The flow examined in this example has some interesting features: Pressure is required to force the liquid through the narrow pores of the porous wall, which results in a 0.75 atm drop over the 10-cm wall thickness, down to pressure 0.45 atm at the exit from the porous wall. Then a radial frictionless flow takes over which reduces velocity to zero and increases pressure to hydrostatic far away from the porous wall. *Thus radial one-phase flows are irrotational following Bernoulli's equation, whereas radial flows in porous media follow Darcy's equation.*

### Example 5.7: Flow in an Annulus

An incompressible liquid flows steadily in the axial direction $z$, through an annulus of inner radius $R_0$ and outer radius $R_1$. Show that the flow is governed by the equation $d(r\tau_{rz})/dr = r\,dp/dz$. Derive the velocity distribution $u_z(r)$ if $dp/dz = \Delta p/\Delta L$ is constant. Then calculate the flow rate $\dot{V}$ and the total frictional retarding force $F$ per unit length.

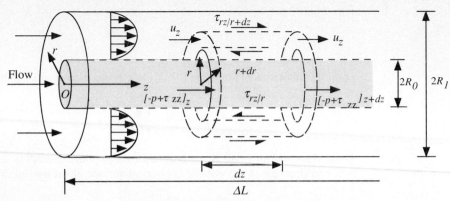

**Figure E5.7.1**   Flow in a rectilinear annulus.

**Solution**   This is a unidirectional steady flow with $u_r = 0$, $u_\theta = 0$, and $u_z \neq 0$. It is also an axisymmetric flow, with $\partial/\partial\theta = 0$. Conservation of mass over the control volume shown in Fig. E5.7.1 yields (see also Example 5.3)

$$\frac{\partial u_z}{\partial z} = 0. \tag{5.7.1}$$

The governing equation of momentum conservation in the $z$-direction can be found by control-volume principles:

$$\frac{d}{dt}(2\pi r\, dr\, dz \rho u_z) = \begin{pmatrix} \text{net convection is zero} \\ \text{because } u_z = u_{z+dz} \end{pmatrix} + \underbrace{(p|_z - p|_{z+dz})2\pi r\, dr}_{\text{net pressure force}}$$

$$+ \underbrace{(-\tau_{zz}|_z + \tau_{zz}|_{z+dz})2\pi r\, dr}_{\substack{\text{net normal force zero} \\ \tau_{zz} = \eta\, \partial u_z/\partial z = 0}} + \underbrace{(\tau_{rz}2\pi r\, dz)_{r+dr} - (\tau_{rz}2\pi r\, dz)_r}_{\text{shear force}}$$

$$+ \begin{pmatrix} \text{zero gravity force,} \\ \text{annulus horizontal} \end{pmatrix}. \tag{5.7.2}$$

Thus, after invoking the definition of a derivative at steady state, the only surviving terms are

$$-\frac{dp}{dz} + \frac{1}{r}\frac{d}{dr}(r\tau_{rz}) = 0. \tag{5.7.3}$$

In unidirectional flows, $dp/dz = \Delta p/\Delta L = $ constant; therefore,

$$r\frac{\Delta p}{\Delta L} = \frac{d}{dr}\left(r\frac{du_z}{dr}\right),$$

$$r\frac{du_z}{dr} = \frac{r^2}{2}\frac{\Delta p}{\Delta L} + c_1,$$

and finally,

$$u_z = \frac{r^2}{4}\frac{\Delta p}{\Delta L} + c_1 \ln r + c_2. \tag{5.7.4}$$

The boundary conditions to determine $c_1$ and $c_2$ are:

At $r = R_0$, $u_z = 0$, and therefore, $0 = \dfrac{R_0^2 \, \Delta p}{4\eta \, \Delta L} + c_1 \ln R_0 + c_2.$

At $r = R_1$, $u_z = 0$ and therefore, $0 = \dfrac{R_1^2 \, \Delta p}{4\eta \, \Delta L} + c_1 \ln R_1 + c_2.$

The result is

$$c_1 = \frac{1}{4\eta} \frac{\Delta p}{\Delta L} \frac{R_1^2 - R_0^2}{\ln(R_0/R_1)} \quad \text{and} \quad c_2 = -\frac{1}{4\eta} \frac{\Delta p}{\Delta L} \left[ R_0^2 + \frac{R_1^2 - R_0^2}{\ln(R_0/R_1)} \ln R_0 \right].$$

Thus the velocity profiles becomes

$$u_z = -\frac{1}{4\eta} \frac{\Delta p}{\Delta L} \left[ R_0^2 - r^2 + (R_1^2 - R_0^2) \frac{\ln(R_0/r)}{\ln(R_0/R_1)} \right]. \tag{5.7.5}$$

The *flow rate* $\dot{V}$ is given by the summation $\sum \dot{V}_i$ of flow rates through each individual ring, as shown in Fig. E5.7.2.

$$\dot{V} = \sum_{i=1}^{n \to \infty} \dot{V}_i = \sum_{i=1}^{N \to \infty} A_i u_z^i = \sum_{i=1}^{N \to \infty} (2\pi r_i \, \Delta r_i) u_z(r_i) \tag{5.7.6}$$

$$= \int_{r=R_0}^{r=R_1} 2\pi r \left[ -\frac{1}{4\eta} \frac{\Delta p}{4L} \left( R_0^2 - r^2 + (R_1^2 - R_0^2) \frac{\ln(R_0/r)}{\ln(R_0/R_1)} \right) \right] dr,$$

The *frictional losses* are obtained by summing the forces acting along the inner and outer cylinders. From the augmented Bernoulli's equation, $\mathcal{F} = (p_1 - p_2)/\rho$. Also, a force balance yields

$$(p_1 - p_2)\pi(R_1^2 - R_0^2) = +\pi_{rz}\big|_{r=R_0} 2\pi R_0 \, \Delta L + \tau_{rz}\big|_{r=R_1} 2\pi R_1 \, \Delta L.$$

Thus

$$\mathcal{F} = \frac{p_1 - p_2}{\rho} = \frac{[+\tau_{rz}\big|_{r=R_0} 2\pi R_0 + \tau_{rz}\big|_{r=R_1} 2\pi R_1] \, \Delta L}{\rho \pi (R_1^2 - R_0^2)} = \frac{p_1 - p_2}{\rho}, \tag{5.7.7}$$

where the shear stresses were calculated by differentiating Eqn. (5.7.5).

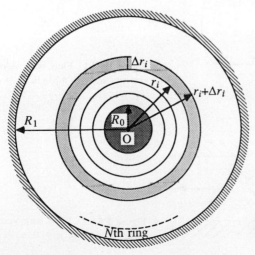

**Figure E5.7.2**  Flow rate through an annulus.

**Example 5.8: Selection of System of Coordinates**

Select the appropriate system of coordinates, among those highlighted in *Appendix A,* for convenient analyses of the following flow situations:

**(a)** *Cylindrical axisymmetric flow:* transferring of natural gas in pipes, hypodermic needle injection, flow under a round jet, film flow down a vertical solid cylinder, annular two-phase flow.

**(b)** *Cartesian flow:* channel flow, ice sliding flow, film down a vertical plate, liquid spreading by knife or blade, flow in a shallow lake induced by wind.

**(c)** *Cylindrical radial flow:* underground oil migration to a well bore, flow across porous wall of cylinder, flow in radially expanding free sheet, sink and source flow.

**(d)** *Spherical radial flows:* liquid flow around growing or collapsing bubble, sink and source flow through porous spherical walls, airflow induced by blast wave.

**(e)** *Torsional flow:* flow around rotating cylinder, flow inside rotating container, azimuthal flow between co- or counterrotating concentric cylinders, vortex flow, spiral flow.

**Solution**

**(a)** *Cylindrical axisymmetric flow* (Fig. E5.8.1):

**(b)** *Cartesian flows* (Fig. E5.8.2):

**(c)** *Cylindrical radial flow* (Fig. E5.8.3):

**(d)** *Spherical radial flow* (Fig. E5.8.4):

**(e)** *Torsional flow* (Fig. E5.8.5):

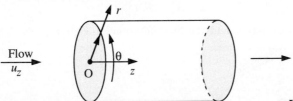

**Figure E5.8.1** Flow with velocity components, $u_r = 0$, $u_\theta = 0$, $u_z = u$.

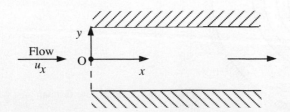

**Figure E5.8.2** Flow with velocity components, $u_x = u$, $y_y = 0$, $u_z = 0$.

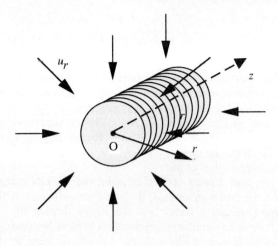

**Figure E5.8.3**   Flow with velocity components, $u_r = u$, $u_\theta = 0$, $u_z = 0$.

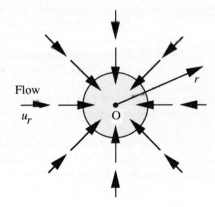

**Figure E5.8.4**   Flow with velocity components, $u_r = u$, $u_\theta = 0$, $u_\phi = 0$.

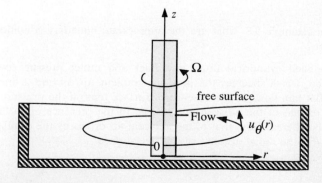

**Figure E5.8.5**   Flow with velocity components, $u_r = 0$, $u_\theta = u$, $u_z = 0$.

**Example 5.9: Velocity Boundary Conditions**

For the steady flows listed in Example 5.8, what are the appropriate boundary conditions?

**Solution**    (a) For natural gas flow (and any other viscous fluid flow) in pipes (and within a hypodermic needle), the boundary conditions are zero velocity at the wall and zero velocity slope at the axis of symmetry.

For flow under a round jet, the velocity derivative is zero at the axis of symmetry, and the shear stress is zero at the free surface in case of adjacent stationary air. If the air moves, the zero shear stress condition is replaced by a liquid velocity identical to that of air.

For film flow down a vertical cylinder, the velocity is zero at the surface of the cylinder. At the free film surface, the shear stress is zero if the air is stationary, or else the liquid velocity is identical to that of the flowing air.

For flow in an annulus, the velocity is zero at the two cylindrical walls.

(b)    For flow in a channel, the velocity is zero at the opposing walls, or equivalently, the velocity is zero at one of the opposing walls and the derivative of the velocity is zero at the plane of symmetry.

For ice sliding, the velocity of the induced thin liquid water film is zero at the lower surface and identical to the speed of the ice at the upper surface, or the shear stress there is identical to that imposed by the weight of the ice.

The same boundary conditions apply to liquid spreading and flow in a lake induced by air.

(c) For cylindrical radial flows either the flow rate, or a known velocity at a specific radial position, are usually applied as boundary conditions.

(d) In collapsing/growing bubbles and cavities, the liquid velocity is identical to that of the moving gas–liquid interface there. In radial flows through porous media, the flow rate is used as a boundary condition. In blast waves the fluid velocity just downstream of the wavefront can be related to the thermodynamics of the explosion.

(e) In torsional flows, the linear azimuthal velocity of liquid adjacent to a rotating surface is identical to the product of the angular speed with the radial distance. In unbounded torsional flows, the velocity away from the center of rotation is negligible.

**Example 5.10: Pressure Boundary Conditions**

For the flows listed in Example 5.8, what are the appropriate boundary conditions for pressure calculation?

**Solution**    (a) For confined cylindrical flows, the inlet and outlet pressure over length $\Delta L$ are specified, which specifies the pressure gradient $\Delta p/\Delta p$ and a linear pressure profile. For flat film flows in the absence of surface tension the pressure is everywhere identical to that of the adjacent air or gas. In the presence of surface tension, the pressure is different from that of the adjacent air or gas by the capillary pressure $\sigma/R$, where $R$ is the radial distance of the free surface.

(b) As in item (a). Also in free films, the pressure is identical to that of the adjacent air or gas even in the presence of surface tension, since for a planar surface $R \rightarrow \infty$.

(c) The pressure far away from the center of the flow is atmospheric or hydrostatic. Occasionally, the presssure may be measured and given at an important radial position, for example at the wall of a well bore.

(d) The pressure far away from the center of the flow is atmospheric or hydrostatic. For growing/collapsing bubbles and cavities, the pressure at the gas–liquid interface may be related to that of the gas. For flow in porous media, the pressure may be given at a specific radial location.

(e) In unbounded torsional flows, the pressure is atmospheric or hydrostatic far away from the center of rotation. The pressure is also identical to the atmospheric pressure at free surfaces, augmented by the appropriate capillary pressure due to any surface tension.

## 5.10 HEAT AND SOLUTE MASS TRANSFER

When the quantity $q$ in Eqn. (5.1) is solute mass or heat, the corresponding equations of change of solute mass and heat result. Solute and heat are transferred by convection when the velocity of the DCV is different from that of its fluid environment; and by diffusion or conduction in case there is a concentration or temperature gradient at the DCV's boundaries, even in fluids at rest. Furthermore, solute mass may be produced or consumed by chemical reaction; and heat may be produced or consumed by exothermic or endothermic chemical reaction and produced from mechanical energy by viscous dissipation. Example 5.11 highlights these mechanisms for selected transfer processes, whereas detailed exposition of heat and solute transfer in static or flowing fluids is given in textbooks on transport phenomena.[15]

### Example 5.11: Heat and Solute Mass Balances

Apply the general conservation equation, Eqn. (5.1), to find the governing equation of change of heat $H$ and solute $s$ in a three-dimensional flowing fluid. The rates of transfer by diffusion of both quantities in the $i$-direction, per unit area per unit time, are governed by

$$\dot{H}_i = -k \frac{dT}{dx_i}, \qquad \dot{m}_c = -\mathcal{D} \frac{dc}{dx_i},$$

where $k$ and $\mathcal{D}$ are appropriate transfer coefficients of heat and solute, respectively; $dT/dx_i$ and $dc/dx_i$ are the corresponding temperature and concentration gradients in the $i$-direction. In addition, due to chemical reaction, there is production of heat and consumption of solute at respective rates $\dot{H}_r$ and $r$ per unit volume per unit time.

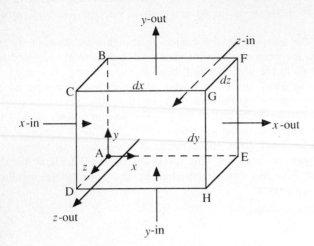

**Figure E5.11** Differential control volume (DCV) for three-dimensional heat and solute transfer.

**Solution**   The general conservation equation states that

$$
\frac{d}{dt}\begin{pmatrix} \text{quantity} \\ \text{conserved} \end{pmatrix} = \begin{pmatrix} \text{rate of convection} \\ \text{of quantity in} \end{pmatrix} - \begin{pmatrix} \text{rate of convection} \\ \text{of quantity out} \end{pmatrix}
$$

$$
\pm \begin{pmatrix} \text{rate of production} \\ \text{or consumption} \end{pmatrix} + \begin{pmatrix} \text{rate of transfer by other mechanisms} \\ \text{of quantity conserved} \end{pmatrix}.
$$
(5.11.1)

**Heat (enthalpy) conservation.** The conserved quantity is heat $H$, the density of the conserved quantity if $\hat{H}$, and the appropriate differential control volume (DCV) is shown in Fig. E5.11. The various terms appearing in Eqn. (5.11.1) are:

*Rate of accumulation:*

$$
\frac{d}{dt}(V\hat{H}) = V\frac{d\hat{H}}{dt} = dx\,dy\,dz\,\frac{d\hat{H}}{dt} \quad \left[\frac{\text{energy}}{\text{time}}\right].
$$

*Rate   of   convection   of   $H$   in:*   The   rate   of   convection   through $[ABCD + ABEF + ADEH]$ is

$$
[u_x\hat{H}\,dy\,dz]_x + [u_z\hat{H}\,dy\,dx]_z + [u_y\hat{H}\,dx\,dz]_y.
$$

*Rate of convection of H out:*

$$
-[u_x\hat{H}\,dy\,dz]_{x+dx} - [u_z\hat{H}\,dx\,dy]_{z+dz} - [u_y\hat{H}\,dx\,dz]_{y+dy}.
$$

*Net rate of convection of H:*

$$
dy\,dz[(u_x\hat{H})_x - (u_x\hat{H})_{x+dx}] + dx\,dy[(u_z\hat{H})_z - (u_z\hat{H})_{z+dz}]
$$

$$
+ dx\,dz[(u_y\hat{H})_y - (u_y\hat{H})_{y+dy}] \quad \left[\frac{\text{energy}}{\text{time}}\right].
$$

*Rate of production:*

$$\mathscr{F}\,dx\,dy\,dz + \dot{H}_r\,dx\,dy\,dz,$$

where $\mathscr{F}$ is loss of mechanical energy by viscous dissipation to heat and $\dot{H}_r$ is the heat of reaction, both per unit volume per unit time.

In addition to convection and conversion of heat, there is transfer of heat by conduction (or diffusion), as follows:

*Rate of diffusion of H in:* The rate of diffusion through [*ABCD* + *ABEF* + *ADEH*] is

$$\left[ -k\frac{\partial T}{\partial x}\bigg|_x dy\,dz - k\frac{\partial T}{\partial z}\bigg|_z dx\,dy - k\frac{\partial T}{\partial y}\bigg|_y dx\,dz \right].$$

*Rate of diffusion of H out:* The rate of diffusion through [*EFGH* + *CDHG* + *BCGF*] is

$$-\left[ -k\frac{\partial T}{\partial x}\bigg|_{x+dx} dy\,dz - k\frac{\partial T}{\partial z}\bigg|_{z+dz} dx\,dy - k\frac{\partial T}{\partial y}\bigg|_{y+dy} dx\,dz \right].$$

*Net rate of diffusion:*

$$\left[ k\frac{\partial T}{\partial x}\bigg|_{x+dx} - k\frac{\partial T}{\partial x}\bigg|_x \right] dy\,dz + \left[ k\frac{\partial T}{\partial z}\bigg|_{z+dz} - k\frac{\partial T}{\partial z}\bigg|_z \right] dz\,dy$$

$$+ \left[ k\frac{\partial T}{\partial y}\bigg|_{y+dy} - k\frac{\partial T}{\partial y}\bigg|_y \right] dz\,dx.$$

All the rates of transfer above are substituted in Eqn. (5.11.1), which yields

$$dx\,dy\,dz\,\frac{d\hat{H}}{dt} = [(u_x\hat{H})_x - (u_x\hat{H})_{x+dx}]\,dy\,dz + [(u_z\hat{H})_z - (u_z\hat{H})_{z+dz}]\,dx\,dy$$

$$+ [(u_y\hat{H})_y - (u_y\hat{H})_{y+dy}]\,dz\,dx + \mathscr{F}\,dx\,dy\,dz \mp \dot{H}_r\,dx\,dy\,dz$$

$$+ \left[ k\frac{\partial T}{\partial x}\bigg|_{x+dx} - k\frac{\partial T}{\partial x}\bigg|_x \right] dy\,dz + \left[ k\frac{\partial T}{\partial z}\bigg|_{z+dz} - k\frac{\partial T}{\partial z}\bigg|_z \right] dx\,dy$$

$$+ \left[ k\frac{\partial T}{\partial y}\bigg|_{y+dy} - k\frac{\partial T}{\partial y}\bigg|_y \right] dz\,dx. \tag{5.11.2}$$

Division by $dV = dx\,dy\,dz$ and use of the definition of derivative in the limit of $dV = dx\,dy\,dz = 0$ in Eqn. (5.11.2) results in

$$\frac{\partial \hat{H}}{\partial t} = -\left[ \frac{\partial(u_x\hat{H})}{\partial x} + \frac{\partial(u_y\hat{H})}{\partial y} + \frac{\partial(u_z\hat{H})}{\partial z} \right] + \mathscr{F} + \dot{H}_r + k\left( \frac{\partial^2 T}{\partial x^2} + \frac{\partial^2 T}{\partial y^2} + \frac{\partial^2 T}{\partial z^2} \right). \tag{5.11.3}$$

Equation (5.11.3) is developed further by invoking the thermodynamic definition,

$$\hat{H} = \rho c_p T, \tag{5.11.4}$$

where $c_p$ is the heat capacity and $T$ is the temperature, and the continuity equation for incompressible fluid,

$$\frac{\partial u_x}{\partial x} + \frac{\partial u_y}{\partial y} + \frac{\partial u_z}{\partial z} = 0. \tag{5.11.5}$$

which is derived by performing total mass balance on the same DCV (check it!). The resulting equation,

$$\underbrace{\rho c_p \frac{\partial T}{\partial t}}_{\text{accumulation}} = \underbrace{-\rho c_p \left[ \frac{\partial T}{\partial x} + u_y \frac{\partial T}{\partial y} + u_z \frac{\partial T}{\partial z} \right]}_{\text{net convection}} + \underbrace{k \left[ \frac{\partial^2 T}{\partial x^2} + \frac{\partial^2 T}{\partial y^2} + \frac{\partial^2 T}{\partial z^2} \right]}_{\text{net diffusion}} + \underbrace{\dot{\mathscr{F}} + \dot{H}_r}_{\text{production}},$$

$$\tag{5.11.6}$$

governs the three-dimensional distribution of temperature $T$.

Equation (5.11.6) is simplified in special cases, as follows:

1. Steady conduction in an inert solid: $\partial T / \partial t = 0$, $u_x = u_y = u_z = 0$, $\dot{\mathscr{F}} = \dot{H}_r = 0$:

$$\frac{\partial^2 T}{\partial x^2} + \frac{\partial^2 T}{\partial y^2} + \frac{\partial^2 T}{\partial z^2} = 0. \tag{5.11.7}$$

2. Steady conduction in one dimension, for example through thin plates:

$$\frac{\partial^2 T}{\partial x^2} = 0. \tag{5.11.8}$$

3. Steady conduction in a unidirectional flow of velocities, $u_x = U$ and $u_y = u_z = 0$, in which case diffusion in the flow direction is negligible:

$$U \frac{\partial T}{\partial x} = k \left( \frac{\partial^2 T}{\partial y^2} + \frac{\partial^2 T}{\partial z^2} \right). \tag{5.11.9}$$

4. Steady incompressible gas flow at constant velocity $U$ with an ongoing endothermic chemical reaction in a plug-flow reactor:

$$U \frac{\partial T}{\partial x} = k \left( \frac{\partial^2 T}{\partial y^2} + \frac{\partial^2 T}{\partial z^2} \right) - \dot{H}_r \tag{5.11.10}$$

**Solute conservation.** By repeating the procedure above and by replacing $\dot{H}_i = -k \, dT/dx_i$ with $(\dot{m}_c)_i = -\mathscr{D} \, dc/dx_i$, $\hat{H}$ with $c$, and $\dot{H}_r$ with $r$, the following solute change equation results:

$$\underbrace{\frac{\partial c}{\partial t}}_{\text{accumulation}} = \underbrace{-u_x \frac{\partial c}{\partial x} + u_y \frac{\partial c}{\partial y} + u_z \frac{\partial c}{\partial z}}_{\text{net convection}} + \underbrace{\mathscr{D} \left( \frac{\partial^2 c}{\partial x^2} + \frac{\partial^2 c}{\partial y^2} + \frac{\partial^2 c}{\partial z^2} \right)}_{\text{net diffusion}} - \underbrace{r}_{\text{consumption}}$$

$$\tag{5.11.11}$$

Equation (5.11.11) is simplified in special cases, as follows:

1. Steady diffusion in inert solids:

$$\frac{\partial^2 c}{\partial x^2} + \frac{\partial^2 c}{\partial y^2} + \frac{\partial^2 c}{\partial z^2} = 0. \tag{5.11.12}$$

2.   Steady diffusion in one-dimensional solid, or through thin membrane, accompanied by chemical reaction of rate $r = kc^b$:

$$\mathscr{D}\frac{\partial^2 c}{\partial x^2} \pm kc^b = 0. \qquad (5.11,13)$$

3.   Steady diffusion across liquid flowing with plug velocity $U$ in the $x$-direction and infinitely wide in the $y$-direction:

$$U\frac{\partial c}{\partial x} = \mathscr{D}\frac{\partial^2 c}{\partial z^2}. \qquad (5.11.14)$$

## PROBLEMS

1.   *Idealized oil well.*[16] An underground porous horizontal cylindrical layer of oil of viscosity of 1 P has a height of $H = 0.05$ mile and an initial radius of $r_2 = 5$ miles (Fig. P5.1). It is surrounded by a large layer of gas whose pressure remains constant at 10 atm. The oil flows radially inward, through the porous well bore and subsequently to the surface, where the pressure is zero, by the vertical pipe of internal radius $r_1 = 1$ ft and length $L = 0.1$ mile.

(a) Derive an expression for the inward radial velocity $u_r$ of the oil as a function of radial position $r$ and the nearly steady flow rate $\dot{V}$, given that the oil–gas interface moves very slowly.

(b) What is the distribution of pressure $p$ in the oil layer as a function of $r$ and $t$ if the velocity follows Darcy's law, $u_r = -(k/\eta)(dp/dr)$, with permeability $k = 0.01$ darcy?

(c) How long will it take for recovery of 20% of the oil?
[*Hint:* Consider two flows, radial and pipe, with a common pressure $p_0 = p(r = r_1) \approx p(r = 0, z = 0)$ at the origin. Notice also that $L \gg H$ and $r_2 \gg r_1$.]

2.   *Boiling of liquids.*[4] Initation of boiling in a liquid starts with formation of small vapor

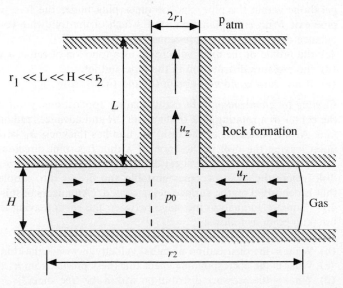

**Figure P5.1**   Oil-well operation.

bubbles having an initial radius $R_0$. The bubble radius $R(t)$ grows with time $t$ at a rate $dR/dt = kR$, where $k$ is a constant.

(a) Derive an expression for $R(t)$.

(b) The bubble becomes unstable and collapses when its volume reaches 10 times its initial volume. What is the time for collapse?

(c) What is the radially outward velocity $u_r$ in the surrounding liquid at distance twice the radius of the bubble at the moment of collapse?

(d) What is the pressure at that point if the bubble is formed at $p = p_0$, its temperature remains constant during growth, and surface tension is negligible such that the vapor pressure is identical to that of the adjacent liquid of negligible viscous stresses?

(e) How would things change in the presence of finite surface tension $\sigma$?

3. *Stabilization of bubble in magnetic liquid.*[17] A spherical bubble in a magnetic liquid evolves to a cylindrical bubble of radius $R_0$ by applying a magnetic field to the liquid, which has density $\rho$. By reducing or increasing the external pressure of the magnetic liquid, the cylindrical bubble starts growing or collapsing at the rate $dR/dt = \pm k$. Obtain expressions for the radial velocity $u_r$ and pressure $p$ in the liquid at a distance $r = 2R$ from the axis of the bubble. You may assume that the bubble is very long, so that end effects are negligible; also, the pressure in the liquid far away from the bubble is $p_\infty = 0$. Consider a vanishingly small viscosity such that all viscous stresses are negligible.

4. *Velocity and pressure in a jet.* Consider horizontal pipe flow of Newtonian liquid of viscosity $\eta = 1$ cP, density $\rho = 0.95$ g/cm³, and surface tension $\sigma = 60$ dyn/cm. The volumetric flow rate is $Q = 0.15$ L/s through the horizontal pipe of internal diameter $D = 10$ cm and length $L = 2$ m. The liquid issues to the atmosphere as a horizontal jet whose diameter is the same as the pipe.

(a) Is the pipe or jet flow laminar or turbulent?

(b) What is the inlet pressure?

The velocity profile under high laminar Reynolds number is everywhere parabolic within the pipe, and becomes plug under the free jet, well away from the pipe exit. With an explanation, state which of the following you can obtain by a mass balance only, and their expressions:

(c) the profile of the axial velocity $u_z$ as a function of radius $r$ in the pipe and jet, or

(d) the pressure distribution in the pipe and jet.

(e) If not, how could you obtain (c) and (d)?

5. *Coating by atomization.*[18] In many coating applications, paint is fed at flow rate $\dot{V}$ at the center of a rotating disk of diameter $2R$ and advances radially in the form of a thin film. At the perimeter of the disk, the film has thickness $h_R \ll R$ and uniform velocity upon leaving the disk in the form of a thin free film directed radially outward. The thickness of this free-going sheet diminishes with distance to zero slope away from the disk, where the sheet becomes unstable and breaks into droplets, forming an aerosol that is deposited on the surface being coated. (Atomizers of this kind are shown in Fig. 4.7). You are to analyze the stage between the sheet leaving the rotating disk and its breakup point as follows:

(a) What is the radial velocity at the perimeter of the disk?

(b) What is the distribution of radial velocity $u_r$ within the thin free sheet?

(c) What is the corresponding sheet thickness profile from $h_R$ to the breakup point?

(d) What is the pressure distribution within the free sheet?

(e) What is a good prototype of a corresponding sink flow?

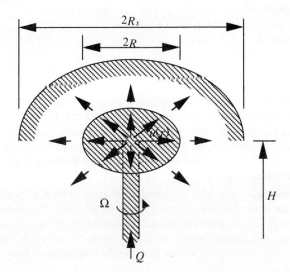

**Figure P5.5**  Coating by atomization.

6. *Compressible flow.* Prove that for steady flow of an incompressible liquid of constant density and viscosity in an insulated pipe, the axial velocity $u_z$ is constant along the pipe; that is, $du_z/dz = 0$. How does the situation change for an ideal compressible gas in which the density is not constant and the viscosity negligible? Can you calculate the pressure distribution in either of the two cases, given the inlet and outlet pressures, $p_0$ and $p_L$, over a measured distance $L$?

7. *Thin film under gravity.* In a heat-exchange arrangement, water flows down the external cylindrical surface of vertical tubes of radius $R = 5$ cm at flow rate $\dot{V} = 0.01$ L/s per tube, in contact with ascending gas at velocity $U = 1$ m/s and uniform pressure everywhere.
   (a) Show that for a flat film the liquid flow is governed by Eqn. (5.31), with $\Delta p/\Delta L = -\rho g$.
   (b) Use the appropriate boundary conditions to arrive at the particular solution. Sketch the velocity profile and comment on its qualitative features.
   (c) What is the thickness of the resulting liquid film?
   (d) What is the interrelation between flow rate and film thickness?

8. *Film flow under gravity.* Consider the arrangement described in Problem 5.7. A slightly different exchange process is now considered, in which the water flows in the form of thin, flat film in the interior of each vertical tube, the core of which is occupied by counter-flowing air at velocity $U = 1$ m/s.
   (a) Address questions (a) to (d) of Problem 5.7 for the flow situation considered here.
   (b) Under what conditions is the net flow rate of liquid zero? Sketch orbits of liquid particles, called *streamlines* under zero net flow rate.
   (c) Under what conditions is the entire pipe interior occupied by liquid?

9. *Cylindrically symmetric unidirectional flows.* The general solution to unidirectional flow along a cylindrically symmetric horizontal geometry is

$$u_z(r) = \frac{1}{4\eta} \frac{\Delta p}{\Delta L} r^2 + c_1 \ln r + c_2.$$

Calculate the exact velocity profile and the flow rate in the following situations:

(a) Flow in a pipe of radius $R_0$.

(b) Flow in an annulus of radii $R_0$ and $kR_0$, with $k > 1$.

(c) Flow of film of thickness $(k - 1)R_0$ on the outside of a vertical cylinder of radius $R_0$, with $k > 0$ and $\rho g = -(\Delta p/\Delta L)$.

(d) How are the three cases related in terms of $k$?

10. *Air-sheared round jet.* An infinitely long horizontal jet of radius $R_0 = 3$ cm is driven by air shear, which results in velocity $u = 10$ m/s at its free surface. Under these conditions, the pressure is everywhere constant and identical to $\sigma/R$, where $\sigma = 72$ dyn/cm is the surface tension. What are the velocity and the shear-stress distributions? If the jet radius varies slightly with distance according to $R = R_0 \exp(-0.01z)$, what are the corresponding velocity, stress, and flow rate? Does Bernoulli's equation apply to these cases?

11. *Bubble growth under constant air mass.* An air bubble collapses within a Newtonian liquid of viscosity $\eta$, with an exponentially decaying pressure $p = p_0 e^{-kt}$, constant air mass $m_0$ and constant temperature $T_0$. Find the velocity distribution in the surrounding liquid at time $t$. Show that the normal stress at the growing interface is constant independent of time. If the normal stresses across the interface are related by

$$-p^G = -p_I^L + \tau_{rr}|_{r=R} - \frac{2\sigma}{R},$$

where $p_I^L$ is the liquid pressure at the interface, how does the liquid pressure $p^L(r)$ evolve with time?

12. *Stack emission of pollutants.*[19] A stack emits $\dot{m} = 0.2$ kg/min of pollutants at a height $h = 20$ m from the ground, where the wind blows horizontally with velocity $v = 10$ m/s. Under these conditions there is a diffusion of pollutants with coefficient $\mathcal{D} = 0.2$ cm²/s to all possible directions away from the stack.

(a) Derive an equation that governs the distribution of pollutants in the area.

(b) Solve the equation in the limiting cases of (i) $v = 0$ (no wind); (ii) $D = 0$ (no diffusion).

[*Hint:* A Cartesian and a spherical system of coordinates are convenient for (a) and (b), respectively. Why?]

13. *Convection of mass and heat.* A stream of Newtonian liquid of viscosity $\eta = 1$ cP, density $\rho = 1$ g/cm³, heat capacity $c_p = 1$ cal/g K, and temperature $T = 50°$C penetrates a surface of area $A = 1$ m² at angle $\theta = 30°$ with velocity $V = 10$ cm/s. Calculate the total rates of mass momentum and heat transfer convected across the surface.

14. *Reaction and diffusion of oxygen in lakes and rivers.* Oxygen from the atmosphere dissolves and diffuses in water at a rate $\dot{m}_D = -\mathcal{D}(dc/dz)$ downward with diffusivity $\mathcal{D} = 0.05$ cm²/s, while being consumed by microorganisms at a rate of $\dot{m}_c = 0.001z$ g/s m³, where $z$ is distance below the free surface. Calculate the concentration distribution in a stationary lake and in a river flowing with constant velocity $u = 1$ m/s, both of depth $h = 2$ m. Comment on the answers. (The diffusion in the direction of the flowing river is negligible.)

15. *Thickness of vertical soap film.* Show that at equilibrium the thickness of a vertical soap film of surface tension $\sigma$ changes along the direction of gravity. Calculate the thickness distribution of a soap film or surface tension $\sigma = 200$ dyn/cm² and density $\rho = 1$ g/cm³ hanging vertically from a horizontal wire of diameter $d = 0.1$ cm and length $l = 20$ cm. When does the film break?

**16.** *Pressurization of granular solids.*[20] The compression of granular solids in horizontal conduits by pistons is governed by the force balance

$$F_x - F_{x+dx} - c\left(\frac{F}{A}\right)kf\,dx = 0,$$

where $F$ is the transmitted compressional force, $C$ the perimeter of the cross-sectional area $A$, $k$ the ratio of radial to axial stresses, and $f$ a coefficient of friction, as shown in Fig. P5.16.

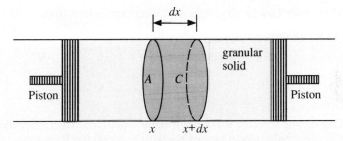

**Figure P5.16**  Pressurization of granular solids.

**(a)** Show that the resulting uniform axial stress distribution is

$$\tau_{xx} = (\tau_{xx})_0 \exp\left(-\frac{fkc}{A}x\right).$$

**(b)** A ram injection molding machine has a 3-in.-diameter barrel, where the ram reciprocates and compresses polymer pellets in order to deliver them to the mold at 20,000 psi. Calculate the volume of pelletized polymer delivered if $f = 0.5$, $k = 1$, and the maximum radial stress the barrel wall can sustain is 30,000 psi.

**17.** *Compression molding.*[21] A simplified compression molding machine is shown in Fig. P5.17. A polymer melt (or molten metal) is compressed between approaching plates due to the motion of the upper plate at velocity $V = -dH/dt$. By choosing an

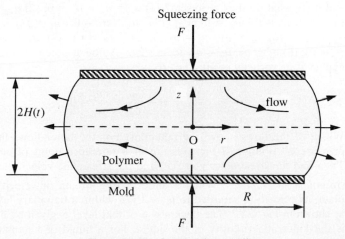

**Figure P5.17**  Compression molding.

appropriate control volume, show that the resulting mass balance equation is

$$\rho \pi r^2 V = 2\rho \pi r H \bar{u}_r = 2\rho \pi r \int_0^H u_r \, dz,$$

and the $r$-momentum equation is

$$\frac{dp}{dr} = \eta \frac{d^2 u_r}{dz^2}.$$

Then show how to obtain the following expressions from these equations:

$$u_r = \frac{1}{2\eta} \frac{dp}{dr} (z^2 - Hz), \qquad p(r) = p_0 + \frac{3\eta V}{H^3} (R^2 - r^2),$$

$$F = \frac{3\eta \pi V R^4}{3H^3} \qquad V = \frac{2H^3 F}{3\pi \eta R^4}.$$

Discuss the physical meaning of these equations. Then calculate the load required to compress polystyrene melt of viscosity $\eta = 600{,}000$ P, in a machine of radius $R = 1$ ft, from an initial thickness of $H_0 = 3$ in. to a 0.3-in. final part within a time interval $\Delta t = 5$ s.

18. *Radial flow between concentric balloons.* An incompressible Newtonian liquid of density $\rho = 1 \, \text{g/cm}^3$ and viscosity $\eta = 1$ cP is contained between two concentric elastic spherical balloons of radii 1 and 1.5 cm, respectively. The inner balloon starts inflating at rate 0.2 cm/s under a constant pressure of $p = 20$ psi. What will happen to the outer balloon if the inflation is strictly radial? What is the resulting pressure distribution in the liquid if the pressure differences across each of the balloons are negligible?

19. *Conservation of mass.* Derive the mass conservation equation of an incompressible liquid flowing in a Cartesian geometry with velocity components $u_x$, $u_y$, and $u_z$. Reduce the equation to one- and two-dimensional flows. What is the appropriate form for flow in rectilinear channel? For flow in a converging or diverging channel? A flow is dynamically admissible if its velocity field satisfies the mass conservation equation (also known as the continuity equation). Which of the following flows are admissible, and if so, what do they represent? (1) $u_x = u_y = u_z = 0$; (2) $u_x = 5 \, \text{m/s}, u_y = u_z = 0$; (3) $u_x = u_y = 0$, $u_z = 10 \, \text{m/s}$; (4) $u_x = u_y = 0$, $u_z = f(x, y)$; (5) $u_x = u_y = 0$, $u_z = f(z)$; (6) $u_x = ax$, $u_y = ay$, $u_z = 0$ $(a \neq 0)$; (7) $u_x = ax$, $u_y = -ay$, $u_z = 0$ $(a \neq 0)$; (8) $u_x = ax, u_y = ay, u_z = -2az$. A flow is potential and Bernoulli's equation applies if its vorticity is zero, i.e., if

$$\omega = \frac{\partial u_i}{\partial x_j} - \frac{\partial u_j}{\partial x_i} = 0.$$

Which of the flows above are potential? For the latter flows find the corresponding pressure distribution if the pressure at a reference point of zero coordinates is $p_0$. Then find the streamlines $y = f(x)$ and the streamline velocity $u = (u_x + y_y)^{1/2}$.

20. *Boundary layer flow.*[22] When a uniform free stream of velocity $V$ overtakes a thin plate, its velocity is retarded to $u_x = Vy/\delta$ within a boundary layer close to the plate, as shown in Fig. E3.7. The thickness $\delta$ of that layer is given by $\delta = 5\sqrt{vx/V}$, where $v$ is the kinematic viscosity of the liquid. For a liquid of kinematic viscosity $v = 1$ St, approaching at a velocity $V = 10 \, \text{m/s}$:

(a) Find the thickness $\delta$ as a function of $x$.

(b) What is the total volume of liquid crossing the line $AB$ of projection $L = 0.2$ m?

(c) What is the normal velocity component $u_y$? Based on $u_y$ and $u_x$, sketch the velocity vector along the line $AB$. Is the direction of flow across $AB$ sketched properly in Fig. P5.7.

**21.** *Radial flows between porous walls.* A liquid of viscosity $\eta$ and density $\rho$ escapes radially at flow rate $Q$ from a porous cylinder of radius $R$ and length $L \gg R$ (Fig. P5.21). The liquid reaches the porous wall of a second cylinder which shares the same axis of symmetry with the first one and has radius $2R$, where the pressure is $p_0 > 0$.

(a) What is the velocity distribution between the two porous walls? Sketch it.

(b) What is the pressure distribution between the two porous walls? Sketch it.

(c) If cavitation occurs at $p = 0$, where does it take place?

(d) What are the velocity and pressure distributions within the outer porous wall of 10 cm thickness?

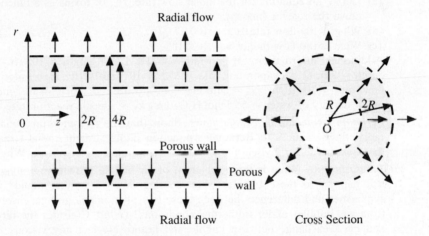

**Figure P5.21**    Flow between coaxial porous cylinders.

**22.** *Radial spherical flow.* A Newtonian liquid of viscosity $\eta$ and density $\rho$ escapes at pressure $p_0$ and flow rate $Q$ m$^3$/min from a spherical porous wall of radius $R$ toward a second spherical wall of radius $4R$ that contains and shares the same origin with the first one. Calculate and plot the velocity and pressure distribution between the two spherical walls. What must be the vapor pressure of the liquid if cavitation is to take place somewhere between the two walls? What must be the velocity and pressure distributions within the smaller porous sphere?

**23.** *Flow with evaporation.*[23] Wetlands are convenient sites of wastewater treatment. The degradation of toxic chemicals (lumped and called species A) in the wastewater is assumed to follow irreversible first-order homogeneous kinetics:

$$A \xrightarrow{\ k_1\ } \text{products}.$$

As the wastewater flows and reacts, it also evaporates at a constant rate $Q$ moles water/h/ft$^2$ from the surface, as shown in Fig. P5.23. None of the toxic species is lost to the air by evaporation. You may assume that the reactor (marsh) is rectangular and that the gentle downhill flow of the water can be modeled as plug flow.

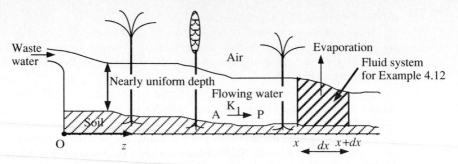

**Figure P5.23**   Flow in wetland with evaporation.

(a) Derive an equation for the molar flow rate, $F_A$, of toxins as a function of distance down the reactor (marsh).

(b) What is the flow rate at $x = 100$ ft?

(c) What is the flow rate at $x = 1000$ ft?

Additional information: $W$ = width = 100 ft, $L$ = length = 1000 ft, $D$ = average depth = 2 ft, $Q$ = evaporation rate = $55.5 \times 10^{-3}$ mol/h/ft$^2$, $v_0$ = entering volumetric flow rate = 100 ft$^3$/h, $c_{A0}$ = entering concentration of toxic = $10^{-5}$ mol/dm$^3$, $\rho_m$ = molar density of water = 55.5 mol $H_2O$/dm$^3$, $k_1$ = specific reaction rate = $10^{-3}$/h.

24. *Bernoulli's equation and radial flows.* Show that the $r$-momentum equation of purely radial flows reduces to Bernoulli's equation along straight radial streamlines. Show that this is true for inviscid flows and for irrotational viscous flows. Why is it so?

25. *Liquid boiling.* What is the mechanism of air-bubble growth in boiling liquid? What will happen to these bubbles when cooling down a boiling liquid? What are the similarities and differences between these two phenomena and gas disolution in liquid from gas bubbles under isothermal conditions? (*Hint:* Consider the direction of heat and mass exchange between bubble and liquid. Neglect any viscous stresses in the liquid.)

26. *Natural flows.* **(a)** What are the most convenient systems of coordinates (i.e., origin, directions, motion) to study the following large-scale processes or flows: river flow; wind; air pollution by stationary source under no-wind conditions; same flow under unidirectional wind; tornados; airflow around train in tunnel; airflow through open tunnel; rainwater flow windshield of speeding car; water penetration in soil; underground storage and recovery of natural gas; blast-wave airflow; rainwater down conical and pyramoidal constructions; flow in wetlands.

    **(b)** For each flow, state clearly the applied boundary conditions of velocity and pressure.

    **(c)** How would you model these large-scale flows in the lab and then rescale your findings to the actual flow?

27. *Streamlines.* Show that the streamlines of channel pipe and annulus flow are straight lines parallel to the wall, those of radial flows identical to the radii, and those of torsional flow concentric circles. What is the streamline velocity and pressure in each class? What is the residence time along each kind of streamlines?

28. *Streamlines.* Show that wall, free surface, and interface lines are streamlines. Why are they called boundary streamlines, [i.e., do they have special properties in addition to those of Eqn. (5.38)]?

# REFERENCES

1. D. M. Himmelblau, *Basic Principles and Calculations in Chemical Engineering*, Prentice Hall, Englewood Cliffs, N.J., 1982.
2. H. S. Fogler, *Elements of Chemical Reaction Engineering*, Prentice Hall, Englewood Cliffs, N.J., 1986.
3. G. K. Batchelor, *An Introduction to Fluid Dynamics*, Cambridge University Press, Cambridge, 1979.
4. L. E. Scriven, On the dynamics of phase-growth, *Chem. Engrg. Sci.*, 10, 1 (1959).
5. A. Afermanesh, S. G. Advani and E. E. Michaelides, A parametric study of bubble growth in low pressure foam molding, *Polym. Engrg. Sci.* 30, 1330 (1990).
6. J. W. Hall and G. M. Wood, Eds., *Cauitation Research Facilities and Techniques*, ASME, New York, 1964.
7. G. F. Hewitt, *Annular flow in two-phase flow and heat transfer in the power process industry* (ed. A. E. Bergles et al.), Hemisphere, New York, 1981.
8. B. V. Deryagin and S. M. Levi, *Film Coating Theory*, Focal Press, New York, 1964.
9. C. Truesdell and W. Noll, *The Non-linear Field Theories of Mechanics*, Handbuch der Physik (ed. S. Flügge), Vol. III/3, Springer-Verlag, Berlin, 1965.
10. R. B. Bird, R. C. Armstrong, and O. Hassager, *Dynamics of Polymeric Liquids*, Vol. 1, *Fluid Mechanics*, Wiley, New York, 1977.
11. T. C. Papanastasiou and A. N. Alexandrou, Isothermal extrusion of non-dilute fiber suspensions, *J. Non-Newtonian Fluid Mech.*, 25, 313 (1987).
12. T. D. Karapantsios, S. V. Paras, and A. J. Karabelas, Statistical characteristics of free falling films, *Internat. J. Multiphase Flow*, 15, 1 (1989).
13. K. Higashitani and A. S. Lodge, Hole pressure error measurements in pressure generated shear flow, *Trans. Soc. Rheol.*, 19, 307 (1975).
14. E. G. Kastrinakis, S. G. Nychas, J. M. Wallace, and W. W. Willmarth, Measurements of streamwise vorticity in a turbulent channel flow, *Bull. APS*, 21, 1237 (1976).
15. R. B. Bird, W. E. Stewart, and E. N. Lightfoot, *Transport Phenomena*, Wiley, New York, 1960.
16. L. P. Dake, *Fundamentals of Reservoir Engineering*, Elsevier, Amsterdam, 1978.
17. R. E. Rosensweig, Magnetic fluids, *Annu. Rev. Fluid Mech.*, 19, 437 (1987).
18. T. C. Papanastasiou, A. N. Alexandrou, and W. Graebel, Thin films in bell sprayers and spin coating, *J. Rheol.*, 32, 485 (1988).
19. K. Wark and C. F. Warner, *Air Pollution: Its Origin and Control*, Harper & Row, New York, 1976.
20. Z. Tadmor and C. G. Gogos, *Principles of Polymer Processing*, Wiley, New York, 1979.
21. T. C. Papanastasiou, C. W. Macosko, and L. E. Scriven, Analysis of lubricated squeezing flow, *Internat. J. Numer. Methods Fluids*, 6, 819 (1986).
22. H. Schlichting, *Boundary Layer Theory* 7th ed., McGraw-Hill, New York, 1979.
23. R. H. Kadlec, Hydrodynamics of wetland treatment systems, in *Constructed Wetlands for Wastewater treatment* (ed. D. A. Hammer), Lewis Publishers, Chelsea, Mich., 1989.

# 6

## Unidirectional Flows

### 6.1 INTRODUCTION

Analytic solutions to the governing equations of fluid flow (i.e., mass and momentum conservation equations) can be found only for limited classes of flows. The following classes of unidirectional flows are in general amenable to analytic approaches (i.e., without the use of computer-aided analysis based on numerical methods):

1. Radial flows (e.g., Figs. E4.16, E5.2, E5.6, P5.1, P5.5, P5.21).
2. Rectilinear pipe, annulus, and channel flows (e.g., Figs. 1.1, E3.13, P3.2a and b, P4.13, 4.2, E5.7.1).
3. Thin-film flows (e.g., Figs. P3.2c, 4.14, P4.33, 5.4).
4. Torsional flows (e.g., Figs. 2.3, 3.10, E5.8.5).

In these flows, there is in general a unique nonzero velocity component that changes with respect to one direction. This velocity component drives the entire flow to its direction (i.e., *the flow is unidirectional*). Unidirectional flows, although simple, are important in a diversity of fluid transferring and processing applications, as demonstrated in Chapters 1 to 5 and in Section 6.5. Prototypes and velocity characteristics of unidirectional flows are listed in Examples 5.6 to 5.10. By *superposition* (i.e., geometrical or vectorial addition) of two or more steady unidirectional flows, more complicated *multidirectional flows* are obtained, as detailed in Section 6.6. In the steady and fully developed flows outlined above, the geometry requires all convective terms to vanish and the governing momentum equation(s) reduces to an ordinary differential equation (ODE) with respect to the unique velocity component which is identical to the streamline velocity. For steady undeveloped flows [e.g., entrance flow, Eqn. (1.37)] or unsteady flows (e.g., initiation of unidirectional flow from rest, Section 6.7). The convective terms cannot be ignored and a partial differential equation (PDE) results. Furthermore, *nearly unidirectional flows* can often be analyzed starting from the solution to the corresponding unidirectional flow. The most important nearly unidirectional flows are *lubrication* and *stretching flows,* which have wide applications in machine operations and materials processing, as discussed in Chapter 8.

The strategy to the analytic solution of *steady unidirectional flows* is the one outlined in Chapter 5. In this chapter, classes of unidirectional flows amenable to common solution approach are presented and analyzed. Selected multidirectional flows obtained by linear superposition of unidirectional flows are also discussed. Finally, the analysis is extended to *transient unidirectional flows,* where velocity and pressure change with space *and* time.

## 6.2 RADIAL FLOWS

Flows belonging to this class have already been addressed in Chapter 5. There may be several types of applications where the radial flow prototype applies:

1. Radial flows of a single fluid, such as flows induced around growing/ collapsing bubbles and cavities (e.g., Example 5.2), and sink and source flows of liquid-free sheets and films (e.g., Fig. P5.5).

2. Radial flows in porous media with applications in oil recovery and hydrology, composite processing (e.g., Section 4.11.3, Example 5.6), and in natural gas storage underground (e.g., as described in this section).

The velocity profile is determined by mass balance on an appropriate control volume with surfaces perpendicular to the radial streamlines. In single-phase radial flows there is either no friction at all (e.g., inviscid flow of fluid of vanishingly small viscosity) or zero net viscous force (e.g., irrotational flow of viscous fluid). In both cases, there is no loss of mechanical energy along streamlines, and therefore the pressure can be calculated easily be means of Bernoulli's equation. According to this equation, the pressure is reduced significantly near sinks and sources to provide for the high velocity there, which may cause cavitation of liquid, in principle. In practice, this is often avoided due to other intervening phenomena.

The growing bubble is perhaps the best prototype of *source radial flow.* It is significant in nucleation, cavitation, and boiling phenomena, where nuclei gas or vapor bubble are formed in liquid fields of low-pressure (and high-temperature) conditions that promote separation of dissolved gas or vapor. Such bubbles start to grow under the low external pressure or due to liquid boiling at the interface and therefore displacement of the liquid by the expending bubble.[1]

In rheometry (Chapter 9) a collapsing air bubble in viscoelastic liquid has been proposed as an extensional rheometer to measure *elongational or extensional viscosity* of viscoelastic liquids,[2] defined as

$$\eta_e = \frac{\tau_{rr} - \tau_{\theta\theta}}{du_r/dr} = \frac{\tau_{rr} - \tau_{\theta\theta}}{\dot{R}}, \tag{6.1}$$

within the induced *radial sink flow* around the bubble, which is extensional of zero vorticity. The extensional viscosity is resistance to stretching or compression of fluid filaments, as opposed to the (shear) viscosity, which is resistance to shearing. The stress difference, $\tau_{rr} - \tau_{\theta\theta}$, is related to the pressure inside the bubble (which is monitored by connecting the bubble to an external pressure chamber through a capillary, as shown in Fig. E5.2) and the hydrostatic pressure far away from the bubble, both of which are measured. The rate of growth/collapse, $\dot{R}$, is monitored by photography for transparent liquids and by dilatometric techniques that measure small volume changes for opaque liquids. The extensional viscosity is extremely important in polymer processing because it controls the developed normal stresses and required processing forces in such common applications as fiber spinning, film casting, and compression molding.

Underground water, oil, and natural gas under pressure migrate in a radial fashion to wells open to the atmosphere, according to the simplified model shown in Fig. P5.1. Oil and natural gas in underground physical or storage reservoirs under high pressure migrate to and are collected in the well and piped to the surface.[3] In this type of radial flow there is significant friction and therefore Bernoulli's equation does not apply. Instead, the pressure distribution is obtained from Darcy's law[4] for flow in porous media,

$$u_r = -\frac{k}{\eta}\frac{dp}{dr}, \tag{6.2}$$

where $k$ is permeability of the porous medium, as introduced in Section 4.3, $\eta$ fluid viscosity, $u_r$ radial velocity, and $dp/dr$ radial pressure gradient. Equation (6.2), unlike Bernoulli's equation, indeed accounts for pressure losses proportional to viscosity and inversely proportional to permeability.

A radial gas flow in porous media arises in underground storage of natural gas.[5] In this process natural gas is injected under pressure through wells that penetrate appropriate porous reservoirs where the gas is stored. Pumped gas advances radially from the well exit in the porous reservoir displacing either air or water through a spherical interface of contact. The radial velocity for constant gas flow rate is given by Eqn. (5.29) for spherical and Eqn. (5.6.3) for cylindrical flows. The pressure distribution is determined by Eqn. (6.2) and the appropriate pressure boundary condition at the advancing front or interface. Such analysis shows that the pumping pressure required to maintain a constant flow rate is inversely proportional to the permeability and porosity and proportional to the pressure at the advancing front (e.g., displacement of another storage system under pressure) or interface with water, which is in addition augmented by capillary pressure due to surface tension. A similar process occurs in cylindrical reservoirs due to gas discharge by a well along the axis of symmetry. Gas is recovered by such systems by lowering the well pressure, which causes the gas to migrate toward the well bore and is then driven to the surface through the well. Thus during storage a source radial flow in a porous medium takes place, whereas during recovery an opposite sink radial flow is initiated. Of course, in these analyses any gas compressibility must be taken into account, as detailed in relative publications.[5,6]

In summary, the velocity profile of any radial flow is obtained by mass balance on an appropriate control volume. For one-phase radial flows the pressure profile is obtained by means of Bernoulli's equation, whereas for radial flow in porous media the pressure is obtained by Darcy's law, given the velocity profiles. The stresses can be determined by means of the constitutive relations of Appendix F, given the velocity profiles.

## 6.3 RECTILINEAR FLOWS

These are usually flows between parallel plates, within channels, pipes, and annuli, and down inclined plates. They are unidirectional; i.e., there is only one nonzero velocity component, which may vary with respect to the other two directions. The flow may be driven by:

1. A boundary moving in the horizontal $i$-direction, in which case the pressure gradient in the flow direction is zero, $dp/dx_i = 0$. This is the *Couette flow*.
2. By a pump, in the horizontal $i$-direction, in which case $dp/dx_i = \Delta p/\Delta L =$ const. within the flow. This is *Hagen–Poiseuille flow*.
3. By gravity alone, in the vertical $i$-direction, in which case $dp/dx_i = 0$ and $\rho g_i \neq 0$. This is *gravity flow*.
4. By any combination or *superposition* of the three.

In these expressions, $\rho g_i$ is a hydrostatic pressure contribution which is present even if the fluid does not flow.

Under steady flow conditions the continuity equation of incompressible fluid (of constant density) becomes

$$\frac{du_i}{dx_i} = 0, \tag{6.3}$$

and therefore

$$u_i = f(x_{j \neq i}) \tag{6.4}$$

for channel-like flows, and

$$u_i = f(r) \tag{6.5}$$

for pipe- and annulus-like flows.

The corresponding momentum equations are

$$\frac{\Delta p}{\Delta L} = \eta \frac{d^2 u_i}{dx_j^2} + \rho g_i \tag{6.6}$$

and

$$\frac{\Delta p}{\Delta L} = \eta \frac{1}{r}\frac{d}{dr}\left(r\frac{du_i}{dr}\right) + \rho g_i,$$ (6.7)

respectively, where $\Delta p/\Delta L$ is a constant pressure gradient.

### 6.3.1 Couette Flow

*Plane Couette flow,* named after Couette, who introduced it in 1890 to measure viscosity, or *drag flow* occurs between two infinitely large and wide plates separated by distance $H$ (Fig. 6.1). One of the plates moves steadily with velocity $V$ relative to the other. The corresponding flow in a pipe is *plug flow.* The *axisymmetric Couette flow* occurs within an annulus with its two boundaries rotating (i.e., not translating) with different velocities. An analysis of plane Couette flow follows.

The procedure of Section 5.7 leads to Eqns. (6.3) and (6.5). For a long channel of $L \gg H$ and away from entrance and exit effects [e.g., Eqn. (1.37)], the flow is similar at any distance $x$, and therefore the velocity $u_x$ is independent of $x$. If $W/H \gg 1$ in addition, the same velocity is independent of $z$ for the same reasons. Thus, the velocity $u_x$ is a function of $y$ alone. The pressure is everywhere identical to the hydrostatic, and therefore its gradient in the horizontal flow direction is zero. Equation (6.6) then becomes

$$\frac{d^2 u_x}{dy^2} = 0,$$ (6.8)

with general solution

$$u_x = c_1 y + c_2,$$ (6.9)

which holds for any Couette flow in a channel, independent of the existing boundary conditions. The general solution is restricted to the specific situation of Fig. 6.1 by invoking the appropriate boundary conditions:

At $y = 0$ (stationary plate), $u_z = 0$, which yields $c_2 = 0$.

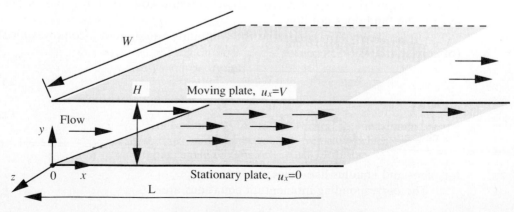

**Figure 6.1**   Plane Couette flow geometry.

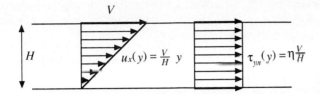

**Figure 6.2** Velocity and shear stress (from upper to lower layer) distribution of Couette flow.

At $y = H$ (moving plate), $u_z = V$, which yields

$$V = c_1 H, \tag{6.10}$$

and therefore $c_1 = V/H$. The constants $c_1$ and $c_2$ are now substituted in the general solution to provide the particular solution to the situation of Fig. 6.2. It is

$$u_z = \frac{V}{H} y, \tag{6.11}$$

which is a linear velocity profile across the gap.

**Macroscopic quantities.**   These may be calculated as follows:

1. Flow rate (per unit width):

$$\frac{\dot{V}}{W} = \int_0^H u_z \, dy = \frac{VH}{2}. \tag{6.12}$$

2. Shear stress distribution: According to the sign convention in the constitutive equations of Appendix F the stress, is from the farther to the closer fluid layer. Therefore,

$$\tau_{yz} = \eta \frac{du_z}{dy} = \eta \frac{V}{H}. \tag{6.13}$$

Shear stress exerted by fluid on upper plate (i.e., from closer to farther):

$$\tau_{yz}^U = -\tau_{yz}\big|_{y=H} = -\eta \frac{V}{H}. \tag{6.14}$$

Shear stress exerted by fluid on lower plate (i.e., from farther to closer):

$$\tau_{yz}^L = \tau_{yz}\big|_{y=0} = \eta \frac{V}{H}. \tag{6.15}$$

Note also that the lower contact surface faces the positive $y$-direction of the chosen system of coordinates, whereas the upper contact surface faces the negative direction.

3. Force per unit width required to move the upper plate:

$$\frac{F}{W} = -\int_0^L \tau_{yz}^U\big|_{y=H} \, dz = \eta \frac{V}{H} L. \tag{6.16}$$

The minus sign here accounts for the fact that the external force must overcome the internal shear stress exerted by the fluid.

4. Vorticity of the flow:

$$\omega = -\frac{du_x}{dy} = -\frac{V}{H} \tag{6.17}$$

is everywhere constant. For this reason, Couette flow is often utilized to study productivity of biological cells under conditions of constant relative rotation of fluid particles under shear stress.

Couette flow occurs between slightly separated (compared to the length and width) moving surfaces. Important examples include lubricated surfaces in relative motion,[7,8] sliding on ice as discussed in Section 1.3, transfer of viscous liquids by gear pumps and extruders,[9] application of thin films on substrates by knife-like devices common in coating processes,[10,11] and experimental environments of constant vorticity for enzyme-producing cells,[12] in fermentation processes,[13] for example.

### 6.3.2 Poiseuille Flow

*Plane Poiseuille flow,* named after the channel experiments by Poiseuille in 1840, occurs when liquid is forced to flow under pressure between infinitely long and wide stationary plates. *Axisymmetric Hagen–Poiseuille flow,* named after Poiseuille's and Hagen's experiments in capillary tubes in 1939, occurs in pipes or annuli under similar conditions. In both cases the flow is driven either mechanically by a pump, which creates a constant pressure gradient $\Delta p/\Delta L$, or by gravity when the channel or pipe is not horizontal. These flows persist up to $Re = \rho \bar{u} D/\eta \approx 2100$, where transition to turbulence starts and the equations developed in Chapter 10 for turbulent flow apply. The axisymmetric Hagen–Poiseuille flow, within the geometry shown in Fig. 6.3, is analyzed below.

The procedure of Section 5.7 leads to Eqn. (6.7),

$$\frac{\Delta p}{\Delta L} = \eta \frac{1}{r}\frac{d}{dr}\left(r\frac{du_z}{dr}\right) + \rho g_z.  \tag{6.18}$$

The differential equation for horizontal flow for which $g_z = 0$ is integrated once,

$$\int_0^r \left(\frac{\Delta p}{\Delta L}\frac{r}{\eta}\right) dr = \int_0^r d\left(r\frac{du_z}{dr}\right),$$

to yield

$$\frac{\Delta p}{\Delta L}\frac{r^2}{2\eta} + c_1 = r\frac{du_z}{dr},$$

and then a second time,

$$\int \left(\frac{\Delta p}{\Delta L}\frac{r^2}{2\eta}\frac{1}{r} + \frac{c_1}{r}\right) dr = \int du_z,$$

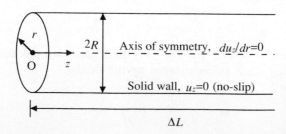

Axis of symmetry, $du_z/dr=0$

Solid wall, $u_z=0$ (no-slip)

$\Delta L$

**Figure 6.3**  Axisymmetric Hagen–Poiseuille flow.

to yield

$$\frac{\Delta p}{\Delta L}\frac{r^2}{4\eta} + c_1 \ln r + c_2 = u_z.$$

The general solution is

$$u_z(r) = \frac{1}{4\eta}\frac{\Delta p}{\Delta L}r^2 + c_1 \ln r + c_2. \tag{6.19}$$

Equation (6.19) represents the general solution to any pipe or annulus, unidirectional, Poiseuille flow. To obtain the particular solution to the problem at hand, the boundary conditions must be utilized:

At $r = 0$, $du_z/dr = 0$; therefore, $c_1 = 0$.

At $r = R$, $u_z = 0$; therefore, $c_2 = \dfrac{1}{4\eta}\dfrac{\Delta p}{\Delta L}R^2$.

The two constants are substituted in the general solution to provide the particular solution

$$u_z = \frac{1}{4\eta}\frac{\Delta p}{\Delta L}(r^2 - R^2), \tag{6.20}$$

which represents a parabolic velocity profile, shown in Fig. 6.4,

**Macroscopic quantities.**    The flow rate, $\dot{V}$, is the summation of individual flow rates $\dot{V}_i$, through annuli of radii $r_i$ and $r_i + \Delta r_i$, and therefore of area $2\pi r_i \Delta r_i$ which make up the cross section of radius $R$ and total area $\pi R^2$:

$$\dot{V} = \sum_{i=1}^{n} \dot{V}_i \simeq \sum_{i=1}^{n} 2\pi r_i \Delta r_i u_z(r_i). \tag{6.21}$$

This approximation is made more and more accurate by forcing the thickness,

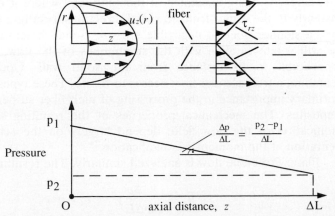

**Figure 6.4**  Velocity, shear stress (from inner to outer layer), slender fiber orientation, and pressure distribution of Hagen–Poiseuille flow.

$\Delta r_i$, to shrink to infinitesimal thickness $dr$, and is exact in the limit of $dr = 0$. Thus

$$\dot{V} = \lim_{\substack{\Delta r_i \to dr \to 0}}^{n \to \infty} \sum_{i=1} 2\pi r_i \, \Delta r_i u_z(r_i) = \int_0^R 2\pi r u_z(r) \, dr, \tag{6.22}$$

where the definition of the integral has been used. For the case at hand,

$$\dot{V} = \int_0^R 2\pi r u_z(r) \, dr = 2\pi \int_0^R r \frac{1}{4\eta} \frac{\Delta p}{\Delta L} (r^2 - R^2) \, dr = -\frac{\pi R^4}{8\eta} \frac{\Delta p}{\Delta L} > 0. \tag{6.23}$$

The shear stress distribution is

$$\tau_{rz} = \eta \frac{du_z}{dr} = \frac{1}{2} \frac{\Delta p}{\Delta L} r, \tag{6.24}$$

and the shear stress exerted by the fluid on the wall (i.e., from closer to farther layer) is

$$\tau_{rz}^w = -\tau_{rz}|_R = -\frac{\Delta p}{\Delta L} \frac{R}{2} > 0. \tag{6.25}$$

Note also that here the contact area faces the negative $r$-direction of the cylindrical system chosen.

The required horizontal force to hold the wall in position is

$$F = -2\pi R \int_0^L \tau_{rz}^w \, dz = \pi \frac{\Delta p}{\Delta L} R^2 L < 0. \tag{6.26}$$

The vorticity of this pressure-driven flow,

$$\omega = -\frac{du_z}{dr} = -\frac{1}{\eta} \frac{\Delta p}{\Delta L} y, \tag{6.27}$$

varies linearly, being maximum at the solid wall where it originates and diffuses constantly to the center from all radial directions, resulting in zero vorticity there. Thus, a slender fiber of negligible cross-sectional dimensions introduced to the flow will rotate to align with the streamlines of the flow, with angular velocity proportional to its distance from the solid wall. Upon alignment with a streamline, a truly slender fiber ceases to rotate. These types of considerations are of primary importance in the processing of melt-fiber suspensions to manufacture composites. The mechanical properties of the resulting solid composite, upon solidification of the flow field, depend largely on the achieved frozen-in fiber orientation at the moment of solidification.[14]

Plane Poiseuille flow is analyzed similarly. The resulting velocity profile is

$$u_z = \frac{1}{2\eta} \frac{\Delta p}{\Delta L} y^2 + c_1 y + c_2, \tag{6.28}$$

and the constants are evaluated by applying the appropriate boundary conditions.

**Linear superposition of flows.** The velocity profile of a Couette–Poiseuille flow is (check it)

$$u_z = \underbrace{\frac{1}{2\eta}\frac{\Delta p}{\Delta L}(y^2 - H^2)}_{\substack{\text{Poiseuille channel due} \\ \text{to pressure } \Delta p/\Delta L}} + \underbrace{\frac{V}{2}(y + 1).}_{\substack{\text{Couette due to motion} \\ \text{of upper plate at } V}} \tag{6.29}$$

It satisfies both the continuity and the momentum equations, and the boundary conditions as well. Superposition of linear unidirectional flows of different directions results in bidirectional (and tridirectional) flows, as discussed in Section 6.6.

### Example 6.1: Gravity-Driven Poiseuille Flow

A device to collect rainwater is designed such that $H_0$ in Fig. E6.1 at any time is minimum, practically $H_0 \approx 0$. The rain is collected by means of a wide slit of $H/W \ll 1$ and of length $L$. What is the time required to collect $V$ amount of water if the roof is always under a thin water layer of $H_0 \approx 0$?

**Solution** This is clearly a plane Poiseulle flow driven by gravity alone, and the procedure of Section 5.7 leads to Eqns. (6.3) and (6.5),

$$\frac{du_z}{dz} = 0 \qquad u_z = f(y) \tag{6.1.1}$$

(no dependence on the $x$-direction, because $W/H \gg 1$), and

$$\frac{\Delta p}{\Delta L} = \eta \frac{d^2 u_z}{dy^2} + \rho g_z. \tag{6.1.2}$$

The pressure difference here is zero because

$$\frac{\Delta p}{\Delta L} = \frac{p_L - p_0}{L} = \frac{p_{\text{atm}} - p_{\text{atm}}}{L} = 0. \tag{6.1.3}$$

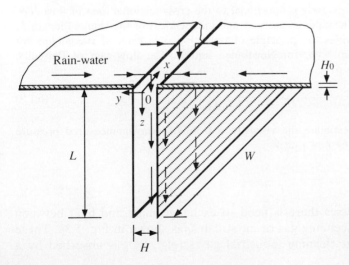

Rain-water

$L$

$H_0$

$W$

$H$

**Figure E6.1** Gravity-driven Poiseuille flow.

The governing differential equation becomes

$$\eta \frac{d^2 u_z}{dy^2} = -\rho g_z,$$

(6.1.4)

which is integrated twice to yield the general solution

$$u_z = -\frac{\rho g_z}{\eta} \frac{y^2}{2} + c_1 y + c_2.$$

(6.1.5)

The boundary conditions determine the two constants.

At $y = 0$, $du_z/dy = 0$; therefore, $c_1 = 0$.

At $y = \pm H$, $u_z = 0$; therefore, $c_2 = \dfrac{\rho g_z}{\eta} \dfrac{H^2}{2}$.

The resulting particular solution is

$$u_z = \frac{\rho g_z}{2\eta}(H^2 - y^2),$$

(6.1.6)

which reveals a parabolic velocity profile in the gap. This flow is equivalent to a mechanically driven plane Poiseuille flow by pressure gradient equivalent to $\Delta p / \Delta L = -\rho g_z$. The flow rate is

$$\dot{V} = W \int_{-H}^{H} \frac{\rho g_z}{2\eta}(H^2 - y^2)\, dy = 2W \int_{0}^{H} \frac{\rho g_z}{2\eta}(H^2 - y^2)\, dy$$

$$= \frac{\rho g_z}{\eta}\left(H^3 - \frac{H^3}{3}\right)W = \frac{2}{3}\frac{\rho g_z}{\eta} H^3 W,$$

(6.1.7)

and therefore the time required to collect volume $V$ of water is

$$t = \frac{V}{\dot{V}} = \frac{3\eta V}{2\rho g_z HWH^2} = \frac{3\nu V}{2 g_z HWH^2}.$$

(6.1.8)

The required time is inversely proportional to the cross-sectional area of flow $HW$, proportional to the kinematic viscosity $\nu$, and independent of the channel length $L$. Equation (6.1.8) provides the principle of operation of a class of rheometers for measuring viscosity, $\eta = \nu\rho$, for Newtonian liquids. The flow rate of Poiseuille flows, for example

$$\dot{V} = -\frac{\pi R^4}{8\eta} \frac{\Delta p}{\Delta L},$$

(6.1.9)

can also be used to estimate the viscosity $\eta$ from the known measured pressure gradient $\Delta p / \Delta L$ and the flow rate $\dot{V}$.

## 6.3.3 Film Flows

In many industrial processes there is need to exchange mass and heat between falling liquid films and ascending gas or air streams, as shown in Fig. 5.3a. These processes are common in cleaning industrial gases (e.g., $SO_2$ is absorbed by a falling solvent[15]), in cooling hot water or preheating air for combustion (heat is removed from falling water to cool air), and in other operations.[16] The fluid mechanics aspects of these processes are essential to estimate the rate of heat or mass exchange. The heat and mass transfer aspects are examined within the context of heat and (solute) mass transfer courses and textbooks.

Figure 5.3a shows a general case of a flat film flow down a vertical wall, in contact with stationary air. Several problems at the end of the chapter deal with a moving airstream. The stationary air of vanishingly small viscosity imposes a vanishingly small shear stress at the interface, taken to be zero, whereas a moving stream of air imposes its velocity on the adjacent liquid film. The thin film has a pressure identical to that of the adjacent air, and therefore pressure gradients in the normal to the free surface direction are everywhere zero. The flow may be driven either by gravity, by air drag (in case of a moving airstream), or by the two combined. Laminar flow may persist in thin films over inclined plates up to Reynolds number 25. For $25 < \text{Re} < 2000$ the film develops ripples, and for $\text{Re} > 2000$ the film becomes turbulent.[17] These limits apply to thin films over vertical cylindrical surfaces of radius much larger than the film thickness. For radius and film thickness of the same order of magnitude these limits are reduced.[18] The fluid mechanics of these films are essential for estimating heat and mass transfer.[19] Flat film flows analyzed in this chapter are without surface tension, the existence of which may disturb film uniformity, as discussed in Chapter 8.

### Example 6.2: Film Flow

Analyze the thin film flow down an inclined plane in contact with stationary air, as shown in Fig. E6.2. The inclination angle of the plane, the flow rate, and the viscosity and density of the liquid are known, whereas its surface tension is negligible.

**Solution**   The procedure of Section 5.8 leads to Eqn. (6.6) with $\Delta p / \Delta L = 0$:

$$0 = \frac{\Delta p}{\Delta L} = \eta \frac{du_z}{dy^2} + \rho g \sin \phi. \tag{6.2.1}$$

Thus, the governing differential equation is

$$\eta \frac{d^2 u_z}{dy^2} = -\rho g \sin \phi, \tag{6.2.2}$$

which has general solution

$$u_z = \frac{-\rho g \sin \phi}{\eta} \frac{y^2}{2} + c_1 y + c_2, \tag{6.2.3}$$

subject to the boundary conditions of no slip along the solid boundary and no stress along the free surface.

At $y = 0$, $u_z = 0$; therefore, $0 = c_2$.
At $y = H$, $du_z/dy = 0$; therefore, $-\rho g \sin \phi / \eta H + c_1 = 0$.

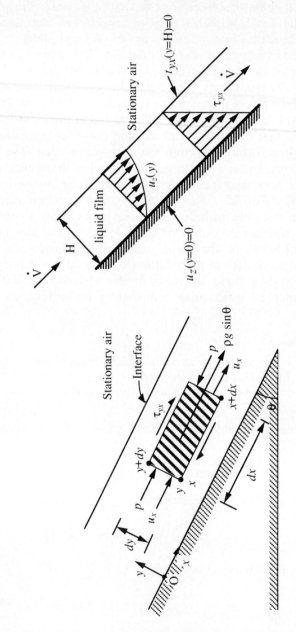

**Figure E6.2** Differential control volume (DCV) for film flow down an inclined plane and resulting velocity and shear stress profiles.

The resulting particular solution is

$$u_z = \frac{\rho g \sin \phi}{\eta}\left(Hy - \frac{y^2}{2}\right), \tag{6.2.4}$$

which represents a semiparabolic velocity profile, as shown in Fig. E6.2.
The flow rate per unit width is

$$\frac{\dot{V}}{W} = \int_0^H u_z\, dy = \frac{\rho g \sin \phi}{\eta}\left[H\frac{y^2}{2} - \frac{y^3}{6}\right]_0^H = \frac{\rho g \sin \phi}{\eta}\frac{H^3}{3}. \tag{6.2.5}$$

Given $\dot{V}/W$, the film thickness $H$ can be estimated, and vice versa.

*Shear stress distribution:*

$$\tau_{yz} = \eta\frac{du_z}{dy} = \rho g \sin \phi (H - y). \tag{6.2.6}$$

*Shear stress at interface:*

$$\tau_{yz}\big|_{y=H} = \rho g \sin \phi (H - y)\big|_{y=H} = 0. \tag{6.2.7}$$

*Wall shear stress:*

$$\tau_{yz}^{P}\big|_{y=0} = \tau_{yz}\big|_{y=0} = \rho g \sin \phi H. \tag{6.2.8}$$

*Force to hold the plane in position:*

$$\frac{F}{W} = -\int_0^L \tau_{yz}\big|_{y=0}\, dy = -\rho g \sin \phi HL. \tag{6.2.9}$$

*Limiting cases:*

1.  Horizontal film: $\sin \phi = 0 \Rightarrow u_z = 0$, no flow.
2.  Vertical film: $\sin \phi = 1 \Rightarrow u_z = (\rho g/\eta)(Hy - y^2/2)$.

    The vorticity of the flow is

$$\omega = -\frac{du_z}{dy} = \rho g \sin \phi (y - H), \tag{6.2.10}$$

and is therefore maximum at the wall and zero at the free surface. Although vorticity generated constantly at the solid wall tends to diffuse away to the free surface, it never makes it there, being swept downstream constantly by convection. This competition between convection and diffusion of vorticity is detailed further in Chapter 7.

## 6.4 TORSIONAL FLOWS

Torsional flows are induced by rotating solid boundaries in contact with liquids. The liquid, due to the no-slip boundary condition, has to follow the motion of the boundary, and therefore a torsional flow is generated. These flows are important in mixing, agitation, and centrifugal separations. They are utilized by commercial

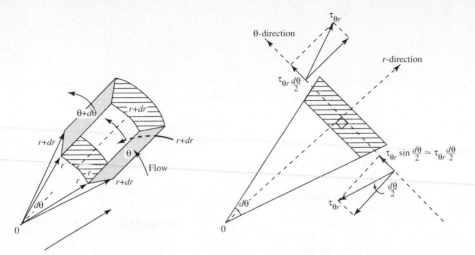

**Figure 6.5**  Control volume for torsional flow, and its projection to the $r\theta$ plane with stress $\tau_{or}$.

viscometers, where the viscosity can be deduced from the torque, $T$, necessary to turn a rod with angular velocity $\Omega$. Torsional flow between concentric cylinders at relative rotation under uniform pressure, called Taylor–Couette flow, has a linear velocity profile similar to plane Couette flow (Fig. E5.8.5). A full analysis of a typical torsional flow follows, following the procedure outlined in Section 5.9.

The geometry is cylindrical, and the flow is axisymmetric with respect to the cylindrical coordinate system. Apart from negligible end effects at the rod bottom, the flow is unidirectional, azimuthal with a unique velocity component $u_\theta \neq 0$, and $u_r = u_z = 0$; there is no radial depletion of fluid, and the free surface attains a fixed configuration at steady rotation. The appropriate control volume for torsional flow is shown in Fig. 6.5 with respect to a cylindrical system of coordinates.

The mass balance equation,

$$\frac{d}{dt}(\rho r\, d\theta\, dr\, dz) = dr\, dz(u_\theta|_\theta - u_\theta|_{\theta+d\theta})\rho, \tag{6.30}$$

under steady conditions yields

$$u_\theta = f(r). \tag{6.31}$$

A momentum balance in the $\theta$-direction on the same control volume results in

$$\frac{d}{dt}(\rho\, u_\theta\, r\, d\theta\, dr\, dz) = dr\, dz(u_\theta^2|_\theta - u_\theta^2|_{\theta+d\theta})\rho$$

$$- (r\, d\theta\, dz\tau_{r\theta})_r + (r\, d\theta\, dz\tau_{r\theta})_{r+dr} + 2\tau_{\theta r}\frac{d\theta}{2}\, dr\, dz - (r\, d\theta\, dr\tau_{z\theta})_z \tag{6.32}$$

$$+ (r\, d\theta\, dr\tau_{z\theta})_{z+dz} - ((-p + \tau_{\theta\theta})\, dr\, dz)_\theta + ((-p + \tau_{\theta\theta})\, dr\, dz)_{\theta+d\theta}.$$

At steady state, dividing by $r\, d\theta\, dr\, dz$ and taking the limit of $d\theta \to 0$, $dz \to 0$, and $dr \to 0$, this equation becomes

$$0 = -\rho\frac{du_\theta^2}{r\, d\theta} + \frac{1}{r}\frac{d}{dr}(r\tau_{r\theta}) + \frac{\tau_{r\theta}}{r} + \frac{d\tau_{z\theta}}{dz} - \frac{1}{r}\frac{dp}{d\theta} + \frac{1}{r}\frac{d\tau_{\theta\theta}}{d\theta}. \tag{6.33}$$

The first term is zero by the mass balance result. The two last terms are zero because of the axisymmetry of the flow. The fourth term is zero by virtue of the constitutive equation, since $u_z = 0$ and $du_\theta/dz = 0$. Thus the only surviving terms are the second and the third and Eqn. (6.33) becomes

$$\frac{1}{r}\left[\frac{d}{dr}(r\tau_{r\theta}) + \tau_{r\theta}\right] = \frac{1}{r}\frac{d}{dr}\left[\frac{1}{r}\frac{d}{dr}(u_\theta r)\right] = 0, \tag{6.34}$$

by means of the appropriate constitutive equation of Appendix F.

**Velocity profile.**    The general solution to Eqn. (6.34), and for that matter to any unidirectional torsional flow, is

$$u_\theta = c_1\frac{r}{2} + \frac{c_2}{r}. \tag{6.35}$$

The particular solution that applies to the flow under consideration here is obtained by invoking the two boundary conditions. If we consider the flow shown in Fig. E5.8.5 in a very wide pool of liquid (i.e., without the presence of the bounding wall), then:

At $r \to \infty$, $u_\theta \to 0$ (liquid at rest); therefore, $c_1 = 0$.

At $r = R$, $u_\theta = R\Omega$ (no-slip); therefore, $c_2 = \Omega R$.

The resulting solution for the velocity is

$$u_\theta = \frac{\Omega R^2}{r}. \tag{6.36}$$

**Pressure profile.**    The remaining two momentum equations, in the $r$- and $z$-directions, can now be utilized to find the pressure profile. Momentum balances on the same control volume in the $r$- and $z$-directions yield

$$-\rho\frac{u_\theta^2}{r} + \frac{\partial p}{\partial r} = 0 \tag{6.37}$$

and

$$-\frac{\partial p}{\partial z} - \rho g = 0 \tag{6.38}$$

respectively. Equations (6.37) and (6.38) combine as

$$p(r, z) = -\rho g z - \frac{\rho R^4 \Omega^2}{2r^2} + c. \tag{6.39}$$

The constant is determined by a free surface pressure boundary condition (in the absence of surface tension here) applied at the contact line of free surface with the rod (e.g., points A, A$^1$ of Fig. E5.8.5):

At $r = R$ and $z = z_0$, $p(R, z_0) = 0$.

Thus,

$$0 = -\rho g z_0 - \rho \frac{\Omega^2 R^4}{2R^2} + c$$

and

$$c = \rho g z_0 + \rho \frac{\Omega^2 R^2}{2}.$$

Then the pressure profile becomes

$$p(r, z) = \rho g(z_0 - z) + \rho \frac{\Omega^2 R^2}{2} \left( 1 - \frac{R^2}{r^2} \right). \tag{6.40}$$

**Free surface profile.**   Is the profile we assumed correct? Is $z_0 \leq z(r)$ for $R \leq r < \infty$? Take another point on the free surface, $B$. Then

$$0 = p(r_B, z_B(r)) = \rho g(z_0 - z_B(r)) + \rho \frac{\Omega^2 R^2}{2} \left( 1 - \frac{R^2}{r_B^2} \right), \tag{6.41}$$

which yields

$$z_B(r) - z_0 = \frac{\Omega^2 R^2}{2g} \left( 1 - \frac{R^2}{r_B^2} \right) > 0. \tag{6.42}$$

Thus, $z_B(r) \gg z_0$ which suggests a parabolic free surface profile accompanied by rod dipping.

Rod climbing or dipping is a very elementary way to distinguish between Newtonian and polymeric, non-Newtonian liquids because Newtonian liquids descend the rotating rod (this analysis), whereas polymeric viscoelastic liquids climb the same rotating rod.[20]

**Calculation of macroscopic quantities.**   Flow rate in $\theta$-direction:

$$\dot{V} = \int_R^\infty \int_0^{z_B(r)} \frac{R^2 \Omega}{r} \, dr \, dz = R^2 \Omega \int_R^\infty \frac{dr}{r} \int_0^{z_B(r)} dz$$

$$= R^2 \Omega \int_R^\infty \left[ z_0 + \frac{\Omega^2 R^2}{2g} \left( 1 - \frac{R^2}{r^2} \right) \right] \frac{dr}{r}$$

$$= \left( R^2 \Omega z_0 + \frac{R^4 \Omega^3}{2g} \right) \int_R^\infty \frac{dr}{r} - \frac{R^6 \Omega^3}{2g} \int_R^\infty \frac{dr}{r^3}$$

$$= \left( R^2 \Omega z_0 + \frac{R^4 \Omega^3}{2g} \right) \infty - \frac{R^6 \Omega^3}{2g} \left( \frac{1}{2R^2} \right)$$

$$\tag{6.43}$$

where the limit of the second integral is found from tables of definite integrals, whereas $\int_R^\infty dr/r \to \infty$. If we assume that the velocity becomes negligible at distance $r = 100R$, then,

$$\dot{V} = \left(R^2\Omega z_0 + \frac{R^4\Omega^3}{2g}\right)\int_R^{100R}\frac{dr}{r} - \frac{R^6\Omega^3}{2g}\int_R^{100R}\frac{dr}{r^3} \approx R^2\Omega z_0 + \frac{4R^4\Omega^3}{g} \qquad (6.44)$$

*Shear stress distribution:*

$$\tau_{r\theta} = \eta\left[r\frac{d}{dr}\left(\frac{u_\theta}{r}\right) + \frac{1}{r}\frac{2u_r}{2\theta}\right] = \eta r\frac{d}{dr}\left(\frac{R^2\Omega}{r^2}\right) = -\frac{2\eta R^2\Omega}{r^2} \qquad (6.45)$$

*Shear stress exerted by liquid on rod* (i.e., from farther to closer layer):

$$\tau_{r\theta}^R = \left(-2\eta\frac{R^2\Omega}{r^2}\right)\Bigg|_{r=R} = -2\eta\Omega \qquad (6.46)$$

*Shear stress exerted by rod on liquid* (i.e., from closer to farther layer):

$$\tau_{r\theta}^0 = -\tau_{r\theta}^R = +2\eta\Omega \qquad (6.47)$$

*Shear stress on tank wall, far away* (i.e., from closer to farther layer):

$$\tau_{r\theta}^T = -\left(-2\eta\frac{R^2\Omega}{r^2}\right)\Bigg|_{r\to\infty} = 0 \qquad (6.48)$$

The torque required to turn the rod is made up by the product of the area of the wetted surface of the rod, the shear stress at $r = R$, and the distance of the sheared surface from the axis of rotation. Thus

$$T = 2\pi z_0 R(2\eta\Omega)R = 4\pi R^2 z_0\eta\Omega. \qquad (6.49)$$

Thus the unknown viscosity of a liquid can be obtained from

$$\eta = \frac{T}{4\pi R^2 z_0\Omega}. \qquad (6.50)$$

The vorticity of the this torsional flow is given by

$$|\omega| = \frac{du_\theta}{dr} = \frac{2\Omega R^2}{r^2} = 2\Omega\frac{R^2}{r^2}. \qquad (6.51)$$

This vorticity is maximum, identical to twice the angular speed $\Omega$ at the solid wall of distance $r = R$ as it should by definition [Eqn. (5.39)], and diffuses constantly to zero value far away, where the liquid remains at rest.

## 6.5 APPLICATIONS OF UNIDIRECTIONAL FLOWS

Annular, multilayer, gear pump and screw extruder, tank draining, thin-film coating, and surface tension gradient-driven flows. In this section, several model unidirectional flows with applications in transferring, characterizing, and processing of liquids are analyzed. The appropriate system of coordinates is shown in the accompanying figures, whereas in most cases the control volume equations are omitted, having been derived earlier in this chapter. For example: For flow in annulus (Example 6.3), and wire coating (Example 6.9), the starting equation is Eqn. (6.7); for two-layer Couette flow (Example 6.4), Eqn. (6.8) is the starting equation; for the gear pump (Example 6.5) and the screw extruder (Example 6.6), the starting equation is Eqn. (6.6); and so on.

### Example 6.3: Translational Flow in an Annulus

An incompressible Newtonian liquid flows through an annulus formed by two concentric cylinders of radii $\lambda R$ and $R$, with $\lambda < 1$. The inner cylinder is stationary, and the outside travels parallel to itself with velocity $V$ (Fig. E6.3). Under these conditions the pressure is constant everywhere.

**(a)** Calculate the velocity profile.

**(b)** Show that in the limiting case of $\lambda \to 1$, the flow is similar to plane Couette.

**(c)** Show that the only possible flow in a pipe of moving wall, at constant pressure, is plug flow.

#### Solution

**(a)** *Velocity profile.* This is an axisymmetric, unidirectional flow, so the $z$-momentum equation reduces to

$$\frac{\Delta p}{\Delta L} = \eta \frac{1}{r}\frac{d}{dr}\left(r \frac{du_z}{dr}\right), \tag{6.3.1}$$

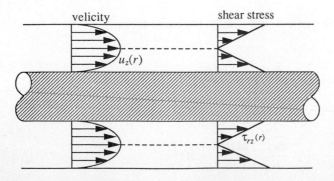

velicity                          shear stress

$u_z(r)$

$\tau_{rz}(r)$

**Figure E6.3**  Velocity and shear stress (from inner to outer layer) distribution of Poiseuille annulus flow.

with general solution

$$u_z = \frac{1}{4\eta}\frac{\Delta p}{\Delta L}r^2 + c_1 \ln r + c_2. \tag{6.3.2}$$

The boundary conditions now must be invoked:

At $r = \lambda R$, $u_z = 0$; therefore, $0 = \dfrac{1}{4\eta}\dfrac{\Delta p}{\Delta L}(\lambda R)^2 + c_1 \ln \lambda R + c_2$.

At $r = R$, $u_z = V$; therefore, $V = \dfrac{1}{4\eta}\dfrac{\Delta p}{\Delta L}R^2 - c_1 \ln R^2 + c_2$.

These two relations yield

$$c_1 = -\frac{V}{\ln \lambda} - \frac{1}{4\eta}\frac{\Delta p}{\Delta L}(\lambda R)^2\frac{1}{\ln \lambda}$$

and

$$c_2 = \frac{1}{4\eta}\frac{\Delta p}{\Delta L}(\lambda R)^2 + V\frac{\ln \lambda R}{\ln \lambda} + \frac{1}{4\eta}\frac{\Delta p}{\Delta L}[(\lambda R)^2 - R^2]\frac{\ln \lambda R}{\ln \lambda}.$$

The particular solution is then obtained by substituting the two latter expressions for $c_1$ and $c_2$ in Eqn. (6.3.2):

$$u_z = \frac{1}{4\eta}\frac{\Delta p}{\Delta L}(r^2 - \lambda^2 R^2) - \frac{\ln r}{\ln \lambda}\left[-V - \frac{1}{4\eta}\frac{\Delta p}{\Delta L}(\lambda^2 R^2 - R^2)\right]$$
$$+ \frac{\ln \lambda R}{\ln \lambda}\left[V + \frac{1}{4\eta}\frac{\Delta p}{\Delta L}(\lambda^2 R^2 - R^2)\right]. \tag{6.3.3}$$

This velocity profile prevails when the entrance and exit pressures are different. In the case $\Delta p/\Delta L = 0$, the velocity profile becomes,

$$u_z = V(\ln \lambda R - \ln r)\frac{1}{\ln \lambda} = V\frac{\ln(\lambda R/r)}{\ln \lambda}. \tag{6.3.4}$$

**(b)** *Limiting case of $\lambda \to 1$ [i.e., $R \approx \lambda R$ (narrow annulus)]:*

$$\lim_{\lambda \to 1} u_z = \lim_{\lambda \to 1} V(\ln \lambda R - \ln r)\frac{1}{\ln \lambda} = \lim_{\lambda \to 1} V\frac{\ln(\lambda R/r)}{\ln \lambda}$$
$$= \lim_{\lambda \to 1} V\frac{\ln(r/\lambda R)}{\ln(1/\lambda)} = \lim_{\lambda \to 1} V\frac{r/\lambda R - 1}{1/\lambda - 1} = V\frac{r/R - \lambda}{1 - \lambda}, \tag{6.3.5}$$

which is plane Couette flow between plates at distances $R$ and $\lambda R$ from the origin.

**(c)** *Limiting case of $\lambda \to 0$ [i.e., $\lambda R = 0$, $R \neq 0$ (pipe)]:*

$$u_z = \lim_{\lambda \to 0} V\frac{\ln(\lambda R/r)}{\ln \lambda} = V\frac{1/\lambda}{1/\lambda} = V, \tag{6.3.6}$$

which is plug flow in pipe, or solid-body translation.

A useful conclusion here is that flows in narrow annuli of gap much smaller than their radii can be analyzed as flows between parallel plates spaced at distance equal to the gap of the annulus. This is convenient because calculations with the Cartesian system of coordinates are generally much simpler than those with the cylindrical system. The accuracy achieved is improved as $\lambda \to 1$.

### Example 6.4: Two-Layer Couette Flow

Two immiscible liquids, $A$ and $B$, of densities $\rho_A > \rho_B$ and viscosities $\eta_A < \eta_B$, flow between two parallel plates of width $W$ and of infinite length (Fig. E6.4). The thicknesses of the layers are $H_A$ and $H_B$, respectively. The lower plate is stationary and the upper plate travels with velocity $V$. Under these conditions the pressure is constant everywhere.

**(a)** Find the velocity distribution in each layer. Then sketch it.

**(b)** Verify the results of (a) by considering a simple limiting case.

**(c)** Calculate the shear force per unit length on the upper plate.

**(d)** What is the shear force per unit length on the lower plate?

### Solution

**(a)** *Velocity profiles.* The governing momentum equations in each liquid are

$$0 = \eta_A \frac{d^2 u_x^A}{dy^2}, \qquad 0 \le y \le H_A \tag{6.4.1}$$

and

$$0 = \eta_B \frac{d^2 u_x^B}{dy^2}, \qquad H_A \le y \le H_A + H_B. \tag{6.4.2}$$

The general solutions are

$$u_x^A = c_1^A y + c_2^A \tag{6.4.3}$$

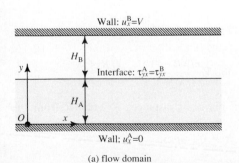

(a) flow domain

(b) velocity and shear stress profiles

**Figure E6.4**  Two-layer Couette flow field and resulting velocity and shear stress (from upper to lower layer) distribution.

and

$$u_x^B = c_1^B y + c_2^B.$$ (6.4.4)

The no-slip boundary conditions are applied first:

At $y = 0$, $u_x^A = 0$; therefore, $c_2^A = 0$.

At $y = H_A + H_B$, $u_x^B = V$; therefore,

$$V_1^B(H_A + H_B) + c_2^B \quad \text{and} \quad c_2^B = V - c_1^B(H_A + H_B).$$ (6.4.5)

Thus, the two velocity profiles become

$$u_x^A = c_1^A y$$ (6.4.6)

and

$$u_x^B = c_1^B[y - (H_A + H_B)] + V.$$ (6.4.7)

At the interface, the following boundary conditions are applied:

At $y = H_A$, $u_x^A = u_x^B$, therefore, $c_1^A H_A = -c_1^B H_B + V$ (6.4.8)

At $y = H_A$, $\tau_{yx}^A = \tau_{yx}^B$; therefore, $\eta_A \left. \dfrac{du_x^A}{dy} \right|_{H_A} = \eta_B \left. \dfrac{du_x^B}{dy} \right|_{H_A}$, (6.4.9)

which substituted in Eqns. (6.4.3) and (6.4.4) yields

$$\eta_A c_1^A = \eta_B c_1^B,$$ (6.4.10)

and finally,

$$c_1^B = \frac{V/H_A}{\eta_B/\eta_A + H_B/H_A}, \qquad c_1^A = \frac{V/H_A}{1 + (H_B/H_A)(\eta_A/\eta_B)}.$$ (6.4.11)

By substituting these constants in Eqns. (6.4.6) and (6.4.7), the velocity profiles in the two layers become

$$u_x^A = \frac{V/H_A}{1 + (H_B/H_A)(\eta_A/\eta_B)} y, \qquad 0 \le y \le H_A$$ (6.4.12)

and

$$u_x^B = \frac{V}{H_A}\left(\frac{H_B}{H_A} + \frac{\eta_B}{\eta_A}\right)[y - (H_A + H_B)] + V, \qquad H_A \le y \le H_A + H_B.$$

(6.4.13)

Checking for consistency:

$$u_x^A(y = 0) = 0, \qquad u_x^B(y = H_A + H_B) = V,$$

$$u_x^A(y = H_A) = \frac{V}{1 + (H_B/H_A)(\eta_A/\eta_B)} = u_x^B(y = H_A)$$

$$= \frac{V}{1 + (H_B/H_A)(\eta_A/\eta_B)}.$$

**(b)** *Limiting case.* For $H_A + H_B = H$ and $\eta_A = \eta_B$ (i.e., one-layer flow),

$$u_x^A = \frac{V}{H_A + H_B} y = \frac{V}{H} y,$$

$$u_x^B = \frac{V}{H_A + H_B}[y - (H_A + H_B)] = \frac{V}{H}(y - H).$$

**(c)** *Stress profiles.* The stress distributions in the two layers are

$$\tau_{yx}^A = \eta_A \frac{du_x^A}{dy} = \eta_A \frac{V/H_A}{1 + (H_B/H_A)(\eta_A/\eta_B)},$$

$$\tau_{yx}^B = \eta_B \frac{du_x^B}{dy} = \eta_A \frac{V/H_A}{(\eta_A/\eta_B)(H_B/H_A) + 1}.$$

Stress exerted by liquid on upper plate:

$$\tau_{yx}^U = -\tau_{yx}^B\big|_{y=H} = -\tau_{yx}^B.$$

**(d)** External force on upper plate ($-y$ surface):

$$F = -\int_0^L \int_0^W \tau_{yx}^U \, dx \, dy = \tau_{yx}^B LW > 0.$$

Stress exerted by liquid on lower surface ($y$ surface):

$$\tau_{yx}^L = \tau_{yx}^A\big|_{y=0} = \tau_{yx}^A.$$

External force on lower plate:

$$F = -\int_0^L \int_0^W \tau_{yx}^A\big|_{y=0} \, dy \, dz \, dx = -\frac{\eta_A V}{H_A}\bigg/\left(1 + \frac{H_B}{H_A}\frac{\eta_A}{\eta_B}\right)WL < 0.$$

### Example 6.5: Gear Pump

The flow shown in Fig. E6.5 is an idealization of a gear pump,[9] with $H \ll W$, so that the flow reversal near the teeth can be neglected, and only the region where the flow is parallel to the walls needs to be analyzed.[21] The liquid recirculates within the tooth, without escape at the considered stage, as depicted in Fig. 4.22.

**(a)** Compute the velocity profile and the pressure gradient of the parallel flow.

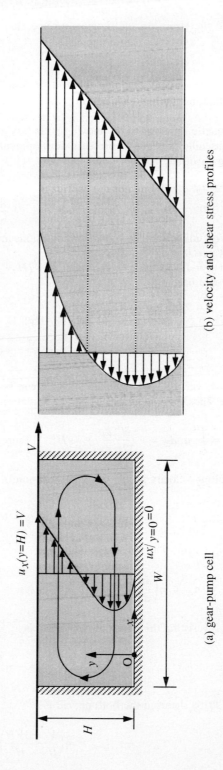

(a) gear-pump cell

(b) velocity and shear stress profiles

**Figure E6.5** Simplified gear pump operation and resulting velocity and shear stress profiles.

**(b)** Compute the stress on the moving surface and the torque applied by a gear pump.

**Solution**

**(a)** *Velocity profile.* Excluding the reversal of flow near the two corners due to drastic fluid momentum change and pressure buildup, the flow is a combination of:

1. Couette flow due to the moving boundary at speed $V$.
2. Poiseuille flow due to pressure development due to the flow reversal.

Thus the governing momentum equation is

$$\frac{dp}{dx} = \eta \frac{d^2 u_x}{dy^2}; \quad \text{therefore,} \quad u_x = \frac{dp}{dx} \frac{1}{2\eta} y^2 + c_1 y + c_2, \quad (6.5.1)$$

with $dp/dx$ unknown. The boundary conditions are:

At $y = 0$, $u_x = 0$; therefore, $c_2 = 0$.

At $y = H$, $u_x = V$; therefore, $c_1 = V/H - (H/2\eta)(dp/dx)$. The velocity profile becomes

$$u_x = \underbrace{\frac{1}{2\eta}\frac{dp}{dx}(y^2 - Hy)}_{\text{Poiseuille flow}} + \underbrace{V\frac{y}{H}}_{\text{Couette flow}}. \quad (6.5.2)$$

However, $dp/dx$ is still unknown; it can be calculated from the zero net flow rate since there is no escape or accumulation of liquid in the cell,

$$\dot{V} = 0 = \int_0^H u_x \, dy = -\frac{H^3}{12\eta}\frac{dp}{dx} + \tfrac{1}{2}VH; \quad \text{therefore,} \quad \frac{dp}{dx} = \frac{6\eta V}{H^2}. \quad (6.5.3)$$

The resulting velocity profile, after substitution of $dp/dx$, is

$$u_x = \frac{3V}{H^2}(y^2 - Hy) + V\frac{y}{H}. \quad (6.5.4)$$

**(b)** *Stress and torque.* The stress distribution is

$$\tau_{yx} = \eta\left(\frac{du_x}{dy}\right)_{y=H} = \eta\left[\frac{V}{H} + \frac{3V}{H^2}(2y - H)\right]_{y=H} = \frac{4V\eta}{H}. \quad (6.5.5)$$

The stress resisting the upper moving plate is

$$\tau_{yx}^P = -\tau_{yx}\big|_{y=H} = -\eta\frac{4V}{H} \quad (6.5.6)$$

and the stress shearing the bottom wall is

$$\tau_{yx}^w = +\tau_{yx}\big|_{y=0} = \eta\left(\frac{V}{H} - \frac{3VH}{H^2}\right) = -\frac{2\eta V}{H}. \quad (6.5.7)$$

The torque per unit width (extending in the $z$-direction) on the gear pump by the liquid is

$$\underbrace{T}_{\text{torque}} = \underbrace{F}_{\text{force}} \underbrace{R}_{\text{distance}} = \underbrace{N}_{\substack{\text{no. of} \\ \text{teeth}}} \underbrace{F_T}_{\substack{\text{force} \\ \text{per tooth}}} R = N \underbrace{W}_{\text{width}} \eta \underbrace{\frac{4V}{HR}}_{\text{stress}}. \qquad (6.5.8)$$

### Example 6.6: Screw Extruder

The model of Fig. E6.5 can be used to analyze the flow in each individual helical channel of a screw extruder,[9] as shown in Fig. E6.6. However, in this case there is continuous transfer of volume at rate $\dot{V}$ along the extrusion direction, due to flow rates $Q = \dot{V}/2\pi R$ from channel to channel, where $R$ is the radius of the extruder, accompanied by pressure buildup, $p_L = p_{i+1} - p_i$, across the width $L$ of each channel.

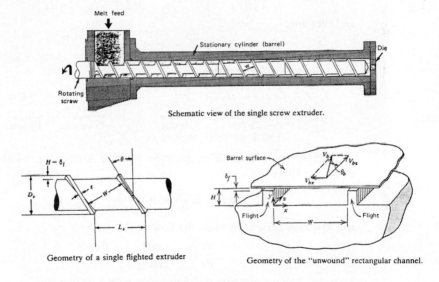

Schematic view of the single screw extruder.

Geometry of a single flighted extruder

Geometry of the "unwound" rectangular channel.

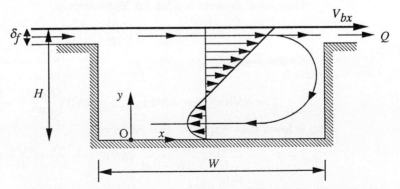

**Figure E6.6**  Simplified screw-extruder operation. The model at the bottom, used in this analysis is the flow seen by an observer (or with respect to a system of coordinates) pinned at the rotating channel bottom. (From Z. Tadmor and C. G. Gogos, *Principles of Polymer Processing*, Wiley, New York, copyright © 1979; reprinted by permission of John Wiley & Sons, Inc.)

**(a)** What are the resulting velocity profiles, pressure buildup, and torque in this case?

**(b)** The liquid from the screw extruder is to be extruded through a narrow slit die of length $l$, width $d$, and gap thickness $2h \ll d$, from die entrance pressure $p_{end} = N(p_{i+1} - p_i)$ to the ambient. What is the speed of extrusion achieved?

**Solution**

**(a)** *Velocity profile.* The flow is similar to that of the gear pump. The difference is the flow rate $\dot{V} \neq 0$. Thus

$$u_x = \frac{1}{2\eta} \frac{dp}{dx}(y^2 - Hy) + V\frac{y}{H}, \tag{6.6.1}$$

$$Q = \int_0^H \left[\frac{1}{2\eta} \frac{dp}{dx}(y^2 - Hy) + V\frac{y}{H}\right] dy = -\frac{H^3}{12\eta} \frac{dp}{dx} + \tfrac{1}{2}VH, \tag{6.6.2}$$

and therefore,

$$\frac{dp}{dx} = \frac{6\eta V}{H^2} - \frac{12\eta Q}{H^3}. \tag{6.6.3}$$

The velocity profile becomes

$$u_x = \underbrace{\frac{1}{2\eta} \frac{p_L}{L}(y^2 - Hy)}_{\substack{\text{Poiseuille flow} \\ \text{(backflow)}}} + \underbrace{V\frac{y}{H}}_{\substack{\text{Couette} \\ \text{forward}}}, \qquad Q = \underbrace{-\frac{H^3}{12\eta} \frac{p_L}{L}}_{\text{Poiseuille}} + \underbrace{\tfrac{1}{2}VH}_{\text{Couette}}. \tag{6.6.4}$$

*Pressure buildup.* The pressure buildup across the width $W$ of each channel is therefore

$$p_W = W\frac{dp}{dx} = W\left(\frac{6\eta V}{H^2} - \frac{12\eta Q}{H^3}\right), \tag{6.6.5}$$

and becomes maximum when the flow rate is zero:

$$\left.\frac{dp}{dx}\right|_{max} = \left.\frac{dp}{dx}\right|_{q=0} = \frac{6\eta V}{H^2}. \tag{6.6.6}$$

Thus, the highest pressure buildup occurs with a gear pump.

The overall pressure buildup for $N$ chambers is

$$p_{end} = Np_W = N\left(\frac{6\eta V}{H^2} - \frac{12\eta Q}{H^3}\right)W. \tag{6.6.7}$$

*Torque required.* The torque per unit length of circumference of the extruder is

$$T = RNW\tau_{yx}^{\text{upper}} = RNW\tau_{yx}\big|_{y=H} = RNW\left(\frac{Hp_W}{2W} + \eta\frac{V}{H}\right), \tag{6.6.8}$$

which is lower than that of the gear pump by

$$T_{\text{pump}} - T_{\text{extr.}} = \left(\frac{4\eta V}{H} - \frac{Hp_W}{2W} - \frac{\eta V}{H}\right)RNW = \left(\frac{3\eta V}{H^2} + \frac{6Q\eta}{H} - \frac{3\eta V}{H}\right)RNW$$

$$= \frac{6Q\eta}{H^2}RNW. \tag{6.6.9}$$

**(b)** *Extrusion speed.* The flow in the slit is pressure-driven flow of velocity profile,

$$u_x^s(y) = \frac{1}{2\eta} \frac{p_{end}}{l}(h^2 - y^2), \tag{6.6.10}$$

and the flow rate is

$$\dot{V} = d \int_{-h}^{h} u_x^s(y)\, dy = \frac{2p_{end}h^3 d}{3\eta l} = 2\pi R Q \delta_f. \tag{6.6.11}$$

Thus, by combining Eqns. (6.6.7) and (6.6.11), the resulting flow rate is

$$Q = \frac{2V}{(6\pi R \delta_f l H^2 / d N h^3 W) + (4/H)} \tag{6.6.12}$$

by an extruder of screw of radius $R$ that has $N$ channels of width $W$ and inclination angle $\theta$, separated by thin walls to minimize pressure drop from one channel to another. It rotates with angular speed $\Omega$ such that $V = \Omega \sin \theta$. The resulting linear speed of extrusion is

$$U_e = \frac{\dot{V}}{2hd} = \frac{2\pi R Q}{2hd}. \tag{6.6.13}$$

Going back, all the interesting quantities can be expressed in terms of the design characteristics, $N$, $\theta$, $d$, $R$, $W$, $h$, $l$, and the speed of rotation $\Omega$.

In these calculations the screw entry pressure and the die exit pressure were taken to be identical, and pressure drop and shear in going from one channel to another and at the die end were neglected. Details on the validity of these assumptions can be found elsewhere.[9,21]

### Example 6.7: Spherical Radial Flow

A Newtonian oil of viscosity $\eta$ escapes steadily from a porous sphere of radius $R$ at a flow rate $\dot{V}$ cm³/s. The flow everywhere outside the sphere is purely radial.

**(a)** Derive the continuity equation, $d(r^2 u_r)/dr = 0$, by using control volume principles.
**(b)** Find the velocity distribution $u_r$ through and outside the porous sphere.
**(c)** What is the flow rate in cm³/s at distance $10R$ from the center of the sphere?
**(d)** Find the pressure distribution $p(r)$ through and outside the porous sphere of permeability $k$.

### Solution

**(a)** Mass balance on a spherical control volume of radius $r$ and thickness $dr$ around the porous sphere yields

$$\underbrace{\frac{\partial}{\partial t}(\rho 4\pi r^2\, dr)}_{\substack{\text{accumulation of} \\ \text{mass in CV}}} = \underbrace{\rho(4\pi r^2 u_r)_r}_{\substack{\text{convection of mass} \\ \text{through boundary at } r}} - \underbrace{\rho(4\pi^2 u_r)_{r+dr}}_{\substack{\text{convection of mass} \\ \text{through boundary at } r+dr}} \tag{6.7.1}$$

For incompressible liquid of constant density and in the limit of $dr \to 0$, Eqn. (6.7.1) reduces to

$$\frac{d}{dr}(r^2 u_r) = 0. \tag{6.7.2}$$

**(b)** Velocity distribution: By integrating Eqn. (6.7.2), we see that

$$\int d(r^2 u_r) = \int 0\, dr, \qquad r^2 u_r = 0 + c, \qquad u_r = \frac{c}{r^2}. \tag{6.7.3}$$

At $r = R$, $4\pi R^2 u_r|_{r=R} = \dot{V}$; therefore, $4\pi R^2 c / R^2 = \dot{V}$ and $c = \dot{V}/4\pi$. Thus

$$u_r = \frac{\dot{V}}{4\pi r^2}. \tag{6.7.4}$$

**(c)** The flow rate is $\dot{V}$ at any distance from the center of the sphere, held constant by a velocity diminishing with radial distance according to Eqn. (6.7.4) through flow area increasing with radial distance.

**(d)** Pressure distribution: The pressure distribution outside the porous sphere along a radial streamline is calculated by means of Bernoulli's equation, which can be shown to be equivalent to the $r$-momentum equation:

$$\frac{p(r)}{\rho} + \frac{u_r^2(r)}{2} = \frac{p_\infty}{\rho} + \frac{u_r^2(r \to \infty)}{2} = \frac{p_\infty}{p}. \tag{6.7.5}$$

Thus

$$p(r) = p_\infty - \rho\,\frac{u_r^2(r)}{2} = p_\infty - \frac{\rho\dot{V}^2}{32\pi^2 r^4}. \tag{6.7.6}$$

The pressure predicted at the surface of the porous sphere is

$$p_s = p(r = R) = p_\infty - \frac{\rho\dot{V}^2}{32\pi^2 R^4}. \tag{6.7.7}$$

The pressure distribution within the porous sphere is governed by Darcy's equation,

$$u_r(r) = -\frac{k}{\eta}\frac{dp}{dr}, \tag{6.7.8}$$

which yields

$$p(r) = \frac{\eta\dot{V}}{12\pi kr} + c, \tag{6.7.9}$$

with the constant determined by using the common pressure value at the sphere surface, given by Eqn. (6.7.7). The resulting pressure distribution is

$$p(r) = p_\infty - \frac{\eta\dot{V}}{12\pi k}\left(\frac{1}{R} - \frac{1}{r}\right) - \frac{\rho\dot{V}^2}{32\pi^2 R^4}. \tag{6.7.10}$$

The distribution of pressure throughout the flow is shown in Fig. E6.7.

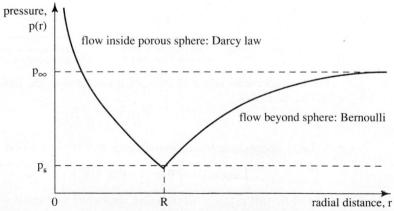

**Figure E6.7** Pressure distribution inside and outside porous sphere having a source at $r = 0$.

**Example 6.8: Tank Drainage**

A cylindrical tank of diameter $D = 3\,\text{m}$ and initial water depth $H_0 = 4\,\text{m}$ is to be drained by means of a pipe of diameter $d = 0.1\,\text{m}$ and length $L = 2\,\text{m}$ (Fig. E6.8). What is the time required for complete drainage? How does this time vary with the length $L$?

**Solution**   The flow in the pipe of diameter $d$ is due to both gravity and hydrostatic pressure. The pressures at the pipe entrance and exit are,

$$p_1 = \rho g H(t) \quad \text{and} \quad p_2 = 0. \tag{6.8.1}$$

The pipe velocity remains nearly constant with time due to slow draining such that the time derivatives $\partial u / \partial t$ are vanishingly small compared to the spatial derivative $\partial u / \partial r$. The momentum equation for this combined flow is

$$\frac{dp}{dz} = -\frac{\rho g H(t)}{L} = \eta \frac{1}{r} \frac{\partial}{\partial r} \left( r \frac{\partial u(r, t)}{\partial r} \right) + \rho g, \tag{6.8.2}$$

which is rearranged to

$$\left( \frac{-\rho g H(t)}{\eta L} - \frac{\rho g}{\eta} \right) r = \frac{\partial}{\partial r} \left( r \frac{du(r, t)}{dr} \right),$$

and then integrated once to

$$\left( \frac{-\rho g H(t)}{\eta L} - \frac{\rho g}{\eta} \right) \frac{r}{2} + \frac{c_1}{r} = \frac{\partial u(r, t)}{\partial r}$$

and integrated again to the general solution,

$$u(r, t) = \left( -\frac{\rho g H(t)}{\eta L} - \frac{\rho g}{\eta} \right) \frac{r^2}{4} + c_1 \ln r + c_2. \tag{6.8.3}$$

Boundary conditions:

At $r = 0$, $\partial u / \partial r = 0$; therefore, $c_1 = 0$.

At $r = R^2$, $u = 0$; therefore, $c_2 = \left( \frac{\rho g H(t)}{\eta L} + \frac{\rho g}{\eta} \right) \frac{R^2}{4}$.

The resulting velocity profile is

$$u(r, t) = \left( \frac{\rho g H(t)}{\eta L} + \frac{\rho g}{\eta} \right) \left( \frac{R^2}{4} - \frac{r^2}{4} \right), \tag{6.8.4}$$

The resulting flow rate is

$$\dot{V}(t) = \int_0^R 2\pi r u \, dr = \left( \frac{\rho g H(t)}{\eta L} + \frac{\rho g}{\eta} \right) 2\pi \int_0^R r \left( \frac{R^2}{4} - \frac{r^2}{4} \right) dr$$

$$= \left( \frac{\rho g H(t)}{\eta L} + \frac{\rho g}{\eta} \right) \frac{\pi}{2} \left( \frac{R^4}{2} - \frac{R^4}{4} \right) = \left( \frac{\rho g H(t)}{L} + \rho g \right) \frac{\pi R^4}{8\eta}. \tag{6.8.5}$$

Mass balance over the entire tank results in

$$\frac{d}{dt}[\pi R_T^2 H(t)] = -\dot{V}(t),$$

$$\pi R_T^2 \frac{dH}{dt} = -\left(\frac{\rho g H(t)}{\eta L} + \frac{\rho g}{\eta}\right)\frac{\pi R^4}{8},$$

$$\int_{H_0}^{H(t)} \frac{dH}{\rho g H(t)/\eta L + \rho g/\eta} = -\int_0^t \frac{R^4}{8R_T^2}\,dt$$

$$\int_{H_0}^{H(t)} \frac{dH}{H/L + 1} = -\frac{\rho g R^4}{8\eta R_T^2}\int_0^t dt, \quad L\ln\left(\frac{H}{L}+1\right)\bigg|_{H_0}^{H(t)} = -\frac{\rho g R^4}{8\eta R_T^2}\bigg|_0^t,$$

and finally,

$$\ln\frac{H(t)/L + 1}{H_0/L + 1} = -\frac{\rho g R^4}{8\eta L R_T^2}t, \qquad t = -\frac{8\eta L R_T^2}{\rho g R^4}\ln\frac{H(t) + L}{H_0 + L}. \tag{6.8.6}$$

The liquid elevation in the tank then descends slowly according to

$$H(t) = (H_0 + L)\exp\left(\frac{-\rho g R^4 t}{R_T^2 8\eta L}\right) - L, \tag{6.8.7}$$

as shown in Fig. E6.8. The time required for complete drainage (i.e., for $H(t) = 0$),

(a)

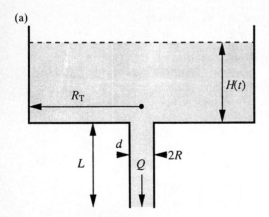

(b)

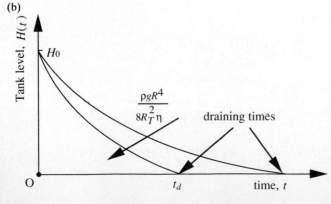

**Figure E6.8**  Drainage of a cylindrical tank and required drainage time $t_d$.

is

$$t_d = \frac{8 \times 0.01 \times 200 \times 22{,}500}{981 \times 625} \ln(3) = 0.587 \ln(3) = 0.66 \text{ min.}$$

## Example 6.9: Two-Phase Annular Flow

In a laminar two-phase, vertical annular flow of a water–steam mixture under gravity alone (Fig. 4.14), the measured mass flow rate within the thin liquid film adjacent to wall of radius $R = 1$ in. is $\dot{m}_L = 50$ g/s. What is the thickness of the thin liquid film and the average velocity?

**Solution**   The velocity $u$ is given by Eqns. (4.112) and (4.113), with $dp/dz = 0$, $\rho_L = 0.96$ g/cm$^3$, $\eta_E = \eta_L = 0.28$ cP, $\rho_G = 0.0006$/cm$^3$, and $\eta_G = 0.013$ cP, at temperature $T = 100°$C. The velocity profile under these conditions reduces to

$$u_L = -\frac{g}{2\eta_L}(\rho_L - \rho_G)(R - \delta)^2 \ln \frac{R}{r} + \frac{\rho_L g}{4\eta_L}(R^2 - r^2), \qquad (6.9.1)$$

and Eqn. (4.116) yields,

$$\begin{aligned}
\dot{m}_L &= -\frac{2\pi g \rho_L(\rho_L - \rho_G)(R - \delta)^2}{2\eta_L} \int_{R-\delta}^{R} r \ln \frac{R}{r} \, dr \\
&\quad + \frac{2\pi \rho_L^2 g}{4\eta_L} \int_{R-\delta}^{R} (R^2 - r^2) r \, dt \\
&= -\frac{\pi g \rho_L(\rho_L - \rho_G)(R - \delta)^2}{2\eta_L}\left[ -\frac{(R-\delta)^2}{2} \ln \frac{R}{R-\delta} - \frac{\delta^4}{4} + \frac{R\delta}{2} \right] \\
&\quad + \frac{\pi \rho_L^2 g}{2\eta_L}\left[ R\delta(R\delta - \delta^2) + \frac{\delta^4}{4} \right].
\end{aligned} \qquad (6.9.2)$$

For thin film of thickness $\delta \ll R$, it is

$$\ln \frac{R}{R - \delta} \approx \frac{\delta}{R}. \qquad (6.9.3)$$

Under this approximation, Eqn. (6.9.2) becomes

$$\dot{m}_L = \frac{\pi \rho_L g}{\eta_L}\left\{ \rho_L\left[ \left(\frac{R\delta}{2}\right)^2 - \frac{R\delta^3}{2} + \frac{\delta^4}{8} \right] + (\rho_L - \rho_G)\left( 2R\delta^3 - \frac{3\delta^2 R^2}{4} + \frac{\delta^5}{2R} - \frac{7\delta^4}{4} \right) \right\} \qquad (6.9.4)$$

and is simplified further to

$$\left(\frac{\dot{m}_L}{2\pi R}\right)\left(\frac{\eta_L}{\rho_L^2 g}\right) = \left(\frac{\rho\delta^2}{4} - \frac{\delta^3}{4}\right) + \left(1 + \frac{\rho_G}{\rho_L}\right)\left(\delta^3 - \frac{3\delta^2 R}{8}\right), \qquad (6.9.5)$$

which can be solved for the thickness $\delta$. For the case considered here, it is

$$\frac{\rho_G}{\rho_L} = \frac{0.006}{0.96} = 0.00061 \ll 1, \qquad (6.9.6)$$

and Eqn. (6.9.5) becomes

$$\delta^3 - \frac{R\delta^2}{6} = \left(\frac{\dot{m}_L}{2\pi R}\right)\left(\frac{4\eta_L}{3\rho_L^2 g}\right), \tag{6.9.7}$$

and for the values given,

$$\delta^3 - 0.423\delta^2 = 0.0026, \tag{6.9.8}$$

which yields

$$\delta = 0.44 \, \text{cm}. \tag{6.9.9}$$

The Reynolds number is

$$\text{Re} = \frac{\rho_L \bar{u}_L \delta}{\eta_L} = \frac{\dot{m}_L}{2\pi R \eta_L} = \frac{50}{2 \times 3.14 \times 2.54 \times 0.0028} = 1120, \tag{6.9.10}$$

which is the upper limit of laminar flow. The average velocity is

$$\bar{u}_L = \frac{\dot{m}_L}{\rho_L 2\pi R \delta} = 7.42 \, \frac{\text{cm}}{\text{s}}. \tag{6.9.11}$$

### Example 6.10: Absorption of $SO_2$ from Exhaust Gases

In an industrial gas cleaning process, an organic solvent of viscosity $\eta = 10 \, \text{cP}$ and density $\rho = 1 \, \text{g/cm}^3$ forms a film of thickness $H = 1 \, \text{mm}$ on a vertical wall. The industrial gas stream ascends (in contact with the solvent) at velocity $V = 5 \, \text{cm/s}$. Under these conditions, the solubility of $SO_2$ is $1 \, \text{g} \, SO_2$ per $100 \, \text{g}$ of solvent. What is the "cleaning rate" of $SO_2$ per unit width of exchange area?

**Solution**    The governing momentum equation in the film is

$$\eta \frac{d^2 u_z}{dy^2} + \rho g = 0, \qquad \frac{d}{dy}\left(\frac{du_z}{dy}\right) = -\frac{\rho g}{\eta}, \qquad \frac{du_z}{dy} = -\frac{\rho g}{\eta} y + c_1,$$

$$u_z = -\frac{\rho g}{\eta}\frac{y^2}{2} + c_1 y + c_2.$$

At $y = 0$, $u_z = 0$; therefore, $c_2 = 0$.

At $y = H$, $u = -V$; therefore, $c_1 = \left(-V + \dfrac{\rho g}{\eta}\dfrac{H^2}{2}\right)\Big/ H = -\dfrac{V}{H} + \dfrac{\rho g H}{2\eta}$.

*Velocity profile:*

$$u_z = \underbrace{\frac{\rho g}{\eta}\left(Hy - \frac{y^2}{2}\right)}_{\substack{\text{gravity flow} \\ \text{due to } g}} + \underbrace{\left(-\frac{V}{H}y\right)}_{\substack{\text{Couette flow} \\ \text{due to } V}}.$$

*Flow rate per unit width:*

$$\dot{V} = \int_0^H u_z \, dy = \frac{2\rho g}{\eta}\frac{H^3}{6} - \frac{VH}{2}.$$

*Mass flow rate per unit width:*

$$\dot{m} = \rho\dot{V} = \frac{\rho^2 g H^3}{3\eta} - \frac{\rho V H}{2}$$

$$= \left(\frac{1^2 \times 981 \times 0.1^3}{3 \times 0.1} - \frac{1 \times 5 \times 0.1}{2}\right) \text{g/s}$$

$$= (3.27 - 0.25) = 3.02 \text{ g/s.}$$

Thus the rate of cleaning is $\dot{m}_{SO_2} = 0.0302$ g of $SO_2$ per second per unit width.

### Example 6.11: Two-Layer Film Show

Two immiscible liquids, with viscosities $\eta_A$ and $\eta_B$ and densities $\rho_A$ and $\rho_B$, are flowing one on top of the other down an inclined plane at angle $\phi$ with the horizontal, at flow rates $Q_A$ and $Q_B$. The upper liquid is in contact with stationary air, as shown in Fig. E6.11, and a zero shear stress is assumed. Calculate the velocity distribution in the two layers.

**Solution**    *Liquid A:*

$$-\frac{dp^A}{dx} + \eta^A \frac{d^2 u_x^A}{dz^2} + \rho^A g \sin\phi = 0, \qquad \frac{du_x^A}{dz} = -\frac{\rho^A g \sin\phi}{\eta A} z + c_1,$$

$$u_x^A = -\frac{\rho^A g \sin\phi}{\eta A}\frac{z^2}{2} + c_1^A z + c_2^A.$$

At $z = 0$, $u_x^A = 0$, $c_2^A = 0$, and

$$u_x^A = -\frac{\rho^A g \sin\phi}{\eta A}\frac{z^2}{2} + c_1^A z.$$

*Liquid B:*

$$u_x^B = -\frac{\rho^B g \sin\phi}{\eta^B}\frac{z^2}{2} + c_1^B z + c_2^B.$$

At $z = H_A + H_B$, $du_x^B/dz = 0$, $c_1^B = \rho^B g \sin\phi \eta^B (H_A - H_B)$. Thus

$$u_x^B = \frac{\rho^B g \sin\phi}{\eta^B}\left[(H_A + H_B)^z - \frac{z^2}{2}\right] + c_2^B, \qquad u_x^A = -\frac{\rho^A g \sin\phi}{\eta^A}\frac{z^2}{2} + c_1^A z.$$

At this point, $c_1^A$ and $c_2^B$ can be calculated by using continuity of velocity and stress at the liquid–liquid interface, $z = H_A$:

$$\tau_{yx}^A = \tau_{yz}^B; \qquad \text{therefore,} \qquad \eta_A \frac{du_x^A}{dz} = \tau_{yx}^A = \tau_{yx}^B = \eta_B \frac{du_x^B}{dz}$$

$$-\rho^A g \sin\phi H_A + c_1^A \eta_A = \rho^B g \sin\phi(H_A + H_B - H_A) + 0$$

and

$$c_1^A = \frac{g \sin\phi(\rho^B H^B + \rho^A H^A)}{\eta^A}.$$

Thus the velocity profile in liquid $A$ becomes

$$u_x^A = \frac{g \sin\phi}{\eta_A}(\rho^B H^B + \rho^A H^A)z - \frac{\rho^A z^2}{2}. \tag{6.11.1}$$

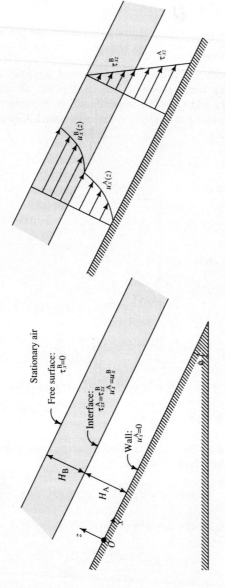

**Figure E6.11** Two-layer film flow and resulting velocity and shear stress profiles.

At $z = H_A$, $u_x^A = u_x^B$; therefore,

$$\frac{g \sin \phi}{\eta^A}\left(\rho^B H^B H^A + \frac{\rho^A H_A^2}{2}\right) = \frac{g \sin \phi}{\eta^B}\left(\rho^B H^A H^A + \frac{\rho^B H^A H^B}{2}\right) + c_2^B.$$

Thus

$$c_2^B = g \sin \phi\left[\frac{\rho^B H^B H^A + \rho^A H_A^2/2}{\eta^A} - \frac{\rho^B H_A^2 + \rho^B(H^A H^B/2)}{\eta^B}\right],$$

and the velocity profile in liquid $B$ becomes

$$u_x^B = \frac{\rho^B g \sin \phi}{\eta^B}\left[(H_A + H_B)z - \frac{z^2}{2}\right]$$

$$\tag{6.11.2}$$

$$+ g \sin \phi\left[\frac{\rho^B H^B H^A + \rho^A(H_A^2/2)}{\eta_A} - \frac{\rho_B H_A^2 + \rho^B(H_A^2/2)}{\eta^B}\right].$$

*Limiting case:* If $\eta^A = \eta^B = \eta$ and $\rho^A = \rho^B = \rho$ (i.e., one liquid),

$$u_x = g \frac{\sin \phi \rho}{\eta}\left[(H^A + H^B) - \frac{z^2}{2}\right].$$

$$\tag{6.11.3}$$

Relations between flow rates $Q_A$ and $Q_B$ and filing thickness $H_A$ and $H_B$ are obtained by means of the equations

$$Q_A = \int_0^{H_A} u_x^A \, dy, \qquad Q_B = \int_{H_A}^{H_A + H_B} u_x^B \, dy.$$

The velocity and the shear stress distribution in the two films are shown in Fig. E6.11. Notice that the velocity and the shear stress are continuous across a planar interface, whereas the velocity derivative is discontinuous. Multilayer flows are common in coating processes, where layers of different liquids are arranged to flow under gravity, one on top of the other over inclined surfaces, eventually being deposited and solidified on the substrate being coated.

### Example 6.12: Film Flow Due to Surface Tension Gradient

A thin liquid film of viscosity $\eta$, density $\rho$, and thickness $\delta$ is on a horizontally moving plane at velocity $V$ to the left (Fig. E6.12). On top, there is nonisothermal gas such that a shear stress gradient, $\sigma' = d\sigma/dx$, results at the interface,[22] due to a surface tension $\sigma(x)$ increasing from left to right,[23] as shown in Fig. E6.12.

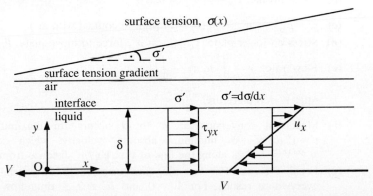

**Figure E6.12** Horizontal film flow induced by nonuniform surface tension, and resulting velocity and stress distribution.

(a) Calculate the velocity profile in the liquid.

(b) Calculate the shear stress profile in the liquid.

(c) Calculate the pressure profile along the film.

(d) What force is required to move the lower boundary?

(e) Under some conditions the flow-rate meter indicates zero flow rate. Explain why.

(f) If a chemical reaction, $A \rightarrow B$, takes place in the liquid film, where does the maximum conversion of $A$ to $B$ take place? (*Hint:* There is no flow or variation perpendicular to your paper plane.)

(g) What are the corresponding answers for uniform surface tension, $\sigma_0$, and $V \neq 0$? For $V = 0$ and $\sigma' \neq 0$? For $V = 0$ and $\sigma = \sigma_0$?

**Solution**  Mass balance yields

$$u_x = f(y).$$

Momentum balance yields

$$-\frac{dp}{dx} + \eta \frac{d^2 u_x}{dy^2} + \rho g_x = 0,$$

where $dp/dx = 0$ (why?), $g_x = 0$ for horizontal flow, and therefore

$$\eta \frac{d^2 u_x}{dy^2} = 0 \quad \text{and} \quad u_x = c_1 y + c_2.$$

The boundary conditions are applied, and the two constants are determined:

At $y = 0$, $u_x = -V$; thus $c_2 = -V$.
At $y = \delta$, $\eta c_1 = \sigma'$, thus $c_1 = \sigma'/\eta$.

The resulting profiles and quantities are:

(a) Velocity profile: $u_x = \dfrac{\sigma'}{\eta} y - V$.

(b) Stress profile: $\tau_{yx} = \eta \dfrac{du_x}{dy} = \sigma'$.

(c) $p = p_{\text{hydr}} = p_{\text{air}} + \rho g (\delta - y)$ (film in contact with air).

(d) Stress on lower plate: $\tau_{xy}\big|_{y=0} = \sigma'$, force to move plate: $F = -\sigma' L$.

(e) Flow rate: $\dot{V} = \displaystyle\int_0^\delta u_x \, dy = \int_0^\delta \left(\frac{\sigma'}{\eta} y - V\right) dy = \frac{\sigma'}{\eta}\frac{\delta^2}{2} - V\sigma'$. Zero net flow rate: $\dot{V} = 0$ therefore, $\dfrac{2\eta V}{\delta^2} = 1$.

(f) Maximum conversion of $A$ to $B$ occurs for maximum residence time, $t = \int_0^L dx/u_x$, or minimum absolute velocity, which occurs at $u_x = \sigma y/\eta - V = 0$ (i.e., along the line of $y = V\eta/\sigma'$ distance from the bottom).

(g) For $\sigma = \sigma_0$ and $V \neq 0$, a plug flow, or a solid body translation, with velocity $V$ everywhere results. For $V = 0$ and $\sigma' \neq 0$, a drag flow of velocity profile $u_x(y) = \sigma' y/\eta$ due to surface tension gradient $\sigma'$ results. The case $V = 0$ and $\sigma = \sigma_0$ and thus $\sigma' = 0$ corresponds to static equilibrium.

## 6.6 SUPERPOSITION OF UNIDIRECTIONAL FLOWS: HELICAL AND SPIRAL FLOWS

Several multidirectional flows are obtained by superposition of unidirectional flows. Each of the unidirectional velocity profiles is a solution to a momentum equation that is linear with respect to a unique velocity component. This precondition is satisfied by all unidirectional flows examined so far, because omission of the nonlinear convective terms always results in a linear ordinary differential equation with respect to a unique streamline velocity. Thus all categories of unidirectional flows examined so far can be superimposed to yield bi- and tridirectional flows. The velocity vector of such multidirectional flows,

$$\mathbf{u} = u_1\mathbf{e}_1 + u_2\mathbf{e}_2 + u_3\mathbf{e}_3, \tag{6.52}$$

is composed from the velocity components $u_1$, $u_2$, and $u_3$ of the superimposed unidirectional flows, each of streamline velocity $u_1$, $u_2$ and $u_3$, respectively, in the corresponding directions defined by the unit vectors $\mathbf{e}_1$, $\mathbf{e}_2$, $\mathbf{e}_3$ (see Appendix A).

**Helical flow.**    Superposition of a Hagen–Poiseuille flow of velocity

$$u_z = \frac{1}{4\eta}\frac{\Delta p}{\Delta L}(r^2 - R^2), \tag{6.53}$$

and of a torsional flow of velocity

$$u_r = \Omega r, \tag{6.54}$$

results in a helical flow of velocity vector

$$\mathbf{u} = u_z\mathbf{e}_z + u_r\mathbf{e}_r = \left[\frac{1}{4\eta}\frac{\Delta p}{\Delta L}(r^2 - R^2)\right]\mathbf{e}_z + \{\Omega r\}\mathbf{e}_r, \tag{6.55}$$

of magnitude

$$u = \sqrt{u_z^2 + u_r^2}, \tag{6.56}$$

as shown in Fig. 6.6. Helical flows are created by rotating nozzles and capillaries

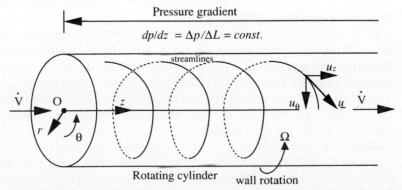

**Figure 6.6**    Superposition of pressure Poiseuille flow and torsional flow results in a helical flow.

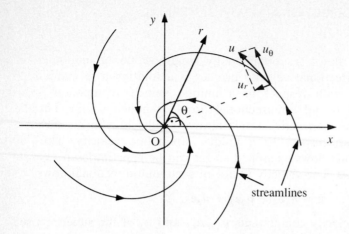

streamlines

**Figure 6.7**  Superposition of sink and torsional flows yields a spiral flow.

to enhance atomization lf liquid jets,[24,25] and in drilling fluid displacement, above its yield stress, in drilling of water and oil wells.[26]

**Spiral flow.**   Superposition of a radial flow of velocity $u_r = \dot{V}/2\pi r$, and a torsional flow of velocity $u_\theta = \Omega r$, results in a vortex flow of velocity

$$\mathbf{u} = u_r\mathbf{e}_r + u_\theta\mathbf{e}_\theta = \frac{\dot{V}}{2\pi r}\mathbf{e}_r + [\Omega r]\mathbf{e}_\theta, \tag{6.57}$$

as shown in Fig. 6.7. Spiral flows are good prototypes of tornados and vortices.[27]

The pressure of multidirectional flows is obtained from the momentum equations given the velocity profiles. In general, it is not a simple superposition of the pressures of the superimposed unidirectional flows.

## 6.7 TRANSIENT FLOWS

Flows considered so far were at steady state, and therefore all time derivatives of velocity in the momentum equation were dropped. This leads to *ordinary differential equations* (ODEs) of the velocity profile $u_x(y)$ with respect to position $y$ to be solved, subject to the appropriate boundary conditions. In flow situations where the velocity may evolve with time, the time derivative cannot be eliminated from the momentum equation. The momentum equation threrfore becomes a *partial differential equation* (PDE) of the velocity $u_x(y, t)$ with respect to position $y$ and time $t$, subject to appropriate boundary and initial conditions. The solution of a PDE yields not only the spatial distribution of the velocity alone, but also how the velocity changes at each position with time. Such transient flow situations include flow development from rest to steady state, flows driven by time-varying pressure gradient or boundary velocity, unstable flows and wavy films, and others. Most of these flows are characterized by generation of vorticity at a boundary set in relative motion with respect to the adjacent fluid. Then the spatial velocity derivative grows progressively from the boundary to the interior of the flow, and so does vorticity according to Eqn. (5.40). It can be shown by dimensionless analysis (Section 1.7.2) and by analytic solution (Example 6.13) that after time $t$ from the moment the boundary is set to motion, vorticity penetrates by diffusion at distance $\delta(t)$, given by

$$\delta(t) = 2\sqrt{vt}, \tag{6.58}$$

where $v = \eta/e$ is the kinematic viscosity. Equation (6.58) also characterizes heat and solute penetration distances by diffusion from their generation sources[9] if $v$ is replaced by the thermal diffusivity $\alpha = k/\rho c_p$ or the mass diffusivity $\mathcal{D}$, respectively. For example, consider a fluid of thermal diffusivity $\alpha$ at initial temperature $T(t = 0) = T_0$ that is suddenly brought in permanent contact with a hot plate at temperature $T(x = 0, t > 0) = T_p > T_0$. After time $t$, heat by conduction will penetrate distance $\delta(t)$ given by Eqn. (6.58) with $v$ replaced by $\alpha$, which causes the local temperature to change from $T_0$ to $T_0 + \varepsilon T_p$, where $\varepsilon$ is a fraction that allows detection of temperature variations. Thus, the three processes of vorticity diffusion, heat conduction, and solute diffusion are dynamically similar and are governed by similar PDEs.

A generic momentum equation of a transient unidirectional flow is

$$\rho \frac{\partial u_i(t, x_j)}{\partial t} = -\frac{dp(t)}{dx_i} + \eta \frac{\partial^2 u_i(t, x_j)}{\partial x_j^2} + \rho g_i. \tag{6.59}$$

The first term represents velocity changes with time and is zero at steady-state situations. The second term represents a time-dependent pressure gradient, which is zero in transient and steady Couette flows and constant in steady pressure Poiseuille flows. Appropriate boundary conditions may include

$$u_i(x_j = 0, t) = V_0 \qquad [V_0 = 0 \text{ for stationary boundary}]$$
$$u_i(x_j = H, t) = V_1(t) \qquad [V_1(t) = V_1 = 0 \text{ for Poiseuille flow}]. \tag{6.60}$$

The system closes with the specification of an initial reference state that the system starts to evolve from. In case of initiation of flow from rest, the appropriate initial condition is

$$u_i(x_j, t = 0) = 0, \tag{6.61}$$

which simply states that at rest the velocity is zero everywhere.

Thus, the transient character of the flow is due to either a time-dependent pressure driving the flow or a boundary set in motion forcing the fluid to move from its initial condition. Following the motion imposed by these conditions that are incompatible with the initial state, the fluid has to evolve to a new state, by developing time-dependent velocity and velocity gradients. Notice that Eqns. (6.59) to (6.61) may represent transient plug, Couette, Poiseuille, and film flows, conforming to a Cartesian system of coordinates. Similar equations describe the corresponding flows conforming to the cylindrical system of coordinates, obtained by adding the velocity time derivative to the already derived steady-state momentum equations.

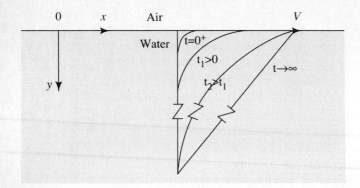

**Figure E6.13**  Initiation of Couette flow in a deep lake by horizontal wind.

The solution of Eqns. (6.59) to (6.61) is much more involved than that of the corresponding steady-state equation, even for the simple case of startup of Couette flow from rest examined in Example 6.13. Solutions to other transient flow situations is studied in more advanced fluid mechanics textbooks.[28]

### Example 6.13: Initiation of Couette Flow

The water of a deep lake of viscosity $\eta$ and density $\rho$ is everywhere at rest. Suddenly, a horizontal wind blows at constant speed $V$, which induces a relative motion within the lake water without altering the planar free surface. Describe the velocity profile development within the lake, extending from the free surface where the velocity is $V$, down to very great depth where the water is practically at rest at all times.

**Solution**  The flow situation and the appropriate system of coordinates are shown in Fig. E6.13. The governing transient momentum and boundary and initial conditions are

$$\frac{\partial u_x(y, t)}{\partial t} = v \frac{\partial^2 u_x(y, t)}{\partial y^2}, \qquad 0 < y < \infty, \quad 0 < t < \infty \qquad (6.13.1)$$

$$u_x(y = 0, t > 0) = V, \qquad u_x(y \to \infty, t) = 0 \qquad (6.13.2)$$

$$u_x(y, t = 0) = 0. \qquad (6.13.3)$$

The qualitative features of the solution sought, also shown in Fig. E6.13, are those of a vorticity diffusion away from the free surface, according to Eqn. (6.58): There is a relation between the lapse time $t$ and the distance $y$ penetrated by vorticity within that time, given by

$$\frac{y}{\sqrt{vt}} = \xi \qquad (6.13.4)$$

which means that the velocity at a point $y_1$ in time $t_1$ will be identical to that of point $y_2$ at time $t_2$, inasmuch as

$$\xi = \frac{y_1}{\sqrt{vt_1}} = \frac{y_2}{\sqrt{vt_2}}. \qquad (6.13.5)$$

Therefore, it is reasonable to assume that the transient velocity profile can be expressed in terms of

$$\xi = \frac{y}{\sqrt{vt}}, \qquad (6.13.6)$$

which incorporates $y$ and $t$ in a combination (suggested by the physics of the flow) rather than on each of $y$ and $t$ separately. Thus

$$u_x(y, t) = u_x(\xi). \tag{6.13.7}$$

The partial derivatives involved in Eqn. (6.13.1) are turned to ordinary derivatives with respect to $\xi$, by the chain rule of differentiation

$$\frac{\partial u_x}{\partial t} = \frac{du_x}{d\xi} \frac{d\xi}{dt} = \frac{du_x}{d\xi}\left(-\frac{y}{2\sqrt{vt^3}}\right)$$

$$\frac{\partial^2 u_x}{\partial y^2} = \frac{\partial}{\partial y}\left(\frac{\partial u_x}{\partial y}\right) = \frac{\partial}{\partial y}\left(\frac{du_x}{d\xi} \frac{1}{\sqrt{vt}}\right) = \frac{d^2 u_x}{d\xi^2} \frac{1}{vt}. \tag{6.13.8}$$

These derivatives are substituted in Eqn. (6.13.1), which results in the ordinary differential equation

$$\frac{d^2 u_x}{d\xi^2} + \frac{\xi}{2} \frac{du_x}{d\xi} = 0. \tag{6.13.9}$$

The boundary and initial conditions are turned now to boundary conditions with respect to $\xi$: The first boundary condition in Eqn. (6.13.2) is equivalent to

$$u_x(\xi = 0) = V. \tag{6.13.10}$$

The second boundary condition in Eqn. (6.13.2) and the initial condition by Eqn. (6.13.3) are represented by the unique condition

$$u_x(\xi \to \infty) = 0. \tag{6.13.11}$$

Thus, by introducing the new independent variable $\xi$, the initial PDE and boundary and initial conditions, Eqns. (6.13.1) to (6.13.3), have been transformed to an ODE and boundary conditions, Eqns. (6.13.9) to (6.13.11), which are nearly trivial to solve. This transformation, called a *similarity transformation*,[29] applies to PDEs on unbounded domains governing momentum or vorticity (and heat and solute) diffusion, and to boundary layers when fluids overtake submerged plates, as discussed in Chapter 7. A precondition for such a transformation to apply is the combination of the initial and one of the boundary conditions to a unique transformed boundary condition [e.g., Eqns. (6.13.2) and (6.13.3) to Eqn. (6.13.11)].
    The solution to Eqn (6.13.9) is obtained by rearranging to the form

$$2\frac{d}{d\xi}\left(\ln\frac{du_x}{d\xi}\right) + \xi = 0, \tag{6.13.12}$$

which is integrated twice to the general solution,

$$u_x(\xi) = c_1 \int_0^\xi \exp\left(-\frac{\xi^2}{2}\right) d\xi + c_2. \tag{6.13.13}$$

Application of the two boundary conditions by Eqns. (6.13.10) and (6.13.11) results in

$$\frac{u_x(\xi)}{V} = 1 - \frac{\int_0^\xi \exp(-\xi^2/2)\,d\xi}{\int_0^\infty \exp(-\xi^2/2)\,d\xi}$$

(6.13.14)

$$\equiv 1 - \frac{2}{\sqrt{\pi}}\int_0^\xi \exp\left(-\frac{\xi^2}{2}\right)d\xi = 1 - \mathrm{erf}\left(\frac{\xi^2}{2}\right) = 1 - \mathrm{erf}\left(\frac{y}{\sqrt{4\nu t}}\right),$$

which is known as the *error function,* arising in many physics and engineering applications, and tabulated in common math handbooks.[30]

From these tabulated values, the velocity profiles of Fig. E6.13 are validated. The velocity $u_x$ reaches $0.01V$ (i.e., the fluid feels the motion) for $\xi = 4$. Thus the vorticity penetration distance defined by Eqn. (6.57) is quantified for this particular flow as

$$\delta(t) = 4\sqrt{\nu t}.$$

(6.13.15)

According to this equation, at time $t = 10\,\mathrm{s}$ after inception of motion at velocity $V$ at the free surface, the velocity reaches $0.01V$ at depth

$$\delta(t = 10\,\mathrm{s}) = 4\sqrt{\nu t} = 4\sqrt{0.01 \times 10} = 1.42\,\mathrm{cm}.$$

For penetration at distance $\delta(t) = 10\,\mathrm{m}$, the time required is

$$t = \frac{\delta^2}{16\nu} = 66.6\,\mathrm{s}.$$

Flow initiation in bounded fields gives rise to linear PDEs that are solvable by *the separation of varibles technique.*[30,31] Figure 6.8 shows velocity development profiles from rest, in Couette and plug flows, due to one or both of the bounding plates set in motion, and in Poiseuille flow due to application of a constant pressure gradient.

Figures E6.13 and 6.8 show that vorticity (end shear stress) propagate through incompressible fluids with finite velocity. It takes a time $t$ for propagation within distance $\delta(t) = c\sqrt{\nu t}$, and in fact for $\nu = 0$, there is no propagation at all, as would be the case in an inviscid gas.

In the radial single-phase flows examined in this chapter, the velocity profile is determined by the continuity equation, which for incompressible fluids is

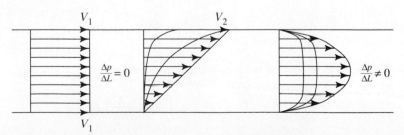

**Figure 6.8**   Initiation of plug, Couette, and Poiseuille flows from rest.

common for steady and transient flows. Any time dependence would be due to boundary motion, felt instantaneously by the entire fluid body. In the bubble growth case, for example, the velocity is

$$u_r(r, t) = \frac{dR}{dt}\frac{r^2}{R^2}, \tag{6.62}$$

and the interface–boundary velocity $v = dR/dt$ is transmitted instantaneously everywhere. The same is true for sink and source flows, where any time dependence comes from an instantaneously transmitted time-dependent flow rate:

$$u_r(r, t) = \frac{\dot{V}(t)}{2m\pi r^m}. \tag{6.63}$$

The corresponding pressure of these radial flows is calculated by the time-dependent $r$-momentum equation,

$$\rho\left(\frac{\partial u_r}{\partial t} + u_r\frac{\partial u_r}{\partial r}\right) = -\frac{\partial p(r, t)}{\partial r} + \rho g_r, \tag{6.64}$$

or alternatively, by the *transient Bernoulli equation,*

$$\Delta\left[\frac{\partial u_r}{\partial t} + \frac{u_r^2}{2} + \frac{p(r, t)}{\rho} + g_r z\right] = 0, \tag{6.65}$$

given the velocity profile $u_r(r, t)$. A prototype of transient torsional flow is described in Problem 6.38 and detailed elsewhere.[32]

## PROBLEMS

1. *Reactor design.* To design continuous flow reactors, the residence time $t_r$ of the reactant in the reactor is of importance because it partially controls the conversion to products. Calculate and plot the residence time achieved as a function of the distance from the symmetry line, $r$ or $y$, for the following reactors: (a) a plug reactor of length $L$, radius $R$, and flow rate $Q$; (b) a pipe reactor of length $L$, radius $R$, and flow rate $Q$; (c) channel-type reactor of length $L$, gap width $2R$, and flow rate per unit width $Q/W$. Calculate the pressure gradient, $\Delta p/\Delta L$, required to drive each of the flows, and then power $P$ required. The reacting mixture has viscosity $\eta$ and density $\rho$ everywhere. Which of the three types of flow yields the most uniform concentration of products?

2. *Viscosity measurement by Couette flow.* The design of this viscometer is based on two parallel plates of dimensions $d \times l$ separated by distance $s$. The liquid of unknown viscosity $\eta$ fills the gap, and the force $F$ on the upper plate, to yield flow rate $Q$, is measured. The flow rate $Q$ is also measured.
   (a) Find the working equation for the viscosity $\eta$ for such a rheometer.
   (b) Construct an appropriate diagram for viscosity reading.
   (c) What is the vorticity of the flow? Is the flow inviscid or irrotational?

3. *Transfer of highly viscous crude oil.* In laminar pipe flows, the required pumping power is $\dot{P} = Q\,\Delta p = QQL(128\eta/D^4) = Q^2 L(128\eta/D^4) = (128\eta Q^2 L/D^4)$. When the viscosity $\eta$ is large, as in the case of crude oil, huge amounts of energy are required,

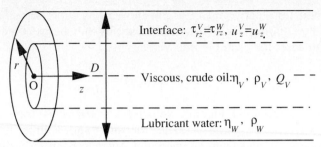

**Figure P6.3**  Lubricated flow in pipe.

which would raise the price for the consumer. To avoid this transferring cost, some oil companies use what is called *lubricated flow*; a lubricant of low viscosity, often water, is injected and flows as a thin film adjacent to the pipe wall, to reduce the friction between the viscous crude oil and the wall.[33] This reduces pressure drop, which results in less power required. The two-layer flow is shown in Fig. P6.3. The pressure is uniform across and linear along the pipe. It is necessary to transfer $Q$ volume of oil per day.

  **(a)** Estimate the power savings, as a function of $Q$, achieved over transfer without lubrication. (*Hint:* The pressure gradient $\Delta p / \Delta L$ will change.)

  **(b)** Show that in the appropriate limit, the two-layer velocity reduces to single-layer pipe flow.

  **(c)** Examine the vorticity distribution across the interface and comment on its form.

  **(d)** Which variables are continuous across the interface? Which are discontinuous?

 **4.** *Recirculation flow.* A reactant is preheated by counterflow of air before entering a reactor at temperature $T_0$, which must always be kept below its degradation temperature $T^*$, by the arrangement shown in Fig. P6.4. The known data are the physical characteristics of the liquid, $\eta$ and $\rho$, the air speed $V$, the film thickness $\delta$, the angle $\phi$, and the temperatures $T_0$ and $T^*$. Suddenly, during operation, the flow of the reactant ceased completely due to an observed accidental increase of the air speed. The production engineer is ordered to shut off the air supply immediately.

  **(a)** Explain what happened to the flow and sketch the streamlines of the flow at that time.

  **(b)** Was his action appropriate? Why?

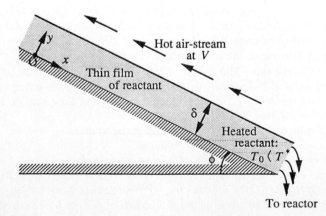

**Figure P6.4**  Heat exchange between falling liquid film and ascending hot air.

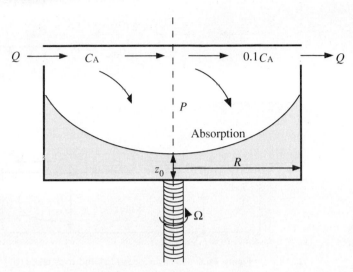

**Figure P6.5**  Absorption reactor.

**(c)** To avoid similar episodes in the future, he equipped the system with a control device to shut off air when the reactant flow rate vanishes. Can you propose a better solution to avoid disruption of process and to save startup cost?

**5.** *Absorption reactor.* The device shown in Fig. P6.5 takes advantage of the extended free surface during rotation to absorb reactant from the airstream of flow rate $Q$ and concentration $c_A$. What should be the rotation rate $\omega$ in order to reduce the exit concentration to $0.1c_A$ if there is absorbtion of $c_A^*$ per unit free surface per time under pressure $p$? How does the concentration at the exit depend on $\omega$, $Q$, and $p$? Will the presence of surface tension, $\sigma$, assist a further reduction of $A$ at the exit?

**6.** *Couette viscometer.* A Couette viscometer consists of an inner cylindrical rod of radius $R_0$ rotating at speed $\Omega$, and a stationary concentric outer cylinder of radius $R_1 > R_0$. A Newtonian liquid of unknown viscosity $\eta$ fills the gap. If the definition of viscosity is $\eta \equiv \tau_w/\dot{\gamma}$, where $\dot{\gamma}$ is the wall shear rate and $\tau_w$ is the wall shear stress, show that the only additional measurement required to determine the viscosity is the torque per unit length of wetted cylinder $T/L$, needed to maintain a constant rotation at angular speed $\Omega$.

**7.** *Superposition of flow*
  **(a)** (*helical flow*). Show that superposition of Couette flow between concentric cylinders of radii $R_0$ and $R_1$ initiated by boundary velocity $V$ in the axial direction, and of torsional flow initiated by rotation of the inner cylinder at angular speed $\Omega$, results in a helical flow. Then calculate the pressure distribution. Does the pressure distribution follow the superposition principle?
  **(b)** (*spiral flow*). It has been suggested that the velocity of a spiral flow is simply the summation of a gravitational and a torsional Couette flow under the same boundary conditions, i.e.,

  $$\mathbf{u}_T = u_G \mathbf{e}_z + u_C \mathbf{e}_\theta = -\frac{\rho g}{\eta}(r^2 - R^2)\mathbf{e}_z + \omega r \mathbf{e}_\theta + \theta \mathbf{e}_r,$$

  where $\mathbf{e}_z$, $\mathbf{e}_\theta$, and $\mathbf{e}_r$ are the unit vectors of the cylindrical system (in a fashion similar to $\mathbf{i}$, $\mathbf{j}$, and $\mathbf{k}$ for the Cartesian system). Prove that the suggestion is true.

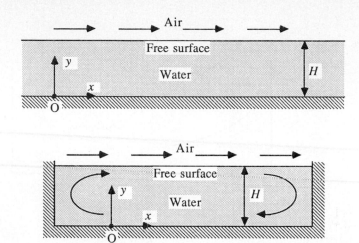

**Figure P6.8**    Air over ocean (a) and over lakes (b).

Then calculate the pressure distribution, and examine whether it follows the superposition principle.

8. *Air over ocean and over lakes* (*Couette vs. Couette–Poiseuille flow*). Air, blowing over infinitely long and wide oceans and over finite bounded lakes and ground cavities with water of viscosity $\eta$ and density $\rho$, is modeled as shown in Fig. P6.8. Under the assumption of planar interface:
   (a) Find the velocity distribution in the two cases.
   (b) Sketch the streamlines.
   (c) What is the stress by the air on the free surface? (Notice that unlike stationary air, the free surface stress is not zero here!)
   (d) What amount of liquid per time is relocated in the two cases?
   (e) What is the stress on the bottom? (Often, bacteria attached to the bottom respond to shear stress by growth and activity.)
   (f) Compare the mixing ability of the two cases in connection with their vorticity distribution (e.g., transfer of oxygen to the bottom).
   (g) Compare these two natural flows to those in gear pumps and in screw extruders.

9. *Recirculating water.* A fascinating exhibition in science museums features water falling on the outside of a tube hanging from the ceiling, but never detaching from it, as shown by Fig. P6.9. In reality, a hidden pump at the upper invisible end of the tube sucks the falling water inside the tube and then releases it on the outside of the tube in order to repeat the cycle. Design a pump to maintain a thin film of water thickness $H = 0.2$ cm on the outside of a tube of internal diameter $d = 5$ cm and wall thickness $\varepsilon = 0.01$ cm and of total length $L = 0.5$ m. Assume that end effects and flow rearrangement there are negligible.

10. *Draining system.* Roof draining systems (number and size of drains) are designed to maintain a minimum water load on the roof under the most intense rainfall. A typical house has height $H = 5$ m and its roof has dimensions $6\,\text{m} \times 12\,\text{m}$. During the heaviest rainfall, rain precipitates at the rate $\hat{Q} = 0.02\,\text{m}^3/\text{m}^2\,\text{min}$. Design a reasonable draining system that will maintain a load of no more than $W = 3$ tons on the roof at any time. What are the resulting flow rates and the momentum transfer at ground impingement? Based on the latter value, redesign the system, if necessary.

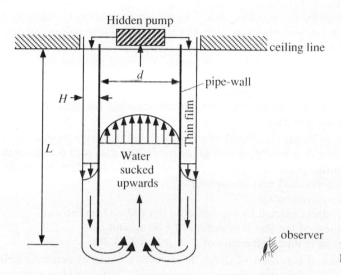

**Figure P6.9**   Water that turns back.

11. *Design of a funnel.* A funnel is to be designed to feed a viscous liquid of density $2 \, \text{g/cm}^3$ and viscosity $\eta = 2800 \, \text{P}$ to a reactor at flow rate $\dot{m} = 10 \, \text{kg/min}$. For a funnel with a stem of diameter $d = 0.3 \, \text{m}$, find suitable ratios of $H/L$ to steadily maintain the flow rate, by gravity alone. Assume that the process is fully controlled by the tube flow. If the resulting ratio $H/L$ appears impractical, what alterations would you implement to come up with a more suitable $H/L$ ratio?

12. *Couette viscometer.* A solid cylinder of radius $R_0 = 1.5 \, \text{cm}$ and height $h = 5 \, \text{cm}$ rotates coaxially inside a fixed cylinder of the same length and of radius $R_1 = 2 \, \text{cm}$. A torque of $10 \, \text{dyn cm}$ is applied to the inner cylinder. Calculate the resulting wall shear rate and the angular speed for water of viscosity $\eta_W = 1 \, \text{cP}$ and glycerin of viscosity $\eta_G = 10 \, \text{P}$. For the two cases, what is the required power which is dissipated by fluid friction?

13. *Capillary rheometer.* Besides the design characteristics (i.e., the radius $R$ and the length $L$), the other measurable quantities are the flow rate $Q$ and the pressure drop $\Delta p$. Show how the shear rate and shear stress at the solid wall can be deduced from $R$, $L$, $Q$, and $\Delta p$ in order to calculate the viscosity $\eta$. Then calculate the viscosity of a liquid that is being forced through a capillary of radius $R = 3 \, \text{mm}$ and length $L = 0.5 \, \text{m}$ by a pressure differential $\Delta p = 20 \, \text{psi}$, which yields the flow rate $Q = 30 \, \text{cm}^3/\text{min}$.

14. *Annular rheometer.* The capillary of Problem 6.13 is replaced by an annulus of radii $R_0$ and $R_1$, and of length $L$. The pressure differential $\Delta p$ to yield a given flow rate $Q$ of a liquid of unknown viscosity is measured. Can the unknown viscosity be estimated from the known values of $R_0$, $R_1$, $l$, $\Delta p$, and $Q$?

15. *Power to move liquids.* Calculate and plot the required power as a function of the hydraulic diameter, the viscosity, and the pressure drop to move liquids in a pipe, an annulus, and a channel of sectional aspect ratio $10:1$. Show that the two expressions $P = Q \, \Delta p$ and $P = Fu$ are equivalent. (Here $P$ is power, $F$ is frictional force, and $u$ is average velocity.)

16. *Free-falling round jet.* An infinitely long round axisymmetric jet of radius $R$ of liquid of viscosity $\eta$ and density $\rho$ falls under gravity in contact with counterflowing air that applies a constant shear stress $\tau_0$ at the free surface.
    **(a)** Show that if $\tau_0 = 0$, the diameter of the jet $2R(z)$ cannot be constant. What is the

resulting thickness under finite values of $\tau_0$? How are these answers different if there is, in addition, surface tension $\sigma \neq 0$? For the remaining questions, assume that $\sigma = 0$ and $\tau_0 = k \neq 0$.

**(b)** What are the velocity and stress distribution within the round jet? Plot it.

**(c)** What is the flow rate? Under what conditions does the flow rate vanish?

**(d)** What is the (force exerted on) the ground?

**(e)** Sketch the pathline of a neutrally buoyant particle entering the cylindrical free surface under zero net flow rate.

17. *Axisymmetric film flow.* A Newtonian liquid of viscosity $\eta$ and density $\rho$ flows down a vertical solid rod of radius $R$ at total flow rate $Q$ (m³/min). The film is in contact with stationary air of uniform pressure. Under these conditions, the film is strictly axisymmetric and of uniform thickness throughout.

**(a)** Calculate the velocity distribution.

**(b)** Calculate the shear stress exerted by the liquid on the rod and on the air.

**(c)** Calculate the thickness of the film $\delta$ in terms of $Q$, $R$, $\eta$, and $\rho$.

**(d)** How is the analysis simplified in the case of $\delta \ll R$ and $\delta \gg R$?

18. *Centrifugation.* A cylindrical container of radius $R$ contains a liquid of density $\rho$ and viscosity $\eta$ and rotates about its axis of symmetry at angular velocity $\Omega$, where the height of the liquid at the wall is $H$.

**(a)** Sketch the flow situation at rotation.

**(b)** Calculate the distribution of the azimuthal velocity $u_\theta$ as a function of $r$.

**(c)** Calculate the pressure distribution $p(r, z)$.

**(d)** Show that the configuration you assumed in (a) is correct. What is the height of the liquid at the axis of rotation?

19. *Annular two-phase flow.* Calculate the resulting film thickness of a vertical two-phase laminar flow of a water–steam system at 100°C flowing under gravity and opposing gas motion at flow rate $\dot{m}_G = 2\,\text{g/s}$ in a vertical pipe of 10 cm diameter. Then plot the resulting velocity distribution within the thin film for gas flow rates up to 10 g/s and comment on the resulting profiles.

20. *Film thickness of steam condensation.* Figure 4.14 may represent ascending steam that condensates and forms a thin liquid film falling along the wall under the influence of gravity and the opposing interfacial stress $\tau_i$, due to an upward pressure gradient, $\Delta p / \Delta z = -100\rho_G g$. Calculate the thickness of the resulting film in the case of zero net water flow rate (i.e., when the mass flow rate of the ascending steam is identical to that of the falling liquid water). Consider a temperature of 100°C throughout, uniform film thickness everywhere, and steady-state conditions. Under what assumptions do these conditions exist, or can they?

21. *Liquid film down exterior cylindrical wall.* The following equation is derived by momentum balance over an annular system, along the lines of the analysis in Section 6.3.2.

$$u_L = \frac{g\rho_L}{4\eta_L}(R^2 - r^2) + \frac{g\rho_L}{2\eta_L}(R + \delta)^2 \ln \frac{r}{R}.$$

It represents the velocity distribution in a liquid film of thickness $\delta$ falling along the outside cylindrical wall of a tube of radius $R$, in contact with condensating steam, common in industrial heat exchange applications. Find the equation that predicts the uniform film thickness $\delta$ in terms of liquid mass flow rate $\dot{m}_L$ under laminar conditions. Then apply to a water system at 100°C, for $\dot{m}_L = 0.05\,\text{kg/s}$ and $R = 1\,\text{in.}$ and compare the results to those of Example 6.9.

**22.** *Thin film flow over inclined plates.* Prove the following expressions for thin-film flow of thickness $H$ over inclined surface of width $W$ and inclination angle $\phi$ with the horizontal:

**(a)** The mess flow rate is

$$\Gamma = \frac{\dot{m}}{W} = \frac{H^3 \rho^2 \cos \phi}{3\eta}.$$

**(b)** The film thickness over vertical surface of width $W$ is

$$H = \sqrt[3]{\frac{3\eta\Gamma}{\rho^2 g}}.$$

**(c)** The Reynolds number is

$$\mathrm{Re} = \frac{4T}{\eta}.$$

**(d)** Sketch the free surface of the film for $\mathrm{Re} < 25$, where the film is everywhere uniform; for $25 < \mathrm{Re} < 2000$, where the film develops ripples; and for $\mathrm{Re} > 2000$, where the film becomes turbulent.

**23.** *Thin-film flow over vertical cylinders*

**(a)** What are corresponding expressions for film flow on the exterior of a vertical cylinder of external cylindrical surface of area $W$, identical to that of Problem 6.22?

**(b)** What are the corresponding expressions for film flow over the interior cylindrical surface of a vertical pipe of diameter identical to the cylinder of part (a)?

**24.** *Vertical annular flow.* Consider vertical flow under gravity in an annulus of gap $H$ identical to the film thickness of Problem 6.22. Calculate the mass flow rate and the Reynolds number when the internal wall cylindrical surface is $0.1W$, $W$, and $10W$. How do they compare with those of Problem 6.22? Explain the differences or similarities between the two.

**25.** *Storage of natural gas.*[5] Natural gas of compressibility factor $z = 0.8$, average molecular weight $M = 16$, and viscosity 0.013 cP is to be stored in a spherical porous reservoir of permeability 100 mD, average porosity $\varepsilon = 0.20$, and radius 1 km, in contact with impermeable rock. The injection pressure is 2500 psia and pressure drop in the vertical well is negligible.

**(a)** What is the resulting storage flow rate time, and how long does it take for the gas to reach the rock?

**(b)** If the initial temperature of reservoir and rock is 130°F and the gas is injected at temperature 60°F, what is the resulting temperature distribution at the time of gas–rock contact?

**26.** *Storage of propane.* Repeat the considerations of Problem 6.25 for propane of average molecular weight $M = 44$, compressibility 0.85, and viscosity 0.02 cP.

**27.** *Storage of natural gas in cylindrical reservoir.* Repeat the considerations of Problem 6.25 for a cylindrical reservoir of the same volume and radius-to-height ratio 5:1.

**28.** *Natural gas discharge from reservoir.*[5] Natural gas stored under the conditions of Problem 6.25 is to be connected to a city gas network.

**(a)** What is the required rejection pressure to maintain gas flow rate 2000 Mft³/day?

**(b)** How long does it take for compete evacuation of gas?

**(c)** Usually, natural gas of molecular weight $M = 16$ is brought inside buildings at

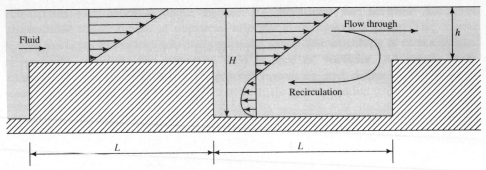

**Figure P6.29**  Slider-pad extruder.

low pressure, typically around 4 in. of water, whereas propane is brought at 11 in. of water. Give an explanation for the difference.

29. *Slider-pad extruder.* The slider pad shown in Fig. P6.29 is an extreme extension of the non-parallel-plate concept for pressurization and delivery of highly viscous liquids.[9] The arrangement is also used for hydrodynamic lubrication because of its good load capacity.

   **(a)** Analyze the flow—wherever possible—of a Newtonian liquid of density $\rho$ and viscosity $\eta$ if a flow rate $Q$ is to be delivered.

   **(b)** Calculate the load capacity, $LC = \int_0^{2L} (p - p_0)\, dx$, of such a device.

   **(c)** Apply to pressurization of a Newtonian polymeric melt of viscosity $\eta = 10^2\,\text{P}$ at delivery $Q = 0.1\,\text{ft}^3/\text{s}$ per unit width by a device of $H = 5\,\text{cm}$, $h = 0.5\,\text{cm}$, $L = 10\,\text{cm}$, and $W = 30\,\text{cm}$.

   **(d)** Where and what are the values of the maximum and minimum pressures, excluding the corner regions? How are the two compared to the pressures of the corner regions?

30. *Flow in an idealized runner system.* Runners are used to feed injection molding machines[9] (Fig. P6.30). Consider a straight tubular runner of length $L$. A melt of constant viscosity $\eta$ is injected at constant pressure into the runner. The melt front progresses along the runner until it reaches the gate that feeds the mold. Under these

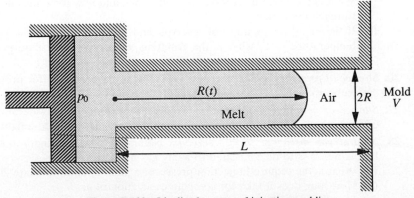

**Figure P6.30**  Idealized runner of injection molding.

conditions the pressure at the inlet of the runner is constant, $p_0$, and at the advancing front is identical to the pressure of the air, i.e., $p = 0$, and therefore the flow rate, $Q(t)$, is a function of time.

(a) Calculate the melt front position and the instantaneous flow rate as a function of time.

(b) How long does it take for the melt to fill first the runner and then the mold of volume $V$?

(c) Complete parts (a) and (b) for a melt of viscosity $\eta = 10^3\,\mathrm{P}$, under pressure $p_0 = 10^6\,\mathrm{N/m^2}$, in a runner of dimensions $R = 5\,\mathrm{cm}$ and $L = 3\,\mathrm{ft}$ which delivers to a mold of volume $1\,\mathrm{m} \times 0.5\,\mathrm{m} \times 0.1\,\mathrm{m}$.

**31.** *Flow in a runner under constant delivery.* The runner of Problem 6.30 is also operated at constant flow rate, $Q$, in which case the pressure at the inlet, $p_0(t)$, is a function of time.

(a) For the runner of Fig. P6.30, calculate how $p_0(t)$ changes with time.

(b) How long does it take to fill the runner and the mold?

(c) Apply to a melt of viscosity $\eta = 10^4\,\mathrm{P}$, under flow rate $Q = 100\,\mathrm{cm^3/s}$, in a runner of dimensions $R = 5\,\mathrm{cm}$ and $L = 3\,\mathrm{ft}$, which delivers to a mold of volume $1\,\mathrm{m} \times 0.5\,\mathrm{m} \times 0.1\,\mathrm{m}$.

**32.** *Combined, spherical radial flows.* Two porous spheres of radii $R$ and $2R$ have their centers a distance $10R$ apart (Fig. P6.32). They generate radial streams of water of total flow rates $Q$ and $2Q$, respectively.

(a) What are the radial velocity profiles, as a function of the radial distance from the centers, in the vicinity of each sphere? (Assume for the moment that each flow pattern can be treated separately.)

(b) Sketch the streamlines, now taking into account the interaction in the intervening space. On which streamline can the velocity be zero at a point?

(c) Where is this stagnation point? That is, where is the composite velocity zero?

(d) If the pressure at the stagnation point is $p^*$, what are the pressures $p_1$ and $p_2$ at the surfaces of each sphere?

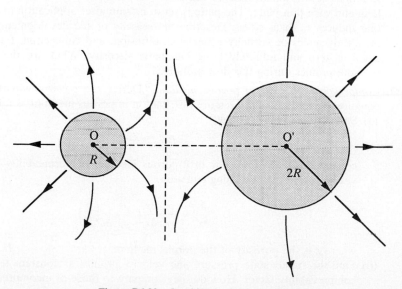

**Figure P6.32**   Combined spherical radial flows.

33. *Artificial porous medium.* Consider an artificial poroous medium made up by $n$ parallel tubes of diameter $d$ and length $L$, cemented together with unpenetrable material such that there are $n$ tube cross sections per square meter of the porous medium. By using Darcy's law for flow through this porous medium and the Poiseuille flow in each individual pipe, find a relation between the permeability of the porous medium and the geometric characteristics of the tubes. Comment on the physical significance of permeability, based on the relation derived.

34. *Spherical porous medium.* Consider an artificial porous spherical shell of thickness $H$ that incorporates $n$ rectilinear cylindrical pores of diameter $r$ and length $H$ pointing to the center of the corresponding sphere of radius $R$, such that a spherical sink flow of radius $R - H$ is enclosed by the shell. Consider sink radial flow through this artificial porous medium at total flow rate $\dot{V}$. By comparing the expressions for Darcy's radial flow in a porous medium and the corresponding expressions for flow through each individual cylindrical pore, derive a relation between the permeability of the artificial porous medium and its geometric characteristics, and comment on the physical significance of permeability.

35. *Bidirectional flow inside a porous cylinder.* Analyze the bidirectional flow inside a vertical pipe of infinite length and thin porous walls of radius $R$ from where the liquid escapes at flow rate $\hat{Q}$ m$^3$/m per meter length. The flow is driven upward by a pressure gradient $\Delta p/\Delta L$ opposing gravity, such that the flow is uniform along the pipe. To account for the effect of the porous wall to the inside flow, assume that he boundary condition to the inside flow is

$$\tau_{rz} = \beta \frac{\partial u_z}{\partial r},$$

where $\beta$ is a wall *slip coefficient.* Then examine limiting cases of $\beta = 0$ and $\beta = \infty$. To what physical situations do they correspond?

36. *Unsteady Couette flow.* Repeat the analysis of Example 6.13 for the following flow situation: A Newtonian fluid of viscosity $\eta$ and density $\rho$ is at rest over an infinitely long and wide thin plate. The plate is set in motion after application of a pulling force that induces $\tau_0$ shear stress. Describe the velocity profile development and comment on the dynamics of vorticity generation, diffusion, and convection. Examine limiting cases of zero and infinitely large kinematic viscosity. What are the corresponding pressure profiles during this flow initiation?

37. *Compressible gas flow in porous medium.*[5] Derive the governing partial differential equation for the transient pressure distribution in porous medium as follows:
    (a) Show that the real gas law,

$$pV = ZnRT,$$

combines with Darcy's law in a porous medium of permeability $\kappa$ for a gas of viscosity $\eta$ to the governing equation,

$$\frac{\partial}{\partial x}\left(\kappa \frac{p}{\eta Z}\frac{\partial p}{\partial x}\right) = \varepsilon \frac{\partial}{\partial t}\left(\frac{p}{Z}\right),$$

where $\varepsilon$ is the porosity of the porous medium.
    (b) Find the steady-state pressure and velocity profiles at constant temperature and compressibility factor. How do they compare to those of incompressible gas flow?
    (c) How could you solve the unsteady-state equation for constant $\kappa$, $\eta$, $\varepsilon$, and $Z$, for

startup of compressible flow in a seminfinite, one-dimensional reservoir, initially at static equilibrium?

**38.** *Unsteady torsional flow.*[32] Consider torsional flow initiation in an infinite pool of incompressible Newtonian liquid of viscosity $\eta$ and density $\rho$, which is at rest, due to a cylindrical rod that starts rotating with angular velocity $\Omega$ at time $t = 0$. Analyze the resulting transient flow, as follows:

(a) Show that the governing momentum equation is

$$\rho \frac{\partial u_\theta}{\partial t} = \eta \frac{\partial}{\partial r}\left[\frac{1}{r}\frac{\partial}{\partial r}(ru_\theta)\right],$$ (P6.38.1)

by adding the time derivative to Eqn. (6.34). By introducing the vorticity,

$$\omega = \frac{\partial u_\theta}{\partial r} + \frac{u_\theta}{r},$$ (P6.38.2)

show that the resulting vorticity equation is

$$\frac{\partial \omega}{\partial t} = \nu\left(\frac{\partial^2 \omega}{\partial r^2} + \frac{1}{r}\frac{\partial \omega}{\partial r}\right),$$ (P6.38.3)

and the boundary and initial conditions are

$$\omega(r, t = 0) = 0,$$
$$\omega(r = R, t > 0) = 2\Omega$$ (P6.38.4)
$$\omega(r \to \infty, t) = 0.$$

(b) Introduce the similarity variable,

$$\xi = \frac{r - R}{\sqrt{2\nu t}},$$ (P6.38.5)

and show that it transforms Eqns. (P6.38.2) to (P6.38.4) to the ordinary differential equation and boundary conditions

$$\frac{d^2\omega}{d\xi^2} + \xi\frac{d\omega}{d\xi} = 0$$

$$\omega(\xi = 0) = 2\Omega$$ (P6.38.6)

$$\omega(\xi \to \infty) = 0.$$

(c) By following the procedure of Example 6.13, show that the resulting solution is

$$u_\theta(r, t) = \frac{\omega(r, t)r}{2} = \Omega r\left[1 - \text{erf}\left(\frac{r - R}{\sqrt{4\nu t}}\right)\right].$$ (P6.38.7)

(d) What is the resulting vorticity penetration distance, and how does it relate to that of Example 6.13?

(e) How would you estimate the pressure distribution and evolution of free surface from planar to that of Fig. E5.8.5?

**39.** *Cessation of Couette flow.* Consider an ongoing Couette flow of Newtonian liquid of kinematic viscosity $\nu$ between two infinitely long and wide plates, the upper of which is moving with velocity $V$ relative to the lower stationary one. At time $t = 0$, the upper moving plate stops moving. Analyze the resulting velocity profile evolution to rest. [First write down the governing momentum equation and the boundary and initial conditions. Then introduce a new velocity $U(y, t) = u_x(y, t) - Vy/H$, which turns the system to that of Example 6.13.]

**40.** *Unsteady radial flow.* An elastic balloon of initial radius $R_0$ connected through a thin tube with a pressure control chamber is stationary within a large bath of Newtonian liquid of viscosity $\eta$ and density $\rho$. At time $t = 0$ the control chamber imposes such a pressure inside the balloon that the balloon starts growing according to

$$R(t) = R_0 \exp(-kt).$$

   **(a)** Calculate the velocity profile development within the surrounding liquid, from rest.
   **(b)** What is the corresponding pressure profile?
   **(c)** Comment on the speed of propagation of the normal velocity or stress imposed by the inflated balloon through the liquid, and compare with that for vorticity or shear stress in Example 6.13.

# REFERENCES

1. L. E. Scriven, On the dynamics of phase growth, *Chem. Engrg. Sci.,* 10, 1 (1959).
2. G. Pearson and S. Middleman, Elongational flow behavior of viscoelastic liquids, *AIChE J.,* 23, 714 (1977).
3. M. Muskat, *Flow of Homogeneous Fluids Through Porous Media,* McGraw-Hill, New York, 1937.
4. H. P. G. Darcy, *Les Fontaines publique de la Ville de Dijon,* Victor Dalmont (Ed.), Paris, 1856.
5. D. L. Katz and R. L. Lee, *Natural Gas Engineering,* McGraw-Hill, New York, 1990.
6. M. H. Cullender, The isochronal performance method of determining the flow characteristics of gas wells, *Trans. AIME,* 204, 137 (1955).
7. N. Tipei, *Theory of Lubrication,* Stanford University Press, Stanford, Calif., 1962.
8. G. K. Miltsios, D. J. Peterson, and T. C. Papanastasiou, Solution of the lubrication problem and calculation of friction of piston ring, *ASME J. Tribol.,* 111, 635 (1989).
9. Z. Tadmor and C. G. Gogos, *Principles of Polymer Processing,* Wiley, New York, 1979.
10. B. V. Deryagin and S. M. Levi, *Film Coating Theory,* Focal Press, New York, 1964.
11. S. S. Hwang, Hydrodynamic analysis of blade coating, *Chem. Engrg. Sci.,* 34, 181 (1979).
12. C. G. Heden, *Fermentation and Enzyme Technology,* Wiley, New York, 1979.
13. J. Y. Oldshue, Fermentation mixing scale-up techniques, *Biotechnol. Bioengrg.,* 8, 24 (1966).
14. A. N. Alexandrou and T. C. Papanastasiou, Nonisothermal extrusion of composite materials, *Internat. Pol. Proc.,* 5, 12 (1990).
15. V. Bontozoglou and A. J. Karabelas, Numerical calculations of simultaneous absorption of $H_2S$ and $CO_2$ in aqueous hydroxide solutions, *Ind. Engrg. Chem. Res.,* 30, 2603 (1991).
16. L. De Leye and G. F. Froment, Rigorous simulation and design of columns for gas absorption and chemical reaction, *Comput. Chem. Engrg.,* 10, 1245 (1980).

17. F. K. Wasden and A. E. Dukler, Insights into the hydrodynamics of free falling wavy films, *AIChE J.,* 35, 187 (1989).

18. T. D. Karapantsios and A. J. Karabelas, Surface characteristics of roll waves on free falling films, *Internat. J. Multiphase Flow,* 16, 835 (1990); Statistical characteristics of free falling films at high Reynolds numbers, *Internat. J. Multiphase Flow,* 15, 1 (1989).

19. F. K. Wasden, Studies of mass and momentum transfer in free falling wavy films, Ph.D. thesis, University of Houston, 1989.

20. R. B. Bird, R. C. Armstrong, and O. Hassager, *Dynamics of Polymeric Liquids,* Vol. 1, *Fluid Mechanics,* Wiley, New York, 1977.

21. M. M. Denn, *Process Fluid Mechanics,* Prentice Hall, Englewood Cliffs, N.J., 1980.

22. L. E. Scriven and C. V. Sternling, On cellular convection driven by surface-tension gradients: Effect of mean surface tension and surface viscosity, *J. Fluid Mech.,* 19, 321 (1964).

23. D. A. Edwards, H. Brenner, and D. T. Wason, *Interfacial Transport Processes and Rheology,* Butterworth-Heinemann, Boston, 1991.

24. W. R. Marshall, Atomization and spray drying, *Chem. Engrg. Prog. Monogr. Ser.,* 2, 50 (1954).

25. T. C. Papanastasiou, A. N. Alexandrou, and W. P. Graebel, Rotating thin films in bell sprayers and spin coating, *J. Rheol.,* 32, 485 (1988).

26. L. C. Walton and S. H. Bittleston, The axial flow of Bingham plastic in a narrow eccentric annulus, *J. Fluid Mech.,* 14, 347 (1984).

27. S. Eskinazi, *Fluid Mechanics and Thermodynamics of Our Environment,* Academic Press, New York, 1985.

28. R. L. Panton, *Incompressible Flow,* Wiley, New York, 1984.

29. A. G. Hansen, *Similarity Analysis of Boundary Value Problems in Engineering,* Prentice Hall, Englewood Cliffs, N.J., 1964.

30. I. S. Sokolnikoff, *Mathematics of Physics and Modern Engineering,* McGraw-Hill, New York, 1966.

31. H. F. Weinberger, *Partial Differential Equations,* Blaisdell, Waltham, Mass., 1965.

32. T. C. Papanastasiou, *Intermediate Fluid Mechanics,* Prentice Hall, Englewood Cliffs, N.J., 1994.

33. D. D. Joseph, Boundary conditions for thin lubrication layers, *Phys. Fluids,* 23 (1980).

# 7

# *Two-Dimensional Laminar Flows: Creeping, Potential, and Boundary Layer Flows*

## 7.1 INTRODUCTION

With the exception of multidirectional flows obtained by superposition of linear unidirectional flows, the flows analyzed in Chapter 6 are *unidirectional* (with just one nonzero velocity component) and *unidimensional* (velocity changes in one dimension only). Although these flows are adequate representations of many industrial applications, most flows are both multidimensional and multidirectional in a way that they cannot be composed by superposition. These include primarily free-stream flows overtaking submerged bodies—or, equivalently, flows induced by traveling bodies in stationary fluids—and confined flows bounded by complex geometries. Two representatives of these classes of flow are shown in Fig. 7.1. In Fig. 7.1a the flow is strongly multidirectional, due to the complex conduit of flow, and in Fig. 7.1b due to deflection of the approaching liquid by a three-dimensional obstacle.

For these flows it is impossible to find a stationary system of coordinates with respect to which there is everywhere only one nonzero velocity component. Still, the methods of analysis of these flows remain based on the same conservation principles, which are, however, more complex when applied to more than one direction with more than one velocity component. Fortunately, these principles lead to universally valid differential equations, which are tabulated and need not be derived each time a new flow situation is to be analyzed.

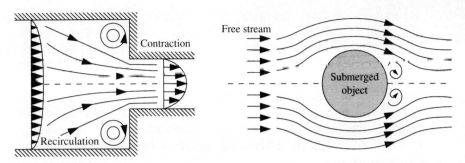

**Figure 7.1**  Multidimensional confined (a), and free-stream flow (b).

For Newtonian liquids, the constitutive equations of Appendix F apply. The conservation equations of mass and momentum, combined with the constitutive equations of Appendix F, lead to equations universally valid for Newtonian fluids, called *Navier–Stokes equations,* in honor of those who derived them.[1,2] These universal conservation equations are the most important and fundamental equations of fluid mechanics and related transport phenomena, since they allow the formulation and analytic or numerical solution of the vast majority of problems arising in physical rate processes, and provide the fundamentals for addressing the rest of them. The Navier–Stokes equations are tabulated in Tables G.1 to G.3 of Appendix G, for Cartesian, cylindrical, and spherical systems of coordinates, respectively. Overviews of the history of development of the Navier–Stokes equations and of alternative forms exist in several review publications.[3]

For non-Newtonian liquids,[4] the corresponding constitutive equations (which are different from Newton's law of viscosity of Appendix F) are used to eliminate the stresses from the corresponding equations, Eqns. (G.2) to (G.4), (G.9) to (G.11), and (G.16) to (G.18) of Tables G.1 to G.3. The resulting equations, which are different from the Navier–Stokes equations in general, govern the flow and processing of non-Newtonian liquids, which is the topic of Chapter 9.

The Navier–Stokes equations apply to laminar and turbulent flows of Newtonian fluids. However, in turbulent flows, due to chaotic and rapid changes of velocity with time, the forms of the equations in Tables G.1 to G.3 are inconvenient. For turbulent flows these equations are averaged over time, which leads to usable *time-smoothed equations,* as discussed in Chapter 10.

## 7.2 NAVIER–STOKES EQUATIONS

The *continuity equations* for an incompressible fluid, Eqns. (G.1), (G.8), and (G.15), are derived from mass balances over the control volume appropriate for each system of coordinates, penetrated by fluid flow with three mutually perpendicular velocity components, $(u_x, u_y, u_z)$, $(u_r, u_\theta, u_z)$, and $(u_r, u_\theta, u_\phi)$, respectively. It states that the volume occupied by a certain mass of incompressible fluid is constant in time and in space, independent of the prevailing local flow conditions. Any attempt to transfer any amount of additional fluid in that volume will result in the displacement of an equal amount of fluid out. The stages of derivation of these equations are those laid down in Chapter 5: positioning of an appropriate system of coordinates, choice of control volume, applications of mass conservation principles according to Eqn. (5.1), where the only mechanism of mass transfer is convection, and transformation of the resulting difference equation to a partial differential equation by taking the control volume as shrinking to zero. A flow represented by its velocity components is *dynamically admissible* (i.e., it can take place and be reproduced in the lab once technical difficulties are overcome) if the components satisfy the corresponding continuity equation.

The *momentum equations in terms of stresses* of Tables G.1 to G.3 are derived according to the principles laid down in Chapter 5: by momentum balances over the appropriate control volume in all the directions of the chosen system of coordinates, for fluid flow with three velocity components that give rise to all possible viscous stress components. These equations apply to any flow of any incompressible fluid, Newtonian or not. They simply state that gradients of pressure, viscous stresses, and body forces (e.g., gravity forces), represented by terms on the right-hand side of each momentum equation, force the fluid to deform and flow at a rate represented by the left-hand side. The *Navier–Stokes equations* are obtained from the momentum equations in terms of stresses, by replacing the viscuos stress components with the appropriate velocity gradients, according to Newton's law of viscosity, Appendix F. These equations are simplified further by utilizing the corresponding continuity equation, as demonstrated in Examples 5.4 and 5.11. The resulting equations apply to flows of incompressible Newtonian fluids alone.

All flows so far analyzed by control volume principles can be also addressed by first simplifying, and solving the appropriate equation(s) among those listed in Tables G.1 to G.3, as will be demonstrated in Section 7.3. For solution of flow problems by the equations of Tables G.1 to G.3, the following procedure is recommended:

1. Identify the geometry of the flow and select the system of coordinates that determines which of the three tables is to be used.
2. Once the appropriate table is selected, identify velocity components which are zero, as well as which derivatives of the nonzero velocity component(s) are zero. Check if the pressure is known a priori.
3. Simplify the continuity equation using the findings of item 2, and see if you can determine the velocity profile by solving this equation alone.
4. If not, select the appropriate momentum equation, usually the one in the main flow direction or direction of the pressure gradient, and use the findings of items 1 and 2 to find the velocity profile equation.
5. Apply the appropriate boundary and initial conditions to find the particular solution that fully determines the velocity profile.

**6.** With all the previous information at hand, solve the remaining momentum equations to calculate the pressure and other velocity components if present. Calculate secondary variables, for example stress, vorticity, and flow rate, as required.

**7.** Regarding the pressure, keep in mind that:

(i) the pressure in the Navier–Stokes equations is the total pressure that includes the hydrostatic and the extra due to flow,

(ii) the gravity components $q_i$ may be positive or negative, if the corresponding positive direction of $x_i$ of the pressure gradient $dp/dx_i$ is towards or away from the Earth's center.

More complicated, classes of two-dimensional laminar flows can also be addressed based on the Navier–Stokes equations by means of the *streamfunction* (e.g., Sections 7.4 and 7.5), by the *potential function* (e.g., Section 7.7), and by *boundary layer theory* (e.g., Section 7.8). Approximate *asymptotic solutions* to the Navier–Stokes equations can be constructed for *nearly unidirectional flows* that include *lubrication* and *stretching* flows, which is the topic of Chapter 8.

# 7.3 APPLICATION OF NAVIER–STOKES EQUATIONS TO UNIDIRECTIONAL FLOWS

Analytic solutions to unidirectional flows, obtained previously by control volume principles, can now be obtained by utilizing and then simplifying the appropriate Navier–Stokes equations. The reader is reminded that according to the principles of Chapters 5 and 6, it was concluded that for the steady unidirectional flows examined the convective terms vanish due to the geometry, and that there was a single velocity component identical to the streamline velocity. In the following examples, pipe, channel, radial, film, and torsional flows, analyzed in Chapters 5 and 6 by control volume methods, are revisited. The purpose is to derive the governing equation based now on the universal Navier–Stokes equations. The solution of the governing equation is done as demonstrated in Chapters 5 and 6.

**Example 7.1: Analysis of Steady Rectilinear Flows by the Navier–Stokes Equations**

Rectilinear flows are unidirectional flows having only one nonzero velocity component with respect to a Cartesian system of coordinates. Thus, Eqns. (G.1) to (G.7) apply, with

$$u_y = u_z = 0 \tag{7.1.1}$$

and

$$u_x = f(y). \tag{7.1.2}$$

Under these conditions, the continuity equation, Eqn. (G.1) becomes

$$\frac{du_x}{dx} = 0, \tag{7.1.3}$$

and the three momentum equations, Eqns. (G.5) to (G.7), become

$$-\frac{dp}{dx} + \eta \frac{d^2 u_x}{dy^2} = 0, \tag{7.1.4}$$

$$\frac{dp}{dy} = 0, \tag{7.1.5}$$

and

$$\frac{dp}{dz} = 0. \tag{7.1.6}$$

For Couette-type flows and planar film flows,

$$\frac{dp}{dx} = 0. \tag{7.1.7}$$

For Poiseuille-type flows,

$$\frac{dp}{dx} = \frac{\Delta p}{\Delta L}. \tag{7.1.8}$$

The velocity profile, $u_x(y)$, is calculated by Eqn. (7.1.4), given the pressure gradient, according to the procedure described in Chapter 6 for Eqn. (6.6).

### Example 7.2: Analysis of Steady Pipe Flow by the Navier–Stokes Equations

Pipe flow is axisymmetric; therefore, Eqns. (G.8) to (G.14) of Table G.2 apply, with

$$\frac{\partial}{\partial \theta} = 0. \tag{7.2.1}$$

In addition, the unidirectional nature of the flow requires that

$$u_r = u_\theta = 0 \tag{7.2.2}$$

and

$$u_z \neq 0. \tag{7.2.3}$$

Furthermore, for a fully developed flow,

$$u_z = f(r). \tag{7.2.4}$$

Under these assumptions, the continuity equation, Eqn. (G.8), becomes

$$\frac{du_z}{dz} = 0, \tag{7.2.5}$$

and the three momentum equations for Newtonian liquid, Eqns. (G.12) to (7.14), become

$$\frac{\partial p}{\partial r} = 0, \tag{7.2.6}$$

$$\frac{\partial p}{\partial \theta} = 0, \tag{7.2.7}$$

and

$$-\frac{dp}{dz} + \eta \frac{1}{r} \frac{d}{dr} \left( r \frac{du_z}{dr} \right) = 0. \tag{7.2.8}$$

Equation (7.2.5) simply states that the velocity is independent of the axial distance, $z$, and Eqns. (7.2.6) and (7.2.7) state that the pressure is uniform over any cross-sectional area of flow. Equation (7.2.8) is similar to Eqn. (6.7), and can therefore be solved accordingly.

**Example 7.3: Analysis of Steady Sink Flow by the Navier–Stokes Equations**

Plane sink flow is axisymmetric and cylindrical; therefore, Eqns. (G.8) to (G.14) apply, with

$$\frac{\partial}{\partial \theta} = 0. \tag{7.3.1}$$

In addition, the unidirectional nature of the flow requires that

$$u_\theta = u_z = 0 \tag{7.3.2}$$

and

$$u_r = f(r) \neq 0. \tag{7.3.3}$$

Under these assumptions, the continuity equation, Eqn. (G.8), becomes

$$\frac{1}{r}\frac{d}{dr}(ru_r) = 0, \tag{7.3.4}$$

which is integrated to

$$u_r = \frac{c}{r} = -\frac{\dot{V}}{2\pi r H} \tag{7.3.5}$$

where $\dot{V} = (2\pi r H)u_r$ is the volumetric flow rate through the cylindrical surface of any fluid cylinder of radius $r$ and height $H$, where the velocity is $u_r$. The pressure is then evaluated using the momentum equation in the $r$-direction, Eqn. (G.12), which under the previous assumptions and the known velocity profile becomes

$$\rho \frac{\dot{V}}{2\pi r}\frac{d}{dr}\left(\frac{\dot{V}}{2\pi r}\right) = -\frac{dp}{dr} + \eta \frac{d}{dr}\left(\frac{1}{r}\frac{d}{dr}(ru_r)\right), \tag{7.3.6}$$

which is integrated to

$$p(r) = p_\infty - \frac{\rho \dot{V}^2}{8\pi^2 r^2}. \tag{7.3.7}$$

**Example 7.4: Analysis of Steady Torsional Flow by the Navier–Stokes Equations**

Torsional flow is axisymmetric and cylindrical; therefore, Eqns. (G.8) to (G.14) apply, with

$$\frac{\partial}{\partial \theta} = 0. \tag{7.4.1}$$

In addition, the unidirectional nature of the flow requires that

$$u_r = u_z = 0 \tag{7.4.2}$$

and

$$u_\theta = f(r) \neq 0. \tag{7.4.3}$$

Under these assumptions, the continuity equation, Eqn. (G.8), is satisfied identically, thus validating the assumptions. The three momentum equations, Eqns. (G.12) to (G.14), become

$$-\rho \frac{u_\theta^2}{2} = -\frac{dp}{dr}, \tag{7.4.4}$$

$$\frac{d}{dr}\left[\frac{1}{r}\frac{d}{dr}(ru_\theta)\right] = 0, \tag{7.4.5}$$

and

$$\frac{dp}{dz} = \rho g_z. \tag{7.4.6}$$

First, the velocity is calculated by Eqn. (7.4.5), and then the pressure distribution is evaluated using Eqns. (7.4.4) and (7.4.6) for the known velocity profile.

### Example 7.5: Analysis of Spherical Flow by the Navier–Stokes Equations

The flow is spherical, and therefore Eqns. (G.15) to (G.21) apply. The flow is purely radial with a unique velocity component

$$u_r \neq 0 \tag{7.5.1}$$

and

$$u_\theta = u_\phi = 0. \tag{7.5.2}$$

In addition, the flow is spherically symmetric, so that

$$\frac{\partial}{\partial \theta} = \frac{\partial}{\partial \phi} = 0. \tag{7.5.3}$$

Thus,

$$u_r = f(r). \tag{7.5.4}$$

Under these assumptions, the continuity equation, Eqn. (G.15), becomes

$$\frac{1}{r^2} \frac{d}{dr}(r^2 u_r) = 0, \tag{7.5.5}$$

which is integrated to

$$u_r = \frac{c}{r^2} = \pm \frac{\dot{V}}{4\pi r^2}, \tag{7.5.6}$$

where $\dot{V}$ is volumetric flow rate through any spherical wall at distance $r$ where the velocity is $u_r$. The momentum equations, Eqns. (G.19) to (G.21), become

$$\rho u_r \frac{du_r}{dr} = -\frac{dp}{dr}, \tag{7.5.7}$$

$$\frac{\partial p}{\partial \theta} = 0, \tag{7.5.8}$$

and

$$\frac{\partial p}{\partial \phi} = 0. \tag{7.5.9}$$

The pressure profile is calculated by Eqn. (7.5.7), given the velocity profile from Eqn. (7.5.6).

## 7.4 STREAMFUNCTION AND CREEPING FLOWS

The complete Navier–Stokes equations for two-dimensional flows (e.g., Fig. 7.1) are rarely solvable analytically. The solution is often facilitated by expressing the two velocity components in terms of a scalar function, which is chosen not only to reduce the number of unknowns, but also to simplify or eliminate one or more of the three governing momentum equations. The most generally applied function is the streamfunction, $\psi$, defined as follows:

*Cartesian flows:*

$$u_x = -\frac{\partial \psi}{\partial y} \Leftrightarrow \psi(x, y) = -\int_0^x u_x \, dy + c(x), \tag{7.1}$$

$$u_y = \frac{\partial \psi}{\partial x} \Leftrightarrow \psi(x, y) = \int_0^y u_y \, dx + c(y). \tag{7.2}$$

*Axisymmetric cylindrical flows with $u_z = 0$, or polar flows:*

$$u_r = -\frac{1}{r}\frac{\partial \psi}{\partial \theta} \Leftrightarrow \psi(r, \theta) = -\int ru_r \, d\theta + c(r), \tag{7.3}$$

$$u_\theta = \frac{\partial \psi}{\partial r} \Leftrightarrow \psi(r, \theta) = \int u_\theta \, dr + c(\theta). \tag{7.4}$$

*Axisymmetric cylindrical flows with $u_\theta = 0$:*

$$u_r = -\frac{\partial \psi}{\partial z} \Leftrightarrow \psi(r, z) = -\int u_r \, dz + c(r), \tag{7.5}$$

$$u_z = \frac{1}{r}\frac{\partial (r\psi)}{\partial r} \Leftrightarrow r\psi(r, z) = \int_0^r ru_z \, dr + c(z). \tag{7.6}$$

*Spherical flow with $u_\phi = 0$:*

$$u_r = -\frac{1}{r^2 \sin \theta}\frac{\partial \psi}{\partial \theta} \Leftrightarrow \psi(r, \theta) = -\int u_r \sin \theta r^2 \, d\theta + c(\theta), \tag{7.7}$$

$$u_\theta = \frac{1}{r \sin \theta}\frac{\partial \psi}{\partial r} \Leftrightarrow \psi(r, \theta) = \int ru_\theta \sin \theta \, dr + c(r). \tag{7.8}$$

All the streamfunction definitions above satisfy the corresponding continuity equation identically, which is therefore dropped in any analysis of flow by streamfunction.

The streamfunction is applied primarily to *creeping flows* defined as having vanishingly small Reynolds number,

$$\mathrm{Re} = \frac{\rho \bar{u} D}{\eta} \ll 1, \tag{7.9}$$

and which are therefore dominated by viscous forces. Under these slow flow conditions, the entire left-hand side of any momentum equation vanishes. In addition, flows with no acceleration or inertial forces (i.e., channel flows) are amenable to streamfunction solution techniques (e.g., Example 7.6). The streamfunction satisfies the continuity equation identically, which is therefore eliminated. The momentum equations are further simplified by appropriate differentiation and subtraction in order to eliminate the pressure: For a two-dimensional creeping Cartesian flow of $u_z = 0$ and $\partial/\partial z = 0$, for example,

Eqn. (G.5) is differentiated with respect to $x$ and Eqn. (G.6) with respect to $y$, and then are subtracted, eliminating the pressure. Then the two velocity components, $u_x$ and $u_y$, are replaced in terms of the streamfunction by Eqns. (7.1) and (7.2), and a single equation, called the *bi-Laplacian equation*, results:

$$\nabla^4\psi = \frac{\partial^4\psi}{\partial x^4} + 2\frac{\partial^4\psi}{\partial x^2\,\partial y^2} + \frac{\partial^4\psi}{\partial y^4} = 0. \qquad (7.10)$$

Thus Eqn. (7.10), supplemented by the appropriate streamfunction boundary conditions [which are easily generated from velocity boundary conditions by means of Eqns. (7.1) and (7.2)] can in principle be solved for $\psi(x, y)$. The velocity is then calculated from $\psi(x, y)$ by means of Eqns. (7.1) and (7.2). Finally, the pressure is computed by integrating the momentum equations, given the velocity components. Examples 7.6 and 7.7 demonstrate this procedure for channel and squeezing flows.

The streamfunction exists for any two-dimensional and axisymmetric steady flow. Equation (7.10) can be solved easily. However, for the many general solutions to Eqn. (7.10) (e.g., $\psi = axy + bx + cy + d$, $\psi = \sum a_n \cos nx + b_n \sin nx$, etc.), there is only one that satisfies the boundary conditions. The latter fact severely inhibits use of the streamfunction.

Besides its mathematical significance, the stream function is a meaningful physical quantity. The equation

$$\psi(x, y) = c \qquad (7.11)$$

represents the *streamlines*, by assigning values to the constant $c$. The streamlines are orbits traced by fluid particles moving with velocity tangential to them, according to Eqns. (7.1) to (7.8). Thus, no mass travels across streamlines because there is no velocity component normal to the streamline. Collections of many streamlines, normal to a cross section of a flow, make up imaginary *streamtubes*, where fluid enters and leaves only axially, as illustrated by Fig. 7.2. The flow rate

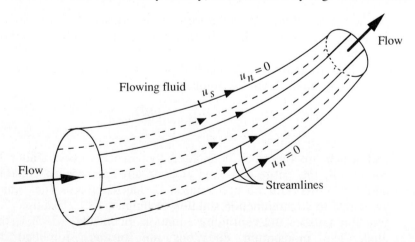

**Figure 7.2**  Streamlines and streamtubes of a cylindrical flow, with streamline velocity $u_s$ and normal velocity $u_n = 0$, across any streamline.

through the streamtube of Fig. 7.2 is given by

$$\dot{V} = \int_0^R 2\pi u_z(r)r\,dr = \int_0^R 2\pi \frac{1}{r}\frac{\partial \psi}{\partial r}r\,dr = 2\pi[\psi(R) - \psi(0)]. \qquad (7.12)$$

Thus, the important conclusion is that the difference between two values of the streamfunction defining any two streamlines is identical to the flow rate between them. In many applications, the locations of the streamlines are essential to calculate deformation and thermal histories of fluid particles, residence time, cavitation, thermal degradation, and others, since the actual conditions experienced by the traveling fluid particle are those along its streamlines.

### Example 7.6: Steady Channel Flow Analysis Using Streamfunctions

Analyze channel Poiseuille flow in a channel of width $H$ under flow rate $\dot{V}$ by means of the appropriate streamfunction (Fig. E7.6).

**Solution**    Equation (7.10) applies here since no acceleration occurs:

$$\frac{\partial^4 \psi}{\partial y^4} + \frac{\partial^4 \psi}{\partial y^2 \partial x^2} + \frac{\partial^4 \psi}{\partial x^4} = 0. \qquad (7.6.1)$$

Since $u_y = \partial\psi/\partial x = 0$ everywhere, the equation reduces to

$$\frac{d^4 \psi}{dy^4} = 0. \qquad (7.6.2)$$

with general solution

$$\psi = c_1\frac{y^3}{6} + c_2\frac{y^2}{2} + c_3 y + c_4. \qquad (7.6.3)$$

The boundary conditions are now applied:

$$y = 0: \quad du_x/dy = -\partial^2\psi/\partial y^2 = 0; \text{ therefore, } c_2 = 0$$
$$y = H: \quad u_x = -\partial\psi/\partial y = 0; \text{ therefore, } c_3 = -c_1 H^2/2.$$

The flow rate requires that

$$\dot{V} = 2\int_0^H u_x(y)\,dy = -2\int_0^H \frac{\partial\psi}{\partial y}\,dy = -2\psi(H) + 2\psi(0)$$

$$= 2\left[\frac{c_1}{6}(-H^3 + 3H^3)\right] = \tfrac{2}{3}c_1 H^3. \qquad (7.6.4)$$

$$y\left(\frac{y^2}{3} - H^2\right) =$$

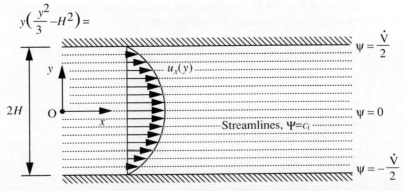

**Figure E7.6**    Streamline pattern for channel.

Thus, $c_1 = 3\dot{V}/2H^3$, and the streamfunction becomes

$$\psi(y) = \frac{3}{12}\frac{\dot{V}}{H^3}y^3 - \frac{3}{4}\frac{\dot{V}}{H}y + c_4. \tag{7.6.5}$$

The streamlines are obtained by defining a reference streamline, $\psi(y = 0) = 0$, for $c_4 = 0$. Thus

$$\psi(y) = \frac{3}{4}\frac{\dot{V}}{H^3}\left(\frac{y^3}{3} - yH^2\right), \tag{7.6.6}$$

which represents straight lines parallel to the axis $y = 0$. Then wall streamlines are

$$\psi(y = \pm H) = \pm\frac{\dot{V}}{2} \tag{7.6.7}$$

and therefore

$$\psi(H) - \psi(-H) = \dot{V} \tag{7.6.8}$$

The velocity $u_x(y)$ is recovered by

$$u_x(y) = -\frac{\partial\psi}{\partial y} = \frac{3}{4}\frac{\dot{V}}{H^3}(y^2 - H^2), \tag{7.6.9}$$

which is independent of the arbitrary constant $c_4$.

The pressure is calculated by means of the momentum equation,

$$\frac{dp}{dx} = \eta\frac{d^2u_y}{dy^2} = \frac{3\eta}{2}\frac{\dot{V}}{H^3}, \tag{7.6.10}$$

which yields the Poiseuille linear pressure profile

$$p(x) = \frac{3\eta}{2}\frac{\dot{V}}{H^3}x + p_0 \tag{7.6.11}$$

of constant pressure gradient

$$\frac{\Delta p}{\Delta L} = \frac{p(L) - p(0)}{L} = \frac{3\eta}{2}\frac{\dot{V}}{H^3}. \tag{7.6.12}$$

The streamline family and velocity profile are shown in Fig. E7.6.

### Example 7.7: Compression Molding Analysis by Streamfunction

A highly viscous Newtonian liquid of density $\rho$ and viscosity $\eta$, and surface tension $\sigma$ is squeezed between two disks of radius $R$ approaching at constant speed $V$, as shown in Fig. P5.17. Analyze the resulting two-dimensional axisymmetric flow by utilizing the appropriate streamfunction.

**Solution**　The appropriate expressions for the streamfunction are

$$u_z = \frac{1}{r}\frac{\partial\psi}{\partial r} \quad \text{and} \quad u_r = -\frac{1}{r}\frac{\partial\psi}{\partial z}, \tag{7.7.1}$$

due to the fact that $\partial/\partial\theta = 0$ because the flow is axisymmetric, and $u_\theta = 0$, due to the radial character of the flow.

The continuity equation,

$$\frac{1}{r}\frac{\partial}{\partial r}(ru_r) + \frac{\partial u_z}{\partial z} = \frac{1}{r}\frac{\partial}{\partial r}\left(\frac{-r}{r}\frac{\partial\psi}{\partial z}\right) + \frac{1}{r}\frac{\partial^2\psi}{\partial r\,\partial z} = 0, \tag{7.7.2}$$

is satisfied identically, so the definitions are consistent. The $z$-momentum and the *r*-momentum equations for creeping squeezing flow, become

$$-\frac{\partial p}{\partial z} + \eta\left[\frac{1}{r}\frac{\partial}{\partial r}\left(r\frac{\partial}{\partial r}\left(\frac{1}{r}\frac{\partial \psi}{\partial r}\right)\right) + \frac{1}{r}\frac{\partial^3 \psi}{\partial r\, \partial z^2}\right] = 0 \tag{7.7.3}$$

and

$$-\frac{\partial p}{\partial r} + \eta\left[-\frac{\partial}{\partial r}\left(\frac{1}{r}\frac{\partial^2 \psi}{\partial r\, \partial z}\right) - \frac{1}{r}\frac{\partial^3 \psi}{\partial z^3}\right] = 0, \tag{7.7.4}$$

respectively. Differentiation of the first with respect to *r*, of the second with respect to *z*, and then subtraction of the two to eliminate the pressure yields

$$\frac{\partial}{\partial r}\left(\frac{1}{r}\frac{\partial}{\partial r}\left(r\frac{\partial}{\partial r}\left(\frac{1}{r}\frac{\partial \psi}{\partial r}\right)\right)\right) + \frac{\partial}{\partial r}\left(\frac{1}{r}\frac{\partial^3 \psi}{\partial r\, \partial z^2}\right) + \frac{\partial}{\partial r}\left(\frac{1}{r}\frac{\partial^3 \psi}{\partial r\, \partial z^2}\right) + \frac{1}{r}\frac{\partial^4 \psi}{\partial z^4} = 0. \tag{7.7.5}$$

A solution of the form $\psi = r^a f(z)$ is sought, for which Eqn. (7.7.5) becomes

$$a(a-2)^2(a-4)r^{a-4}f(z) + 2a(a-2)r^{a-3}f''(z) + r^{a-1}f^{IV}(z) = 0. \tag{7.7.6}$$

with

$$f(z) \neq 0 \quad \text{and} \quad f'(z) = 0. \tag{7.7.7}$$

In Eqn. (7.7.6) it must be $a \geq 1$ ($a < 0$, predicts infinite velocity at $r = 0$, and $a = 0$ predicts velocity independent of *r*). Also, $a \neq 1$; otherwise, Eqn. (7.7.6) predicts $f(z) = f''(z) = f^{IV}(z) = 0$, and therefore $\psi = 0$ and $u_z = u_r = 0$. The same is true for $a = 3$. Thus, $a = 2$, which yields $f^{IV} = 0$ [without $f(z) = f''(z) = 0$], with general solution

$$f(z) = az^3 + bz^2 + cz + d. \tag{7.7.8}$$

The boundary conditions determine the four constants:

1. $z = 0$: $u_z = 0$, thus $f = 0$; therefore, $c = 0$.
2. $z = H$: $u_z = -V$, thus $f = -V$; therefore, $b = -V/H - dH^2 - aH$.
3. $z = 0$: $\tau_{zr} = 0$, thus $f'' = 0$; therefore, $a = 0$.
4. $z = H$: $u_r = 0$, thus $f' = 0$; therefore, $d = V/2H^3$.

Thus,

$$f(z) = \frac{V}{2H^3}(z^3 - H^2 z) - \frac{V}{H}z. \tag{7.7.9}$$

The streamfunction is therefore

$$\psi = r^2\left[\frac{V}{2H^3}(z^3 - H^2 z) - \frac{V}{H}z\right], \tag{7.7.10}$$

with corresponding velocities,

$$u_r = -\frac{1}{r}\frac{\partial \psi}{\partial z} = -\frac{rV}{2H^3}(3z^2 - H^2) - \frac{V}{H}r, \tag{7.7.11}$$

$$u_z = \frac{1}{r}\frac{\partial \psi}{\partial r} = \frac{V}{2H^3}(z^3 - H^2 z) - \frac{V}{H}z. \tag{7.7.12}$$

The pressure distribution is obtained by first substituting Eqn. (7.7.10) into the two streamfunction momentum equations, Eqns. (7.7.3) and (7.7.4), resulting in

$$-\frac{\partial p}{\partial z} + \eta\left(0 + \frac{6V}{3H^3}z\right) = 0 \tag{7.7.13}$$

and

$$-\frac{\partial p}{\partial r} + \eta\left(0 - \frac{6V}{2H^3}r\right) = 0, \tag{7.7.14}$$

respectively. Integration of these two equations yields

$$p(r, z) = \frac{6V}{2H^3}\left(\frac{z^2}{2} - \frac{r^2}{2}\right) + c. \tag{7.7.15}$$

The constant $c$ is evaluated from the boundary condition:

At $r = R$ and $z = 0$: $p = p_{\text{atm}} = 0$; therefore, $c = 6VR^2/4H^3 + p_0$. Thus the pressure becomes

$$p(r, z) = \frac{6V}{2H^3}\left(\frac{R^2 - r^2}{2} + \frac{z^2}{2}\right) + p_0. \tag{7.7.16}$$

Other important creeping flows include slow flows overtaking submerged bodies or suspended particles, that lead to Stokes' law of drag force,[5] discussed in several applications of Chapters 3 and 4; processing of highly viscous glasses and polymers;[6] flows in the vicinity of corners, wedges, and other geometries that decelerate flow;[7] and slow flows in converging and diverging channels and dies common in a diversity of material processing,[8] as analyzed in Section 7.5.

## 7.5 FLOWS IN CONVERGING OR DIVERGING CHANNELS: CREEPING, LUBRICATION, REVERSING, AND RADIAL FLOWS

The prototype geometry of a diverging flow is shown in Fig. 7.3. The flow is best described with respect to a polar system of coordinates, positioned as shown in

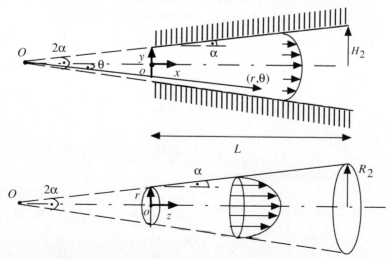

**Figure 7.3**   Diverging or converging channel and pipe flows and polar system of coordinates.

Fig. 7.3. The two velocity components are expressed in terms of the streamfunction by means of Eqns. (7.3) and (7.4). Since for these flows $u_\theta = 0$, Eqn. (7.4) shows that the streamfunction is only a function of $\theta$, and the velocity components are

$$u_r(r, \theta) = -\frac{1}{r}\frac{d\psi(\theta)}{d\theta} = -\frac{g(\theta)}{r}, \quad u_\theta = 0, \quad u_z = 0, \tag{7.13}$$

where $g(\theta) = d\psi/d\theta$ was used. These expressions are substituted in Eqns. (G.9) and (G.10), which are then differentiated with respect to $\theta$ and $r$, respectively, and then subtracted to eliminate the pressure. The resulting governing streamfunction equation is

$$2g\frac{dg}{d\theta} - \nu\frac{d^3g}{d\theta^3} - \nu\frac{dg}{d\theta} = 0, \tag{7.14}$$

with boundary conditions

$$g(\theta = \pm\alpha) = 0 \tag{7.15}$$

and

$$\dot{V} = \int_{-\alpha}^{\alpha} ru_r\, d\theta = \int_{-\alpha}^{\alpha} g\, d\theta, \tag{7.16}$$

where $\dot{V}$ is the volumetric flow rate and $\alpha$ angle in *radians*, with $\dot{V} > 0$ for diverging and $\dot{V} < 0$ for converging flow.

Equations (7.14) to (7.16) are made dimensionless by means of the dimensionless variables,

$$\xi = \frac{\theta}{\alpha}, \quad f = -\frac{g}{\dot{V}}, \quad \mathrm{Re} = \frac{\dot{V}}{\nu}, \tag{7.17}$$

which turn the original equations to

$$2\alpha\, \mathrm{Re}\, f\frac{df}{d\xi} + \frac{d^3f}{d\xi^3} + 4\alpha^2\frac{df}{d\xi} = 0, \quad -1 < \xi < 1, \tag{7.18}$$

the boundary conditions to

$$f(\xi = -1) = f(\xi = 1) = 0 \tag{7.19}$$

and the flow rate constraint to

$$\int_{-1}^{1} \alpha f = 1. \tag{7.20}$$

Equations (7.19) to (7.20) may represent the following cases of converging or diverging flows.

**Creeping flow.**    At a vanishingly small Reynolds number,

$$\mathrm{Re} = \frac{\dot{V}}{\nu}, \tag{7.21}$$

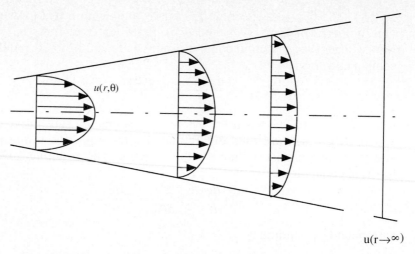

**Figure 7.4** Velocity profiles in diverging creeping flow. In converging flow the velocity profiles are opposite to those shown.

the first term of Eqn. (7.18) is dropped and the resulting solution is

$$u(r, \theta) = -\frac{g(\theta)}{r} = \pm \frac{\dot{V}}{r} \frac{e^{-2\alpha} + e^{-2\alpha} - e^{2\theta} - e^{-2\theta}}{(2\alpha - 1)e^{2\alpha} + (1 - 2\alpha)e^{-2\alpha}}, \qquad (7.22)$$

which represents parabolic velocity profiles that diminish with distance, as shown in Fig. 7.4. Equation (7.22) predicts that the velocity vanishes only at the two boundaries, which means that there cannot be any flow reversal under creeping conditions, since the velocity profile preserves the same sign and therefore direction for all values of $\theta$.

**Lubrication flows.**   Under small channel inclination such that

$$\tan \alpha \approx \alpha \ll 1, \qquad \alpha \, \mathrm{Re} \ll 1, \qquad (7.23)$$

the flow obtained resembles that taking place in lubrication applications of slightly inclined surfaces in relative motion.[9] The first and the third terms of Eqn. (7.18) drop out, and the resulting velocity profile is

$$u_r(r, \theta) = \frac{3\dot{V}}{4\alpha^3}(\alpha^2 - \theta^2)\frac{1}{r}. \qquad (7.24)$$

Notice that according to the criteria of Eqn. (7.23), lubrication flows are not necessarily creeping flows. Lubrication flows are good prototypes for studying not only lubrication applications, but applications of thin liquid films on solid substrates or coatings,[10] and are examined in detail in Chapter 8.

**Reverse flow at high Reynolds numbers.**   At a high laminar Reynolds number,

$$\alpha \, \mathrm{Re} \gg 1, \qquad (7.25)$$

an approximate solution of Eqn. (7.18) can be obtained by asymptotic analysis:

$$u(r, \theta) = -\frac{g(\theta)}{r} = \frac{f\dot{V}}{r} = \frac{\dot{V}}{2\alpha r}\left[\left(1 - \frac{\theta^2}{\alpha^2}\right) + \frac{1}{\alpha \, \text{Re}}\left(5\frac{\theta^4}{\alpha^4} - \frac{\theta^6}{\alpha^6} - 4\frac{\theta^2}{\alpha^2}\right)\right]. \qquad (7.26)$$

This equation predicts that the velocity vanishes not only at the two boundaries where $\theta = \pm\alpha$, but also at angles given by

$$\left(\frac{\theta^*}{\alpha}\right)^2 = 2 \pm \sqrt{4 - 30\alpha \, \text{Re}}, \qquad (7.27)$$

which reveals reversal of flow, depending on the sign and magnitude of $\alpha$ Re: A purely diverging flow of $\alpha$ Re $> 0$ is possible if $4 > 30\alpha$ Re; otherwise, there will be reversals of flow indicated by vanishing velocity at angles predicted by Eqn. (7.27). A purely converging flow at infinitely large $\alpha$ Re $< 0$ is not possible according to Eqn. (7.27) and flow reversals will occur, as predicted by the same equation. Solutions to converging and diverging flows at finite $\alpha$ Re have been obtained in terms of elliptic integrals by Jeffery (1915) and Hamel (1917) and reported elsewhere.[7]

**Radial flow.**    As the inclination of the channel becomes larger and larger the flow becomes more and more purely radial of zero vorticity, except within thin boundary layers near the two walls, the flow within which is dominated by vorticity. The velocity profile of the radial bulk flow is given by Eqn. (7.26) in the limit of $\theta/\alpha \ll 1$:

$$u_r(r) \approx \frac{\dot{V}}{2\alpha r}. \qquad (7.28)$$

Notice that Eqn. (7.28) does not satisfy the two boundary conditions at the two walls. Indeed, this radial flow does not extend down to the walls, where vorticity-rich, thin boundary layers intervene between the radial flow and the solid walls. Such situations of envelopes adjacent to solid boundaries rich in vorticity, surrounded by irrotational flows, arise frequently and are modeled by *boundary layer* and *potential flow* theories, respectively, as described in Sections 7.6 and 7.7.

## 7.6 VORTICITY DYNAMICS

Creeping flows are usually associated with the presence of boundaries and obstacles to flow where vorticity is generated and diffuses away, infecting the flow. Away from flow barriers, the vorticity is unable to penetrate by diffusion being swept away by the large velocity that dominates flows at high Reynolds numbers. Such flows free of vorticity are termed *irrotational*. This is due to the fact that fluid particles do not rotate with respect to each other, (or, equivalently, bouyancy neutral slender fibers do not rotate with respect to the flow streamlines).

According to these vorticity dynamics, flow around the sphere shown in Fig. 7.1 is creeping at a low Reynolds number, known as *Stokes' flow*. In Stokes-like flows, vorticity generated at solid boundaries diffuses away and infects the entire flow. Equivalently, vorticity penetrates by diffusion to nearly infinite distances left unprotected by slow convection at low Reynolds numbers. At high Reynolds numbers, the flow is of dual character: Within an envelope around the sphere whose thickness shrinks as the Reynolds number increases, viscous stresses and vorticity generated at the boundary continue to dominate and give rise to a *boundary layer flow*, detailed in Section 7.8. Outside the boundary layer, the flow is irrotational, free of vorticity that cannot travel beyond the boundary layer, since it is swept away by high convection at high Reynolds numbers.

The same is true for the converging/diverging channel flows examined in Section 7.5. As the inclination increases, a radial, potential flow of zero vorticity is established in the core, whereas thin boundary layers dominated by vorticity intervene between the potential radial flow and the two walls.

The concepts of *vorticity* and *irrotationality* have been introduced for the unidirectional flows studied in Chapters 5 and 6. These concepts are generalized here to two-dimensional flows. The vorticity is a vector $\boldsymbol{\omega}$ perpendicular to the plane defined by the two velocity components and has magnitude $\omega$ defined by velocity derivatives, as follows:

*Cartesian flows of $u_z = 0$:*

$$\boldsymbol{\omega} = \omega \mathbf{k}, \qquad \omega = \frac{\partial u_x}{\partial y} - \frac{\partial u_y}{\partial x} \tag{7.29}$$

*Axisymmetric cylindrical flows with $u_z = 0$, or polar flows:*

$$\boldsymbol{\omega} = \omega \mathbf{e}_z, \qquad \omega = \frac{1}{r}\frac{\partial u_r}{\partial \theta} - \frac{\partial u_\theta}{\partial r} \tag{7.30}$$

*Axisymmetric cylindrical flows with $u_\theta = 0$:*

$$\boldsymbol{\omega} = \omega \mathbf{e}_\theta, \qquad \omega = \frac{\partial u_z}{\partial r} - \frac{\partial u_r}{\partial z} \tag{7.31}$$

*Spherical flows with $u_\phi = 0$:*

$$\boldsymbol{\omega} = \omega \mathbf{e}_\phi, \qquad \omega = \frac{1}{r}\frac{\partial}{\partial r}(r u_\theta) - \frac{1}{r}\frac{\partial u_r}{\partial \theta} \tag{7.32}$$

The vectors $\mathbf{k}$, $\mathbf{e}_z$, $\mathbf{e}_\theta$, and $\mathbf{e}_\phi$ have unit magnitude and directions parallel to those of the corresponding axis, in the Cartesian, polar, cylindrical, and spherical system of coordinates (e.g., Appendix A). Notice that the vorticity of a two-dimensional flow is in the direction of the zero-velocity component. A positive value of vorticity denotes clockwise rotation and a negative value otherwise.

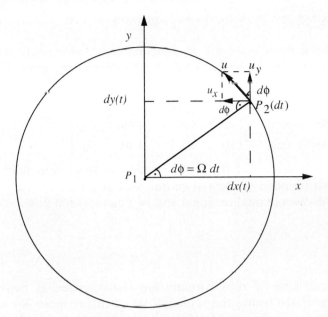

**Figure 7.5**  Instantaneous rotation of fluid particle $P_2$ about fluid particle $P_1$ at angular velocity $\Omega$ and vorticity $\omega = 2\Omega$.

The physical significance of vorticity as a driving mechanism of fluid rotation is assessed by studying a typical rotational motion, depicted in Fig. 7.5. The figure shows two fluid particles, $P_1$ and $P_2$, traveling with the flowing fluid. A Cartesian and a polar system of coordinates pinned at and traveling with particle $P_1$ are introduced. In a generalized flow, $P_2$ is seen by an observer pinned at $P_1$ to rotate about particle $P_1$ with a certain angular velocity $\Omega$. If a small time interval $dt$ is considered, the angular velocity can be assumed constant. In reference to Fig. 7.5, the following trigonometric relations hold:

$$d\phi = \Omega\,dt, \quad dx = -R\sin(d\phi) \approx -R\,d\phi, \quad dy = R\cos(d\phi) \approx R. \quad (7.33)$$

At time $dt$, particle $P_2$ is at position $P_2(dt)$ with velocity vector $\mathbf{u}(dt)$ of magnitude $u = \Omega R$. Its Cartesian components are

$$u_x(dt) = -u\sin(d\phi) \approx -u\,d\phi, \quad u_y(dt) = u\cos(d\phi) \approx u. \quad (7.34)$$

Derivatives of velocity components are computed by means of the chain rule of differentiation:

$$\frac{\partial u_x}{\partial y} = \cos(d\phi)\frac{d\phi}{dy} = \frac{u}{R} = \Omega, \quad \frac{\partial u_y}{\partial x} = -u\sin(d\phi)\frac{d\phi}{dx} = -u\frac{dy}{R}\frac{1}{dy} = -\Omega. \quad (7.35)$$

The vorticity of this fundamental rotational motion is then

$$\boldsymbol{\omega} = \left[\frac{\partial u_x}{\partial y} - \frac{\partial u_y}{\partial x}\right]\mathbf{k} = 2\Omega\mathbf{k}, \quad (7.36)$$

a vector perpendicular to the sheet plane and of magnitude twice the angular velocity. Therefore, torsional and solid-body rotational flows are rich in vorticity, whereas radial and plug flows are free of vorticity. Also, since relative rotation of particles is induced by velocity gradients, any inviscid flow is rid of vorticity.

In three-dimensional flows there are in general three components of the vorticity vector, each perpendicular to its two defining velocity components. The vorticity vector of a three-dimensional Cartesian flow is defined by a matrix determinant,

$$\boldsymbol{\omega} = \mathbf{i}\left[\frac{\partial u_z}{\partial y} - \frac{\partial u_y}{\partial z}\right] - \mathbf{j}\left[\frac{\partial u_z}{\partial x} - \frac{\partial u_x}{\partial z}\right] + \mathbf{k}\left[\frac{\partial u_y}{\partial x} - \frac{\partial u_x}{\partial y}\right]. \tag{7.37}$$

where $\mathbf{i}$, $\mathbf{j}$, $\mathbf{k}$, are the unit vectors of the Cartesian system (e.g., Appendix A). Similar expressions exist for other systems of coordinates. It is easy to verify that Eqn. (7.37) includes the cases of unidirectional and two-dimensional flows.

## 7.7 POTENTIAL FLOWS

Irrotational and inviscid flows of zero vorticity are characterized as *potential flows*: As will be demonstrated below, the velocity components of these flows can be expressed in terms of a single *potential function* that represents the inviscid conservative forces driving a potential flow. One way to introduce the potential function is by means of Eqns. (7.29) to (7.32): For potential flows these equations must be identically zero, which defines the potential function in a fashion similar to the use of the continuity equation to introduce the streamfunction. Based on Eqns. (7.29) to (7.32), the following potential functions are defined:

*Cartesian flows of $u_z = 0$:*

$$u_x = -\frac{\partial\phi}{\partial x} \Leftrightarrow \phi(x, y) = -\int u_x\, dx + c(y), \tag{7.38}$$

$$u_y = -\frac{\partial\phi}{\partial y} \Leftrightarrow \phi(x, y) = -\int u_y\, dy + c(x) \tag{7.39}$$

*Axisymmetric flows of $u_z = 0$, or polar flows:*

$$u_r = -\frac{\partial\phi}{\partial r} \Leftrightarrow \phi(r, \theta) = -\int u_r\, dr + c(\theta), \tag{7.40}$$

$$u_\theta = -\frac{1}{r}\frac{\partial\phi}{\partial\theta} \Leftrightarrow \phi(r, \theta) = -\int r u_\theta\, d\theta + c(r) \tag{7.41}$$

*Axisymmetric cylindrical flows of $u_\theta = 0$:*

$$u_r = -\frac{\partial\phi}{\partial r} \Leftrightarrow \phi(r, z) = -\int u_r\, dr + c(z), \tag{7.42}$$

$$u_\theta = -\frac{\partial\phi}{\partial\theta} \Leftrightarrow \phi(r, z) = -\int u_z\, dz + c(r) \tag{7.43}$$

*Spherical flows with $u_\phi = 0$:*

$$u_r = -\frac{\partial \phi}{\partial r} \Leftrightarrow \phi(r, \theta) = -\int u_r\, dr + c(\theta), \tag{7.44}$$

$$u_\theta = -\frac{1}{r}\frac{\partial \phi}{\partial \theta} \Leftrightarrow \phi(r, \theta) = -\int r u_\theta\, d\theta + c(r) \tag{7.45}$$

It is easy to see that these expressions substituted in the appropriate equations, Eqns. (7.29) to (7.32), yield zero vorticity.

By replacing the velocity components with the corresponding potential function expressions, Eqns. (7.29) to (7.32), the continuity equation for a potential flow reduces to the corresponding *Laplace equation* that includes the scalar potential function:

$$\nabla^2 \phi = \frac{\partial^2 \phi}{\partial x^2} + \frac{\partial^2 \phi}{\partial y^2} = 0, \tag{7.46}$$

$$\nabla^2 \phi = \frac{\partial^2 \phi}{\partial r^2} + \frac{1}{r}\frac{\partial \phi}{\partial r} + \frac{1}{r^2}\frac{\partial^2 \phi}{\partial \theta^2} = 0, \tag{7.47}$$

$$\nabla^2 \phi = \frac{\partial^2 \phi}{\partial r^2} - \frac{1}{r}\frac{\partial \phi}{\partial r} + \frac{\partial^2 \phi}{\partial z^2} = 0, \tag{7.48}$$

$$\nabla^2 \phi = \frac{1}{r}\frac{\partial}{\partial r}\left(r^2\frac{\partial \phi}{\partial r}\right) + \frac{1}{r^2 \sin \theta}\frac{\partial}{\partial \theta}\left(\frac{\partial \phi}{\partial \theta}\sin \theta\right) = 0. \tag{7.49}$$

The corresponding momentum equations degenerate to the Bernoulli equation[11] along streamline distance $ds$,

$$\rho\, u\frac{du}{ds} + \frac{dp}{ds} + \rho g\frac{dz}{ds} = 0, \tag{7.50}$$

which is integrated to give

$$\Delta\left(\frac{u^2}{2} + \frac{p}{\rho} + gz\right) = 0 \tag{7.51}$$

along streamlines of arc length $s$. Equation (7.50) assigns physical meaning to the potential function, when rearranged in the form

$$\frac{d\phi}{ds} = -u = 2\sqrt{p + \rho gz}, \tag{7.52}$$

which indicates that the potential function is the potential energy possessed by the fluid due to gravity or pressure, which can be converted into velocity and kinetic energy according to the velocity–potential relation along a streamline,

$$u = -\frac{d\phi}{ds}. \tag{7.53}$$

Because of this property, the potential function represents lines of

$$\phi(x, y) = c, \tag{7.54}$$

perpendicular to the streamlines, by assigning values to $c$ based on a chosen datum. Potential lines are lines crossed perpendicularly by fluid particles traveling at velocity proportional to pressure and/or elevation differences between adjacent potential lines according to Eqn. (7.52). Thus, given the location of the potential lines, the location of the streamlines can be determined, and vice versa.

**Solution of potential flow equations.** The existence of the potential function for those flows that are irrotational greatly facilitates solution of the Navier–Stokes equations: First, Laplace's equation is solved for $\phi(x, y)$, which satisfies the appropriate boundary conditions. These boundary conditions must be consistent with the physics of the potential flow; i.e., no-slip kind of boundary conditions that imply generation of vorticity are meaningless. Meaningful boundary conditions for potential flow include specification of a datum potential level and/or of potential first-order derivatives equivalent to velocity components. Then the velocity components, $u_x$ and $u_y$, are recovered by means of Eqns. (7.38) to (7.45). Finally, the pressure is computed by means of Eqn. (7.51).

Unfortunately, in addition to the difficulties encountered in solving flow problems using the streamfunction, potential functions (which are solutions to the Laplace equations) exist only for the limited class of potential flows. These are in general flows away from solid boundaries, free of vorticity. Even for such flows that are known a priori to be potential, it is not always possible to find the potential function. However, potential functions for a diversity of potential flows have been computed and are tabulated in potential theory textbooks.[12,13] Some are shown in Table 7.1. Example 7.8 demonstrates the use of this type of tabulated information to solve potential flow problems.

**TABLE 7.1** POTENTIAL FUNCTION AND STREAMFUNCTION FOR SELECTED POTENTIAL FLOWS

| Geometry of flow | Potential and streamfunction[a] |
|---|---|
| Uniform stream | $\phi = (u_x x + u_y y),\ u_x^2 + u_y^2 = V^2$ <br> $\psi = -u_x y + u_y x$ |
| Stagnation flow | $\phi = \pm A(x^2 - y^2)$ <br> $\psi = \pm Axy$ |

**TABLE 7.1**—(*Continued*)

| Geometry of flow | Potential and streamfunction[a] |
| --- | --- |
| Hyperbolic flow | $\phi = \pm Axy$<br>$\psi = \pm A(x^2 - y^2)$ |

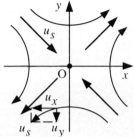

Spiral flow $\qquad\qquad\qquad\qquad\qquad\qquad$ $\phi = -A\theta + c \ln r$
$\psi = -A \ln r + C\theta$

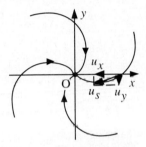

Sink and source flows $\qquad\qquad\qquad$ $\phi = \pm c \ln(x^2 + y^2)^{1/2}$
$\qquad\qquad\qquad\qquad\qquad\quad = \pm c \ln r$
$\qquad\qquad\qquad\qquad\qquad\quad \psi = \pm C\theta$

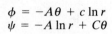

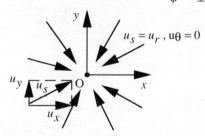

Free vertex flow $\qquad\qquad\qquad\qquad$ $\phi = \pm A \tan^{-1} \dfrac{y}{x}$

$\qquad\qquad\qquad\qquad\qquad\qquad = \pm A\theta$

$\qquad\qquad\qquad\qquad\qquad\quad \psi = \pm A \ln r$

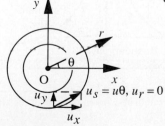

**TABLE 7.1**—(*Continued*)

| Geometry of flow | Potential and streamfunction[a] |
|---|---|
| Flow around cylinder | $\phi = u_\infty \left( r + \dfrac{R^2}{r} \right) \cos \theta$ |
| | $\psi = u_\infty \left( r - \dfrac{R^2}{r} \right) \sin \theta$ |

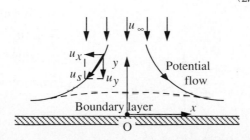

| Flow around sphere | $\psi = \dfrac{u_\infty R}{2^2} \sin^2 \theta - \dfrac{u_\infty r^2}{2} \sin^2 \theta$ |
|---|---|
| | $\phi = -\dfrac{u_\infty R^3}{2r^2} \cos \theta - u_\infty r \cos \theta$ |

| Stagnation flow | $\phi = -u_\infty \left( \dfrac{R^3}{2r^2} \cos \theta + r \cos \theta \right)$ |
|---|---|
| | $\psi = u_\infty \left( \dfrac{R^3}{2r} \sin^2 \theta - \dfrac{r^2}{2} \sin^2 \theta \right)$ |

[a] In all figures, $u_s$ is the streamline velocity, tangent to the streamline.

**Example 7.8: Orthogonal Stagnation Flow**

Analyze the flow obtained by two colliding liquid planar jets at flow rate $Q$ each (Fig. E7.8). Consider only the region away from the capillaries such that jet-exit effects fade out.

**Solution**    Assume that the field to be analyzed, enclosed by the dashed line in Fig. E7.8, is far away from the jet exits. Since the entire field under consideration is away from solid boundaries, the flow *may be* potential, satisfying Laplace's equation,

$$\nabla^2\phi = \frac{\partial^2\phi}{\partial x^2} + \frac{\partial^2\phi}{\partial y^2} = 0. \tag{7.8.1}$$

The boundary conditions are:

1.  $y = 0$: $u_y = -\partial\phi/\partial y = 0$ (symmetry about the $x$-axis).
2.  $x = 0$: $u_x = -\partial\phi/\partial x = 0$ (symmetry about the $y$-axis).

A solution that satisfies both the Laplace equation and the boundary conditions is

$$\phi = c(x^2 - y^2), \tag{7.8.2}$$

for which the velocity components are

$$u_x = -\frac{\partial\phi}{\partial x} = -2cx \tag{7.8.3}$$

and

$$u_y = -\frac{\partial\phi}{\partial y} = 2cy. \tag{7.8.4}$$

(a)

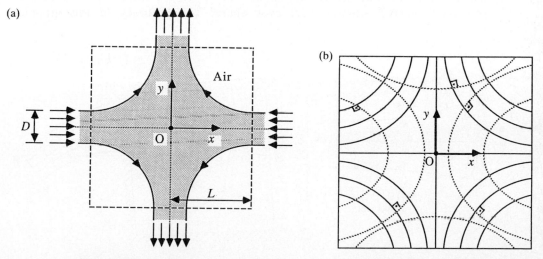

(b)

**Figure E7.8**  Collision of opposing sheet jets, and selected resulting streamlines and potential lines.

The constant $c$ is calculated from the flow rate,

$$Q = Du_x|_L = -2DcL; \quad \text{therefore, } c = -\frac{Q}{2DL}. \quad (7.8.5)$$

The pressure *along* streamlines is calculated by the momentum equations, which for potential flow degenerate to Bernoulli's equations is

$$\frac{p(x = 0)}{\rho} = \frac{p(x, y)}{\rho} + \frac{4c^2(x^2 + y^2)}{2}, \quad (7.8.6)$$

where $c = -Q/2DL$. The stream function can be calculated by means of Eqns. (7.1) and (7.2), which yield

$$\psi(x, y) = cxy + c_1. \quad (7.8.7)$$

The lines $\psi(x, y) = cxy + c_1 = m$ or $y = k/x$ represent streamlines by assigning values to $k$. The lines $\phi(x, y) = c(x^2 - y^2) = l$ or $y = nx$ are the potential lines, perpendicular to the streamlines. Families of these lines are plotted in Fig. E7.8. Orthogonal stagnation flow is now receiving attention from rheologists as a means of producing extensional flow,[14] to measure elongational viscosity.

**Graphical solution of potential flow problems.** Because of their special properties, streamlines and potential lines can be constructed graphically and a potential flow reproduced from them, without knowledge of its potential function. This graphical procedure,[15] illustrated in Fig. 7.6, starts by identifying fixed selected streamlines, which in this case are the boundary lines and the free surface line. Then, internal streamlines are sketched, such that the flow rates through each pair of them are identical, which is achieved by placing these streamlines at equal distances in places where the flow is fully developed (i.e., along the line $AA'$). Then the equipotential lines are sketched perpendicular to the streamlines, such that curved squares are formed by streamlines and potential lines. These two families of streamlines and potential lines are updated until reasonably curved squares exist everywhere. The accuracy of this graphical

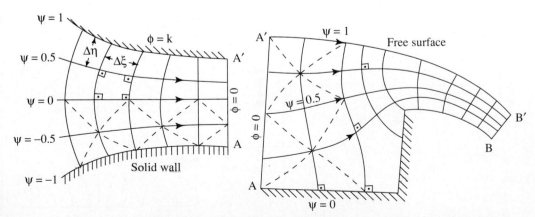

**Figure 7.6** Graphical construction of streamlines and potential lines for confined and free-surface potential flows. (From Ref. 15, by permission.)

method is evaluated by checking that the diagonals of each square are mutually perpendicular. As the tessellation becomes finer and finer, the curved squares become truly squares with straight sides. Once this graphical construction is brought into the required accuracy, the streamline velocity is computed by

$$u = \frac{\Delta\psi}{\Delta\eta},\tag{7.55}$$

or, equivalently, by

$$\frac{u_1}{u_2} = \frac{\Delta\xi_1}{\Delta\xi_2} = \frac{\Delta\eta_1}{\Delta\eta_2},\tag{7.56}$$

where $\xi$ and $\eta$ are as shown in Fig. 7.6. This kind of graphical construction is applied in large-scale hydraulic flows where excessive accuracy is not required.[15]

## 7.8 BOUNDARY LAYER FLOW

As demonstrated earlier in this chapter, high Reynolds number flows away from solid boundaries are virtually potential, free of vorticity. Potential flows cannot extend to the vicinity of solid boundaries because vorticity is generated at a solid boundary and diffuses away. The velocity of the fluid has to change from a finite value, away from the boundary, to the velocity of the boundary, which is different (zero for stationary boundaries). This velocity change occurs in a fluid envelope adjacent to the boundary, dominated by large velocity gradients and therefore vorticity, as shown in Fig. 7.7. This vorticity-rich envelope is called a *boundary layer*. The appropriate forms of the governing momentum equations are entirely different from those of the surrounding potential flow. These boundary layer equations were first derived and studied by Prandtl[16] in connection with

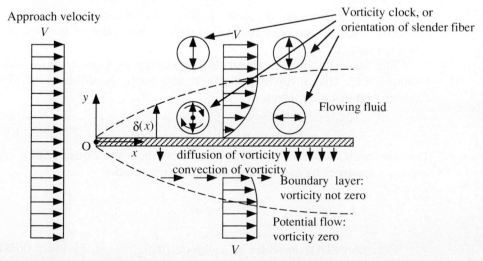

**Figure 7.7**   Boundary layer flow over flat plate.

free-stream flow over a plate. Figure 7.7 illustrates the scenario when an approaching fluid stream at univorm velocity $V$ overtakes a plate. The arrows denote vorticity indicators traveling with the fluid. The envelope is contaminated by vorticity, whereas the potential field is free of vorticity. At steady state, vorticity is generated steadily at the solid boundary and diffuses steadily away from the boundary at the rate

$$\dot{\omega}_D = v \frac{\partial^2 \omega}{\partial y^2} = v \frac{\partial^2}{\partial y^2} \left( \frac{\partial u_y}{\partial x} - \frac{\partial u_x}{\partial y} \right) \approx v \frac{\partial^3 u_x}{\partial y^3}, \tag{7.57}$$

while being swept downstream by convection at the rate,

$$\dot{\omega}_C = u_x \frac{\partial w}{\partial x} = u_x \frac{\partial}{\partial x} \left( \frac{\partial u_y}{\partial x} - \frac{\partial u_x}{\partial y} \right) \approx u_x \frac{\partial^2 u_x}{\partial x \partial y}. \tag{7.58}$$

The thickness of the boundary layer, $\delta(x)$, shown in Fig. 7.7, is defined as the distance from the plate where the diffusion and the convection rates are of the same order of magnitude, that is, where

$$O\left[ v \frac{\partial^3 u_x}{\partial y^3} \right] = O\left[ u_x \frac{\partial^2 u_x}{\partial x^2} \right], \tag{7.59}$$

and thus

$$O\left[ v \frac{V}{\delta^3(x)} \right] = O\left[ \frac{V}{\delta(x)} \frac{V}{x} \right], \tag{7.60}$$

which yields

$$\delta(x) = k \sqrt{\frac{vx}{V}}, \tag{7.61}$$

with $k$ being a constant of order 1. Thus the order of the thickness of the boundary layer is approximated by means of an *order-of-magnitude analysis*, based on vorticity dynamics.

The flow situation is similar to the startup of Couette flow examined in Example 6.13, that involves generation and unsteady diffusion of vorticity to distance $\delta(t)$ in time $t$, given by

$$\delta(t) = 4\sqrt{vt}. \tag{7.62}$$

The similarity between the two cases described by Eqns. (7.61) and (7.62) becomes apparent by substituting the kinematic relation

$$t = \frac{x}{V}. \tag{7.63}$$

Equation (7.61) provides an estimate of the boundary layer thickness and can be combined with appropriate mass and momentum balances to derive the boundary layer velocity, stress, and pressure profiles. This is an approximate, yet accurate boundary layer analysis, known as the *von Kármán method,*[17] after the person who pioneered it. An exact solution to the boundary layer flow over plate is the *Blasius solution,*[18] which, however, is much more complicated. The two approaches are summarized below.

### 7.8.1 Von Kármán's Approximate Boundary Layer Method

The velocity parallel to the plate is assumed to be nearly parabolic and can therefore be represented by

$$u_x = ay^2 + by + c. \tag{7.64}$$

The boundary conditions determine the constants $a$, $b$, and $c$:

**1.** $y = 0$: $u_x = 0$; thus $c = 0$.
**2.** $y = \delta(x)$: $u_x = V$; thus $b = V/\delta(x) - a\delta(x)$.
**3.** $y = \delta(x)$: $\partial u_x/\partial y = 0$; thus $b = -2a\delta(x)$.

From these equations, $c = 0$, $a = -V/\delta^2(x)$ and $b = 2V/\delta(x)$, resulting in

$$\frac{u_x}{V} = -\left(\frac{y}{\delta(x)}\right)^2 + 2\left(\frac{y}{\delta(x)}\right), \tag{7.65}$$

or in dimensionless form,

$$u_x^* = -\xi^2 + 2\xi, \tag{7.66}$$

where the dimensionless variables, $u_x^* = u_x/V$ and $\xi = y/\delta(x)$, have been used. Other kinds of velocity representation (e.g., exponential) are usable inasmuch as they satisfy the boundary conditions.

To complete the analysis, the thickness of the boundary layer, $\delta(x)$, approximated up to a constant by Eqn. (7.61) and appearing in the velocity profile of Eqn. (7.65), must be determined. This is easily done by means of mass and momentum balances over control volumes, extending from the plate to the potential field, as shown in Fig. 7.8. The approximate method for boundary layers above is called the von Kármán approximate method.

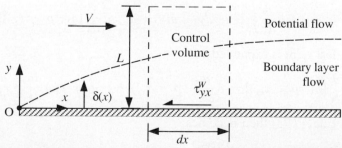

**Figure 7.8**  Mass and momentum balance in boundary layer over a flat plate (von Kármán's method).

The steady-state mass balance yields

$$\rho \int_0^{\delta(x)} u_x(y)\,dy - \rho \int_0^{\delta(x+dx)} u_x(y)\,dy + \dot{m}\,dx = 0, \tag{7.67}$$

where $\dot{m}$ is the rate of mass transfer to the control volume through the boundary layer line $AB$. This equation is rearranged in the form

$$\dot{m}\,dx = \rho \int_0^{\delta(x+dx)} u_x\,dy - \rho \int_0^{\delta(x)} u_x\,dy. \tag{7.68}$$

and in the limit of $dx \to 0$,

$$\dot{m} = \rho \lim_{dx \to 0} \frac{\int_0^{\delta(x+dx)} u_z\,dy - \int_0^{\delta(x)} u_x\,dy}{dx}. \tag{7.69}$$

By invoking the definition of the derivative, Eqn. (7.69) becomes

$$\dot{m} = \rho \frac{d}{dx} \int_0^{\delta(x)} u_x\,dy. \tag{7.70}$$

Along similar lines, a momentum balance results in

$$\rho \int_0^{\delta(x)} u_x^2\,dy - \rho \int_0^{\delta(x+dx)} u_x^2\,dy + \dot{m}V\,dx$$
$$- \tau_{yx}^w\,dx - [\delta(x)(-p + \tau_{xx})]_x + [\delta(x)(-p + \tau_{xx})]_{x+dx} = 0, \tag{7.71}$$

which can be rearranged in the form

$$\lim_{dx \to 0} \rho \frac{\int_0^{\delta(x+dx)} u_x^2\,dy - \int_0^{\delta(x)} u_x^2\,dy}{dx} - \dot{m}V + \tau_{yx}^w$$
$$+ \lim_{dx \to 0} \frac{(\delta p)_{x+dx} - (\delta p)_x}{dx} - \lim_{dx \to 0} \frac{(\delta \tau_{xx})_{x+dx} - (\delta \tau_{xx})_x}{dx} = 0. \tag{7.72}$$

Then, by invoking the definition of derivatives it results in

$$\rho \frac{d}{dx} \int_0^{\delta(x)} u_x^2\,dy - \dot{m}V + \tau_{yx}^w - \frac{d}{dx}[\delta(-p + \tau_{xx})] = 0. \tag{7.73}$$

The last term is expanded to

$$\frac{d}{dx}[\delta(-p + \tau_{xx})] = \delta \frac{d}{dx}(-p + \tau_{xx}) + (-p + \tau_{xx})\frac{d\delta}{dx} \simeq \frac{d\delta}{dx}(-p), \tag{7.74}$$

and Eqn. (7.73) then becomes

$$\rho \frac{d}{dx} \int_0^{\delta(x)} u_x^2\,dy - \dot{m}V + \tau_{yx}^w + p\frac{d\delta}{dx} = 0, \tag{7.75}$$

which is the governing *integral momentum equation.*

Thus, the characteristic approximate equations for steady, laminar boundary layers are

$$\delta(x) = k\sqrt{\frac{vx}{V}}, \tag{7.76}$$

$$\frac{u_x}{V} = -\frac{y^2}{\delta(x)^2} + \frac{2y}{\delta(x)}, \tag{7.77}$$

$$\dot{m} = \rho\frac{d}{dx}\int_0^{\delta(x)} u_x\,dy = 0, \tag{7.78}$$

and

$$\rho\frac{d}{dx}\int_0^{\delta(x)} u_x^2\,dy - \dot{m}V + \eta\left.\frac{\partial u_x}{\partial y}\right|_w + p\frac{d\delta}{dx} = 0. \tag{7.79}$$

These equations can be solved to determine the unknown constant of proportionality $k$. Then $\delta(x)$ is calculated, which, upon insertion into the velocity profile fully determines the approximate velocity $u_x$. The stress at the wall, $\tau_{yx}^w$, and the total drag are evaluated by

$$\tau_{yx}^w = \eta\left.\frac{\partial u_x}{\partial y}\right|_{y=0} \tag{7.80}$$

and

$$F_D = \int_0^L \tau_{yx}^w\,dx, \tag{7.81}$$

respectively.

### 7.8.2 Blasius's Exact Boundary Layer Analysis

Blasius solved the entire set of differential Navier–Stokes equations by a *similarity transformation* (e.g., Example 6.13). The two independent variables $x$ and $y$ in the governing continuity and momentum equations are first replaced by a single variable, $\xi$, defined by

$$\xi = \frac{y}{\delta(x)} = y\sqrt{\frac{V}{vx}}, \tag{7.82}$$

Then a solution of the form

$$u_x^* = \frac{u_x}{V} = f(\xi) \tag{7.83}$$

is sought, as demonstrated in Example 6.13. This reduces the equations to a single nonlinear ordinary differential equation, which Blasius solved as an initital value problem. The details of his analysis are presented in textbooks with extensive

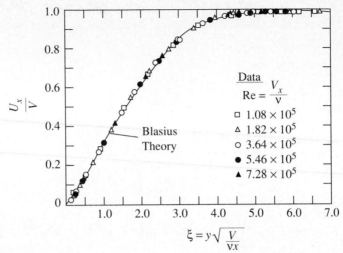

**Figure 7.9** Blasius's exact solution for boundary layer over a flat plate (From H. Schlichting, *Boundary Layer Theory*, McGraw-Hill, New York, copyright 1960; by permission of McGraw-Hill.)

boundary layer coverage.[19] The results of his analysis are tabulated in tables and figures similar to Fig. 7.9. Figure 7.9 gives the dimensionless velocity profile, $u_x/V$, in terms of the similarity variable, $\xi = y\sqrt{V/vx}$. The diagram shows the dimensionless velocity, $u^* = u_y/V$, as a function of elevation $y$ from the plate and distance $x$ downstream from the leading edge, and of the kinematic viscosity, $v = \eta/\rho$, of the fluid, incorporated in the similarity variable, $\xi = y/\sqrt{V/2vx}$. The boundary layer thickness, $\delta(x)$, is taken where the velocity approaches 99% of the potential velocity $V$. From Fig. 7.9, this occurs approximately at $\xi = 4.86$. Thus,

$$\delta(x) = 4.86\sqrt{\frac{vx}{V}}. \tag{7.84}$$

Other useful quantities can be evaluated as follows:

**1.** Local shear stress, along the boundary:

$$\tau_w = \tau_{yx}|_{y=0} = \eta \frac{\partial u_x}{\partial y}\bigg|_{y=0} = \eta\sqrt{\frac{V^3}{2vx}} f''(\xi = 0) = 0.332(\rho V^2)\sqrt{\frac{v}{Vx}} \tag{7.85}$$

**2.** Drag on both sides of finite length $L$ and width $w$:

$$F_D = 2w\int_0^L \tau_w\, dx = (1.328\sqrt{V^3\rho\eta L})w \tag{7.86}$$

**3.** Transverse velocity component:

$$u_y = -\int_0^y \frac{\partial u_x}{\partial x}\, dx = \sqrt{\frac{vV}{2x}}[\xi f'(\xi) - f(\xi)] \tag{7.87}$$

Example 7.9 illustrates the use of Fig. 7.9 and Eqns. (7.84) to (7.87) to solve boundary layer flow problems over plates.

**Example 7.9: Boundary Layer Flow Analysis by Blasius's Exact Solution**

Consider flow over a plate of length $L = 4\,\text{m}$ and width $W = 3\,\text{m}$ with water of denstiy $\rho = 1\,\text{g/cm}^3$ and viscosity $\eta = 1\,\text{cP}$ approaching at velocity $V = 1\,\text{cm/s}$.

(a) What is the profile of the resulting boundary layer? What is the boundary layer thickness at distances $x = 0$, $x = 0.5L$, and $x = L$ from the leading edge of the plate?

(b) What is the velocity profile at elevations $y = 0$, $y = 0.5\delta(x)$ and $y = \delta(x)$ over the plate? What is the velocity value at the point of Cartesian coordinates $[x = 0.5L, y = 0.5\delta\ (x = 0.5L)]$?

(c) What is the total drag froce exerted by the water on both faces of the plate?

(d) What are the corresponding answers for the same flow of air of viscosity $\eta = 0.02\,\text{cP}$ and density $\rho = 0.001\,\text{g/cm}^3$?

**Solution**    A. *Water boundary layer:* The maximum Reynolds number is

$$\text{Re} = \frac{\rho V L}{\eta} = \frac{1 \times 1 \times 400}{0.01} = 4 \times 10^4,$$

and therefore the flow is laminar over the entire plate.

(a) Thickness of boundary layer, by Eqn. (7.84):

$$\delta(x) = 4.86\sqrt{\frac{vx}{V}} = (0.486x^{1/2})\ \text{cm}.$$

Thus $\delta(x = 0) = 0$ $\delta(x = 0.5L) = 6.87\,\text{cm}$ and $\delta(x = L) = \delta_{\max} = 9.72\,\text{cm}$.

(b) For $y = 0$, $0.5\delta(x)$, and $\delta(x)$, the similarity variable $\xi$ is 0, 2.43, and 4.86, respectively. For these values of $\xi$, the diagram of Fig. 7.9 yields values of $u_x/V$ equal to 0, 0.72, and 0.99, respectively. To satisfy these values, an admissible velocity profile is

$$\frac{u_x(y)}{V} = a\xi^2 + b\xi + c = 0.88\xi^2 + 1.88\xi = 1.88\left(\frac{y}{0.486\sqrt{x}}\right) + 1.88\left(\frac{y}{0.486\sqrt{x}}\right)^2.$$

Thus

$$u_x = [x = 0.5L, y = 0.5\delta(0.5L)] = 0.72V = 0.72\,\text{cm/s}.$$

(c) Drag on both sides, by Eqn. (7.86):

$$F_D = (1.328\sqrt{V^3\rho\eta L}) = (1.328\sqrt{1 \times 1 \times 0.01 \times 400}) \times 300 = 796.8\,\text{dyn}.$$

**B.** *Air boundary layer:* Laminar flow persists in this case, too:

$$\mathrm{Re} = \frac{1 \times 10^{-3} \times 1 \times 400}{2 \times 10^{-4}} = 2000.$$

The values for air corresponding to (a), (b), and (c) are:

**(a)** Boundary layer thickness:

$$\delta(x) = 4.86 \sqrt{\frac{0.2x}{1}} = (2.17x^{1/2}) \text{ cm}$$

**(b)** Velocity profile:

$$\frac{u_x(y)}{V} = a\xi^2 + b\xi + c = 1.88\left(\frac{y}{2.17\sqrt{x}}\right) - 0.88\left(\frac{y}{2.17\sqrt{x}}\right)^2$$

**(c)** Drag force:

$$F_D = (1.328 \times \sqrt{1 \times 10^{-3} \times 2 \times 10^{-4} \times 400}) \times 300 = 3.563 \text{ dyn}$$

**C.** *Comparisons water–air boundary layers*: Air laminar boundary layer persists at higher Reynolds number, has larger thickness, and results in a much lower drag force than the corresponding water boundary layer.

The preceding analysis provides an exact solution to boundary layer flow over plates and similar slender bodies. Approximate solutions to the same flows are obtained by means of Eqns. (7.76) to (7.79). These approximate solutions are significantly simpler and not much less accurate than Blasius's solution, as demonstrated in Example 7.10.

### Example 7.10: Boundary Layer Analysis by Blasius's Exact Solution and by von Kármán's Approximation

Assume that the velocity profile under the boundary layer is linear, given by $u_x/V = y/\delta$. Solve von Kármán's equations, Eqns. (7.76) to (7.79), with this velocity profile and compare the results with Blasius's solution, given by Eqns. (7.84) to (7.87).

**Solution** By substituting the velocity prople $u_x/V = y/\delta$ in Eqn. (7.79) and integrating from $y = 0$ to $y = \delta$, we find that

$$\delta \frac{d\delta}{dx} = \frac{6v}{V}, \tag{7.10.1}$$

which is integrated to

$$\frac{\delta^2}{2} = \frac{6vx}{V_\infty} \quad \text{or} \quad \delta = \left(\frac{12vx}{V_\infty}\right)^{1/2} = 3.46\left(\frac{vx}{V_\infty}\right)^{1/2}. \tag{7.10.2}$$

Comparing this result with Blasius's exact solution for the same boundary layer, Eqn. (7.84), we find that this approximate solution gives a boundary layer thickness that is about 3.46/4.86, or approximately 70%, of the correct one.

The drag coefficient is computed as

$$C_D = \frac{\eta(V/\delta)}{\frac{1}{2}\rho V^2} = \frac{2\nu}{V}\left(\frac{V}{12\nu x}\right)^{1/2} = 0.577\left(\frac{\nu}{Vx}\right)^{1/2}. \tag{7.10.3}$$

By comparing this value with the drag coefficient based on Blasius's solution, we find that this approximate solution gives a drag coefficient of 0.557/0.664, or about 87% of the correct solution. In the same way, all the other properties of the boundary layer can be computed for this assumed velocity profile or for any other assumed one. The important conclusion here is that this very simple profile gives reasonably accurate results with an expenditure of much less effort than in the Blasius solution. These results become more accurate as more sophisticated, admissible velocity profiles are used, for example the one given by Eqn. (7.66).

### 7.8.3 Turbulent Boundary Layers over Plates

The boundary layer becomes turbulent at a local Reynolds number of approximately

$$\mathrm{Re} = \frac{Vx}{\nu} > 10^5. \tag{7.88}$$

Beyond this value, the Blasius and von Kármán laminar flow solutions both fail. Experimental observations and dimensional analysis suggest that the approximations

$$\delta = 0.37x\left(\frac{\nu}{Vx}\right)^{1/5} \tag{7.89}$$

and

$$F_D = 2C_D\tfrac{1}{2}\rho V^2 A \qquad C_D = 0.072\left(\frac{\nu}{Vx}\right)^{1/5} \tag{7.90}$$

apply for turbulent boundary layers over flat surfaces. In Eqn. (7.90), $C_D$ is the drag coefficient introduced in Section 4.6, and $A$ the boundary area wetted by the fluid. Since the criterion is, by Eqn. (7.88), based on distance $x$ downstream from the leading edge, there may be cases of a laminar boundary layer followed by a turbulent boundary layer over the same plate, as demonstrated in Example 7.11.

**Example 7.11: Turbulent Boundary Layer Flow**

Consider water flow over a plate, as described in Example 7.9.

**(a)** What is the maximum velocity $V_c$ allowed for laminar flow throughout?

**(b)** If the free-stream velocity is $2V_c$, where does the transition to turbulent flow occur? What is the velocity along this transition line?

**(c)** Calculate the boundary layer thickness profile throughout the plate for cases (a) and (b).

**(d)** Calculate the total drag force on the plate for cases (a) and (b).

**Solution**

**(a)** The maximum allowed velocity is

$$V_c = \frac{10^5 \eta}{\rho L} = \frac{10^5 \times 10^{-2}}{1 \times 400} = 2.5 \text{ cm/s}.$$

**(b)** For $V = 2V_c = 5$ cm/s, transition to turbulent flow occurs at

$$x_L = \frac{10^5 \, \eta}{\rho V} = \frac{10^5 \times 10^{-2}}{1 \times 5} = 200 \text{ cm}, \qquad \delta(x) = 4.86 \sqrt{\frac{0.01 \times 200}{5}} = 3.07 \text{ cm}$$

The velocity is given by

$$\frac{u_x(y)}{V} = 1.88\xi - 0.88\xi^2 = 1.88\left(\frac{y}{0.217\sqrt{x}}\right) - 0.88\left(\frac{y}{0.217\sqrt{x}}\right)^2 = 0.614y - 0.093y^2$$

**(c)** In the first case,

$$\delta^I_{(x)} = 4.86 \sqrt{\frac{0.01x}{2.5}} = (0.307\sqrt{x}) \text{ cm}, \qquad 0 < x < 400.$$

In the second case,

$$\delta^{II}_{(x)} = \begin{cases} \delta^{II}_L(x) = 4.86 \sqrt{\dfrac{0.01x}{5}} = (0.217\sqrt{x}) \text{ cm}, & 0 < x < 200. \\[3ex] \delta^{II}_T(x) = 0.37x\left(\dfrac{v}{Vx}\right)^{1/5} = 0.107x^{4/5} \text{ cm}, & 200 < x < 400. \end{cases}$$

**(d)** Case I:

$$F^I_D = (1.328\sqrt{V^3_c\rho\eta L})w = (1.328\sqrt{2.5^3 \times 1 \times 0.01 \times 400}) \times 300 = 3149.6 \text{ dyn}.$$

Case II:

$$F^{II}_D = (1.328\sqrt{V^3\rho\eta x_L})w + 2 \times 0.072\left(\frac{v}{V(L - x_L)}\right)^{1/5} \rho V^2(L - x_L)w$$

$$= (6299 + 21{,}600) \text{ dyn}.$$

## 7.8.4 Boundary Layer over Nonslender Bodies

The potential flow may itself be affected by the presence of a nonslender body or by the presence of the growing boundary layer. The pressure gradient, which is zero in the boundary layer in slender bodies, is not zero in general:

$$\frac{dp}{dx} = V\frac{dV}{dx} \neq 0. \tag{7.91}$$

The procedure is one of trial and error until the assumed velocity of the potential region satisfies all the equations. In other cases, the velocity of the potential region—which takes into account the presence of the body—may be known, in which case the analysis is straightforward and proceeds as in the case of a slender body (e.g., potential flow around a cylinder).

## 7.9 COMPUTATIONAL FLUID MECHANICS

When none of the preceding approaches apply to the flow under consideration, the Navier–Stokes equations may be discretized and solved numerically by finite difference and/or finite element techniques.[20,21] In the finite element method, the unknown velocities, $u_x$ $u_y$, and $u_z$ and the unknown pressure $p$ (and the unknown domain dimension $h$, in case of free surface flows) are expanded in terms of known basis functions in space, $\phi_k(x, y, z)$, and unknown coefficients, $(u_i)_k$ and $p_k$, that represent velocities and pressure values at $k$ selected grid points of the domain:

$$u_i = \sum_k (u_i)_k \phi_k(x, y, z), \qquad i = x, y, z \tag{7.92}$$

$$p = \sum_k p_k \phi_k(x, y, z). \tag{7.93}$$

The unknown coefficients, $(u_i)_k$ and $p_k$, are determined by substituting these expansions in the continuity and momentum equations, and minimizing the interpolation error. Minimization of interpolation can be done by several methods, for example the method of *least squares,* which minimizes the difference or "distance" between the exact solution and the interpolated one.

The appearance and evolution of fast computers, supercomputers, and machines of parallel processing allow the replacement of often slow and expensive laboratory experiments with *numerical experiments*: Data are supplied to an appropriate mathematical problem of a real flow or process. The computer solves the mathematical problem by a numerical algorithm placed into its memory by the user, providing quick and relatively inexpensive answers. Graphic display hardware and software enhance further these results for design and optimization. Computers can also be used to study flows and processes where tractable mathematical models do not exist to start with, for example molecular motion, orientation of fibers or macromolecules in complex flow fields, and others. In this case, the flow field with its population is input to the computer's memory and allowed to rearrange position according to fundamental physical laws, for example Newton's law of motion. The resulting motion and arrangement in space and time define statistically average properties of interest. This method of computer-aided analysis is called *simulation*.

## 7.10 EXPERIMENTAL FLUID MECHANICS

When all other approaches fail to produce a solution, physical experiment is often the last resort. However, in many cases experiments are costly and time consuming and often demand the application of highly sophisticated techniques. Pressure transducers, based on thrust on deformable membranes, measure pressure and stress.[22] Thermal and optical methods,[23] such as hot-wire assemometry and laser-doppler velocimetry (LDV), measure local velocity components and evolution with time. Free surface and film thickness can be measured by mechanical probes of high resolution,[24] by section photography,[25] and by wire conductance methods.[26] Experiment, of course, may serve a diversity of other purposes not necessarily associated with flow solution, such as determinations of flow parameters, scaling down large flows, investigation of mechanisms of flow, and many others. Experiments are also important to test the accuracy of theoretical analysis; the ultimate test of any theoretical or numerical analysis is always the real world.

## PROBLEMS

1. *Radial flows with the Navier–Stokes equations.* Repeat the solution to Problems:
   (a) 5.1 oil well;
   (b) 5.2 boiling of liquids;
   (c) 5.21 radial flow between porous walls.
   Use the appropriate Navier–Stokes equations. Find both the velocity and the pressure distributions.

2. *Rectilinear flows with the Navier–Stokes equations.* Repeat the solution to Problems:
   (a) 6.3 transfer of highly viscous crude oil;
   (b) 6.4 recirculation flow;
   (c) 6.7a superposition of flow: helical flow;
   (d) 6.17 axisymmetric film flow.
   Use the appropriate Navier–Stokes equations. Find both the velocity and the pressure distributions.

3. *Torsional flows with the Navier–Stokes equations.* Repeat the solution to Problems:
   (a) 6.6 Couette viscometer;
   (b) 6.7b spiral flow;
   (c) 6.18 centrifugation.
   Use the appropriate Navier–Stokes equations. Find both the velocity and the pressure distributions.

4. *Cone-and-plate rheometer.* The geometrical description and working equations of this

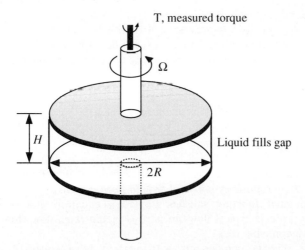

T, measured torque

$\Omega$

H

Liquid fills gap

2R

**Figure P7.5**  Parallel-plate viscometer.

device to measure viscosity are given in Section 9.2, based on the assumption that the velocity profile is $u_\theta(r, z) = \Omega rz/H(r) = \Omega rz/r \tan \beta \simeq \Omega z/\beta$, where $\Omega$ is the angular velocity, $\beta$ the cone angle, $z$ the distance from the lower stationary plate, and $r$ the radial distance. Show that this velocity profile is admissible, and find the corresponding pressure distribution. (*Hint:* See Fig. 9.5.)

**5.** *Parallel-plate rheometer.* This device consists of a lower stationary plate and an upper rotating plate with angular velocity $\Omega$, both of radius $R$, set a small distance $H$ apart (Fig. P7.5). Show that under these conditions, the resulting velocity profile is $u_\theta(r, z) = \Omega rz/H$. What is the implied pressure distribution $p(r, z)$?

**6.** *Squeezing flow at constant velocity.* A liquid of viscosity $\eta$ and density $\rho$ is being squeezed slowly between two disks of radius $R$, initially set a distance $2H$ apart, approaching slowly at constant squeezing speed $V$ and $-V$ with respect to the midplane of symmetry. Analyze the flow by means of the appropriate Navier–Stokes equations and calculate:
**(a)** the velocity and pressure distributions;
**(b)** the squeezing force required.

**7.** *Squeezing flow at constant force.* A liquid of viscosity $\eta$ and density $\rho$ is being squeezed slowly between two disks of radius $R$, initially set at distance $2H$ apart, approaching under constant external squeezing forces $F/2$ and $-F/2$ on the two disks. Analyze the flow by means of the appropriate Navier–Stokes equations and calculate:
**(a)** the velocity and pressure distributions;
**(b)** the squeezing speed achieved.

**8.** *Potential flow past a sphere.* The velocity potential of free-stream laminar flow past a sphere of radius $a$ at high Reynolds number is

$$\phi = V \cos \theta \left( r + \frac{a^3}{2r^2} \right),$$

$$\psi = -\tfrac{1}{2} V r^2 \sin^2 \theta \left( 1 - \frac{a^3}{r^3} \right).$$

**(a)** Show that the potential function $\phi$ satisfies the spherical Laplace's equation as well as the appropriate boundary conditions.

**(b)** What is the expression for the streamfunction?

**(c)** What are the resulting velocity and pressure distribution?

**(d)** Show that these potential flow profiles do not extend down to the solid boundary. (*Hint:* Examine the boundary velocity and resulting drag force on sphere.) Why is it so?

9. *Potential flow in porous media.* The velocity components of a two-dimensional fluid flow in porous media follows Darcy's law:

$$u_x = -\frac{k}{\eta}\frac{\partial}{\partial x}(p + \rho g_x z), \qquad u_y = -\frac{k}{\eta}\frac{\partial}{\partial y}(p + \rho g_y z). \qquad (P7.9.1)$$

**(a)** What is a potential-like function for this flow? Is the flow really potential? If not, how do you satisfy the terminology *potential flow in porous media*?

**(b)** Show that the potential function satisfies Laplace's equation and boundary conditions that may arise in typical flows in porous media (e.g., flow from earth level down to an impermeable rock).

**(c)** The corresponding velocity components with respect to a spherical system centered within the porus medium are

$$u_r = -\frac{k}{\eta}\frac{\partial p}{\partial r}, \qquad u_\theta = -\frac{k}{\eta}\frac{\partial p}{\partial \theta}. \qquad (P7.9.2)$$

Show that the pressure $p$ obeys Laplace's equation,

$$\sin\theta \frac{\partial}{\partial r}\left(r^2 \frac{\partial p}{\partial r}\right) + \frac{\partial}{\partial \theta}\left(\frac{\partial p}{\partial \theta}\sin\theta\right) = 0. \qquad (P7.9.3)$$

10. *Water penetration in soil.* Consider flow of water in soil of given permeability $k$, penetrating from surface to depth $2H$. A long cylindrical cavity of radius $R \ll H$ runs horizontally at depth $H$. Calculate the resulting pressure distribution around the cavity for the following types of cavities:

**(a)** cavity filled with water, offering no resistance to water penetration;

**(b)** cavity replaced by pipe, impermeable to penetration;

**(c)** cavity filled with natural gas at pressure $p_g$, impermeable to penetration;

**(d)** cavity filled with crude oil of viscosity 10 times that of water.

[*Hint:* Examine profiles of the form

$$p(r, \theta) = \left(c_1 r + \frac{c_2}{r^2}\right)\cos\theta,$$

where $r$ is the radial distance and $\theta$ angle from the horizontal plane of symmetry.]

11. *Tornados.* A tornado may be idealized as a potential vortex with a rotational "eye" or core that behaves approximately as a solid body. A rough rule of thumb is that the radius of the eye is on the order of 100 ft. How does the pressure vary along the ground around the eye? For a tornado with a maximum wind velocity of 100 miles per hour, what is the maximum drop in pressure? Show how this underpressure may inflict damages (e.g., lifting the roof) to a house in the tornado's eye.

12. *Film flow analysis by streamfunction.* A Newtonian liquid of density $\rho$ and viscosity $\eta$ flows at a rate $Q$ (per unit width) down an inclined plate of infinite width, length $L$, and angle $\theta$ with the horizontal. The falling liquid forms a thin film of depth $H$ whose free surface is in contact with the surrounding still air. (a) What drives the flow?

(b) What are the pressure gradients $\partial p/\partial x$, $\partial p/\partial y$, and $\partial p/\partial z$? (c) Where is the vorticity zero? Why? (d) Show that $\partial\psi/\partial x = 0 \Leftrightarrow \psi = c$ along a $y = c$ curve. (e) Without using the momentum equations, show that $u = u(y)$. Solve the flow by the streamfunction formulation, as follows: (f) Select the equations and boundary conditions (g) Find the vorticity distribution. (h) Find the streamline pattern. (i) Sketch the streamlines $\psi = c$ and vorticity lines $\omega = c$. (j) Find the velocities and show that the assumptions made above were correct. (k) What is the relation between the pressures at two points on the same streamline? Apply the relation to the free surface and to the solid wall. (l) What is the vorticity distribution along the wall? Along the free surface? From these, can you deduce the pressure profile $p(x)$?

13. *Potential flows.* Consider the two-dimensional flow field against a plate given by the equations $u_x = 4x$ and $u_y = -4y$. Show that this flow satisfies the equation of continuity. Calculate the vorticity of the flow. What is the pressure distribution? A two-dimensional incompressible flow field is described by the equations $v_\theta = \Omega r$ and $v_r = 0$, in which $\Omega$ is constant. Sketch this flow and show that it satisfies the continuity equation. Also, calculate the vorticity to the flow and the pressure distribution. What is $\Omega$?

14. *Bubble rise in potential flow.*[7,27,28] Figure P7.14 shows the cross section through a large bubble rising between two vertical plates, or equivalently, water overtaking the same bubble. Far back from its nose, the bubble is cylindrical with radius $R - d$. When $d$ is not too small and surface tension and viscosity effects are negligible, the flow is potential and the liquid velocity across the narrow annulus is uniform.
  (a) Show that in this case the uniform velocity is given by

$$u(y) = \sqrt{2gy} \tag{P7.14.1}$$

with respect to the shown Cartesian system.
  (b) By appropriate liquid mass balance show that

$$\frac{d}{R} = 1 - \left(1 - \frac{V}{\sqrt{2gy}}\right)^{1/2} \tag{P7.14.2}$$

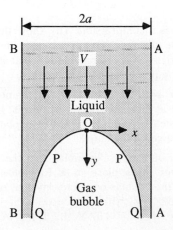

**Figure P7.14**  Two-phase potential flow.

and for $x \gg R$ it becomes

$$\frac{d}{R} = \sqrt{\frac{V^2}{8gy}}.$$  (P7.14.3)

(c) Experimental observations show that the rising velocity is given by[27]

$$V = 0.48\sqrt{gR}.$$  (P7.14.4)

What is the resulting $d/R$ ratio? Does this ratio validate the assumptions made a priori of potential flow?

15. *Bubble rise in viscous flow.* In Fig. P7.14, when $d$ is not so small and vorticity generated at the wall contaminates the entire liquid annular flow, the velocity profile becomes parabolic. In this case, the resulting volumetric flow rate through the annular liquid film, falling under gravity while being dragged in the opposite direction by the velocity $V$, is given by

$$\dot{V} = 2\pi R\left(Vd + \frac{gd^3}{3\nu}\right).$$  (P7.15.1)

(d) Show that with $V$ given by Eqn. (P7.14.4), the resulting liquid film thickness is

$$\frac{d}{R} \approx 0.90\left(\frac{\nu^2}{R^3g}\right).$$  (P7.15.2)

*Application:* A common situation that can be modeled by these equations is water draining out of a vertical tube in the form of an annular sheet around an air pocket occupying the core of the tube at its exit.

(e) What is the resulting sheet thickness for pipe radii 1, 10, and 100 cm?

16. *Airflow over a flat plate.* Air at 100°F is flowing over a flat plate that is 1 ft wide. It has $\nu_{100°F} = \eta/\rho = [3.96 \times 10^{-7}\,\text{lb}_f \cdot \text{s/ft}^2/0.079\,\text{lb}_m/\text{ft}^3] = 1.89\,\text{ft}^2/\text{s}$, and uniform approach velocity $V = 7.2\,\text{ft/s}$. Estimate: (1) the boundary layer thickness 1 ft from the leading edge, and (2) the total drag force on the plate down to 2 ft from the leading edge. At what length downstream does the layer become turbulent?

17. *Flow analysis by streamfunction.* A flow is characterized by the streamfunction

$$\psi = 3ax^2y - ay^3.$$

(a) Show that the flow is incompressible. (b) Show that the flow is irrotational. (c) Show that the magnitude of the velocity at any point in the flow field depends only on its distance from the origin of coordinates. (d) What is the pressure in terms of the pressure at the stagnation point? (e) Sketch the flow.

18. *Smooth plate moving in water.* A ship is towing a smooth plate 1 ft wide and 200 ft long through still water at 68°F at a speed of 50 ft/s. Determine the boundary layer thickness at the end of the plate and the drag on the plate. (*Hint:* Re = ?) How would you calculate the drag force on the ship?

19. *Laminar boundary layer over plate.* Assuming that the transition from a laminar to a turbulent boundary layer takes place at a Reynolds number of $10^6$, what is the maximum thickness for the laminar boundary layer over a flat plate for (a) air flowing at 10 ft/s ($\nu = 1.43 \times 10^{-4}\,\text{ft/s}$); (b) water flowing at 10 ft/s ($\nu = 1.08 \times 10^{-5}\,\text{ft}^2/\text{s}$); (c) glycerin flowing at 10 ft/s ($\nu = 8.07 \times 10^{-3}\,\text{ft}^2/\text{s}$)? All fluids are at 68°F.

**20.** *Use of Blasius's solution to determine normal velocity.* From Blasius's solution (Fig. 7.9), find the normal velocity component $u_y$, as follows: Starting with the mass balance equation, prove that

$$u_y = \int_0^y \frac{du_x}{dx}\,dy = \frac{V}{2(Vx/\nu)^{1/2}} \int_0^\xi \frac{d(u_x/V)}{d\xi}\,\eta\,d\xi. \qquad \text{(P7.20.1)}$$

Then rearrange in the form

$$\frac{u_y}{V}\left(\frac{Vx}{\nu}\right)^{1/2} = \frac{1}{2}\int_0^\xi \frac{d(u_x/V_\infty)}{d\xi}\,\xi\,d\xi. \qquad \text{(P7.20.2)}$$

**(a)** Show that $u_y(y = 0)$ and $u_y(y \to \infty) \to 0$.
**(b)** Is it true that the vorticity is maximum where $u_y = 0$? What is the vorticity as $y \to \infty$? Justify your answer.
**(c)** Explain how would you utilize the diagram of Fig. 7.9 to estimate $u_y$ given by Eqn. (P7.20.2).
**(d)** From Eqn. (P7.20.2) alone, can you conclude the direction and order of magnitude of $u_y$?

**21.** *River flow.* A river of width $W$ meters and depth $D$ meters flows steadily with a total mass flow rate of $\dot{m}$ kg/s. The width is much bigger than the depth (i.e., $W \gg D$). The viscosity and density of the water are $\eta$ and $\rho$, respectively. Split the flow into meters of width of potential flow and meters of width of boundary layer flow. Then show all necessary and sufficient boundary conditions for both flows. At what distance downstream do the two boundary layers join? [If you need to use a diagram, show how (sketch it) and represent any numerical values by constants.]

**22.** *Incompressible Newtonian flows.* Complete the following table, in which $u_x$ is the principal velocity in the $x$-direction.

| | True for incompressible Newtonian flows? | | | | |
| --- | --- | --- | --- | --- | --- |
| Assertion | Potential | Shear | Stretching | Lubrication | Boundary layer |
| $\nabla \cdot \bar{u} = 0$ | | | | | |
| $\psi$ exists | | | | | |
| $\phi$ exists | True | False | True | False | False |
| $\nabla^2\psi = 0$ | | | | | |
| $\partial u_x/\partial x \ll \partial u_x/\partial y$ | | | | | |
| $\partial u_x/\partial x \gg \partial u_x/\partial y$ | | | | | |
| Vorticity $w = 0$ | | | | | |
| $\Delta(p/\rho + u^2/2 + gz) = 0$ | | | | | |
| Material lines rotate | | | | | |
| Torsional flow is | | | | | |

**23.** *Two-dimensional incompressible flow.* In two-dimensional incompressible flow, the fluid velocity components are given by $u_x = x - 4y$ and $u_y = -y - 4x$.
**(a)** Show that this is an admissible flow and obtain the expression for the stream function.
**(b)** If the flow is potential, also obtain the expression for the velocity potential.

(c) Sketch the flow.

(d) If the pressure at $x = y = 0$ is $p_0$, find the pressure distributions $p(x)$ and $p(y)$ along the x- and y-axis, respectively.

24. *Flow of crude oil over a flat plate.* Crude oil at 70°F ($v = 10^{-4} \, \text{ft}^2/\text{s}$, $\rho = 1.66 \, \text{lb/ft}^3$) with a free-stream velocity of 10 ft/s flows past a thin flat plate that is 4 ft wide and 6 ft long in a direction parallel to the flow.

(a) Determine and plot the boundary layer thickness and the shear stress distribution along the plate.

(b) For the conditions of part (a), determine the resistance on each side of the plate.

(c) If the velocity distribution across the boundary layer is $u_x/V = y/\delta(x)$, what is the normal velocity, $u_y$, directed away from the plate? Sketch $u_x$ and $u_y$ vs. $y$ and $x$.

25. *Radial flows.*

(a) Show that all single-fluid radial flows (plane, cylindrical, and spherical) are potential flows. For each class, find the stream function and the potential function. Then calculate the vorticity and the pressure distributions.

(b) Show how the potential theory can be used to study radial flows in porous media. For each flow calculate the velocity and pressure distribution.

(c) Repeat the considerations of parts (a) and (b) to steady compressible flow.

26. *Heat conduction similarities to potential flow.*[29] Heat is produced steadily across a 100-cm-thick slab. The ends of the slab are at 150 and 350°C so that the midpoint temperature is 250°C. A 10-cm-diameter cyclindrical hole is drilled around the centerline of the slab.

(a) Sketch the isotherms of the heat conduction.

(b) By heat balance over the appropriate control volume, derive the governing equation of steady conduction. What are the appropriate boundary conditions?

(c) Show that the solution with respect to a polar system centered at the center of the hole is

$$\tau(r, \theta) = A\left(r + \frac{r^2}{r}\right)\sin\theta + B,$$

and define the constants $A$ and $B$.

(d) To what potential flow situation does this problem correspond? What are the answers corresponding to (a), (b), and (c)?

(e) What are the streamfunction, potential function, velocity components, and vorticity of the flow?

27. *Pressure of converging/diverging flows.* Calculate the pressure distribution of the flow of Section 7.5, given the computed velocity profiles for creeping flow. Compare to those of rectilinear flow over the same channel length at the same flow rate $\dot{V}$. (You may need to convert from a polar to a Cartesian system.)

28. *Von Kármán boundary layer.* Repeat the analysis of Example 7.10 by assuming a quadratic velocity profile,

$$\frac{u_x}{V} = -\frac{y^2}{\delta^2(x)} + 2\frac{y}{\delta(x)}.$$

(a) Show that this velocity profile satisfies the appropriate boundary conditions.

(b) Calculate the predicted boundary layer thickness and drag force.

(c) Compare the part (b) values to those of Example 7.10 and to Blasius's exact analysis, and comment on the accuracy of von Kármán's approximate solution in connection with the velocity profiles assumed.

**29.** *Expotential boundary layer.*

(a) Evaluate the velocity profile

$$u_x = V(1 - e^{-ay/\delta(x)})$$

as a boundary layer velocity profile. What is the value of $a$?

(b) Calculate the predicted boundary layer thickness and drag force.

(c) How do they compare with Blasius's exact solution?

(d) What are the values of (a), (b) and (c) for flow of water and air at 1 atm and 20°C at velocity $V = 2$ m/s over a 10-m-long plate?

**30.** *Drag force on flying objects.* Calculate the shear force experienced by a flying bird at velocity 40 miles/h, an aircraft at velocity 300 miles/h, and a rocket at 1000 miles/h. Consider a similar bird, aircraft, and rocket minus wings of relative sizes 1:10,000:1000 and a reasonable bird body. Repeat for a reasonable body car running at 60 miles/h.

# REFERENCES

1. G. C. Stokes, On the theories of the internal friction of fluids in motion, and of the equilibrium and motion of elastic solids, *Trans. Cambridge Philos. Soc.*, 8, 287 (1845).

2. C. Truedell, *Essays in the History of Mechanics,* Springer-Verlag, New York, 1968.

3. S. Whitaker, The development of fluid mechanics in chemical engineering in *One Hundred Years of Chemical Engineering* (ed. N. A. Peppas), Kluwer Academic, New York, 1989,

4. R. B. Bird, R. C. Armstrong, and O. Hessager, *Dynamics of Polymeric Liquids*; Vol. 1, *Fluid Mechanics*, Wiley, New York, 1977.

5. G. C. Stokes, On the effect of the internal friction of fluids on the motion of pendulums, *Trans. Cambridge Philos. Soc.*, 9, 8 (1851).

6. Z. Tadmor and G. C. Gogos, *Principles of Polymer Processing*, Wiley, New York, 1979.

7. G. K. Batchelor, *An Introduction to Fluid Dynamics,* Cambridge University Press, Cambridge, 1979.

8. S. Middleman, *Fundamentals of Polymer Processing,* McGraw-Hill, New York, 1977.

9. O. Reynolds, On the theory of lubrication and its application to Mr. Beauchamp Tower's experiments including an experimental determination of the viscosity of olive oil, *Philos. Trans. Roy. Soc. London*, 177, 157 (1886).

10. B. V. Deryagin and S. M. Levi, *Film Coating Theory,* Focal Press, New York, 1964.

11. T. Carmady and H. Kobus, *Hydrodynamics by Daniel Bernoulli and Hydraulics by Johann Bernoulli,* Dover Publications, New York, 1968.

12. H. F. Weinberger, *Partial Differential Equations,* Blaisdell, Walthm, Mass., 1969.

13. I. S. Sokolnikoff and R. M. Redheffer, *Mathematics of Physics and Modern Engineering,* McGraw-Hill, New York, 1966.

14. K. J. Mikkelsen, C. W. Macosho, and G. G. Fuller, Opposed jets: an extensional rheometer for low viscosity liquids, *Proc. Internat. Congr. Rheol.*, Sydney, 1988.

15. G. K. Noutsopoulou, *Lectures of Theoretical and Applied Hydraulics,* National Technical University of Athens, Athens, Greece, 1975.

16. L. Prandtl, Veber Fluessigkeitsbewegung mit kleiner Reibung (Concerning fluid movements with small friction) in *Vier Abhandlungen zur Hydrodynamic,* Goettingen, 1927.

17. T. von Kármán and H. S. Tsien, Boundary layers in compressible fluid, *J. Aeronaut. Sci.*, 5, 227 (1938).

18. H. Blasius, Grenzschichten in Flüssigkeiten mit kleiner Reibung, *Z. Math. Phys.*, 56 1 (1908) (transl. NACA TM 1256).

19. H. Schlichting, *Boundary Layer Theory* 7th ed. McGraw-Hill, New York, 1979.

20. B. A. Finlayson, *Nonlinear Analysis in Chemical Engineering,* McGraw-HIll, New York, 1980.

21. J. J. Connor and C. A. Brebbia, *Finite Element Techniques for Fluid Flow,* Newnes-Butterworth, Boston, 1976.

22. K. Higashitani and A. S. Lodge, Hole pressure error measurements in pressure generated shear flow, *Trans. Soc. Rheol.,* 19, 307 (1975).

23. R. J. Goldstein, *Fluid Mechanics Measurements,* Hemisphere, New York, 1983.

24. J. J. von Rossum, Viscous lifting and drainage of liquids, *Appl. Sci. Res.,* A7, 121 (1958).

25. W. Merzkirch, *Flow Visualization,* Academic Press, New York, 1974.

26. A. S. Telles and A. E. Dukler, Statistical characteristics of thin, vertical, wavy, liquid films, *Ind. Engrg. Chem. Fundam.,* 9, 412 (1970).

27. R. S. Brodkey, *The Phenomena of Fluid Motion,* Addison-Wesley Series in Chem. Engng., Ohio State University, Columbus, Ohio, 1967.

28. R. Clift, J. R. Grace, and M. E. Weber, *Bubbles, Drops, and Particles,* Academic Press, New York, 1978.

29. H. M. Schey, *Div, Grad, Curl and All That*: *An Informal Text on Vector Calculus,* W. W. Norton, New York, 1973.

# 8

## Nearly Unidirectional Flows: Lubrication and Stretching Flows

## 8.1 INTRODUCTION: LUBRICATION VS. STRETCHING FLOWS

Flows having a dominant velocity component are *nearly unidirectional*. The other velocity component(s) are vanishingly small compared to the driving component. The magnitude of the dominant component may change with respect to more than one direction, in general, but there is always a direction of dominant change. The most important flows that can be analyzed under this approximation are *lubrication flows* and *stretching flows*.

*Lubrication flows*[1] are flows between nearly parallel walls of a small inclination $\alpha$ with respect to each other, as well as thin film flows under nearly planar interfaces. The dominant velocity component is nearly parallel to the walls or the interface, and changes significantly only with respect to the perpendicular direction(s). Thus, the necessary conditions for lubrication flow in the $x_1$-direction are

$$|u_1| \gg |u_2| + |u_3| \tag{8.1}$$

and

$$\left| \frac{\partial u_1}{\partial x_1} \right| \ll \left| \frac{\partial u_1}{\partial x_2} \right| + \left| \frac{\partial u_1}{\partial x_3} \right|. \tag{8.2}$$

Important operational flows such as journal-bearing[2] and piston-ring lubrications[3] of engines, and processing flows such as application of thin films or coating[4] and multilayer extrusion[5] can be analyzed as lubrication flows by the procedure highlighted in Section 8.2.

*Stretching flows*[6] are free-surface flows of small interface curvature $dh/dx_1 \ll 1$, nearly free of vorticity. The dominant velocity component is nearly parallel to the free surface, and changes significantly only with respect to its direction. Thus, the necessary conditions for stretching flow in the $x_1$-direction are

$$|u_1| \gg |u_2| + |u_3| \tag{8.3}$$

and

$$\left|\frac{\partial u_1}{\partial x_1}\right| \gg \left|\frac{\partial u_1}{x_2}\right| + \left|\frac{\partial u_1}{\partial x_3}\right|. \tag{8.4}$$

Important processing flows such as fiber spinning,[7] film casting[8] and film blowing of polymers,[9] and sheet and wire casting of metals,[10] can be analyzed as stretching flows according to the procedure presented in Section 8.3. Other types of stretching flows are induced around spherical cavities and bubbles that are growing or collapsing in liquids,[11] which have been examined in Chapters 5 and 6.

Figure 8.1 shows some prototype geometries of lubrication and stretching flows:

(a) *Journal-bearing* flow is a good example of confined lubrication flows in lubricants between solid surfaces in relative motion. As will be shown later, the small inclination gives rise to pressure that serves to keep the two surfaces apart, and therefore solid–solid contact is prevented. In this type of lubrication flow, the geometry is usually known, and the pressure developed and liquid–solid friction occurring need to be determined.

(b) *Coating flow* is a good prototype to study continuous deposition of thin liquid films on top of fast-moving substrates, as shown in Fig. 8.1. The coated substrate is subsequently driven through ovens, where the wet liquid film solidifies, forming a permanent solid layer. In this type of lubrication flow, the pressure distribution is known, being imposed by the external pressure and the capillary pressure, and the film thickness profile, including the final target thickness, need to be determined. Such coating operations are used to protect metals from corrosive environments, to laminate paper and other materials, and to deposit magnetic and photographic layers on tapes.

(c) *Film casting* and its axisymmetric analog *fiber spinning* are good prototypes to study manufacturing of films and fibers by drawing molten glass and metals and polymer melts extruded through capillary dies. Upon extrusion, the film or fiber is cooled by blowing air and solidifies by the time it is wound up at the takeup end. In this kind of stretching flow, of primary importance is the relation between the drawing force and the final thickness achieved, given the extrusion rate and the geometry of the arrangement.

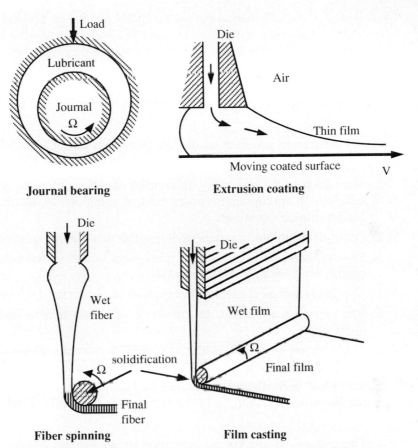

**Figure 8.1**  Typical lubrication and stretching flows: (a) journal-bearing lubrication, (b) extrusion coating, (c) fiber spinning, and (d) film casting.

## 8.2 LUBRICATION FLOW EQUATIONS

The equations governing lubrication flows can be derived and studied from simplified mass and momentum balances over appropriate and differential control volume (DCV), as discussed in Chapter 5. The same equations can be obtained by simplifying the general Navier–Stokes equations (NSEs) of Appendix G, utilizing the features of lubrication flows and Eqns. (8.1) and (8.2), as detailed in Ref. 12. In both cases, the resulting lubrication equations lead to the celebrated *Reynolds lubrication equation*:[1]

$$R(h, p, \text{Ca}, \text{St}) = 0, \qquad (8.5)$$

Equation (8.5) represents a relation between $h$, $p$, Ca, and St, where $h$ is the thickness profile of the narrow channel or of the thin film, $p$ is the pressure distribution, $\text{Ca} = \eta V/\sigma$ is the capillary number entering with film flows alone, and $\text{St} = \rho g D^2/\eta V$ is the Stokes number entering in nonhorizontal flows.

Equation (8.5) can be solved to find:

(a) The pressure distribution $p(x)$ and related quantities, such as load capacity, friction and wear, cavitation, etc., when the thickness $h(x)$ is known. The most typical applications are in journal-bearing-type lubrications[2,3] and other machine parts lubrications.

(b) The thickness $h(x)$ when the pressure is known. The most typical applications are in the formation of thin films in lubrication and in coating.[4,5]

The derivation and solution of Eqn. (8.5) goes through the following stages:

1. Derivation of the governing differential equations by mass and momentum balances on appropriately chosen control volumes—or, alternatively, by the Navier–Stokes equations.

2. Order-of-magnitude analysis to derive the lubrication equations.

3. The solution of the differential lubrication equation and related boundary conditions to find the velocity profile.

4. The integral mass conservation equation to derive the Reynolds equation.

5. The surface tension and the curvature of a thin film to find the pressure distribution.

6. The solution of the Reynolds equation to find the film thickness or the pressure distribution.

7. Simplified perturbation techniques to find limiting asymptotic solutions to the Reynolds equation, since the full equation is rarely solvable analytically.

These stages of analysis accompanied by application examples follow. The equations of *lubrication approximation* for flows in *nearly* rectilinear channels or pipes of *nearly* cylindrical walls can be derived intuitively from the equations of flow in rectilinear channels and pipes. The equations that govern flows in rectilinear channels and pipes are the continuity or mass conservation equation, which demands constant flow rate:

$$u_z = 0, \qquad \frac{\partial u_x}{\partial x} = 0 \quad \Leftrightarrow \quad u_x = f(z), \qquad (8.6)$$

where $x$ is the flow direction and $z$ its perpendicular direction; and the conservation of linear momentum in the flow direction equation,

$$\frac{dp}{dx} = \eta \frac{d^2 u_x}{dz^2}, \qquad (8.7)$$

which under a constant pressure gradient $dp/dx$ predicts linear shear stress and parabolic velocity profiles.

When one or both walls are in slight inclination, $\alpha$, the same governing equations are expected to hold, but the velocity may now be locally a weak function of $x$, of order $\alpha$ because now there is a barrier to flow analogous to $\alpha$, known to cause pressure buildup (e.g., Examples 6.5 and 6.6). The most obvious difference is the pressure gradient: In the case of a lubrication flow, which may be accelerated or decelerated in a converging or diverging channel, respectively, $dp/dx$ is not constant along the channel. As illustrated by Fig. 8.2, this pressure gradient of lubrication flow depends on $h(z)$ and is therefore a function of $x$. The

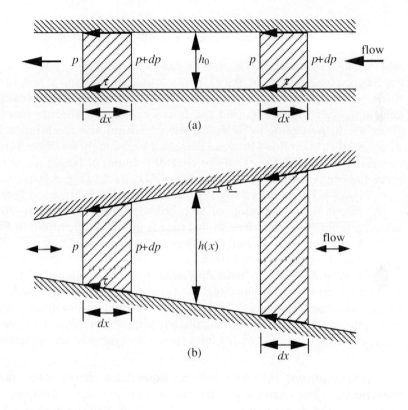

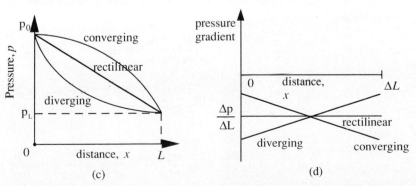

**Figure 8.2**  Momentum balances: (a) in a rectilinear flow, $h_0\, dp = 2\tau\, dx$; (b) in lubrication flow, $h(x)\, dp(x) = 2\tau(x)\, dx$; (c) resulting pressure distribution. (From Ref. 12, by permission.)

pressure distribution in rectilinear and lubrication flows is shown in Fig. 8.2. The velocity profile of a lubrication flow satisfies the equations

$$\frac{\partial u_x}{\partial x} + \frac{\partial u_z}{\partial z} = 0 \tag{8.8}$$

and

$$\frac{\partial p}{\partial x} = \eta \frac{\partial^2 u_x}{\partial z^2}, \tag{8.9}$$

*and unlike rectilinear flows, it is a function of both x and z.* Equation (8.9) expresses conservation of linear momentum, or Newton's law of motion, that there is no accumulation of momentum in a control volume because there is no substantial net convection, and the forces capable of altering momentum are in equilibrium. According to Newton's law of motion, the acceleration is vanishingly small in lubrication flows because there is a vanishingly small net force acting on a control volume. The forces on the control volume, of height $dx$, are net pressure force $(dp/dx)A(x)$ and shear stress force $A(x)\,dx\,\tau_{xy}$ (Fig. 8.2b).

Figure 8.3 illustrates a slightly different kind of lubrication flow, where there is significant relative motion of one boundary with respect to the other. The underlying mechanism of flow in this case is more complex than in Poiseuille flow. First, the moving wall on one side sweeps fluid into a narrowing passage through the action of viscous shear forces, which gives rise to a local velocity profile of Couette type $u_x = Vy/h$, with flow rate $\dot{V} = Vh/2$. Because $\dot{V}$ is constant by continuity and $h(x)$ is diminishing, the flow sets up a pressure gradient to supply a Poiseuille component, which redistributes the fluid and maintains a constant flow rate. This mechanism leads to significant pressure buildup, as shown in Fig. 8.3. This pressure is responsible for load capacity in lubrication applications.

**Development of lubrication equations from the Navier–Stokes equations.** The lubrication equations can be developed by an order-of-magnitude and dimensional analysis of the full, two-dimensional Navier–Stokes equations,

$$\frac{\partial u_x}{\partial x} + \frac{\partial u_x}{\partial z} = 0, \tag{8.10}$$

$$\rho\left(\frac{\partial u_x}{\partial t} + u_x\frac{\partial u_x}{\partial x} + u_z\frac{\partial u_z}{\partial z}\right) = -\frac{\partial p}{\partial x} + \eta\left(\frac{\partial^2 u_x}{\partial x^2} + \frac{\partial^2 u_x}{\partial z^2}\right), \tag{8.11}$$

and

$$\rho\left(\frac{\partial u_z}{\partial t} + u_x\frac{\partial u_z}{\partial x} + u_z\frac{\partial u_z}{\partial z}\right) = -\frac{\partial p}{\partial z} + \eta\left(\frac{\partial^2 u_z}{\partial x^2} + \frac{\partial^2 u_z}{\partial z^2}\right), \tag{8.12}$$

where $x$ is the direction of flow and $z$ is the gapwise direction, for which the geometry is shown in Fig. 8.3.

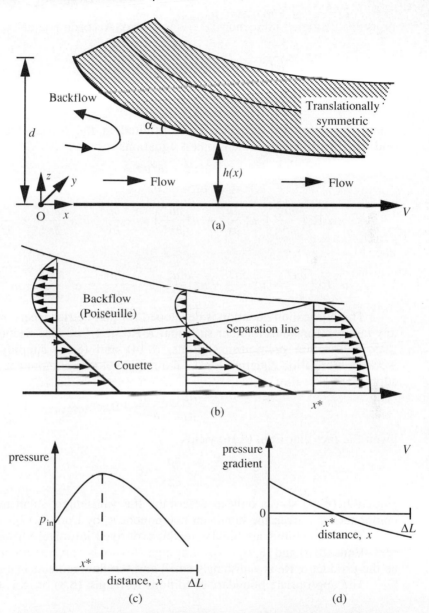

**Figure 8.3**   (a) Geometry of a two-dimensional lubrication flow; (b) the velocity profiles along the channel are mixtures of Couette and Poiseuille; (c) the backflow gives rise to significant pressure buildup. (From Ref. 12, by permission.)

There are several good reasons to work with dimensionless equations and variables: To reduce the dependence of the solution to minimum dimensionless numbers; to simplify the equations judging from the relative magnitude of a dimensionless number with respect to unity; and to scale up experiments to real applications of the same dimensionless numbers. To achieve these goals, the procedure outlined in Section 1.7.2 is followed. Accordingly, define

$$x^* = \frac{x}{L}, \qquad z^* = \frac{z}{\alpha L}, \qquad t^* = \frac{tV}{L}, \qquad h^* = \frac{h}{\alpha L},$$

$$u_x^* = \frac{u_x}{V}, \qquad u_z^* = \frac{u_z}{\alpha V}, \qquad p^* = \frac{p}{\eta(V/\alpha^2 L)}. \tag{8.13}$$

These dimensionless variables are substituted in the Navier–Stokes equations, and the corresponding dimensionless equations result:

$$\frac{\partial^* u_x^*}{\partial x^*} + \frac{\partial^* u_z^*}{\partial z^*} = 0, \tag{8.14}$$

$$\alpha \, \mathrm{Re}\left(\frac{\partial u_x^*}{\partial t^*} + u_x^* \frac{\partial u_x^*}{\partial x^*} + u_z^* \frac{\partial u_x^*}{\partial z^*}\right) = -\frac{\partial p^*}{\partial x^*} + \alpha^2 \frac{\partial^2 u_x^*}{\partial x^{*2}} + \frac{\partial^2 u_x^*}{\partial z^{*2}}, \tag{8.15}$$

and

$$\alpha^3 \, \mathrm{Re}\left(\frac{\partial u_z^*}{\partial t^*} + u_x^* \frac{\partial u_z^*}{\partial x^*} + u_z^* \frac{\partial^2 u_z^*}{\partial z^{*2}}\right) = -\frac{\partial p^*}{\partial z^*} + \alpha^4 \frac{\partial^2 u_z^*}{\partial x^{82}} + \alpha^2 \frac{\partial^2 u_z^*}{\partial z^{*2}}. \tag{8.16}$$

The lubrication equation is developed for geometries where $\alpha \ll 1$, and for any laminar Reynolds number such that $\alpha \, \mathrm{Re} \ll 1$ (see also Section 7.5). Under these lubrication prerequisites, Eqns. (8.14) and (8.15) simplify to yield the working lubrication equation for the dominant velocity component (with asterisks suppressed hereafter),

$$-\frac{dp}{dx} + \frac{\partial^2 u_x}{\partial z^2} = 0, \tag{8.17}$$

given the fact that Eqn. (8.16) yields

$$\frac{\partial p}{\partial z} = 0. \tag{8.18}$$

Equation (8.14) serves only to determine the vanishingly small normal velocity component $u_z$, given the dominant component $u_x$ by Eqn. (8.17).

These equations are similar to those derived intuitively from channel flow [e.g., Eqns. (8.8) and (8.9)]. Note that high Reynolds numbers are allowed as long as the product $\alpha \, \mathrm{Re}$ is vanishingly small and the flow remains laminar.

The appropriate boundary conditions for Eqns. (8.8) or (8.17) are:

At $z = 0$, $u_x = V$ (no-slip boundary condition).
At $z = h$, $u_x = 0$ (slit flow: no-slip boundary condition).
At $z = h$, $\tau_{zx} = 0$ (thin film: no shear stress at free surface).

Under these conditions, the solution to Eqn. (8.17) is

$$u_x = -\frac{1}{2\eta}\frac{dp}{dx}(zh - z^2) + V - \frac{z}{h}V \qquad \text{(slit flow)}, \tag{8.19}$$

$$u_x = \frac{1}{2\eta}\frac{dp}{dx}(2zh - z^2) + V \qquad \text{(film flow)}. \tag{8.20}$$

### 8.2.1 Reynolds Lubrication Equation

Mass must be conserved in any differential control volume of length $dx$ and height $h(x)$, identical to the film thickness. Therefore,

$$-\dot{V}_{x+dx} + \dot{V}_x = dx\,\frac{dh}{dt}, \tag{8.21}$$

which states that the net mass convection in the control volume is being used to increase the volume at rate $d(dx\,dh)/dt$, where $dx$ and $dh$ are the width in the flow direction and the height of the volume, respectively. Rearrangement yields

$$-\frac{d\dot{V}}{dx} = \frac{dh}{dt}, \tag{8.22}$$

which for confined and film flows reduces to

$$\frac{d}{dx}\left(-\frac{1}{2\eta}\frac{dp}{dx}\frac{h^3}{6} + \frac{hV}{2}\right) = -\frac{dh}{dt} \qquad \text{(slit flow)} \tag{8.23}$$

and

$$\frac{d}{dx}\left(-\frac{1}{\eta}\frac{dp}{dx}\frac{h^3}{3} + hV\right) = -\frac{dh}{dt} \qquad \text{(film flow)}, \tag{8.24}$$

respectively.

### 8.2.2 Confined Lubrication Flows

The steady-state form of Eqn. (8.23),

$$\frac{d}{dx}\left(-\frac{1}{2\eta}\frac{dp}{dx}\frac{h^3}{6} + \frac{hV}{2}\right) = 0, \tag{8.25}$$

is integrated to

$$-\frac{1}{2\eta}\frac{dp}{dx}\frac{h^3}{6} + \frac{hV}{2} = \dot{V}. \tag{8.26}$$

Thus, the pressure distribution is

$$p(x) = p_0 + 6\eta V \int_0^x \frac{dx}{h^2(x)} - 12\eta\dot{V}\int_0^x \frac{dx}{h^3(x)}, \tag{8.27}$$

where

$$\dot{V} = \frac{p_0 - p_1}{12\eta\int_0^L h^{-3}(x)\,dx} + \frac{V\int_0^L h^{-2}(x)\,dx}{2\int_0^L h^{-3}(x)\,dx}. \tag{8.28}$$

A lubrication layer will generate a positive pressure and thus a load capacity, normal to this layer, only when the layer is arranged so that the relative motion of the two surfaces tends to drag fluid by viscous stresses from the wider to the narrower end of the layer. For a rectilinear wall described by the equation $h(x) = d - \alpha x$ (Example 8.5), the load capacity $W$ is

$$W = \int_0^L (p - p_0)\, dx = \frac{6\eta V}{\alpha^2} \left[ \log \frac{d}{d - \alpha L} - 2 \left( \frac{\alpha L}{2d - \alpha L} \right) \right]. \qquad (8.29)$$

The resulting liquid-wall frictional force that is responsible for wearing of lubrication surfaces is computed by

$$F = \int_0^L [\tau_{zx}]_{z=h(x)}\, ds = -\int_0^L \eta \frac{\partial u_x}{\partial z}\, ds = -\int_0^L \left( \frac{h}{2} \frac{dp}{dx} - \eta \frac{V}{h} \right) = \frac{\alpha}{2} W. \qquad (8.30)$$

Thus the load capacity is of order $\alpha^{-2}$, whereas the shear of friction is of order $\alpha^{-1}$. Therefore, the load/friction ratio increases with $\alpha^{-1}$. To enhance this load-to-friction ratio further, shear-thinning polymeric additives are added to lubricant oils. These additives reduce viscosity and therefore wear according to Eqn. (8.30), and more important, give rise to viscoelastic normal stresses that act as additional pressure and increase load capacity according to Eqn. (8.29). Shear-thinning and normal viscoelastic stress effects are discussed in Chapter 9, on non-Newtonian fluids.

The most important applications of the lubrication theory for confined flows are in the lubrication of journal bearing[2] and piston-ring[3] systems of engines. Other confined flows that can be studied by means of the same lubrication equations include wire coating,[13] roll coating,[14,15] and many polymer applications.[6,16] The solutions to these problems follow the procedure outlined above, starting from Eqn. (8.22). The flow rate is usually given by

$$\dot{V} = V h_f, \qquad (8.31)$$

where $\dot{V}$ is the speed of withdrawal or production and $h_f$ is the final film thickness.

The boundary condition on the pressure at the outlet may vary. When the outlet pressure is known, then $p(L) = p_L$; otherwise, good approximate boundary conditions are $[dp/dx]_{x=L} = 0$ introduced by Reynolds,[1] or $p = -[R\,Ca]^{-1}$, where Ca is the capillary number and $R^{-1}$ the curvature of the exit thin film,[17] or $[dp/dx]_{x=L} = 2h_L^{-2}$ introduced by Prandtl.[18] Several outlet boundary conditions were evaluated by finite element analysis.[19] At the inlet, common pressure boundary conditions are $p(x = 0) = p_0$, $[dp/dx]_{x=0} = q$, and $[dp/dx]_{x=0} = \kappa p$.

### 8.2.3 Thin Film Lubrication Flows

In converging lubrication flows there is a pressure buildup due to the inclination $\alpha$, which may give a rise in backflow of some of the entering fluid. The pressure is then used effectively to support loads and keep moving surfaces apart to reduce friction and wear. In thin-film lubrication flows, any pressure buildup is due to surface tension.

The steady-state form of Eqn. (8.24),

$$\frac{d}{dx}\left(-\frac{1}{\eta}\frac{dp}{dx}\frac{h^3}{3} + Vh\right) = 0,  \tag{8.32}$$

is integrated to

$$-\frac{1}{\eta}\frac{dp}{dx}\frac{h^3}{3} + VH = \dot{V} = Vh_f.  \tag{8.33}$$

The film thickness profile, $h(x)$, is not known. However, the pressure drop $dp/dx$ can be deduced from the surface tension by means of the Young–Laplace equation[17] under the lubrication approximation that the slope, $dh/dx$, is much less than unity:

$$-p = \frac{\sigma(d^2h/dx^2)}{[1 + (dh/dx)^2]^{1/2}} \simeq \sigma\frac{d^2h}{dx^2}.  \tag{8.34}$$

Here $h(x)$ is the elevation of the free surface from the substrate which coincides with the $x$-axis, and $\sigma$ is the surface tension of the liquid. Then

$$-\frac{dp}{dx} = \sigma\frac{d^3h}{dx^3},  \tag{8.35}$$

which substituted in Eqn. (8.33) yields

$$\frac{\sigma}{\eta}h^3\frac{d^3h}{dx^3} + 3hV = Vh_f,  \tag{8.36}$$

which is rearranged in the form

$$h^3\frac{d^3h}{dx^3} + 3\,\text{Ca}(h - h_f) = 0.  \tag{8.37}$$

Equation (8.36) is highly nonlinear and cannot be solved analytically.

The most important applications of the thin-film lubrication equations are films with surface tension, nonisothermal films, dip and extrusion coating, and wetting and liquid spreading.[20] A similar class of problems includes centrifugal spreading, common in bell sprayers[21] and in spin coating.[22] It must be re-emphasized here that the film analysis presented here applies to uneven films at low Reynolds numbers of free surface nonuniformities and waves of small amplitude to wavelength, such that lubrication analysis applies. Nonlinear traveling free surface waves that tend to develop in all films at high Reynolds numbers is still the topic of ongoing research.[23,24]

**Example 8.1: Vertical Dip Coating**

An example of thin lubrication film formed under the combined action of gravity, surface tension, and viscous drag arises in dip coating,[12,16] shown in Fig. E8.1. This method of coating is employed to coat metals with anticorrosion materials and to laminate paper and polymer films. The substrate is being withdrawn at speed $V$ from a liquid bath of density $\rho$, viscosity $\eta$, and surface tension $\sigma$. The analysis must predict the final coating thickness as a function of the processing conditions (withdrawal speed), the physical characteristics of the liquid ($\rho$, $\eta$, and $\sigma$), and the geometry ($W$ and $L$).

**Solution**   The governing momentum equation with respect to the Cartesian system of coordinates shown is

$$-\frac{dp}{dz} + \eta \frac{\partial^2 u_z}{\partial y^2} - \rho g = 0. \tag{8.1.1}$$

The boundary conditions are

$$at \quad y = 0, \qquad u_z = V, \tag{8.1.2}$$

and

$$at \quad y = H(z), \qquad \tau_{zy} = \eta \frac{\partial u_z}{\partial y} = 0. \tag{8.1.3}$$

The particular solution for the velocity profile is

$$u_z = \frac{1}{\eta}\left(\frac{dp}{dz} + \rho g\right)\left(\frac{y^2}{2} - Hy\right) + V. \tag{8.1.4}$$

The resulting Reynolds equation is

$$-\frac{1}{\eta}\left(\frac{dp}{dz} + \rho g\right)\frac{H^2}{3} + VH = Q = VH_f, \tag{8.1.5}$$

where $H_f$ is the final coating thickness.

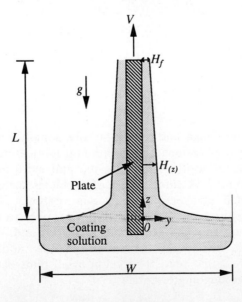

**Figure E8.1**  Dip coating: A coated plate is being withdrawn from a coating solution. A final thin film or coating results on the plate under the combined action of gravity, surface tension, and drag by the moving substrate. (From Ref. 12, by permission.)

The pressure gradient, related to the slope of the film by the Young–Laplace equation,[4]

$$\frac{dp}{dz} = -\sigma \frac{d^3H}{dz^3},$$  (8.1.6)

is substituted in Eqn. (8.1.5) to yield the final Reynolds equation,

$$\frac{1}{\eta}\left(\frac{d^3H}{dz^3} - \rho g\right)\frac{H^3}{3} + V(H - H_f) = 0.$$  (8.1.7)

The Reynolds equation is rearranged in the form

$$\frac{H^3}{3}\frac{d^3H}{dz^3} - \frac{\rho g}{\sigma}\frac{H^3}{3} + \frac{V\eta}{\sigma}(H - H_f) = 0.$$  (8.1.8)

By identifying the dimensionless capillary number,

$$\mathrm{Ca} = \frac{V\eta}{\sigma},$$  (8.1.9)

and the Stokes number,

$$\mathrm{St} = \frac{\eta V}{\rho g H_f^2},$$  (8.1.10)

Eqn. (8.1.8) becomes

$$\frac{H^3}{\mathrm{Ca}}\frac{d^3H}{dz^3} - \frac{1}{\mathrm{St}}\frac{H^3}{H_f^2} + 3(H - H_f) = 0,$$  (8.1.11)

which can be solved directly for the following limiting cases:

1. *Negligible surface tension* $(\mathrm{Ca} \to \infty)$. Equation (8.1.11) reduces to the third-order algebraic equation

$$\frac{1}{\mathrm{St}}H^3 - 3H_f^2 H + 3H_f^3 = 0.$$  (8.1.12)

In the limit of $\mathrm{St} \to 0$ (e.g., vertical molten metal flow), the only solution is $H = 0$ (i.e., no coating). In the limit of $\mathrm{St} = \infty$ (i.e., a horizontal arrangement), the solution is $H = H_f$ (i.e., plug flow). For finite values of St, the solution is independent of $z$, which predicts a flat film throughout.

2. *Infinitely large surface tension* $(\mathrm{Ca} \to 0)$. Equation (8.1.11) reduces to

$$\frac{d^3H}{dz^3} = 0,$$  (8.1.13)

with general solution

$$H(z) = c_1\frac{z^2}{2} + c_2 z + c_3,$$  (8.1.14)

along with the boundary conditions, $H(z = 0) = W/2$, $H(z = L) = H_f$, and $(dH/dz)_{z=L} = 0$. The solution is

$$H(z) = \left(\frac{W - 2H_f}{L^2}\right)\left(\frac{z^2}{2} - zL\right) + \frac{W}{2},$$  (8.1.15)

which is a parabolic film thicknss.

3.  *Finite surface tension* $(0 < Ca < k)$. Equation (8.1.11) is cast in the form

$$H^3\left(\frac{d^3H}{dz^3} - St \cdot Ca\,\frac{1}{H_f^2}\right) + 3\,Ca(H - H_f) = 0, \qquad (8.1.16)$$

with no apparent analytic solution. For a special case of horizontal coating $(St = \infty)$, and since usually $H_f/W \ll 1$, the transformation

$$H^* = \frac{H}{W}, \qquad z^* = \frac{z}{W}, \qquad (8.1.17)$$

reduces Eqn. (8.1.16) to

$$(H^{*3})\frac{d^3H^*}{dz^{*3}} + 3\,Ca\left(H^* - \frac{H_f}{W}\right) = 0. \qquad (8.1.18)$$

Equation (8.1.18) predicts that near the inlet, where $H^* = 1$, the film decays with rate depending on the Ca. Near the other end, where $H^* = H_f/W$, the film becomes flat, surface tension becomes unimportant, and therefore the slope is zero. Equation (8.1.18) can be solved asymptotically by perturbation techniques.[25]

It must be noted that the analysis presented here may not apply in the vicinity of $z = 0$, where the flow may become truly two-dimensional, and for the free surface curved to such slope, the lubrication approximation does not apply anymore. In such cases a complete solution is obtained by matching the two-dimensional solution in the vicinity of $z = 0$ to the lubrication solution away from $z = 0$.

### 8.2.4 Oseen's Generalized Lubrication Equation

A better approximation to the Navier–Stokes equations is Oseen's equation,

$$Re\,\frac{\partial u_x}{\partial x} = -\frac{dp}{dx} + \frac{\partial^2 u_x}{\partial y^2}, \qquad Re = \frac{Vh_L}{\nu}, \qquad (8.38)$$

in case of a boundary moving with velocity $V$ (e.g., Fig. 8.3), and

$$Re = \frac{\bar{u}(x)h(x)}{\nu} = \frac{\dot{V}}{\nu} \qquad (8.39)$$

in case of converging or diverging channel of stationary walls. In case of converging or diverging stationary pipe or annulus, the corresponding equations are

$$Re\,\frac{\partial u_z}{\partial z} = -\frac{dp}{dz} + \frac{1}{r}\frac{\partial}{\partial r}\left(r\frac{\partial u_z}{\partial r}\right), \qquad Re = \frac{\dot{V}}{\eta\bar{R}\nu}, \qquad \bar{R} = \frac{R_0 + R_L}{2}. \qquad (8.40)$$

These generalized lubrication equations are extensions of the Stokes-like lubrication equations [Eqns. (8.8) and (8.9)] to include any significance of inertia terms, $Re(\partial u_i/\partial x_i)$, away from solid boundaries. This extension from Stokes to Oseen equations was first applied to free-stream creeping flow overtaking submerged spheres and cylinders[26] (e.g., Section 4.6).

Results predicted by Eqns. (8.38) to (8.40) are identical to those predicted by the lubrication equations for a Re < 0.01. Improved results are predicted by Eqns. (8.38) to (8.40) for 0.01 < a Re < 0.1, whereas both kind of equations fail for a Re > 0.1. Equations (8.38) and (8.40) predict increased or decreased Re absolute values of pressure gradient for converging or diverging conduits, respectively. This is consistent with the requirements of mechanical energy conservation, in the presence of differences in the kinetic energy term $\text{Re}(\partial u_i / \partial x_i)$.

All the foregoing differences are small. The most important difference between the two kinds of equations is in the *stability of steady lubrication flow.* A system may be removed from its steady flow by velocity or pressure fluctuations from their steady values, due to one-time or delta-function-like variations of flow rate and vibrations of system boundary. The velocity and pressure fluctuations may or may not decay with time, leading the flow back to or away from its steady state. The steady flow is *stable* in the first case and *unstable* in the second. The lubrication equation predicts always stable confined flows and stable or unstable film flows depending on the capillary number alone. The Oseen equation predicts stable or unstable confined flows depending on the Reynolds number, and stable or unstable film flows depending on both the capillary and the Reynolds number. The latter patterns agree with experimental observations and numerical analysis predictions of the complete Navier–Stokes equations.[19]

Some details on the principles of stability analysis are presented in Section 10.2.1 and Example 10.1, as they apply to stable laminar flow and unstable laminar flow, or turbulent flow. The important subject of stability analysis is the topic of several publications, for fluid flows[27,28] and for processing in general.[29,30]

## 8.3 STRETCHING FLOW EQUATIONS

The dominant deformation in stretching flows is the stretching or compression of liquid filaments. The most elementary stretching flow is the one shown in Fig. 8.4, where a cylindrical filament of a highly viscous polymeric liquid is stretched

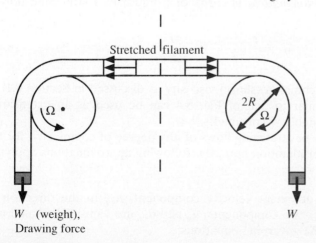

Stretched filament

$\Omega$

$2R$

$\Omega$

W  (weight),
Drawing force

W

**Figure 8.4**  Elementary form of stretching flow.

under the normal stress created by the two hanging weights and/or by the rotation of the two drums. The filament is elongated in the axial direction and compressed in all radial directions. The force (weight here) is a normal *drawing force*. The rotating drum accelerates the liquid from a vanishingly small velocity near the middle of the filament to a linear velocity,

$$u_L = R\Omega,  \tag{8.41}$$

at the takeup ends. The ratio of the velocities between two points along a filament a distance *l* apart,

$$D_R = \frac{u_L}{u_0},  \tag{8.42}$$

is commonly called the *draw ratio*. The drawing force relates to the stresses by

$$W = F = \int_0^{d/2} 2\pi r(-p + \tau_{zz})\,dr = \int_0^{d/2} 2\pi r(\tau_{zz} - \tau_{rr})\,dr = \pi \frac{d^2}{4}(\bar{\tau}_{zz} - \bar{\tau}_{rr}),  \tag{8.43}$$

where the interfacial relation under negligible surface tension,

$$-p + \tau_{rr} = \frac{\sigma}{d} = 0,  \tag{8.44}$$

has been used to eliminate the pressure.

The *extension rate* is simply the derivative of the velocity

$$\dot{\varepsilon} = \frac{\partial u_z}{\partial z}  \tag{8.45}$$

along the filament. The *elongational viscosity* of a liquid in a stretching flow is defined by

$$\eta_e = \frac{\tau_{zz} - \tau_{rr}}{\dot{\varepsilon}} = \frac{4F}{\pi d^2 \dot{\varepsilon}}  \tag{8.46}$$

and is different from the common (shear) viscosity as discussed in Section 6.1. In fact, an arrangement similar to that of Fig. 8.4 can be used as an extensional rheometer, restricted to highly viscous melts.[31]

Although there may be stretching flows of any degree of complexity, for the stretching flows under consideration here, the following approximations apply:

1. There is only one dominant velocity component, $u_z$, in the direction of flow. Thus, the other two components, $u_\theta$ and $u_r$, are vanishingly small and are dropped from the governing equations.

2. The dominant velocity component is virtually uniform over any cross-sectional area of flow (i.e., $\partial u_z / \partial r = 0$).

3. Shear stress is vanishingly small, and normal stresses are virtually uniform over any cross-sectional area of flow [i.e., $(\tau_{ij})_{i \neq j} = 0$ and $\partial \tau_{ii} / \partial r = 0$].

4. The slope of the free surface (or of the thickness of the film or filament) in the flow direction is small, and therefore surface tension effects are not important. This is especially evident in highly viscous liquids, where viscous forces are dominant, and therefore $\tau_{rr} = p$ [Eqn. (8.44)].

Under these conditions, the governing continuity and momentum equations for Cartesian geometry in film stretching flows reduce to

$$\frac{du_z}{dz} + \frac{du_x}{dx} = 0 \tag{8.47}$$

and

$$\rho u_z \frac{du_z}{dz} + \frac{dp}{dz} - \frac{d\tau_{zz}}{dz} - \rho g = 0, \tag{8.48}$$

respectively. The same equations for cylindrical geometry of fiber-like stretching reduce to

$$\frac{du_z}{dz} = -\left(\frac{du_r}{dr} + \frac{u_r}{r}\right) \simeq -2\frac{du_r}{dr}, \tag{8.49}$$

since $u_r$ is very small and nearly uniform on each cross section, and

$$\rho u_z \frac{du_z}{dz} + \frac{dp}{dz} - \frac{d\tau_{zz}}{dz} = -\rho g = 0 \tag{8.50}$$

for filament or fiber drawing.

The pressure $p$ is related to the radial stress $\tau_{xx}$ or $\tau_{rr}$ by the interfacial boundary conditions

$$-p + \tau_{rr} = \sigma\left(\frac{1}{R(z)} + \frac{d^2R}{dz^2}\right) \simeq \frac{\sigma}{R(z)}, \tag{8.51}$$

and under negligible surface tension by

$$-p + \tau_{rr} = \sigma \frac{d^2h}{dz^2} \simeq 0. \tag{8.52}$$

The thickness of the film $h(z)$, or the radius of the fiber $R(z)$, are related to the corresponding velocities by the mass conservation equations,

$$\dot{V} = W u(z) h(z) \tag{8.53}$$

for film or sheet, and

$$\dot{V} = \pi R^2(z) u(z) \tag{8.54}$$

for fiber. Here $\dot{V}$ is the liquid extrusion flow rate and $W$ is the width of the film.

### 8.3.1 Spinning of Newtonian Fibers

The fiber-spinning process is an important industrial operation in the manufacturing of synthetic fibers. The process is shown schematically in Fig. P2.26. A polymer melt is extruded continuously through a die and is taken up downstream, at a higher speed, at the desired diameter. The wet fiber zone consists of two regions. The first is known as the *extrudate-swell* region, caused by the transition from confined to free-surface flow. The second, succeeding the first region, is a *draw-down* region in which the liquid is drawn downward to form a filament. For long filaments, of length $L$ from the die to the takeup end such that $L/D_0 \gg 1$, the extrudate-swell region becomes unimportant to the final fiber diameter and required tension, and accurate results can be obtained by considering a one-dimensional stretching flow throughout, as demonstrated in Examples 8.2 and 8.3.

**Example 8.2: Derivation of Fiber-Spinning Equations**

Derive the governing equations for the process shown in Fig. P3.18b and described in Section 8.3.1.

**Solution**   The governing equation is

$$-\rho u_z \frac{du_z}{dz} - \frac{dp}{dz} + \frac{d\tau_{zz}}{dz} + \rho g = 0, \tag{8.2.1}$$

or

$$\frac{d}{dz}\left( -\rho \frac{u_z^2}{2} - p + \tau_{zz} \right) = -\rho g. \tag{8.2.2}$$

The equation is averaged over the cross-sectional area by

$$\frac{1}{\pi R^2} \int_0^{R(z)} \frac{d}{dz}\left( -\rho \frac{u_z^2}{2} - p + \tau_{zz} \right) 2\pi r \, dr = -\rho g. \tag{8.2.3}$$

By applying Leibnitz's integration formula, Eqn. (8.2.3) becomes

$$\frac{d}{dz} \int_{r=0}^{r=R(z)} \left[ -\rho \frac{u_z^2}{2} - p + \tau_{zz} \right] 2\pi r \, dr - \frac{dR}{dz} \left[ -\rho \frac{u_z^2}{2} - p + \tau_{zz} \right]_{r=R(z)} 2\pi R = -\pi R^2 \rho g. \tag{8.2.4}$$

Using the definition of a cross-sectional area averaged quantity,

$$\bar{q} = \frac{\displaystyle\int_0^{R(z)} q 2\pi r \, dr}{\pi R^2(z)}, \tag{8.2.5}$$

in the integral term of Eqn. (8.2.4) and Eqn. (8.2.2) in its boundary term results in

$$\frac{d}{dz}\left[\pi R^2(z)\left(-\rho\frac{\bar{u}_z^2}{2} - \bar{p} + \bar{\tau}_{zz}\right)\right] \approx -\pi R^2 \rho g. \tag{8.2.6}$$

Conservation of mass requires that

$$\dot{V} = \pi R^2(z)\bar{u}_z(z), \tag{8.2.7}$$

and the free-surface condition at $r = R(z)$ defines the pressure $p$ by

$$-p + \tau_{rr} = \frac{\sigma}{R(z)}. \tag{8.2.8}$$

Equations (8.2.7) and (8.2.8) are used to eliminate the pressure and the radius, $R(z)$, from Eqn. (8.2.6). The resulting equation in terms of the average velocity $\bar{u}_z$ is

$$\dot{V}\frac{d}{dz}\left[-\frac{\rho\bar{u}_z(z)}{2} + \frac{\bar{\tau}_{zz} - \bar{\tau}_{rr}}{\bar{u}_z(z)} + \sigma\sqrt{\frac{\pi}{\dot{V}\bar{u}_z(z)}}\right] + \frac{\rho g}{\bar{u}_z(z)} = 0, \tag{8.2.9}$$

which is integrated to

$$-\rho\frac{\bar{u}_z(z)}{2} + \frac{\bar{\tau}_{zz} - \bar{\tau}_{rr}}{\bar{u}_z(z)} + \sigma\sqrt{\frac{\pi}{\dot{V}\bar{u}_z(z)}} + \int_0^z\frac{\rho g\, dz}{\bar{u}_z(z)} = c. \tag{8.2.10}$$

The stress difference is expressed in terms of the derivative of the velocity according to Eqn. (8.46), which reduces Eqn. (8.2.10) to

$$3\eta\frac{1}{\bar{u}_z(z)}\frac{d\bar{u}_z(z)}{dz} + \sigma\sqrt{\frac{\pi}{\dot{V}\bar{u}_z(z)}} + \int_0^z\frac{\rho g\, dz}{\bar{u}_z(z)} - \frac{\rho}{2}\bar{u}_z(z) = c. \tag{8.2.11}$$

Equation (8.2.11) states that viscous stresses $3\eta\, d\bar{u}_z(z)/dz$, surface tension forces $\sigma\sqrt{\pi\dot{V}\bar{u}_z(z)}$, and gravity forces $\rho gz/\bar{u}_z(z)$ acting on the fiber result in acceleration $\rho u_z(z)/2$. Equation (8.2.11) can also be derived by a direct force balance on a cylindrical control volume, identical to a short fiber of radius $R(z)$ and of differential length $dz$. This procedure is demonstrated in Example 8.7 and Fig. E8.7 for film casting.

In Chapter 9, the fiber-spinning process of non-Newtonian liquids, common in the polymer industry, is highlighted. Equation (8.2.9) is the starting point. However, the viscoelastic stress difference $\bar{\tau}_{zz} - \bar{\tau}_{rr}$, must be replaced by an appropriate non-Newtonain consititutive equation, which leads to equations significantly more complicated than Eqn. (8.2.11).

## Example 8.3: Solution of Fiber-Spinning Equations

Analyze the fiber-spinning process for a Newtonian polymeric melt of known viscosity $\eta$, density $\rho$, and surface tension $\sigma$, extruded through a capillary of diameter $D_{\text{die}}$ at flow rate $\dot{V}$, and taken up at distance $L$ downstream. Investigate the relation between applied tension and resulting final fiber diameter.

**Solution** The governing equation is

$$3\eta \frac{1}{u(z)} \frac{du(z)}{dz} + \sigma \sqrt{\frac{\pi}{\dot{V}u(z)}} + \int_0^z \frac{\rho g\, dz}{u(z)} - \frac{\rho u(z)}{2} = c. \tag{8.3.1}$$

This is a highly nonlinear differential equation and cannot be solved analytically. However, it is instructive to study limiting cases, as follows:

**(a)** *polymer fiber at draw ratio* $D_R = u_L/u_0$. In this case the viscosity is large enough that viscous forces are dominant and Eqn. (8.3.1) reduces to

$$3\eta \frac{1}{u(z)} \frac{du}{dz} = c \tag{8.3.2}$$

with boundary conditions $u(z = 0) = u_0 = \dot{V}/4^2$ and $u(z = L) = u_0 D_R = u_L$. The resulting solution is

$$u(z) = u_0 \exp\left(\frac{z}{L} \ln \frac{u_L}{u_0}\right). \tag{8.3.3}$$

**(b)** *Metallic fiber under gravity alone.* In this case the temperature is high and the viscosity small such that gravity, surface tension, and inertial forces control and Eqn. (8.3.1) reduces to

$$\sigma \sqrt{\frac{\pi}{\dot{V}u}} + \int_0^z \frac{\rho g\, dz}{u} - \frac{\rho u}{2} = c \tag{8.3.4}$$

subjected to the boundary condition $u(z = 0) = u_0$. The resulting solution is

$$\sigma \sqrt{\pi} \left[u(z) - u_0\right] = \frac{\rho}{2}\left[u^2(z) - u_0^2\right] = \rho g(z - z_0), \tag{8.3.5}$$

which shows that the fiber is drawn by its own weight $\rho g(z - z_0)$, and that surface tension suppresses its diameter.

**(c)** *Glass fiber at draw ratio* $D_R$. In this case, in addition to viscous forces, surface tension may become important, in which case Eqn. (8.3.1) reduces to

$$3\eta \frac{1}{u(z)} \frac{du}{dz} + \sigma \sqrt{\frac{\pi}{\dot{V}u}} = c_1. \tag{8.3.6}$$

The solution is given explicitly by

$$4\sqrt{u(z)} + c_1 \ln[\sqrt{u}\,(z) - c_1] = -\frac{\sigma\pi}{6\eta\dot{V}} + c_2 \tag{8.3.7}$$

with the two constants $c_1$ and $c_2$ determinable by two boundary conditions. Notice that the solution depends on the capillary number, which is a ratio of viscous to surface tension forces.

**(d)** *Determination of fiber diameter.* The radius of the fiber is recovered from the calculated velocity profiles by means of the continuity equation, which for the case of the first example yields

$$R(z) = \sqrt{\frac{\dot{V}}{\pi u(z)}} = R_0 \exp\left(-\frac{z}{2L} \ln D_R\right). \tag{8.3.8}$$

**(e)** *Determination of drawing force or tension.* The corresponding drawing force that is required to achieve a certain draw ratio, $D_R$, and thus certain spinnability, is obtained by integrating the normal stresses over the cross section of the fiber at the takeup end:

$$F = \int_0^{R(L)} 2\pi r(-p + \tau_{zz}) \, dr = \int_0^{R(L)} 2\pi r(\tau_{zz} - \tau_{rr}) \, dr = 6\pi\eta \int_0^{R(L)} \left(\frac{du}{dz}\right)_{z=L} r \, dr$$

$$= 6\pi\eta \frac{R^2(L)}{2} \frac{u_L}{L} \ln \frac{u_L}{u_0} = \frac{3\eta \dot{V}}{L} \ln \frac{u_L}{u_0}. \qquad (8.3.9)$$

# 18.4 ADDITIONAL EXAMPLES: SPIN COATING, JOURNAL BEARING, WIRE COATING, FILM AND SHEET CASTING

As discussed earlier, lubrication flows are good models to study processes where materials flow in the form of thin layers or films, adjacent to solid boundaries, as in journal-bearing applications, in flows through converging/diverging dies, and in coating of solid substrates. Stretching flows are good models for processes where materials are subjected to elongation or squeezing under negligible shearing, as in fiber spinning, sheet casting, and film blowing. Some representative applications follow.

### Example 8.4: Spin Coating

In magnetic disk manufacturing, an important stage is the application of a thin film on a fast-rotating disk.[32] The disk is fed continually at its center at flow rate $\dot{V}$. The liquid spreads outward, forming a thin film of thickness $H(r)$ diminishing with radial distance $r$, due to the fast rotation of the disk at an angular speed $\Omega$, as shown in Fig. E8.4. A similar process takes place in bell sprayers,[22] where liquid atomizes upon leaving the bell, forming an aerosol, as discussed in Section 4.10.

(a) Derive the Reynolds equation that governs the film thickness profile. [The film $H(r)$ varies only slightly with the radius $r$, and the surface tension is negligible.]

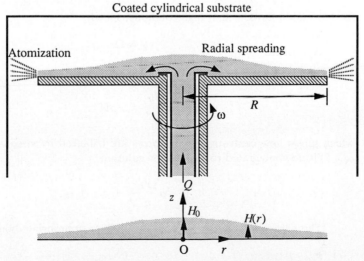

**Figure E8.4**   Arrangement and film profile in spin coating.

(b) Solve for the velocity distribution and the thickness profile. What is the final coating at the edge of the disk?

(c) Write the expression for the film thickness. How does it relate to film thickness down an inclinded perpendicular plane under gravity (e.g., Example 6.2)?

In a subsequent stage of processing, the supply of liquid stops and the nonuniform film is allowed to level off, under the ongoing rotation.

(d) What is the governing equation of film leveling?

(c) How does the film thickness evolve with time? Will it ever reach a planar film? If not, how could such a film be obtained?

**Solution**

*Film application stage*
A steady-state mass balance on a cylindrical shell around the center and of height identical to the film thickness yields

$$\frac{1}{r}\frac{\partial}{\partial r}(\rho r u_r) = 0, \qquad r u_r = c(z), \qquad u_r = \frac{c(z)}{r}. \tag{8.4.1}$$

So the velocity is not fully determined from the continuity equation. We need to proceed to the momentum equation in the direction of motion, which yields

$$\rho\left(u_r\frac{\partial u_r}{\partial r} - \frac{u_\theta^2}{r}\right) = \eta\frac{\partial^2 u_r}{\partial z^2}. \tag{8.4.2}$$

Furthermore,

$$\frac{\partial u_r}{\partial z} \gg \frac{\partial u_r}{\partial r} \tag{8.4.3}$$

for lubrication flows, and at high rotational speeds,

$$u_\theta \simeq \Omega r. \tag{8.4.4}$$

Under these assumptions, Eqn. (8.4.2) becomes

$$\eta\frac{\partial^2 u_r}{\partial z^2} = -\rho\Omega^2 r, \tag{8.4.5}$$

which shows how centrifugal body forces are balanced by viscous forces. Equation (8.4.5) is now integrated to the general solution

$$u_r = -\rho\frac{\Omega^2 r}{\eta}\frac{z^2}{2} + c_1 z + c_2. \tag{8.4.6}$$

The particular solution is found by applying appropriate boundary conditions:

1. At $z = 0$, $u_r = 0$; therefore, $c_2 = 0$.

**2.** At $z = H(r)$, $\partial u_r / \partial z = 0$; therefore, $c_1 = \rho \Omega^2 r H(r) / \eta$.

**(a)** The velocity profile is

$$u_r = \frac{\rho \Omega^2 r}{\eta} \left( Hz - \frac{z^2}{2} \right) z. \tag{8.4.7}$$

The film thickness is found by means of the constant flow rate:

$$
\begin{aligned}
\dot{V} &= 2\pi r \int_0^{H(r)} \frac{\rho \Omega^2 r}{\eta} \left( Hz - \frac{z^2}{2} \right) dz \\
&= \frac{2\pi \rho \Omega^2 r^2}{\eta} \left( \frac{H^3}{2} - \frac{H^3}{6} \right) = \frac{2\pi \rho \Omega^2 r^2}{\eta} \frac{H^3}{3}.
\end{aligned} \tag{8.4.8}
$$

**(b)** The resulting thickness is

$$H(r) = \sqrt[3]{\frac{3\eta \dot{V}}{2\pi \rho \Omega^2 r^2}} = c r^{-2/3}. \tag{8.4.9}$$

**(c)** The film thickness down a perpendiculat plate of width $W$ (Example 6.2), is $H = (3\eta \dot{V} / \rho \, gW)^{1/3}$. Thus, film spreading in spin coating is due to the centrifugal body force $F_C = \rho \Omega^2 r$, whereas in a film falling under gravity, the spreading is due to the gravity body force $F_g = \rho \, g$:

$$H(r) = \sqrt[3]{\frac{3\eta \dot{V}}{(2\pi r)(\rho \Omega^2 r)}} \quad \text{and} \quad H = \sqrt[3]{\frac{3\eta \dot{V}}{\rho \, gW}}. \tag{8.4.10}$$

*Film leveling stage*
**(d)** The governing transient Reynolds equation is obtained by mass balance over the same control volume with its height allowed to evolve with time:

$$2\pi r \frac{\partial}{\partial t} [H(r, t)] \, dr \cdot 2\pi r H(r, t) = [\dot{V}]_r - [\dot{V}]_{r+dr}. \tag{8.4.11}$$

By dividing by $dr$ and then taking the limit $dr \to 0$, this difference equation changes into the partial differential equation,

$$2\pi r \frac{\partial H(r, t)}{\partial t} = -\frac{\partial \dot{V}}{\partial r} = -\frac{2\pi \rho \dot{V}^2}{3\eta} \frac{\partial}{\partial r} (r^2 H^3). \tag{8.4.12}$$

**(e)** A particular solution to Eqn. (8.4.12) of the form $H(t, r) = A r^a t^b$ is sought which upon substitution in Eqn. (8.4.12) results in

$$H(r, t) = \sqrt{\frac{3\eta}{4\rho \Omega^2}} \frac{1}{\sqrt{t}}. \tag{8.4.13}$$

Thus as time progresses, the film thickness diminishes and becomes uniform.
The complete solution to Eqn. (8.4.12) can be shown to be,[33]

$$\frac{H(r, t)}{H_0(r_0)} = \left[ 1 + \frac{4\rho \Omega^2 H_0^2(r_0)}{3\eta} t \right], \qquad \frac{r}{r_0} = \left[ 1 + \left( \frac{4\rho \Omega^2 H_0^2(r_0)}{3\eta} \right) t \right],$$

which predicts that no matter what the initial thickness profile $H_0(r_0)$ is, the film will eventually level off.

Recent reviews of the spin-coating process, including aspects of non-Newtonian rheology and accompanied by solvent evaporation[32,34] as well as air drag at high rotational speeds,[35] may alter the predicted thickness profiles significantly.

### Example 8.5: Journal-Bearing Lubrication

In the journal-bearing systems shown in Fig. E8.5, 10W/30 oil of viscosity $\eta = 2.0\,\text{P}$ and density $\rho = 50\,\text{lb}_\text{m}/\text{ft}^3$ is used. The journal is of radius $R = 0.5\,\text{ft}$ and rotates at $\Omega = 200\,\text{rpm}$. The gap between journal and bearing after being mapped to an almost rectilinear channel has dimensions $L = 5\,\text{in.}$, $d_1 = 0.015\,\text{in.}$, and $d_2 = 0.013\,\text{in.}$, Calculate the following:

**(a)** the maximum pressure and its location along $L$;

**(b)** the total load capacity;

**(c)** the friction on the two plates;

**(d)** the torque to turn the journal;

**(e)** the power required to turn the journal.

Comment on the relative magnitude by comparing your values with common units (e.g., maximum pressure to atmospheric pressure).

**Solution**   The inclined wall line is given by the equation

$$d(x) = d_1 - \frac{d_1 - d_2}{L}x = d_1 - ax, \qquad a = \frac{d_1 - d_2}{L}. \tag{8.5.1}$$

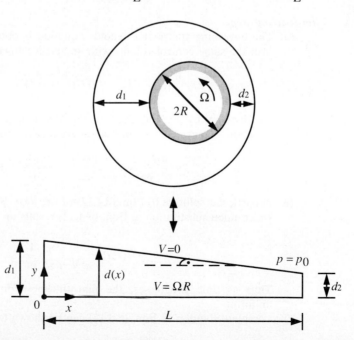

**Figure E8.5**  Journal-bearing lubrication and corresponding nearly rectilinear geometry.

With this profile, Eqn. (8.27) yields

$$-p_0 = \frac{6\eta}{\alpha}\left[V\left(\frac{1}{d(x)} - \frac{1}{d_1}\right) - \dot{V}\left(\frac{1}{d(x)} - \frac{1}{d_1^2}\right)\right] \tag{8.5.2}$$

and Eqn. (8.28) yields

$$\dot{V} = \frac{Vd_1d_2}{d_1 + d_2}. \tag{8.5.3}$$

Then Eqn. (8.5.9) becomes

$$p(x) = p_0 + \frac{6\eta V}{a(d_1 + d_2)}\left[\left(\frac{d_1}{d(x)} - 1\right)\left(1 - \frac{d_2}{d(x)}\right)\right]. \tag{8.5.4}$$

The maximum pressure occurs for $x = x^*$ such that $[dp/dx]_{x=x^*} = 0$ and $[d^2p/dx^2]_{x=x^*} < 0$. Thus

$$\frac{dp}{dx}\bigg|_{x=x^*} = \frac{6\eta V}{d_1 + d_2}[(d(x) - d_2)d_1 - (d_1 - d(x^*))d_2]\frac{1}{d^3(x^*)} = 0, \tag{8.5.5}$$

which yields

$$d(x^*) = \frac{2d_1d_2}{d_1 + d_2}, \qquad \frac{x^*}{L} = \frac{d_1}{d_1 + d_2}. \tag{8.5.6}$$

It is easy to show that $[d^2p/dx^2]_{x=x^*} < 0$, and therefore Eqns. (8.5.5) and (8.5.6) define the position $d(x^*)$ or $x^*$, where the pressure becomes maximum. This maximum value is

$$p_{max} = p\left(x = x^* = \frac{d_1L}{d_1 + d_2}\right) = \frac{6\eta VL}{4d_2d_1}. \tag{8.5.7}$$

Other quantities are calculated by means of Eqns. (8.29) and (8.30), given Eqns. (8.5.1) to (8.5.7), as follows.

**(a)** *Maximum pressure*:

$$p_{max} = \frac{6\eta VL}{4d_2d_1} = \frac{6 \times 2/14.88 \times 200/60 + 5/12}{4 \times (0.015 \times 0.013)/144} = 2 \times 10^5 \text{ lb}_f/\text{ft}^2,$$

$$d_{max} = \frac{2d_1d_2}{d_1 + d_2} = \frac{2 \times 0.015 \times 0.013}{12 \times 0.028} = 0.001161 \text{ ft} \times 0.0139 \text{ in.},$$

$$z_{max} = \frac{d_{max} - d_1}{d_2 - d_1}L = \frac{0.013961 - 0.0150}{-0.002} = 0.55L.$$

**(b)** *Total load capacity*:

$$\ln\frac{d_1}{d_2} = \ln\frac{0.015}{0.013} = 0.1431, \qquad \frac{d_1 - d_2}{d_1 + d_2} = \frac{0.002}{0.018} = 0.071428,$$

$$\frac{\eta V}{a} = \frac{(2/15)200.1}{0.002/5} = \frac{2.67}{0.0004} = 6.6 \times 10^4 \text{ lb}_f/\text{ft},$$

$$\int_0^L (p - p_0)\,dx = \frac{6 \times 6.6 \times 10^4}{4 \times 10^{-4}}(0.1431 - 0.1428) = 2.4 \times 10^5 \text{ lb}_f/\text{ft}.$$

**(c)** *Friction on the plates:*
*Upper plate:*

$$\int_0^L \eta\left(\frac{du}{dy}\right)_{y=d} dx = \frac{2\eta v}{a}3\frac{d_1 - d_2}{d_1 + d_2} - \ln\frac{d_1}{d_2}$$

$$= 1.36 \times 10^4(0.2242 - 0.1431) = 1.1 \times 10^3 \, \text{lb}_f/\text{ft}.$$

*Lower plate:*

$$\int_0^L \eta\left(\frac{du}{dy}\right)_{y=0} dx = \frac{2\eta v}{a}3\frac{d_1 - d_2}{d_1 + d_2} - 2\ln\frac{d_1}{d_2}$$

$$= 1.36 \times 10^4(0.2242 - 0.2862) \times -1.42 \times 10^3 \, \text{lb}_f/\text{ft}.$$

Regarding the friction, note that:
1. Its value on the two plates is not symmetric.
2. It is positive on the upper and negative on the lower plate, according to our convention.
3. It is much less than the total load capacity:

$$\frac{\text{Load capacity}}{\text{friction}} \simeq 200.$$

**(d)** *Torque to turn the journal:*

$$T = \underbrace{(-1.42 \times 10^3) \, \text{lb}_f/\text{ft}}_{\substack{\text{force per unit length} \\ \text{circumference}}} \quad \underbrace{(2\pi \times 0.5) \, \text{ft}}_{\substack{\text{length of} \\ \text{circumference}}} \quad \underbrace{0.5 \, \text{ft}}_{\substack{\text{distance from} \\ \text{axis of rotation}}}$$

$$= 2.23 \times 10^3 \, \text{lb}_f \, \text{ft}.$$

**(e)** *Power to turn the journal:* For channel-like flows, it is convenient to write $P = Q \, \Delta p$. For torsional flows, the power can also be calculated by means of

$$P = Fu_\theta = F\Omega R = T\Omega,$$

which yields

$$P = (2.23 \times 10^3) \, \text{lb}_f \, \text{ft} \, \frac{200}{60}\frac{1}{\text{s}} = 7.43 \times 10^3 \frac{\text{lb}_f \, \text{ft}}{\text{s}} = \frac{7.43 \times 10^3}{1818} \, \text{hp} = 4.088 \, \text{hp}.$$

## Example 8.6: Wire Coating

The equipment shown in Fig. E8.6 is utilized for wire coating. The pressure at the inlet and outlet of the die is atmospheric, $p^0$.

**(a)** Analyze the flow and compute the velocity and pressure profiles.

**(b)** Plot the resulting pressure distribution and compare it to those of rectilinear and diverging annulus.

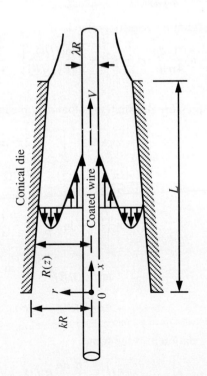

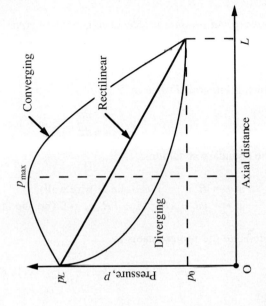

**Figure E8.6** Wire coating die and resulting pressure profile.

**Solution**

**(a)** *Velocity and pressure profiles.* The governing equation is

$$-\frac{dp}{dz} + \eta \left[ \frac{1}{r} \frac{d}{dr} \left( r \frac{du_z}{dr} \right) \right] = 0, \tag{8.6.1}$$

which is integrated twice to

$$u_z = \frac{r^2}{4\eta} \frac{dp}{dz} + c_1 \ln r + c_2. \tag{8.6.2}$$

The boundary conditions,

at $r = R$, $u_z = V$ (no-slip at wire wall)
at $r = (R - kR)z/L + kR$, $u_z = 0$ (no-slip at die wall)

determine the two constants as

$$c_2 = V - \frac{R^2}{4\eta} \frac{dp}{dz} - c_1 \ln R$$

and

$$c_1 = \left[ -\frac{1}{4\eta} \frac{dp}{dz} \left( \left( (\lambda R - kR) \frac{z}{L} + kR \right)^2 - R^2 \right) - V \right] \times \frac{\ln(\lambda R - kR)}{R} \frac{z}{L} + kR.$$

The resulting velocity profile is

$$u_z \frac{1}{4\eta} \frac{dp}{dz} (r^2 - R^2) + \frac{\ln(r/R)}{\ln[R(z)/R]} \left[ -\frac{1}{4\eta} \frac{dp}{dz} (R^2(z) - R^2) - V \right] + V. \tag{8.6.3}$$

The pressure distribution is found by means of the known flow rate:

$$\dot{V} = 2\pi R H_f = \int_R^{R(z)} u_z 2\pi r \, dr$$

$$= \frac{\pi}{2\eta} \frac{dp}{dz} \left[ \frac{R^4(z)}{4} - \frac{R^2 R^2(z)}{2} - \frac{R^4}{4} - \frac{(R^2(z) + R^2)}{\ln[R(z)/R]} \right.$$

$$\times \left. \left( \frac{R^2(z)}{2} \ln \frac{R(z)}{R} - \frac{R^2(z)}{4} - \frac{R^2}{4} \right) \right] \tag{8.6.4}$$

$$- \frac{V}{\ln[R(z)/R]} \left( \frac{R^2(z)}{2} \ln \frac{R(z)}{R} - \frac{R^2(z)}{4} - \frac{R^2}{4} \right)$$

$$+ 2\pi \frac{V}{2} (R^2(z) - R^2).$$

The relation is of the form

$$2\pi R H_f = \frac{\pi}{2\eta} \frac{dp}{dz} R^4 f_1(k, \lambda, z) + V R^2 f_2(k, \lambda, z), \tag{8.6.5}$$

with $f_1$ and $f_2$ known functions. Thus

$$\frac{dp}{dz} = \frac{2\pi R H_f - V R^2 f_2(k, \lambda, z)}{(\pi/2\eta) R^4 f_1(k, \lambda, z)} \tag{8.6.6}$$

and

$$p(z) = p^0 + \frac{4\eta H_f}{R^3} \int_0^z \frac{1}{f_1} dz - \frac{2\eta V}{\pi R^2} \int_0^z \frac{f_2}{f_1} dz. \tag{8.6.7}$$

**(b)** The pressure distributions for converging rectilinear and diverging dies are plotted in Fig. E8.6. The nonmonotonic pressure of the converging die is due to backflow, as suggested by the shown velocity profile.

### Example 8.7: Film and Sheet Casting

Repeat the analysis of Example 8.2 for the case of film casting shown in Fig. 8.1.

**(a)** Derive the corresponding equation that governs the velocity profile.

**(b)** Solve the resulting equation for negligible inertia, surface tension, and gravity effects.

### Solution

**(a)** *Derivation of governing equations.* An appropriate control volume of width $W$, height $dz$, and thickness identical to that of the film is chosen, as shown in Fig. E8.7. A force or momentum balance on the control volume produces the governing momentum equation,

$$\frac{\partial}{\partial t}(\rho u h W \, dz) = (h W \rho u^2)_z - (h W \rho u^2)_{z+dz} + p W h|_z$$

$$- p W h|_{z+dz} - \tau_{zz} W h|_z + \tau_{zz} W h|_{z+dz} + \rho g W h (dz)$$

$$- 2(W + h)\sigma|_z + 2(W + h)\sigma|_{z+dz}, \tag{8.7.1}$$

which is rearranged to a differential form by invoking the definition of the

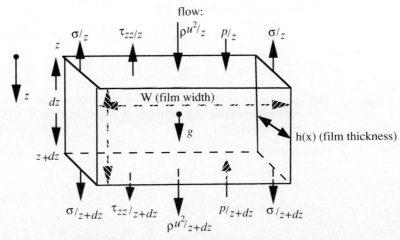

flow:

**Figure E8.7**  Control volume in film casting.

derivative:

$$-\rho W \frac{\partial}{\partial z}(hu^2) - W\frac{\partial}{\partial z}(ph) + W\frac{\partial}{\partial z}(\tau_{zz}h) + \rho gWh + 2\sigma\frac{\partial h}{\partial z} = 0.$$

(8.7.2)

The thickness $h$ is eliminated in favor of velocity $u$, by means of the flow rate equation,

$$\dot{V} = h(z)Wu(z),$$

(8.7.3)

and the resulting equation is

$$\rho\frac{du}{dz} = \frac{d}{dz}\left(\frac{-p + \tau_{zz}}{u}\right) + \rho g\frac{1}{u_z} + 2\sigma\frac{d}{dz}\left(\frac{1}{u}\right).$$

(8.7.4)

At the free surface, $-p + \tau_{xx} = 2H\sigma \approx 0$; therefore, $p = \tau_{xx}$ and Eqn. (8.7.4) becomes

$$\rho\frac{d}{dz} = \frac{d}{dz}\left(\frac{\tau_{zz} - \tau_{xx}}{u}\right) + \rho g\frac{1}{u} + 2\sigma\frac{d}{dz}\left(\frac{1}{u}\right).$$

(8.7.5)

When inertia, surface tension, and gravity effects are negligible, the equation becomes

$$\frac{d}{dz}\left(\frac{\tau_{zz} - \tau_{xx}}{u}\right) = 0.$$

(8.7.6)

This equation applies to both Newtonian and non-Newtonian liquids. To obtain the Newtonian film casting, Newton's law of viscosity is invoked:

$$\tau_{zz} = \eta\frac{\partial u_z}{\partial z} \quad \text{and} \quad \tau_{xx} = \eta\frac{\partial u_x}{\partial x} = -\eta\frac{\partial u_z}{\partial z}$$

(8.7.7)

and Eqn. (8.7.6) becomes

$$\frac{d}{dz}\left(2\frac{1}{u}\frac{du}{dz}\right) = 0, \qquad \frac{1}{u}\frac{du}{dz} = c_1, \qquad \ln u = c_1 z + c_2.$$

(8.7.8)

The boundary conditions are:

At $z = 0$, $u = u_0$; therefore, $u_0 = c_2$.
At $z = L$, $u = u_L$; therefore, $\ln u_L = c_1 L + \ln u_0$, $c_1 = (1/L)$ $\ln(u_L/u_0)$.

(b) The velocity profile is first obtained by applying the two boundary conditions in order to determine the constants $c_1$ and $c_2$ in Eqn. (8.7.7). The following results are obtained:

*Velocity profile:*

$$u(z) = u_0 \exp\left(\frac{z}{L}\ln\frac{u_L}{u_0}\right).$$

(8.7.9)

*Normal stress difference profile:*

$$\tau_{zz} - \tau_{xx} = 2\frac{du}{dz} = 2u_0\frac{1}{L}\ln\frac{u_L}{u_0}\exp\left(\frac{x}{L}\ln\frac{u_L}{u_0}\right) = 2u(z)\frac{1}{L}\ln D_R.$$

(8.7.10)

*Draw ratio:*

$$D_r \equiv \frac{u_L}{u_0}. \tag{8.7.11}$$

*Drawing force:*

$$F = W \int_0^{h(x)} (\tau_{xx} - \tau_{xx})\, dx = W h(x) 2u(z) \frac{1}{L} \ln D_R = 2\eta \frac{\dot{V}}{L} \ln D_R. \tag{8.7.12}$$

The similarities between these results and their axisymmetric analog fiber spinning examined in Examples 8.2 and 8.3 are obvious. Cases of film casting of non-Newtonian liquid are examined in Chapter 9.

# PROBLEMS

1. *Flow in a converging annulus.* Analyze the flow in a slightly converging linear annulus of the following dimensions: inner and outer radii at its inlet, $R$ and $kR$, respectively; inner and outer radii at its outlet, $R$ and $(k - \varepsilon)R$, respectively; and length of its inner cylindrical surface, $L$, The inlet pressure is $p_0$ and the outlet atmospheric, and the resulting flow rate is $Q$ for a Newtonian liquid of density $\rho$ and viscosity $\eta$.

2. *Flow around a conical rotometer.*[36] Flow around conical objects, as shown by Fig. P8.2, is utilized to measure flow rates by estimating the resulting pressure drop and the known dimensions, weight, and equilibrium position of the cone. Analyze the lubrication flow between the containing cylinder and its concentric cone under the assumption that the geometry is such that the lubrication equations apply. What is a necessary condition for lubrication flow?

3. *Flow in sinusoidal channel.* The gap of an infinitely wide and long sinusoidal channel varies according to the relation $H(x) = H_0 + \alpha \sin(2\pi x/L)$, where $H_0$ is the gap at the

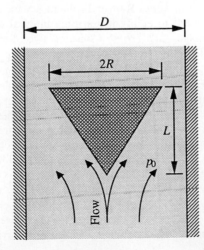

**Figure P8.2**  Measurement of flow rate by a rotometer.

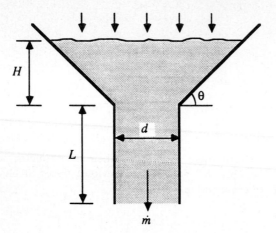

**Figure P8.5**   Flow in a conical hoop.

inlet, $\alpha$ is the amplitude of the sinusoidal walls, $x$ is the direction of flow, and $L$ is the channel length. The pressure at the inlet is $p_0$ and the channel discharges to the open atmosphere, water of viscosity $\eta$, and density $\rho$ at flow rate $Q$.
(a) What is the necessary condition for the lubrication approximation to apply?
(b) Find the pressure distribution along the channel.
(c) What is likely to happen under conditions that violate the criterion for the lubrication approximation?

4. *Film flow over rough surface.* A thin film of Newtonian liquid of density $\rho$ and viscosity $\eta$ flows down a rough, inclined plane of length $L$ at angle $\phi$ with the horizontal, at flow rate $Q$. The roughness of the plane can be approximated by $y(z) = \alpha \sin(k\pi x/L)$. For an opaque liquid of surface tension $\sigma$, will the configuration of the free surface reveal the plane roughness? Is the answer different for negligible surface tension?

5. *Flow in a conical hoop.* Analyze the entire flow in the conical hoop shown in Fig. P8.5, for a Newtonian liquid of density $\rho$ and viscosity $\eta$, when the hoop is always filled with liquid. The angle $\theta$ is such that lubrication approximation applies.

6. *Flow in a discharging conical hoop.* Repeat Problem 8.5 for $t > 0$ when the hoop is filled at time $t = 0$ and is left to empty within time interval $\Delta t$. What is the required time for complete drainage, $\Delta t$? (*Hint:* Consider two flows, a pipe flow and a lubrication flow, and match flow rates. Also, $\tan \theta \approx \theta \approx 0$.)

7. *Thin film flow down a conical surface.*[37] In connection with gas-absorption studies, it is decided to explore several arrangements in systems of simple geometry. One system that seems attractive is that in which a liquid jet impinges on the point of a cone and flows down along the cone as a thin liquid film. The cone has base diameter $D$ and opening angle $\theta$. Obtain an approximate expression for the film thickness profile $h(x)$ with $x$, the distance along the conical surface. You can neglect surface tension, so that pressure is hydrostatic in the liquid film. How will things be different with finite surface tension? (*Hint:* Although the geometry is cylindrical, it is convenient to use the Cartesian system and the corresponding equations. Why?)

8. *Wire coating.* Wire is advanced along the axis of a cylindrical converging die by tension pulling it at a velocity $V$. The converging annular space between the wire of radius $R$ and the die of exit radius $H_0$ is always filled. The die-wire arrangement is identical to

that of Example 8.6. The coating film decays from the $H_0$ to the final target film $H_\infty$, under the competing actions of surface tension and drag by velocity $V$, within a distance $L$. The physical characteristics of the coating liquid are density $\rho$, viscosity $\eta$, and surface tension $\sigma$. The pressure due to surface tension is $\sigma/H(z)$ along the film, where $H(z)$ is the local radius (distance from the $z$-axis) of the free surface. Analyze the flow between $0 < z < L$, as follows:

(a) What is the velocity profile? What type of flows does it incorporate? What is the force $F$ needed to advance the wire at a velocity $V$?

(b) Derive the Reynolds equation and its appropriate boundary conditions as a function of the physical characteristics $v = \eta/\rho$ and $\sigma$, the velocity $V$, and the geometry $H_0$, $H_\infty$, $R$, and $L$.

(c) Solve the Reynolds equation in the limiting cases of capillary numbers $\text{Ca} = 0$ and $\text{Ca} = \infty$.

(d) Show that in the common case of $H_0/H_\infty \approx 1$, the film thickness decays exponentially with distance, at rate depending on the capillary number.

9. *Film flow down concave surface.* A film of Newtonian liquid of density $\rho$, viscosity $\eta$, and surface tension $\sigma$ is formed when the liquid flows at a total rate $Q$ down a vertical, concave cylindrical column of height $H$, base radii $R_0$, and half-height radius $R_1$. The liquid film wets the entire concave cylindrical surface of the column and the flow is axisymmetric.

(a) What is the equation that gives the radius $R(z)$ of the curved surface of the cylinder in terms of the distance $z$ measured downward from the origin indicated? You may assume that it is of the form $R = c_1 + c_2 z + c_3 z^2$.

(b) What is a necessary and sufficient condition for lubrication theory to apply?

(c) Assuming that the pressure in the surrounding air is zero, the pressure inside the liquid film is given by $p(z) = -\sigma/H(z)$. What is the velocity profile $u_z(r, z)$?

(d) Derive the Reynolds equation for this flow, and simplify by observing that $H(z) \approx R(z)$, and thus $\ln[H(z)/R(z)] \approx 0$.

(e) Solve the Reynolds equation for both zero and infinite surface tension. Explain qualitatively the behavior of the free surface in both cases.

10. *Pressurization in polymer processing.*[38] A simple device for pressurization involves two parallel plates in relative motion, transferring the in-between liquid to a vertical barrier equipped with a discharge valve. Show that this device can give rise to a pressure rise. Examine the pressure buildup and flow rate achieved under *closed discharge and open discharge* conditions, when the valve is fully closed and fully open, respectively. Calculate the maximum pressure gradient achieved by such a device for a polymeric melt of approximately constant viscosity, $\eta = 1000\,\text{P}$, when the upper plate travels parallel to itself at velocity $V = 1\,\text{m/s}$, delivering $0.02\,\text{m}^3/\text{s}$ of polymer.

11. *Converging channel flow.* Consider the steady flow of an incompressible Newtonian liquid of viscosity $\eta$ and density $\rho$ in a planar converging channel. The channel is symmetric about its centerplane (the $xz$-plane at $y = 0$). The length of the channel is $L$, and the gap distance between the channel walls is given by $h(x)$, where $h(x = 0) = h_1$ and $h(x = L) = h_2$.

(a) For what conditions would a lubrication approximation be valid, in this geometry?

(b) Assuming that the lubrication approximation is valid, derive an expression for the fluid velocity as a function of the pressure gradient in the channel and the physical properties of the fluid.

(c) Derive an expression for the pressure profile in terms of the channel dimensions, given that the fluid pressure is $p_0$ at $x = 0$ and $p_L$ at $x = L$.

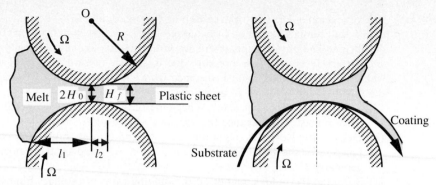

**Figure P8.12**   Roll calendering and coating.

(d) Compare your velocity and pressure profiles along the channel to those characteristics of the steady flow of an incompressible Newtonian liquid between two parallel flat plates.

12. *Rotating rolls: calendering*[16] *and roll coating.*[15] Figure P8.12 shows a simplified version of two co-rotating rolls which is used extensively to form plastic sheets (calendering) or to apply thin films (coating) on substrates. The two identical rolls, of radius $R$ and lateral width $W$, are set a minimum distance $2H_0$ apart and rotate with frequency $N$. The melt fed contacts both rolls at distance $l_1$ upstream from the nip and detaches in the form of a sheet of thickness $2H_f$ at location $l_2$ downstream from the nip.

   (a) What is the relation between the nip gap $H_0$ and the final thickness $H_f$?
   (b) What is the speed of production of thin film?
   (c) Derive the lubrication equations for a Newtonian liquid of density $\rho$ and viscosity $\eta$. First find the velocity distribution, then formulate the Reynolds equation and the required pressure boundary condition.
   (d) Where does the maximum pressure occur? Sketch the pressure profile.
   (e) Sketch velocity profiles at selected positions in $l_1 < x < l_2$.

13. *Blade coating.*[16,39] A schematic view of blade coating is depicted in Fig. P8.13.
   (a) Calculate the maximum pressure if the pressure at the inlet and the outlet is zero.
   (b) Show that the resulting coating thickness is $H_f = H_0/(1 + H_0/H_1)$.
   (c) What are the required shear and normal forces to keep the blade in place?

14. *Calender design.* It is desired to manufacture a 1-m-wide and 0.2-m-thick PVC film at

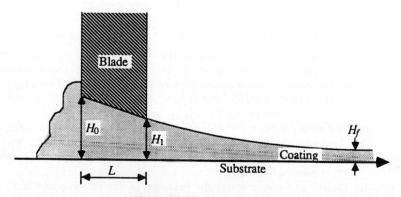

**Figure P8.13**   Blade coating.

a production rate of 1.2 m³/h. Design an appropriate calendering facility and state the appropriate operating conditions. Consider the viscosity of PVC constant at the temperature of processing. The prototype arrangement of calendering is shown in Fig. P8.12 and summarized in Problem 8.12.

15. *Film casting.* Solve the film casting equation of Example 8.7 in selected limiting cases of the parameters involved: Examine cases of free-falling metallic sheet under gravity alone, drawn polymeric film, and drawn glass film.

16. *Casting of polymer film and steel sheet.* Consider a common casting arrangement of die gap $h_0 = 1$ cm and width $W = 1$ m extruding at rate $\dot{V} = 3$ L/s a film taken up at distance $L = 2$ m downstream by a rotating drum of radius $R = 15$ cm. At what angular velocity of drum $\Omega$ films of target thickness $h_L = 0.1$ cm are obtained for the following materials:
    (a) Polymer melt of viscosity $\eta = 1000$ P, density $\rho = 0.95$ g/cm³ and surface tension $\sigma = 55$ dyn/cm.
    (b) Sheet of steel of viscosity 0.05 cP, density 8 g/cm³, and surface tension 1200 dyn/cm.

17. *Free-falling impinging fiber.* A fiber emerges from a capillary of radius $R$ at flow rate $Q$ and falls freely under gravity in a liquid bath located a distance $L$ downstream. The governing equation for Newtonian liquid of density $\rho$, viscosity $\eta$, and surface tension $\sigma$ is Eqn. (8.2.11). Solve this equation under the appropriate boundary conditions at the impinging end for selected limiting cases of the involved parameters. Repeat for a free-falling impinging sheet.

18. *Extrusion coating.* Liquid of viscosity $\eta$, density $\rho$, and surface tension $\sigma$ is extruded at flow rate $Q$ and deposited onto a substrate moving with velocity $V$. The die is at distance $H^0$ over the substrate and the coating attains a practically constant thickness at distance $L$ downstream.
    (a) What is the final film thickness?
    (b) What is the velocity profile?
    (c) What is the resulting Reynolds equation?
    (d) Make the equation dimensionless, identify the resulting dimensionless numbers, and solve for limiting values of these numbers.

19. *Fiber-spinning boundary conditions.*[40] Fiber spinning of Newtonian liquids under dominant viscous and gravity forces is governed by the one-dimensional equation,

$$\frac{d}{dz}\left[3\eta \frac{du_z}{dz} \pi R^2(z)\right] + \rho g \pi R^2(z) = 0,$$

where $R(z)$ is the radius of the fiber. Bring the equation to its dimensionless form,

$$u_z^* \frac{d}{dz}\left(\frac{1}{u_z^*}\frac{du_z^*}{dz^*}\right) = -\frac{\rho g L^2}{3\eta V} = -\text{St},$$

where $V$ is the inlet velocity and $L$ is the length of the fiber. Show how the solution,

$$u_z^* = Ve^{-c_2 x^*} + \frac{\text{St}}{z}z^{*2},$$

may be obtained. What do each of the two terms represent? What are acceptable boundary conditions at the other end of the fiber, and what kind of spinning do they represent (e.g., free falling fiber, drawn fiber)? Justify the limiting value of $u_z^*(z^* \to \infty)$. Show that the fiber will never attain a constant diameter, and explain the

physical significance of the fact. What is the solution for fiber drawn by velocity
$u_z = V_L$ at $z = L$ (or $z^* = 1$)?

**20.** *Unsteady film flow with thinning viscosity.* A liquid of viscosity $\eta = \eta_0/(1 + k\partial u_z/\partial y)$
and density $\rho$ forms a nonuniform film $\delta(z)$ on a vertical wall under gravity and in
contact with still air, such that $\rho g/\eta_0\delta \ll 1$, in which case the velocity profile is

$$u_z(y) = \frac{\rho g}{2\eta_0}[1 + m(\delta - y)](2\delta y - y^2).$$

**(a)** What is $m$ here?
**(b)** Derive the unsteady Reynolds lubrication equation and boundary and initial
conditions for films of your choice (e.g., film of constant thickenss at inlet, film of
constant slope at outlet, etc.).
**(c)** Make your equations dimensionless and identify the resulting dimensionless
numbers.
**(d)** Construct perturbation solutions in the limits of $m \to 0$ and $m \to \infty$, and comment
on the physical significance of your solution(s).
**(e)** Is there any analytic solution to your case by similarity transformation, $\xi = ct^a z^b$?
**(f)** Is there any wave solution to your equations of the form $\delta(t, z) = \delta(\xi = z + ct)$?
What is $c$?
**(g)** Compare the results of parts (d) and (e) and/or (f).

**21.** *Cylinder-plate lubrication.* Consider an infinitely long cylinder of radius **a**, slowly
translating at speed $U$ in the $x$-direction parallel to a flat plate in a quiescent fluid. The
minimum separation distance is $\delta$. Show that as $\delta \to 0$, the distance between plate and
cylindrical surface is

$$h(x) = \delta + \frac{x^2}{2\mathbf{a}}.$$

**(a)** In the lubrication limit ($\delta \to 0$), find the velocity and pressure profiles generated
by such motion.
**(b)** Derive the Reynolds equation for the pressure profile.
**(c)** Using the Reynolds equation and the appropriate boundary conditions, determine
the pressure and velocity profiles in the gap between the surfaces. Plot the
pressure profile as a function of $x$.

(Provided by S. Bike, University of Michigan, fall 1989, by permission.)

**22.** *Spin-coating topography.*[41] In spin coating, one of the stages involves leveling of a thin
film of initial thickness $h^0(t)$ on top of a disk of radius $R$ and roughness approximated
by $\hat{h}(r) = a \sin(\pi r/R) \ll h(r, t)$, where $h(r, t)$, is the thickness at time $t$, rotating at
angular velocity $\Omega$.
**(a)** Under what conditions or assumptions does the lubrication approximation hold?
**(b)** Show that under lubrication conditions the resulting equations are

$$-\rho r\Omega^2 = -\frac{\partial p}{\partial r} + \eta \frac{\partial^2 u_r}{\partial z^2} \qquad \text{(P8.22.1)}$$

and

$$0 = -\frac{1}{r}\frac{\partial p}{\partial \theta} + \eta \frac{\partial^2 u_\theta}{\partial z^2}. \qquad \text{(P8.22.2)}$$

**(c)** What are the appropriate boundary conditions under surface tension $\sigma(r)$?

**(d)** Solve for the velocity profile and show that the resulting flow rates in the two directions are

$$q_r = \left(\rho r\Omega^2 - \frac{\partial p}{\partial r}\right)\frac{(h - \hat{h})^3}{3\eta} + \frac{\partial \sigma}{\partial r}\frac{(h - \hat{h})^2}{2\eta} \tag{P8.22.3}$$

and

$$q_\theta = -\frac{1}{r}\frac{\partial p}{\partial \theta}\frac{(h - \hat{h})^3}{3\eta} + \frac{1}{r}\frac{\partial \sigma}{\partial r}\frac{(h - \hat{h})^2}{2\eta} \tag{P8.22.4}$$

and therefore the resulting Reynolds equation is

$$\frac{\partial h}{\partial t} = -\frac{1}{r}\frac{\partial}{\partial r}(rq_r) - \frac{1}{r}\frac{\partial}{\partial \theta}(q_\theta). \tag{P8.22.5}$$

**(e)** The pressure in Eqns. (P8.22.3) and (P8.22.4) is eliminated by means of the normal stress condition at the free surface,

$$-p + 2\eta\frac{\partial u_r}{\partial z} = \sigma\left[\frac{1}{r}\frac{\partial}{\partial r}\left(r\frac{\partial h}{\partial r}\right) + \frac{1}{r^2}\frac{\partial^2 h}{\partial \theta^2}\right] \simeq \frac{\sigma}{r}\frac{\partial h}{\partial r}, \tag{P8.22.6}$$

to yield the final form of the Reynolds equation, which is what?

**(f)** Solve the Reynolds equation for $h^0(r) = h_0$, $\sigma(r) = k$, analytically or asymptotically for small and large $a$. How will things change in case of $k = 0$? Explain the physical significance of your solution(s).

**23.** *Wave propagation in a basin.* Show that the Reynolds lubrication equation

$$\frac{\partial h}{\partial t} \simeq gh\frac{\partial^2 h}{\partial x^2} \tag{P8.23.1}$$

when inertia, $\delta^2(AV)/\partial x^2 \ll 1$, and slope, $g(\partial h/\partial x)^3 \ll 1$, are small, by combining the kinematic equation,

$$\frac{\partial h}{\partial t} + \frac{\partial h}{\partial x}(hV) = 0, \tag{P8.23.2}$$

and the momentum equation,

$$\frac{\partial}{\partial t}(hV) + \frac{\partial}{\partial x}\left(hV^2 + \frac{gh^2}{2}\right) = gh\tan\phi - c_f\frac{V^2}{2}S, \tag{P8.23.3}$$

where $V$ is the nearly rectlinear velocity, $A$ the area of the cross section of the flow, $\phi$ the inclination angle, $c_f$ a friction coefficient at the bottom, and $S$ the wetted bottom area.

**(a)** Show how Eqns. (P8.23.2) and (P8.23.3) are derived.

**(b)** Solve Eqn. (P8.23.1) with appropriate conditions of your choice, and comment on the physical significance of the solution.

## REFERENCES

1. O. Reynolds, Papers on mechanical and physical aspects, *Philos. Trans. Roy. Soc.*, 177, 157 (1886).
2. D. D. Fuller, *Theory and Practice of Lubrication for Engineers*, Wiley, New York, 1984.

3. G. K. Miltsios, D. J. Peterson, and T. C. Papanastasiou, Solutions of the lubrication problem and calculation of friction of piston ring, *J. Tribol. ASME,* 111, 635 (1989).
4. B. V. Deryagin and S. M. Levi, *Film Coating Theory,* Focal Press, New York, 1964.
5. N. A. Anturkar, T. C. Papanastasiou, and J. O. Wilkes, Lubrication theory for *n*-layer thin-film flow with applications to multilayer extrusion and coating, *Chem. Engrg. Sci.,* 45, 3271 (1990).
6. J. R. A. Pearson, *Mechanics of Polymer Processing,* Elsevier Applied Science, London, 1985.
7. M. M. Denn, Continuous drawing of liquids to form fibers, *Annu. Rev. Fluid Mech.,* 12, 365 (1980).
8. P. J. Luccbest, E. H. Roberts, and S. J. Kurtz, Reducing draw resonance in LLDPE film resins, *Plastics Eng.,* May 1985.
9. C. D. Han and J. Y. Park, Studies on blown film extrusion. II. Analysis of deformation and heat transfer processes, *J. Appl. Polym. Sci.,* 19, 3277 (1975).
10. S. Kalpakjian, Manufacturing Engineering and Technology, Addison-Wesley, Reading, Mass., 1989.
11. G. Pearson and S. Middleman, Elongational flow behavior of viscoelastic liquids, *AIChE J.,* 23, 714 (1977); 722 (1977).
12. T. C. Papanastasiou, Lubrication flows, *Chem. Engrg. Ed.,* 24, 50 (1989).
13. M. M. Denn, *Process Fluid Mechanics,* Prentice Hall, Englewood Cliff, N.J., 1980.
14. K. Adachi, T. Tamura and R. Nakamura, Coating flows in a nip region and various critical phenomena, *AIChE J.,* 34, 456 (1988).
15. D. J. Coyle, C. W. Macosko, and L. E. Scriven, Film splitting flows in forward roll coating, *J. Fluid Mech.,* 171, 183 (1986).
16. S. Middleman, *Fundamentals of Polymer Processing,* McGraw-Hill, New York, 1977.
17. J. C. Coyne and H. G. Elrod, An exact asymptotic solution for a separating film, *J. Lubrication Technol.,* Oct., 651 (1969).
18. L. Prandtl, General discussion on lubrication, *Proc. Inst. Mech. Engr.,* 182, 104 (1937).
19. N. E. Bixler and L. E. Scriven, Robin conditions for open-flow boundaries, *Bull. Amer. Phys. Soc.,* 25, 1079 (1980).
20. G. F. Teletzke, H. T. Davis and L. E. Scriven, How liquids spread on solids, *Chem. Engrg. Commun.,* 55, 41 (1988).
21. A. H. Lefebvre, *Atomization and Sprays,* Hemisphere, New York, 1989.
22. T. C. Papanastasiou, A. N. Alexandrou, and W. P. Graebel, Rotating thin films, *J. Rheol.,* 32, 485 (1988).
23. F. K. Wasden and A. E. Dukler, Insights into the hydrodynamics of free falling wavy films, *AIChE J.,* 35, 187 (1989).
24. T. D. Karapantsios and A. J. Karabelas, Surface characteristics of roll waves on free falling films, *J. Multiphase Flow,* 16, 835 (1990).
25. M. Van Dyke, *Perturbation Methods in Fluid Mechanics,* Parabolic Press, Stanford, Calif., 1964.
26. C. W. Oseen, Über die Stokessche Formel und über die verwandte Aufgabe in der Hydrodynamik, *Arkiv. Mat. Astron. Fysik,* 6 29 (1910).
27. P. G. Drazin and W. H. Reid, *Hydrodynamic Stability,* Cambridge University Press, New York, 1981.
28. N. R. Anturkar, T. C. Papanastasiou and J. O. Wilkes, Linear Stability analysis of multilayer plane Poiseuille flow, *Phys. Fluids A,* 2, 530 (1990).
29. M. M. Dens, *Stability of Reaction and Transport Processes,* Prentice Hall, Englewood Cliffs, N.J., 1975.

30. T. C. Papanastasiou and W. J. Maier, Dynamics of biodegradation of 2,4-dichlorophenoxyacetate in the presence of glucose, *Biotechnol. Bioengrg.*, 25, 2337 (1983).

31. J. Meissner, Development of a universal extensional rheometer for the uniaxial extension of polymer melts, *Trans. Soc. Rheol.*, 16, 405 (1972).

32. S. A. Jenekhe and S. B. Schudt, Coating flows of non-Newtonian fluids on a flat rotating disk, *Ind. Engrg. Chem. Fundam.*, 23, 425 (1984); 23, 432 (1984).

33. A. G. Emslie, F. T. Bonner, and L. G. Peck, Flow of viscous liquid on a rotating disk, *J. Appl. Phys.*, 29, 858 (1958).

34. D. E. Bornside, C. W. Macosko, and L. E. Scriven, On the modeling of spin-coating, *J. Imaging Technol.*, 13, 122 (1987).

35. T. J. Rehg and B. G. Higgins, The effects of inertia and interfacial shear on film flow on a rotating disk, *Phys. Fluids*, 31, 1988 (1360).

36. R. H. Perry, *Perry's Chemical Engineer's Handbook*, 6th ed., McGraw-Hill, New York, 1984.

37. R. B. Bird, W. E. Stewart, and E. N. Lightfoot, *Transport Phenomena*, Wiley, New York, 1960.

38. Z. Tadmor and C. G. Gogos, *Principles of Polymer Processing*, Wiley, New York, 1979.

39. F. R. Pranckh and L. E. Scriven, Elastohydrodynamics of blade coating, *AIChE J.*, 36, 587 (1990).

40. C. J. S. Petrie, *Elongational Flows: Aspects of the Behavior of Model Elastoviscous Fluids*, Pitman, London, 1979.

41. P. C. Sukanek, A model for spin coating with topography, *J. Electrochem. Soc.*, 10, 3019 (1989).

# 9

# Rheology
# and Flows
# of Non-Newtonian Liquids

## 9.1 INTRODUCTION

The preceding chapters have dealt exclusively with Newtonian liquids such as gases, water, water-based solutions, most inorganic liquids, and in general, liquids of small and stiff molecules of isotropic resistance to motion. Most industrial liquids are non-Newtonian, such as liquids of polymeric origin, suspensions and emulsions, food sauces, paints, coatings, and others. These liquids are made primarily of elastic macromolecules that exhibit three distinct properties: (1) They can deform substantially under normal stresses, being elongated or compressed; (2) they can orient themselves to align along streamlines, reducing resistance to flow and viscosity. The mechanics and flow of non-Newtonian matter are detailed in numerous publications in the fields of non-Newtonian fluid mechanics, rheology, and polymer processing. In this chapter we examine important non-Newtonian effects in channel, film, lubrication, and stretching flows. A summary on rheology, rheometry, and constitutive equations that are prerequisites to non-Newtonian flows is also presented.

## 9.2 NEWTONIAN VS. NON-NEWTONIAN LIQUIDS

The (shear) viscosity of a liquid is proportional to the resistance exhibited by its molecules to align with the flow and to deform under shear (and normal) stress:

$$\underbrace{\frac{du_x}{dy}}_{\substack{\text{relative flow or} \\ \text{deformation, shear rate}}} = \underbrace{\frac{1}{\eta}}_{\substack{\text{fluidity or} \\ \text{inverse viscosity}}} \underbrace{\tau_{yx}}_{\text{stress}}. \tag{9.1}$$

Newtonian liquids exhibit constant viscosity independent of shear rate and of type of flow, shear, or extensional. Equivalently, Newtonian liquids exhibit uniform resistance to flow independent of flow conditions. Non-Newtonian liquids exhibit viscosity which changes with the shear rate and with the type of deformation: In extensional flows, they exhibit elongational viscosities in no-apparent relation to their (shear) viscosity, as discussed in Section 9.3. Equivalently, non-Newtonian liquids exhibit different resistance to flow, at different shear and extensional rates. Non-Newtonian liquids, in general, can be:

1. *Shear thinning,* or thixotropic when their viscosity decreases (*thins*) with increasing shear rate.
2. *Shear thickening,* or rheopectic when their viscosity increases (*thickens*) with increasing shear rate.
3. *Viscoplastic or Bingham plastics,* which flow only if subjected to a shear stress bigger than a characteristic stress, the *yield stress.* Beyond this point, they may behave as Newtonian, shear-thinning, or shear-thickening liquids. Below their yield stress, these liquids behave like elastic solids.

   Figure 9.1 illustrates the behavior of the viscosity of several rheologically different liquids and the resulting stress and viscosity as functions of the shear rate. $\tau_0$ is the value of the yield stress of a viscoplastic liquid.

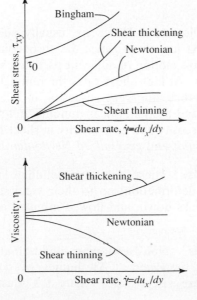

**Figure 9.1**  Shear stress and shear viscosity of several rheological classes of liquids.

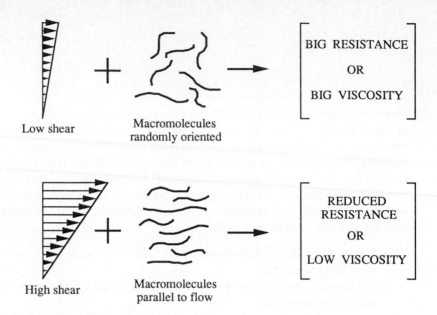

**Figure 9.2**   Viscosity is resistance of molecules to flow.

**Shear-thinning liquids.**   Most non-Newtonian liquids consist of macro-molecules or particles and droplets, suspended in solvents. These molecules or particles are randomly oriented under no flow conditions or at low shear rates and therefore exhibit large resistance to flow and thus high viscosity. At large shear rates accompanied by large vorticity and local angular velocity [e.g., Eqn. (2.37)], these molecules rotate and orient themselves parallel to the flow. In this way they exhibit small resistance to flow and thus reduced viscosity. Figure 9.2 illustrates this situation.

**Shear-thickening liquids.**   These are usually solid-particle suspensions and droplet emulsions. At low shear rates, the suspending medium serves as a lubricant film to the relative motion of adjacent particles, so the resistance to flow is small, as is the viscosity. At high shear rates the lubrication fails, and therefore resistance to flow and thus apparent viscosity increase, due to suspended particle–particle contact and friction. Shear-thinning and shear-thickening liquids exhibit non-Newtonian viscosity and no elasticity in shear flows, and are therefore called *viscous inelastic liquids* and *generalized Newtonian liquids.*

**Viscoplastic liquids.**   These exhibit a solidlike structure and therefore vanishingly small elastic deformation, up to their yield stress. Beyond the yield stress, their structure breaks down, and flow is initiated accompanied by viscous stress and deformation.

**Viscoelastic liquids.**   Non-Newtonian liquids that exhibit both viscosity and elasticity are called *viscoelastic.* All liquids of polymeric origin (melts, solutions, suspensions) are viscoelastic. Figure 9.3 illustrates the growth and

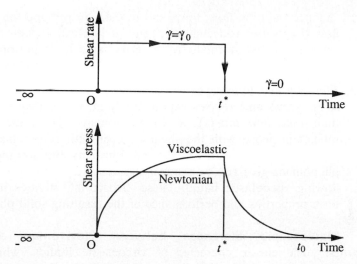

**Figure 9.3** Stress development in Newtonian and viscoelastic liquids. The relaxation time is $\lambda$.

relaxation of shear stress for a viscoelastic liquid and Newtonian liquid, after imposition and subsequent elimination of a constant shear rate. For Newtonian liquids the deformation and the stress are in phase, for viscoelastic liquids they are out of phase, and therefore viscoelastic liquids may continue to be under stress even under zero deformation or shear rate. The viscous character of viscoelastic liquids is controlled by their ability to orient themselves differently under different flow conditions, which gives rise to shear thinning. The elastic character is controlled by the flexibility and ability of elastic macromolecules to respond to shear and extensional deformations.

Common natural and industrial liquids are classified as (1) Newtonian: air, water, oil, glycerin, aqueous solutions; (2) shear thinning: polymeric melts and polymeric solutions; (3) shear thickening: suspensions and emulsions; (4) Bingham plastics: paint, ketchup, mayonnaise, fiber suspensions; and (5) viscoelastic: liquids of polymeric origin.

Important fundamental differences between Newtonian and non-Newtonian liquids include:

1. *Viscosity.* The viscosity of Newtonian liquids is constant. The viscosity of non-Newtonian liquids depends on the shear rate and may increase or decrease with shear rate. Some non-Newtonian liquids exhibit *yield stress* in addition. Thus, shear-thinning liquids reduce friction and wear while shear-thickening liquids increase them.[1] Furthermore, viscoelastic liquids exhibit large *elongational viscosity,* independent of (shear) viscosity.

2. *Normal stress.* The normal stress difference (e.g., $\tau_{xx} - \tau_{yy}$) of Newtonian and viscous inelastic liquids in viscometric flows (where there is only shear deformation, e.g., channel and pipe flow) is zero. Non-Newtonian, viscoelastic liquids exhibit nonzero normal stresses in viscometric flows. In

channel and pipe flow, they tend to push away opposing walls, in torsional flow they cause rod climbing,[2] and in lubrication flow they enhance load capacity.[3] These normal stresses grow large in extensional or elongational flows (e.g., radial and stretching flows), due to large elongational viscosity.

**3.** *Elasticity and time-dependent stresses.* Due to elastic macromolecules, the stress grows and relaxes exponentially at a rate inversely proportional to their *relaxation time(s)*, $\lambda$. In Newtonian liquids, the stress grows and relaxes in phase with the shear rate, or with zero relaxation time. These principles are illustrated in Fig. 9.3. Elasticity and accompanying relaxation phenomena give rise to residual stresses and strain upon solidification of a flowing viscoelastic liquid. These "frozen-in" stresses influence the long-term properties and performance of the resulting solid plastic.[4]

**Macroscopic dierences between Newtonian and viscoelastic liquids.** The elastic character of viscoelastic liquids, which induces time-dependent growth and relaxation of shear and normal stresses, gives rise to fascinating macroscopic phenomena.[5] Some of these are shown in Fig. 9.4. In (a), the viscoelastic liquid climbs the rotating rod under the action of normal

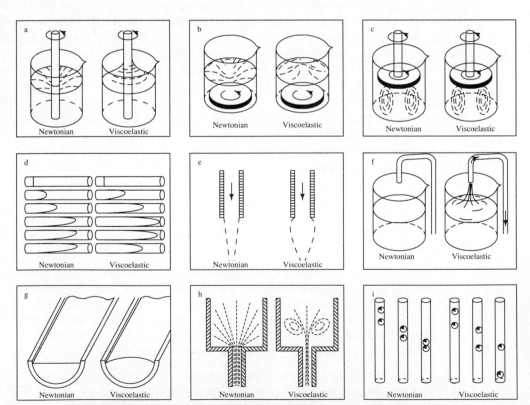

**Figure 9.4** Unusual flows of viscoelastic liquids, in comparison to the corresponding flows of Newtonian liquid. (From Ref. 5, by permission.)

viscoelastic stresses, overcoming the underpressure which causes the Newtonian liquid to descend the rod. The secondary flow in (c) and the elevation of the free surface in (b) are also due to the action of viscoelastic normal stresses, which are not present with Newtonian liquids. In (d), the elimination of the driving pressure gradient causes the Newtonian motion to cease instantaneously. However, the viscoelastic motion tends to be reversed in order to bring the viscoelastic liquid to its initial state of a flat-zero velocity profile, due to memory effects. In (e), viscoelastic jets tend to swell much more than their Newtonian counterparts, due to the release of the high-normal viscoelastic stresses. In (f), the viscoelastic liquid siphons across the gap due to the ability of the elastic macromolecules to sustain stretching, transmitting tension upstream. In (g) the action of normal stresses raises the free surface of a flowing viscoelastic liquid, and in (h) large vortices are induced in contraction flows of viscoelastic liquids at low Reynolds numbers. Finally, in (i), the first of the two spheres travels faster in a viscoelastic liquid due to its shear-thinning viscosity, which reduces friction, but the second sphere follows at a lower speed due to the relaxation of the stresses, created by passage of the first sphere.

## 9.3 RHEOLOGY AND RHEOMETRY

The name *rheology* originates from the Greek words *rheo,* which stands for "flow," and *logos,* which stands for "science." Rheology investigates the fundamental relation between stress and deformation during flow,[6] for example

$$\underbrace{\tau_{yx}}_{\text{stress}} = \underbrace{\eta}_{\substack{\text{material} \\ \text{property}}} \quad \underbrace{\frac{du_x}{dy}}_{\text{deformation}}, \tag{9.2}$$

which is called the *constitutive equation.* Thus the constitutive equation is an equation of state in nonequilibrium situations. The corresponding equation of state under equilibrium, for example the ideal gas law,

$$\underbrace{p}_{\text{stress}} = \underbrace{\frac{RT}{M}}_{\substack{\text{material} \\ \text{property}}} \quad \underbrace{\rho}_{\text{deformation}}, \tag{9.3}$$

can also be considered as a stress-deformation constitutive relation under static conditions; indeed, pressure is a normal stress and density describes the strain that arranges matter in space.

*Rheometry* is the experimental part of rheology. It deals with finding appropriate ways to measure deformation under imposed stress, and vice versa,

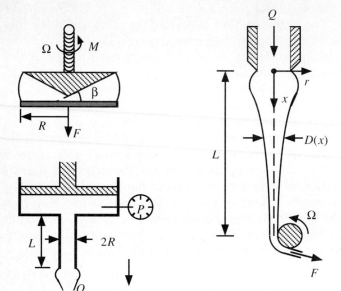

**Figure 9.5** Widely applied rheometers: (a) cone-and-plate, (b) capillary, and (c) fiber-spinning.

in order to formulate a working constitutive equation. Investigations in this area have produced several commercially available rheometers to measure viscosity, normal stresses, relaxation times, and elongational viscosity. Three widely used rheological instruments are the cone-and-plate, capillary, and fiber-spinning rheometers, as shown in Fig. 9.5 and described below.

**Cone-and-plate rheometer.**   This device is used to measure the shear viscosity and the normal stresses at several shear rates, as well as the relaxation time of viscoelastic liquids. The flow, which is initiated when the upper cone rotates over the bottom stationary plate, with the liquid filling the gap, gives rise to a known shear rate, uniform along the liquid-cone contact wall,

$$\dot{\gamma} = \frac{\Omega}{\beta}, \tag{9.4}$$

where $\Omega$ is the speed of rotation and $\beta$ the cone angle. The developed shear and normal stresses within the liquid act against the upper cone or the lower plate and the resulting total normal thrust, $F$, and the torque, $T$, are recorded by sensitive transducers connected with the cone and/or the plate. These are related to the shear and the normal stress at the cone wall by

$$\tau_{12} = \frac{3T}{2\pi R^3}, \qquad \tau_{11} - \tau_{22} = \frac{2F}{\pi R^2}, \tag{9.5}$$

where $R$ is the radius of the cone. The shear viscosity, at shear rate $\dot{\gamma}$, is then

$$\eta(\dot{\gamma}) = \frac{\tau_{12}}{\dot{\gamma}} = \frac{3T}{2\pi R^3 \dot{\gamma}} = \frac{3T\beta}{2\pi R^3 \Omega}, \tag{9.6}$$

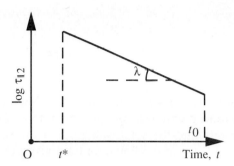

**Figure 9.6**  The relaxation time, $\lambda$, of a viscoelastic liquid is obtained by measuring the stress decay, $\tau_{12}(t)$, after elimination of a constant shear rate, as represented by Fig. 9.3.

and the *first normal stress coefficient*, $N_1(\dot{\gamma})$, is

$$N_1(\dot{\gamma}) = \frac{\tau_{11} - \tau_{22}}{\dot{\gamma}^2} = \frac{2F}{\pi R^2 \dot{\gamma}^2} = \frac{2F\beta^2}{\pi R^2 \Omega^2}. \tag{9.7}$$

Cone-and-plate rheometers theoretically can perform measurements down to vanishingly small shear rates, but are limited in practice to moderate shear rates by material failure or instability. Measurements at very high shear rates are usually done by means of capillary rheometers.

The relaxation time, $\lambda$, is measured by performing the experiment described by Fig. 9.3 and recording the duration of the exponential decay of stress as shown by Fig. 9.6. This is done by exposing the liquid to steady rotation following the upper rotating cone (Fig. 9.5), and then forcing it to come to an abrupt rest by halting the rotation. Relaxation time(s) are also determined by sinusoidal rotation with small strain amplitude and increasing frequency and recording the in-phase (viscous) and out-of-phase (elastic) stresses. This is also a way to distinguish and quantify the viscous and the elastic contribution to stress of a viscoelastic lqiuid.[7]

In flows of viscoelastic liquids, the dimensionless *Deborah number* is defined by[4]

$$\text{De} = \frac{\lambda}{L/\bar{u}} \qquad \left[\frac{\text{relaxation time}}{\text{residence time}}\right], \tag{9.8}$$

where $L/\bar{u}$, residence time, represents the effective local elasticity that persists for that residence time.

The dimensionless *Weissenberg number*, defined by

$$W_s = \frac{\tau_{11} - \tau_{22}}{\tau_{12}} = \lambda\dot{\gamma} \qquad \left[\frac{\text{elastic force}}{\text{viscous force}}\right] \tag{9.9}$$

represents the relative magnitude of the viscoelastic normal stresses (associated with the elastic character) to the magnitude of the viscoelastic shear stresses (associated with the viscous character).[2] When the Maxwell mode is used (Table 9.1) in channel and pipe flows, then $\text{De} = 3W_s$, by comparing the two defining equations.[3]

The ratio of viscosity $\eta$ to the relaxation time $\lambda$ is the *coefficient* or *modulus* of the *monodisperse viscoelastic liquid*

$$g = \frac{\eta}{\lambda}, \qquad \eta = g\lambda,$$

where $\eta$ is the *zero shear viscosity*. In case of a *polydisperse viscoelastic liquid* of many relaxation times $\lambda_i$, the zero shear viscosity is

$$\eta = \sum_i g_i \lambda_i. \tag{9.10}$$

**Capillary rheometer.** This instrument utilizes Poiseuille-type flow in a capillary under known pressure difference, $\Delta p$ and flow rate $Q$. Under these flow conditions, both the shear rate and the solid wall, $\dot{\gamma}$, and the shear stress there, $\tau_{12}$, are independent of the nature of the liquid and are given by

$$\dot{\gamma} = \frac{Q}{\pi R^3}\left[3 + \frac{d(\ln Q)}{d(\ln \Delta p)}\right], \qquad \tau_{12} = \frac{\Delta p}{\Delta L}\frac{R}{2}, \tag{9.11}$$

where $R$ and $L$ are the radius and length of the capillary, respectively. Thus the viscosity, $\eta(\dot{\gamma})$, can be measured at high shear rates induced by high flow rates in thin capillaries. Normal stress and relaxation time can also be deduced by more elaborate calculations.

**Fiber-spinning rheometer.** The cone-and-plate rheometers and the capillary rheometers are shear rheometers. They can be used to evaluate and predict the behavior of non-Newtonain materials only in shear flows and deformations. However, important industrial applications such as spinning of fibers, casting of films, coating, and compression molding are dominated by elongational stresses. These stresses have no apparent relation to the shear stresses and therefore need to be measured and formulated. Elongational or extensional stresses are proportional to an *elongational viscosity*, which is much higher than the corresponding (shear) viscosity of the material. This is due to the fact that viscoelastic macromolecules resist stretching or compression much more drastically than they resist shearing between each other. The elongational viscosity is defined by

$$\eta_e = \frac{\tau_{11} - \tau_{22}}{\dot{\varepsilon}}, \qquad \dot{\varepsilon} = \frac{du_1}{dx_1}, \tag{9.12}$$

where $\tau_{11} - \tau_{22}$ is normal stress difference arising due to imposition of a *constant* extension rate $\dot{\varepsilon}$ in the direction of flow.

In axisymmetric fiber-spinning rheometers, the diameter of the fiber, $D(x)$, is measured by mechanical micrometers or by projection and photography. This is adequate to calculate the extension rate $\dot{\varepsilon}(x_1^*)$ at position $x_1^*$ along the fiber, by

$$\dot{\varepsilon}(x_1^*) \equiv \frac{du_1}{dx_1} = \frac{d}{dx_1}\left[\frac{4Q}{\pi D^2(x_1)}\right]_{x_1=x_1^*} = -\frac{8Q}{\pi D^3(x_1^*)}\frac{dD(x_1)}{dx_1}\bigg|_{x_1=x_1^*}, \tag{9.13}$$

where $Q$ is the flow rate of the fiber. The stress difference along the fiber is given by

$$[\tau_{11} - \tau_{22}]_{x_1=x_1^*} = \frac{4F}{\pi D^2(x_1^*)}, \tag{9.14}$$

where $F$ is the tension required to draw the fiber. The stress difference is measured by connecting the orifice, called a spinneret, to a tensiometer. The elongational viscosity, at extension rate $\dot{\varepsilon}(x_1^*)$, is then

$$\eta_e(\varepsilon) = \frac{[\tau_{11} - \tau_{22}]_{x=x_1^*}}{\dot{\varepsilon}(x_1^*)} = \frac{FD(x^*)}{Q} \left| \frac{dD(x)}{dx_1} \right|_{x_1=x_1^*}. \tag{9.15}$$

Fiber spinning appears to be the only feasible method to measure elongational viscosity of liquids of low shear viscosity. Yet *apparent elongational viscosities* measured by fiber spinning according to Eqn. (9.15) are in general different from truly elongational viscosities defined by Eqn. (9.11). This is due to the extension rate $\varepsilon(x_i^*)$ varying along the fiber of a fiber-spinning rheometer, and therefore the prerequisites of the defining equation, Eqn. (9.11), are not met. The difference between apparent fiber spinning and true elongational viscosity is vanishingly small at low Deborah number, De $= \lambda \bar{u}/L = \lambda \dot{\varepsilon} < 1$, and grows large for De $\gg 1$, where resistance of macromolecules to stretching gives rise to large elongational stresses.[8] For highly viscous polymeric melts, sheet and filament stretching (e.g., Fig. 8.5) is also applicable.[9]

Viscoelastic melts are usually extension thickening, which gives rise to high elongational stresses during processing. Viscoelastic solutions are usually extension thinning. At low extension rates the elongational viscosity of all liquids approaches the Newtonian viscosity, which is three times the corresponding shear viscosity. This limit, known to rheologists as the Trouton limit,[10] is illustrated in Fig. 9.7.

**Example 9.1: Working Equations for Rheometers**

Derive the working equations for cone-and-plate [Eqns. (9.4) to (9.7)] and capillary rheometers [Eqn. (9.11)].

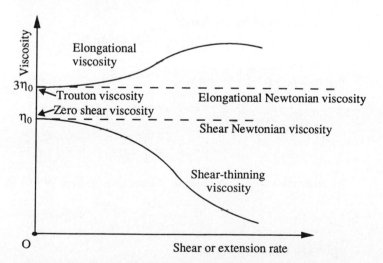

**Figure 9.7**   Shear and elongational viscosity of a typical polymer melt.

**Solution** *Cone-and-plate rheometer* (Fig. 9.5). For small angle $\beta$ and opening $H(r)$, the induced torsional flow can be approximated by

$$u_\theta = \Omega r \frac{z}{H(r)}, \qquad (9.1.1)$$

where $\Omega r$ is the linear azimuthal velocity at distance $r$ from the axis of rotation, and $H(r)$ is the gap at radial position $r$. The wall shear rate is then

$$\dot\gamma = \left.\frac{du_\theta}{dz}\right|_{z=H(r)} = \Omega\frac{r}{H(r)} = \frac{\Omega}{\tan\beta} \simeq \frac{\Omega}{\beta}, \qquad (9.1.2)$$

since for $\beta \to 0$, $\sin\beta \simeq \tan\beta \simeq \beta$.

The torque necessary to sustain the rotation is

$$T = \int_0^R 2\pi \frac{r}{\cos\beta} r\tau_{z\theta}\, dr = 2\pi \frac{R^3}{3}\frac{1}{\cos\beta}\tau_{z\theta} \simeq \tfrac{2}{3}\pi R^3 \tau_{z\theta},$$

and therefore

$$\tau_{z\theta} = \frac{3T}{2\pi R^3}. \qquad (9.1.3)$$

*Capillary rheometer.* The governing momentum equation, in reference to Fig. 9.5 is

$$r\frac{P}{L} + \frac{d}{dr}(r\tau_{rz}) + \rho g = 0, \qquad (9.1.4)$$

which is integrated to

$$\tau_{rz} = -\frac{\Delta p r}{\Delta L 2} + \frac{c}{r}, \qquad \frac{\Delta p}{\Delta L} = \frac{P}{L} + \rho g. \qquad (9.1.5)$$

At $r = 0$, $\tau_{rz} = 0$ (symmetry condition), and therefore $c = 0$. Thus

$$\tau_{rz} = -\frac{\Delta p}{\Delta L}\frac{r}{2}, \qquad (9.1.6)$$

and at the wall of the capillary,

$$\tau_{rz}\big|_{r=R} = -\frac{\Delta p}{\Delta L}\frac{R}{2}. \qquad (9.1.7)$$

The flow rate is

$$Q = \int_0^R 2\pi r u_z\, dr = \int_0^R \pi \frac{d}{dr}(r^2)u_z\, dr$$

$$= \pi r^2 u_z\big|_0^R - \pi\int_0^T \frac{du_z}{dr}r^2\, dr = \pi\int_0^R \dot\gamma r^2\, dr. \qquad (9.1.9)$$

By replacing $r$ with $\tau_{rz}$ by means of Eqn. (9.1.6), Eqn. (9.1.9) becomes

$$\frac{Q}{\pi R^3} = \frac{1}{(\Delta p/\Delta L)^3}\int_0^{\Delta p/\Delta L} \dot\gamma \tau_{rz}^2\, d\tau_{rz}, \qquad (9.1.10)$$

which can be shown to yield[2]

$$\dot\gamma = \frac{Q}{\pi R^3}\left[3 + \frac{d(\ln Q)}{d(\ln \Delta p)}\right] \qquad (9.1.11)$$

**Fiber-spinning rheometer.**    The working equations are Eqns. (9.13) to (9.15). Collection of data and reduction are described in Problem 9.23.

Notice that in both Example 9.1 and in the theory behind the fiber-spinning rheometer, the governing expressions were derived without the use of a constitutive equation. The necessary and sufficient conditions that a rheometer must fulfill are:

1. Working equations independent of constitutive equations.
2. Measurable deformation and stress.

## 9.4 CONSTITUTIVE EQUATIONS FOR NON-NEWTONIAN LIQUIDS

The phenomena illustrated by Fig. 9.4 cannot be explained by the Navier–Stokes equations. Since they are based on the simple *Newtonian law,*

$$\tau_{yx} = \eta \left( \frac{du_x}{dy} + \frac{du_y}{dx} \right), \tag{9.16}$$

where $\eta$ is the constant Newtonian viscosity, they apply only to flows of Newtonian fluids. Different constitutive equations must be employed, which incorporate the fundamental microscopic differences between Newtonian and non-Newtonian molecules that give rise to different macroscopic phenomena. Some of the derived or designed constitutive equations are tabulated in Table 9.1 and explained as follows.

**Viscous inelastic liquids.**    These are liquids that exhibit shear-thinning or shear-thickening viscosity, but no normal stresses in viscometric flows and no elasticity or memory. They are also called *generalized Newtonian liquids* because their constitutive equation is similar to Newton's law of viscosity; however, the

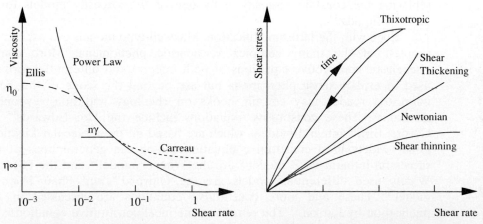

**Figure 9.8**    Viscosity models for viscous inelastic liquids, and hysteresis loop characteristic of thixotropic liquids.

viscosity itself is a function of the shear rate, as shown in Fig. 9.8. The shear dependence of viscosity has been expressed by several empirical relations obtained by fitting rheological data.[2] The three widely used relations are the power-law model, the Ellis model, and the Carreau model, as recorded in Table 9.1.

**Thixotropic and Rheopectic liquids.** These are liquids that possess structure that reforms with the flow or deformation and therefore the viscosity changes not only with the shear rate alone, but also with the duration of deformation. In thixotropic liquids the viscosity and shear stress increase with time, and in rheopectic liquids they decrease. This kind of behavior leads to a kind of hysteresis loop on the curve of shear stress vs. shear rate as shown in Fig. 9.8. This kind of behavior can not be described by any form of equations of Table 9.1. A profound class of thixotropic liquids are thermosetting plastic resins[3] and oil well drilling mud.[12]

**Bingham plastic liquids.** These are materials that exhibit a solid-like structure unless sheared by an external stress higher than a specific stress, the yield stress, which is characteristic of the material. Once initiation of flow occurs, the material behaves as a Newtonian or as a viscous-inelastic liquid, as shown by Fig. 9.9. The appropriate constitutive equation is the ideal Bingham model tabulated in Table 9.1, and modified continuous representations of it.[11]

**Viscoelastic liquids.** These are liquids that may exhibit both shear-thinning or shear-thickening viscosity and elasticity and memory. The simplest viscoelastic model is the Maxwell model, which is based on the assumption that a viscoelastic liquid exhibits both viscous resistance to flow, measured by its viscosity $\eta$, and elastic resistance to deformation, measured by its relaxation time $\lambda$. The two resistances are connected in series or in parallel, as shown by Fig. 9.10, to produce the overall stress-deformation constitutive equation given in Table 9.1. These constitutive equations predict time dependence and elastic behavior but no shear thinning or thickening. This deficiency is reconciled by replacing the constant viscosity $\eta$ by one of the viscosity models for viscous inelastic liquids.[13]

Even with the latter modification, Maxwell-type models can explain only the simplest of the many complex viscoelastic phenomena. More complicated viscoelastic constitutive equations of both integral and differential forms can be used to explain these phenomena but are beyond the scope of this book. The interested reader may consult books on rheology and non-Newtonian fluid mechanics. These constitutive equations include the Doi–Edwards[14] and the Curtiss–Bird[15] integral models, which are based on the molecular kinetic theory. Other widely used constitutive equations along the general lines of the very successful integral BKZ model[16] are the Wagner model[17] and the PSM model.[18] Widely used differential models are the Oldroyd[19] and Phan–Thien Tanner model.[20] These and more constitutive equations are discussed in a recent publication by Larson.[21] The pursuit of the ideal constitutive equation is still one of the most active research areas in rheology.

**TABLE 9.1**   CONSTITUTIVE RELATIONS OF COMMON LIQUIDS IN CARTESIAN SYSTEM

| Rheological class | Constitutive equation | Representative liquids |
|---|---|---|
| Newtonian: | $\tau_{ij} = \eta\left[\dfrac{du_i}{dx_j} + \dfrac{du_j}{dx_i}\right]$  $\eta = 0 = \text{const.}$ | Water and inorganic liquids and solutions, lubricants, glycerin, honey |
| Viscous inelastic: | Same as Newtonian, with $\eta$ given by:  Power law: $\eta(\dot{\gamma}) = \eta\left\|\dfrac{du_i}{dx_j}\right\|^n$  Ellis: $\eta(\dot{\gamma}) = \eta\left[1 + \eta\left\|\dfrac{du_i}{dx_j}\right\|^n\right]^{-1}$  Carreau:  $\eta(\dot{\gamma})\dfrac{\eta(\dot{\gamma}) - \eta_\infty}{\eta_0 - \eta_\infty} = \left[1 + \eta\left\|\dfrac{du_i}{dx_j}\right\|^n\right]^{-1}$ | Food sauces, paints, slurries, polymer melts and solutions, suspensions and emulsions, fiber suspensions |
| Viscoplastic: | Solid structure, no-flow:  $\dfrac{du_i}{dx_j} = 0 \quad \text{if } \tau_{ij} < \tau_y$  Flow:  $\tau_{yx} = \left[\text{sign}\left(\dfrac{du_i}{dx_j}\right)\right]\tau_y + \dfrac{du_i}{dx_j} \quad \text{if } \|\tau_{xy}\| > \tau_y$  ($\eta$ as in inelastic liquids) | Paints, emulsions, suspensions, ketchup, mayonnaise, pastes, filled rubber, fiber suspensions, blood, electrorheological fluids, concrete |
| Viscoelastic: | $\tau_{xx} + \lambda\left(u_x\dfrac{\partial\tau_{xx}}{\partial x} + u_y\dfrac{\partial\tau_{xx}}{\partial y}\right.$  $\left. - 2\tau_{xx}\dfrac{\partial u_x}{\partial x} - 2\tau_{xy}\dfrac{\partial u_x}{\partial y}\right) = 2\eta\,\dfrac{\partial u_x}{\partial x}$  $\tau_{xy} + \lambda\left(\dfrac{\partial u_x}{\partial x}\tau_{xy} + \dfrac{\partial u_y}{\partial y}\tau_{xy} - \tau_{xx}\dfrac{\partial u_y}{\partial x}\right.$  $\left. - \tau_{yy}\dfrac{\partial u_x}{\partial y}\right) = \eta\left(\dfrac{\partial u_x}{\partial y} + \dfrac{\partial u_y}{\partial x}\right)$  $\tau_{yy} + \lambda\left(\dfrac{\partial u_x}{\partial x}\tau_{yy} + \dfrac{\partial u_y}{\partial y}\tau_{yy}\right.$  $\left. - 2\tau_{xy}\dfrac{\partial u_y}{\partial x} - 2\tau_{yy}\dfrac{\partial u_y}{\partial y}\right) = 2\eta\,\dfrac{\partial u_y}{\partial y}$ | Melts, solutions, emulsions and suspensions of polymeric origin |
| Elastic: | $\tau_{ij} = g\displaystyle\int_0^x \dfrac{\partial u_i\,dx_i}{\partial x_j\,u_j}$  $g = \text{elasticity modulus}$ | Elastometers, rubber products, fiber suspensions |

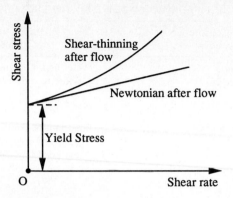

Figure 9.9   Stress of Bingham liquids.

**Elastic liquids.**    These are viscoelastic materials of infinitely large relaxation time and therefore of perfect, nonfading memory. Important classes are elastomers like rubber and concentrated fiber suspensions. The latter are becoming increasingly important in the manufacture of composite plastics reinforced with fibers. A recently advanced constitutive equation for these materials is the Dinh–Armstrong integral equation[22] based on Batchelor's[23] fiber suspension theory, and the kinetic theory applied to stiff fiber molecules.

To study complex viscoelastic flows, the constitutive equation is combined and solved along with the conservation equation.[24] This combination results in highly nonlinear, often integrodifferential systems of equations, which are difficult to solve numerically even with the most sophisticated computers. This is another active research area in rheology and polymer processing, which advances with the appearance of more sophisticated computational means.

## 9.5 FLOWS OF NON-NEWTONIAN LIQUIDS: COUETTE, POISEUILLE, FILM, LUBRICATION, AND STRETCHING FLOWS

The constitutive equations of Table 9.1 are combined with the stress momentum equations of Appendix G in a study of flows of non-Newtonian liquids. Some of

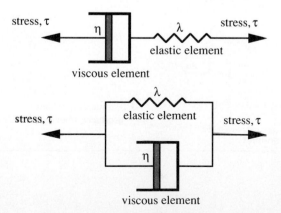

Figure 9.10   Elementary string and dashpot Maxwell-type viscoelastic models.

the flows analyzed previously for Newtonian liquids, including Couette, Poiseuille, film, lubrication, and casting are revisited in Examples 9.2 to 9.8.

### 9.5.1 Unidirectional Flows of Shear Thinning or Thickening and Viscoelastic Liquids

Examples 9.2 to 9.4 examine and compare unidirectional flows of Newtonian, shear thinning or thickening, and viscoelastic liquids.

**Example 9.2: Couette Flow of Non-Newtonian Liquids**

The governing momentum equation for any incompressible liquid in Couette flow is

$$\frac{d\tau_{yx}}{dy} = 0, \tag{9.2.1}$$

which is specialized to rheologically different liquids by means of the constitutive equation as follows.

*Newtonian liquid.* The corresponding constitutive equation,

$$\tau_{yx} = \eta \frac{du_x}{dy}, \tag{9.2.2}$$

is substituted into Eqn. (9.2.1) to produce the governing equation restricted to Newtonian liquid,

$$\frac{d^2 u_x}{dy^2} = 0. \tag{9.2.3}$$

The general solution of Eqn. (9.2.3) is

$$u_x = c_1 y + c_2. \tag{9.2.4}$$

The boundary conditions, which are independent of the rheological class of the liquid, are:

At $y = 0$, $u_x = 0$; thus $c_2 = 0$.
At $y = H$, $u_x = V$; thus $c_1 = V/H$.

The particular solution is then

$$u_x = \frac{V}{H} y, \tag{9.2.5}$$

which represents a linear velocity profile. The resulting stress distribution is

$$\tau_{yx} = \eta \frac{du_x}{dy} = \eta \frac{V}{H}, \tag{9.2.6}$$

and the shear force required to move the upper plate is

$$F = -\int_0^L -\tau_{yx}\, dx = \int_0^L \eta \frac{V}{H} dx = \eta \frac{V}{H} L > 0. \tag{9.2.7}$$

The resulting flow rate and average velocity are

$$\dot{V} = \int_0^H u_x\, dy = \int_0^H \frac{V}{H} y\, dy = \frac{VH}{2}, \qquad \bar{u} = \frac{\dot{V}}{H} = \frac{V}{2}. \tag{9.2.8}$$

*Viscous inelastic liquid.* Here we select the power law to represent the viscosity:

$$\tau_{yx} = \eta \left| \frac{du_x}{dy} \right|^n \frac{du_x}{dy}. \tag{9.2.9}$$

The resulting momentum equation for a power-law liquid is, by substituting Eqn. (9.2.9) into Eqn. (9.2.1),

$$\frac{d}{dy} \left[ \left| \frac{du_x}{dy} \right|^n \frac{du_x}{dy} \right] = 0. \tag{9.2.10}$$

The shear rate, $du_x/dy$, is always positive, and therefore Eqn. (9.2.10) is integrated to

$$\left( \frac{du_x}{dy} \right)^{n+1} = c_1,$$

and then to

$$u_x = c_1^{1/(n+1)} y + c_2, \tag{9.2.11}$$

which yields the particular solution by imposing the boundary conditions:

At $y = 0$, $u_x = 0$; thus $c_2 = 0$.
At $y = H$, $u_x = V$; thus $c_1 = (V/H)^{n+1}$.

The resulting solution is

$$u_x = \frac{V}{H} y, \tag{9.2.12}$$

which is identical to Eqn. (9.2.6) for Newtonian liquid. However, the resulting stress is

$$\tau_{yx} = \eta \left| \frac{du_x}{dy} \right|^n \frac{du_x}{dy} = \eta \left( \frac{du_x}{dy} \right)^{n+1} = \eta \left( \frac{V}{H} \right)^{n+1}, \tag{9.2.13}$$

which is lower than that of Newtonian liquid if the liquid is shear thinning ($n < 0$) and higher if the liquid is shear thickening ($n > 0$). The required force follows the same pattern,

$$F = -\int_0^L -\tau_{yx} \, dx = \eta \left( \frac{V}{H} \right)^{n+1} L. \tag{9.2.14}$$

The flow rate is identical to that achieved by the Newtonian liquid.

*Maxwell viscoelastic liquid.* The third-component constitutive equation of Table 9.1 for viscoelastic liquid yields $\tau_{yy} = 0$. The second-component constitutive relation becomes

$$\tau_{xy} = \tau_{yx} = \eta \frac{du_x}{dy}. \tag{9.2.15}$$

This is combined with Eqn. (9.2.1) to produce the governing equation,

$$c = \eta \frac{du_x}{dy}, \tag{9.2.16}$$

which yields a velocity profile identical to the Newtonian profile, and thus an identical flow rate, too. The shear stress predicted by this model is also identical to the Newtonian shear stress. However, the normal stress difference, $N_1 = \tau_{xx} - \tau_{yy}$, may not be zero. In fact, from the remaining component of the constitutive equation,

$$\tau_{xx} - 2\lambda \tau_{xy} \frac{du_x}{dy} = 0, \tag{9.2.17}$$

it turns out that

$$\tau_{yy} = 0, \qquad \tau_{xx} = 2\lambda \left( \eta \frac{V}{H} \right) \frac{V}{H} = 8 \left( \frac{\eta}{\lambda} \right) \left( \frac{\lambda V}{2H} \right)^2 = 8g \, \text{De}^2 = 2g(\lambda \dot{\gamma})^2 = 2g \, \text{Ws}^2, \tag{9.2.18}$$

which predicts the normal stress difference,

$$\tau_{xx} - \tau_{yy} = 8g \, \text{De}^2 = 2g \, \text{Ws}^2. \tag{9.2.19}$$

By contrast, Newtonian and viscous inelastic liquids yield invariantly zero normal stress difference under similar flow conditions. The normal stress predicted by Eqn. (9.2.19) is proportional to the relaxation time. This normal stress difference acts against the wall and should be taken into account when designing piping systems and dies to transfer or process viscoelastic liquids. In lubrication applications this extra pressure-like normal stress is taken advantage of to support a load (Section 8.2).

   *Maxwell liquid with shear-thinning viscosity.* If a power-law viscosity is introduced in the Maxwell model, the resulting constitutive equations yield $\tau_{yy} = 0$, and

$$\tau_{yx} = \eta \frac{du_x}{dy} = \eta \left( \frac{du_x}{dy} \right)^{n+1} \tag{9.2.20}$$

and

$$\tau_{xx} = 2\lambda \tau_{yx} \frac{du_x}{dy}. \tag{9.2.21}$$

The velocity and stress components in a Couette flow are

$$\tau_{yx} = \eta \left( \frac{du_x}{dy} \right)^{n+1} = \eta \left( \frac{V}{H} \right)^{n+1}, \qquad u_x = \frac{V}{H} y, \qquad \tau_{xx} = 2\lambda \eta \left( \frac{V}{H} \right)^{n+2}. \tag{9.2.22}$$

   Several important conclusions can be drawn from the study of Couette flow, which is a good prototype of drag processes and operations such as gear pumps, screw extruders, lubrication applications, and tranfer of liquids by belts:

1. The kinematics, (i.e., the velocity and the shear-rate profiles) are independent of shear thinning and elasticity.

2. The shear stresses and required forces are independent of elasticity but do depend strongly on shear thinning or shear thickening.

3. Normal stresses may develop only with elastic liquids.

   As discussed in Section 8.2, advantage is taken of 2 by adding shear-thinning additives in lubricants to reduce shear and wear. Simultaneously, advantage is taken of 3 since these polymeric additives develop normal viscoelastic stresses capable of supporting loads.

### Example 9.3: Steady Flow of Non-Newtonian Liquids in Pipe

The continuity relation for any incompressible liquid demands that

$$\frac{du_z}{dz} = 0, \tag{9.3.1}$$

and therefore the velocity $u_z$ is constant along $z$. The resulting momentum equation becomes

$$\frac{\Delta p}{\Delta L} = \frac{dp}{dz} = \frac{1}{r}\frac{d}{dr}(r\tau_{rz}), \tag{9.3.2}$$

which is integrated to

$$\tau_{rz} = \frac{r}{2}\frac{\Delta p}{\Delta L} + \frac{c_1}{r}. \tag{9.3.3}$$

The constant $c_1$ is determined by the symmetry condition that at $r = 0$, $\tau_{rz} = 0$, and thus $c_1 = 0$. Therefore, the governing momentum equation for the incompressible liquid becomes

$$\tau_{rz} = \frac{r}{2}\frac{\Delta p}{\Delta L}, \tag{9.3.4}$$

where $\Delta p/\Delta L$ is the constant pressure gradient (the assumption of a constant pressure gradient will be evaluated later).

*Newtonian liquid.* The Newtonian constitutive equation,

$$\tau_{rz} = \eta\frac{du_z}{dr}, \tag{9.3.5}$$

substituted in Eqn. (9.3.4), leads to the general solution

$$u_z(r) = \frac{r^2}{4\eta}\frac{\Delta p}{\Delta L} + c_2. \tag{9.3.6}$$

The boundary condition,

$$u_z(r = R) = 0, \tag{9.3.7}$$

determines the constant $c_2$, which produces the familiar parabolic profile

$$u_x = \frac{1}{4\eta}\frac{\Delta p}{\Delta L}(r^2 - R^2). \tag{9.3.8}$$

The shear stress is

$$\tau_{rz} = \frac{r}{2}\frac{\Delta p}{\Delta L}, \tag{9.3.9}$$

and the flow rate and average velocity are

$$\dot{V}_N = \int_0^R 2\pi r u_z\, dr = -\frac{\pi}{2\eta}\frac{\Delta p}{\Delta L}\int_0^R (r^2 - R^2)r\, dr = \frac{\pi R^4}{8\eta}\frac{\Delta p}{\Delta L},$$

$$\bar{u}_N = \frac{\dot{V}_N}{\pi R^2} = \frac{R^2}{8\eta}\frac{\Delta p}{\Delta L}. \tag{9.3.10}$$

The normal stress difference is zero.

*Viscous inelastic liquid of power-law viscosity.* The power-law viscosity relation inserted in to Eqn. (9.3.4) results in the governing momentum equation

$$\eta \left| \frac{du_z}{dr} \right|^n \frac{du_z}{dr} = \frac{r}{2} \frac{\Delta p}{\Delta L}, \tag{9.3.11}$$

and therefore Eqn. (9.3.4) is rearranged in the form

$$-\eta \left( -\frac{du_z}{dr} \right)^n \left( -\frac{du_z}{dr} \right) = -\eta \left( -\frac{du_z}{dr} \right)^{n+1} = \frac{r}{2} \frac{\Delta p}{\Delta L}, \tag{9.3.12}$$

by observing that $du_z/dr < 0$. The latter equation is integrated to

$$u_z = -\left( -\frac{1}{2\eta} \frac{\Delta p}{\Delta L} \right)^{1/(n+1)} \frac{r^{[1/(n+1)]+1}}{[1/(1+n)]+1} + c_2. \tag{9.3.13}$$

The constant $c_2$ is determined by the boundary condition $u_z(r = R) = 0$, which results in the velocity profile for power-law liquids in pipe flow,

$$u_z = \frac{1}{1 + 1/(n+1)} \left( -\frac{1}{2\eta} \frac{\Delta p}{\Delta L} \right)^{1/(n+1)} [R^{[1/(n+1)]+1} - r^{[1/(n+1)]+1}]. \tag{9.3.14}$$

Notice that the Newtonian velocity profile is recovered in the limit of $n = 0$, as it should. The shear stress profile is given by Eqn. (9.3.9). The normal stress difference is zero.

The resulting flow rate and average velocity under the constant pressure gradient are

$$\dot{V}_p = \int_0^R 2\pi r u_z(r) \, dr$$

$$= \left( \frac{-2\pi}{1 + 1/(n+1)} \right) \left( -\frac{1}{2\eta} \frac{\Delta p}{\Delta L} \right)^{1/(n+1)} \left[ \frac{R^{[1/(1+n)]+3}}{3 + 1/(1+n)} - \frac{R^{[1/(1+n)]+3}}{2} \right]_{r=0}^{r=R}$$

$$= \left( \frac{\pi}{3 + 1/(1+n)} \right) \left( -\frac{1}{2\eta} \frac{\Delta p}{\Delta L} \right)^{1/(1+n)} R^{[1/(1+n)]+3}, \tag{9.3.15}$$

$$\bar{u}_p = \frac{\dot{V}_p}{\pi R^2} = \frac{1+n}{3+4n} \left( -\frac{R}{2\eta} \frac{\Delta p}{\Delta L} \right)^{1/(1+n)} R^{(3+3n)/(1+n)}$$

which yields the corresponding Newtonian flow in the limit of $n = 0$ and $k = \eta$, as it should. The important conclusion here is that shear-thinning liquids yield a higher flow rate than do their Newtonian counterparts under the same driving pressure gradient, due to the fact that their resistance to flow decreases. The opposite is true for shear-thickening liquids.

*Maxwell viscoelastic liquid.* The governing stress equation, Eqn. (9.3.2), is combined with the constitutive equation of Table 9.1,

$$\tau_{rz} = \eta \frac{du_z}{dr}, \tag{9.3.16}$$

where $u_r = 0$ and $\partial/\partial z = 0$ due to unidirectionality, to yield the governing momentum equation for pipe flow of Maxwell liquid,

$$\frac{r}{2}\frac{\Delta p}{\Delta L} = \eta \frac{du_z}{dr}. \tag{9.3.17}$$

This equation yields a velocity profile identical to the Newtonian one, and thus an identical flow rate. However, the normal stress at the upper wall given by the constitutive equation,

$$\tau_{zz} - 2\lambda\tau_{rz}\frac{du_z}{dr} = 0, \qquad \tau_{rr} = 0, \tag{9.3.18}$$

is

$$\tau_{zz} = \frac{2\lambda}{\eta}\left(\frac{R}{2}\frac{\Delta p}{\Delta L}\right)^2 = 32\lambda^2\left(\frac{\eta}{\lambda}\right)\left(\frac{R}{8\eta}\frac{\Delta p}{\Delta L}\right)^2 = 32g(\bar{u}\lambda)^2$$

$$= 32g\,\mathrm{De}^2 = 2\lambda\eta\dot{\gamma}_w^2 = 2g\,\mathrm{Ws}^2, \tag{9.3.19}$$

drastically different from the Newtonian result of zero. The other normal stress, $\tau_{rr}$, is zero for both Newtonian and Maxwell liquids.

*Maxwell liquid with power-law viscosity.* It can be shown that the velocity profile, the flow rate, and the average velocity are identical to those given by Eqns. (9.3.14) and (9.3.15), and the shear rate and shear stress to those given by Eqns. (9.3.12) and (9.3.9). In addition, there is wall normal stress given by

$$\tau_{zz} = 2\lambda\tau_{rz}\frac{du_z}{dr} = 2\lambda\eta\left(-\frac{R}{2\eta}\frac{\Delta p}{\Delta L}\right)^{[1/(n+1)]+1} = 2g^{n+1}(-\lambda\dot{\gamma}_w)^{n+2} = 2g^{n+1}\,\mathrm{Ws}^{n+2}$$

$$= 2\lambda\eta\frac{\bar{u}_p^{n+2}}{R^{3(n+2)}}\left(\frac{1+n}{4+3n}\right)^{n+2} = 2g^{n+1}\,\mathrm{De}^{n+2}\left(\frac{1+n}{4+3n}\right)^{n+2}. \tag{9.3.20}$$

In contrast to Couette flow, shear thinning yields a velocity profile and flow rate different from the corresponding Newtonian and shear stress identical to that of Newtonian liquid under the same pressure gradient. As in Couette flow, elasticity induces nonzero normal stresses, proportional to the relaxation time, $\lambda$.

### Example 9.4: Film Flows of Non-Newtonian Liquids

The governing momentum equation is

$$\frac{d\tau_{yz}}{dy} + \rho g\cos\theta = 0. \tag{9.4.1}$$

*Newtonian liquid.* The solution for Newtonian liquid is

$$u_x(y) = -\frac{\rho g\cos\theta}{\eta}\frac{y^2}{2} + c_1 y + c_2. \tag{9.4.2}$$

The constants, $c_1$ and $c_2$, are determined from the two boundary conditions,

$u_x(0) = 0$ and $du_x/dy|_{y=H} = 0$, which yield $c_2 = 0$ and $c_1 = \rho g \cos \theta H$, and the velocity profile becomes

$$u_x(y) = -\frac{\rho g \cos \theta}{\eta}\left(\frac{y^2}{2} - Hy\right). \tag{9.4.3}$$

The flow rate, $\dot{V}$, is related to the thickness of the film, $H$, by

$$\dot{V} = \int_0^H u_x(y)\,dy = \frac{\rho g \cos \theta}{\eta}\frac{H^3}{3}, \tag{9.4.4}$$

which is rearranged in the form

$$H = \left(\frac{3Q\eta}{\rho g \cos \theta}\right)^{1/3}. \tag{9.4.5}$$

The shear stress distribution is

$$\tau_{yx} = \eta\frac{du_x}{dy} = \rho g \cos \theta(H - y), \tag{9.4.6}$$

which yields zero shear stress at the free surface (at $y = H$), and

$$\tau_{yx}|_{y=0} = H\rho g \cos \theta \tag{9.4.7}$$

at the inclined plane. The normal stress difference, $N_1 = \tau_{xx} - \tau_{yy}$, is zero.

*Viscous inelastic liquid with power-law viscosity.* The stress momentum equation becomes

$$\eta\left|\frac{du_x}{dy}\right|^n\frac{du_x}{dx} = \rho g \cos \theta(H - y). \tag{9.4.8}$$

Since $du_x/dy > 0$, and $du_x/dy = 0$ at the free surface, Eqn. (9.4.8) is integrated to

$$u_x = \left(-\frac{\rho g \cos \theta}{\eta}\right)^{1/(1+n)}\frac{[(H - y)^{[1/(1+n)]+1} - H^{[1/(1+n)]+1}]}{[1/(1 + n)] + 1}, \tag{9.4.9}$$

which includes the Newtonian profile in the limit of $n = 0$ and $k = \eta$, as it should. The flow rate, $\dot{V}$, is related to the thickness of the film, $H$, by

$$\dot{V} = \int_0^H u_x\,dy = \left(\frac{-\rho g \cos \theta}{\eta}\right)^{1/(1+n)}\left[\frac{H_y^{[1/(1+n)]+2}}{([1/(1 + n)] + 1)([1/(1 + n)] + 2)}\right.$$
$$\left.- \frac{H^{[1/(1+n)]+1}y}{[1/(1 + n)] + 1}\right]_0^H = \left(\frac{\rho g \cos \theta}{\eta}\right)^{1/(1+n)}(H)^{[1/(1+n)]+2}\left(\frac{1}{[1/(1 + n)] + 2}\right). \tag{9.4.10}$$

The thickness of the film, $H$, is related to the flow rate by

$$H = \left[\left(\frac{1}{1 + n} + 2\right)\dot{V}\left(\frac{\eta}{\rho g \cos \theta}\right)^{1/(1+n)}\right]^{1/[1/(1+n)+2]}, \tag{9.4.11}$$

which includes the Newtonian case in the limit of $n = 0$ as it should.

*Maxwell viscoelastic liquid.* The governing equation is obtained by combining the stress form of the momentum equation with the constitutive equation:

$$-\rho g \cos \theta (y - H) = \eta \frac{du_x}{dy}, \tag{9.4.12}$$

which yields a velocity profile, flow rate, and film thickness identical to those of Newtonian liquid of viscosity $\eta$. The shear stress is also identical to that of Newtonian liquid. The normal stress is obtained by means of the constitutive relation

$$\tau_{xx} - 2\lambda \tau_{xy} \frac{du_x}{dy} = 0, \tag{9.4.13}$$

which yields

$$\tau_{xx} = \frac{2\lambda}{\eta} [\rho g \cos \theta (H - y)]^2. \tag{9.4.14}$$

*Maxwell viscoelastic liquid with power-law viscosity.* The stress form of the momentum equation,

$$\frac{d\tau_{yx}}{dy} + \rho g \cos \theta = 0, \tag{9.4.15}$$

combines with the constitutive relation

$$\tau_{yx} = \eta \left| \frac{du_x}{dy} \right|^n \frac{du_x}{dy}, \tag{9.4.16}$$

to yield the governing equation

$$-\rho g \cos \theta (y - H) = \eta \left( \frac{du_y}{dy} \right)^{n+1}. \tag{9.4.17}$$

Equation (9.4.17) predicts a velocity profile and film thickness identical to those of a purely shear-thinning liquid [Eqs. (9.4.9) and (9.4.11)]. However, the normal stress difference in this case is not zero:

$$\tau_{xx} = 2\lambda \tau_{xy} \frac{du_x}{dy} = \frac{2\lambda}{1/\eta (1 + n)} [\rho g \cos \theta (H - y)]^{[1/(1+n)]+1}. \tag{9.4.18}$$

Thus, shear thinning results in a thinner film than results from its Newtonian counterpart at the same flow rate. Elasticity does not alter the velocity profile or the film thickness, but gives rise to normal viscoelastic stresses that depend on the magnitude of relaxation time.

## 9.5.2 Unidirectional Flows of Viscoplastic Liquids

Viscoplastic liquids exhibit flow characteristics not encountered in other classes of liquids: Due to the existence of the yield stress, there may be three different flow situations: If the externally applied stress (e.g., by pressure difference, or gravity, or spreading force) is less than the yield stress then there is no flow at all.

Otherwise, there may be flow everywhere, or there may be stationary islands of zero velocity adjacent to stationary walls or traveling islands of plug velocity carried by adjacent flowing liquid.

### Example 9.5: Flows of Viscoplastic Liquid

The governing momentum equation in rectilinear channel flows,

$$\frac{dp}{dx} = \frac{d\tau_{yx}}{dy}, \tag{9.5.1}$$

combines with the ideal Bingham plastic model,

$$\frac{du_x}{dy} = 0 \quad \text{if } \tau_{yx} < \tau_y, \tag{9.5.2}$$

and

$$\tau_{yx} = \tau_y + \eta \left| \frac{du_x}{dy} \right| \quad \text{if } |\tau_{yx}| > \tau_y, \tag{9.5.3}$$

to describe Couette, Poiseuille, and film flows, as follows.

*Couette flow.* The pressure gradient is zero, and therefore the shear stress is constant over the cross section of flow. Thus, in case of $\tau_{yx} = c_1 > \tau_y$, Eqn. (9.5.3) becomes

$$\tau_y + \eta \frac{du_x}{dy} = c_1 \tag{9.5.4}$$

and

$$u_x = \frac{c_1 - \tau_y}{\eta} y + c_2. \tag{9.5.5}$$

The lower plate is stationary [i.e., $u_x(y = 0) = 0$], and thus $c_2 = 0$. The upper plate is traveling with speed $u_x(y = H) = 0$, and thus $c_1 = \tau_y + (V\eta/H)$. The velocity profile then is

$$u_x = \frac{V}{H} y, \tag{9.5.6}$$

and the shear stress is

$$\tau_{yx} = \tau_y + \frac{\eta V}{H}. \tag{9.5.7}$$

Thus, once the upper plate is set in motion with finite speed $V$, the liquid must yield and flow. To achieve this, a shearing force per unit length,

$$\frac{F}{L} = \tau_y + \eta \frac{V}{H}, \tag{9.5.8}$$

is required. Thus, yielding and flow will occur only when an external drag or shear stress larger than the yield stress:

$$\frac{F}{L} > \tau_y, \tag{9.5.9}$$

is applied. Otherwise, the liquid will remain still and the upper plate stationary (if slip does not occur).

*Poiseuille channel flow.* The governing equation is

$$\frac{du_x}{dy} = 0 \quad \text{if } \tau_{yx} \le \tau_y$$

and

$$\tau_{yx} = \left[\text{sign}\left(\frac{du_x}{dy}\right)\right]\tau_y + \frac{du_x}{dy} \quad \text{if } |\tau_{yx}| > \tau_y. \tag{9.5.10}$$

The shear stress may not be constant over the cross section of flow; close to the solid wall, the shear rate and the shear stress are high and flow is likely to occur. Away from the wall and close to the symmetry line, the shear rate and the shear stress become smaller and smaller, and regions of solid-like liquid may be induced. Thus, the flow pattern of Fig. E9.5a appears appropriate. The flow in the regions $D \ge y \ge H$ has a velocity profile with negative derivative, and therefore Eqn. (9.5.10) becomes

$$\tau_{yx} = -\tau_y + \frac{du_x}{dy} = \frac{\Delta p}{\Delta L}y, \tag{9.5.11}$$

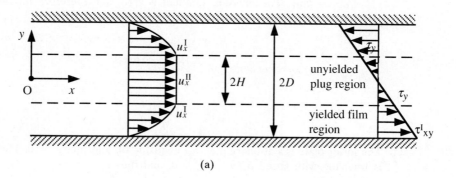

(a)

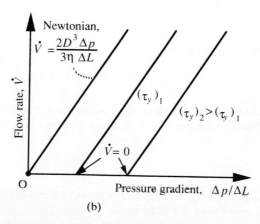

(b)

**Figure E9.5** Pressure driven channel flow of Bingham liquid: (a) velocity and stress profiles; (b) pressure vs. flow rate.

which is integrated to

$$u_x(y) = \frac{1}{2\eta}\frac{\Delta p}{\Delta L} y^2 + \frac{\tau_y}{\eta} y + c. \qquad (9.5.12)$$

At $y = D$, $u_x = 0$. Thus, $c = -(\tau_y D/\eta) - (\Delta p/\Delta L)(D^2/2\eta)$, and the velocity profile, in $D \geq y > H$, is

$$u_x'(y) = \frac{\Delta p}{\Delta L}\frac{1}{2\eta}(y^2 - D^2) - \frac{\tau_y}{\eta}(D - y) \qquad (9.5.13)$$

and in $0 \leq y \leq H$ becomes

$$u_x'' = u_x'(y = H) = \frac{1}{2\eta}\frac{\Delta p}{\Delta L}(H^2 - D^2) - \frac{\tau_y}{\eta}(D - H). \qquad (9.5.14)$$

The location of the yield line, $H$, is still unknown, but it can be determined by the common slope, zero, of the velocity at the yield line: Equation (9.5.13) gives

$$\left.\frac{du_x'}{dy}\right|_{y=H} = \frac{1}{2\eta}\frac{\Delta p}{\Delta L}2H + \frac{\tau_y}{\eta} = 0,$$

and the location of the yield line is

$$H = -\frac{\tau_y}{\Delta p/\Delta L}, \qquad (9.5.15)$$

defined by the ratio of the yield stress, which resists the flow to the pressure gradient that drives the flow. The same relation could be obtained by equating the driving pressure force to the resisting yield force on a control volume of length $\Delta L$ and height $2H$ across the plane of symmetry: $2H \Delta p = 2\tau_y \Delta L$. Thus, the ratio of pressure to yield forces determines the extent of liquid flow:

$$\frac{\tau_y}{|\Delta p/\Delta L|} \begin{cases} \geq D, & \text{no flow at all} \\ <D, & \text{yielded and unyielded regions} \\ = 0, & \text{complete flow.} \end{cases}$$

The flow rate achieved under a pressure gradient $\Delta p/\Delta L$ of a liquid of yield stress $\tau_y$ is

$$\frac{\dot{V}}{2} = \int_0^H u_x'' \, dy + \int_H^D u_x' \, dy = \frac{1}{2\eta}\frac{\Delta p}{\Delta L}(H^2 - D^2)H - \frac{\tau_y}{\eta}(D - H)H$$

$$+ \frac{1}{2\eta}\frac{\Delta p}{\Delta L}\left(\frac{D^3 - H^3}{3} - (D^3 - D^2 H)\right) - \frac{\tau_y}{\eta}\left(D^2 - DH - \frac{D^2 - H^2}{2}\right)$$

$$= \frac{1}{3\eta}\frac{\Delta p}{\Delta L}(H^3 - D^3) - \frac{\tau_y}{2\eta}(D - H)^2. \qquad (9.5.16)$$

Several observations are in order:

1.  In the limit of $\tau_y = 0$ (and therefore $H = 0$, too), the Newtonian case is recovered.
2.  Figure E9.5b shows the flow rates achieved at several levels of yield stress by several pressure gradients. Notice that for every yield stress $\tau_0$, there is a characteristic pressure gradient below which there is no flow at all.
3.  It is important to realize that the solid-like region at the center of the channel

does not stay still; it is transferred in a plug-like fashion by the two adjacent flowing layers at velocity $u_x'' = u_x'(y = H)$ and shear stress $\tau_{yx} \le \tau_y$.

4. Flows of Bingham plastic fluids in pipes can be analyzed by the same procedure; the flow is split into yielded and unyielded regions and the appropriate equations are applied.

5. Films of Bingham plastic fluids of thickness $\delta$ over inclined planes at angle $\phi$ are made up by a lower layer of thickness $\delta_L$ where the shear stress exceeds the yield stress and the velocity $u_z^L(y)$ parabolic, and an upper layer of thickness $\delta_v = \delta - \delta_L$ where the shear stress is less than the yield stress and the velocity $u_z^v = u_z^L(y = \delta_L)$ is plug. The thickness $\delta_v$ is defined by

$$\delta_v = \frac{\tau_y}{\rho g \sin \phi},\tag{9.5.17}$$

and the relation between flow rate and film thickness is

$$\frac{\dot{V}}{w} = \int_0^{\delta_L} u_z^L(y)\, dy + \int_{\delta_L}^{\delta} u_z^v\, dy = \int_0^{\delta_L} u_z^L(y)\, dy + \delta_v u_z^v.\tag{9.5.18}$$

## 9.5.3 Stretching Flows of Shear Thinning or Thickening and Viscoelastic Liquids

The film-casting and fiber-spinning stretching flows, analyzed in Section 8.3 for Newtonian liquid, are not extended to the above non-Newtonian liquids. The basic difference between Newtonian and non-Newtonian analyses is the constitutive equation that is used to express the normal stresses in the stress–momentum equation, Eqns. (8.2.10) and (8.7.5).

**Example 9.6: Film Casting of Non-Newtonian Liquid**

The governing momentum equation is (from Chapter 8)

$$\frac{\tau_{xx} - \tau_{yy}}{u_x} = c_1.\tag{9.6.1}$$

*Newtonian liquid.* The normal stresses are

$$\tau_{xx} = 2\eta \frac{du_x}{dx}, \qquad \tau_{yy} = 2\eta \frac{du_y}{dy} = -\frac{du_x}{dx}.\tag{9.6.2}$$

Equation (9.6.1) then yields

$$u_x = c_2 e^{-(c_1/4\eta)z}.\tag{9.6.3}$$

The boundary conditions $u_x(x = 0) = u_0$ and $u_x(x = L) = u_L$ determine $c_2 = u_0$ and $c_1 = (4\eta/L)\ln(u_L/u_0)$, and the solution becomes

$$u_x = u_0 \exp\left(\frac{x}{L}\ln\frac{u_L}{u_0}\right).\tag{9.6.4}$$

The thickness profile of the film is

$$H(x) = \frac{\dot{V}}{Wu_0} \exp\left(-\frac{x}{L}\ln\frac{u_L}{u_0}\right) = H_0 \exp\left(-\frac{x}{L}\ln\frac{u_L}{u_0}\right),\tag{9.6.5}$$

where $\dot{V}$ is the flow rate, $W$ the width of the film, $H_0$ the initial thickness of the film, and $L$ the length of the film casting zone. The force required to draw the film is

$$F = [WH(\tau_{xx} - \tau_{yy})]_{x=L} = \left[WH(4\eta)\frac{du_x}{dx}\right]_{x=L} = 4\eta \frac{\dot{V}}{L}\ln\frac{u_L}{u_0}.\tag{9.6.6}$$

*Viscous inelastic liquid.* An alternative extension thinning or thickening. A power-law constitutive equation is

$$\tau_{xx} = 2\eta \left|\frac{du_x}{dx}\right|^n \frac{du_x}{dx}, \qquad \tau_{yy} = 2\eta \left|\frac{du_x}{dx}\right|^n \frac{du_y}{dy} = -2\eta \left|\frac{du_x}{dx}\right|^n \frac{du_x}{dx}. \qquad (9.6.7)$$

The resulting momentum equation is

$$4\eta \left|\frac{du_x}{dx}\right|^n \frac{du_x}{dx} = cu_x. \qquad (9.6.8)$$

Since the velocity increases monotonically, Eqn. (9.6.8) becomes

$$\left(\frac{du_x}{dx}\right)^{n+1} = \frac{cu_x}{4\eta} = c_1 u_x,$$

which is integrated to

$$u_x = (c_1 x + c_2)^{(1+n)/n}. \qquad (9.6.9)$$

At $x = 0$, $u_x = u_0$, and thus $c_2 = u_0^{n/(1+n)}$.
At $x = L$, $u_x = u_L$, and thus $c_1 = (u_L^{n/(1+n)} - u_0^{n/(1+n)})/L$.

Thus the velocity profile is

$$u_x = u_0\left[\left(\left(\frac{u_L}{u_0}\right)^{n/(1+n)} - 1\right)\frac{x}{L} + 1\right]^{(n+1)/n}. \qquad (9.6.10)$$

The film thickness is

$$H(x) = H_0\left[\left(\left(\frac{u_L}{u_0}\right)^{n/(1+n)} - 1\right)\frac{x}{L} + 1\right]^{-(n+1)/n}. \qquad (9.6.11)$$

The drawing force required is

$$F = [WH(x)(\tau_{xx} - \tau_{yy})]_{x=L} = W\left[\frac{\dot{V}}{Wu_L}(4k)\left(\frac{du_x}{dx}\right)^{n+1}\right]_{x=L}$$

$$= \frac{4k\dot{V}u_0^n}{L^{n+1}}\left(\frac{n+1}{n}\right)^{n+1}\left[\left(\frac{u_L}{u_0}\right)^{n/(n+1)} - 1\right]^{n+1}. \qquad (9.6.12)$$

In the Newtonian limit of $n = 0$, Eqn. (9.6.13) reduces to Eqn. (9.6.6).

*Viscoelastic Maxwell liquid.* The shear stress $\tau_{yx}$ and transverse velocity gradients $du_x/dy$ are negligible and the Maxwell equations of Table 9.1 reduce to

$$\tau_{xx} + \lambda\left(u_x\frac{d\tau_{xx}}{dx} - 2\tau_{xx}\frac{du_x}{dx} - u_x\frac{d\tau_{zz}}{dz}\right) = 2\eta\frac{du_x}{dx} \qquad (9.6.13)$$

$$\tau_{yy} + \lambda\left(u_x\frac{d\tau_{yy}}{dx} - 2\tau_{yy}\frac{du_y}{dy} - u_x\frac{d\tau_{yy}}{dx}\right) = 2\eta\frac{du_y}{dy}. \qquad (9.6.14)$$

It is also

$$\frac{du_y}{dy} = -\frac{du_x}{dx} \qquad (9.6.15)$$

by the continuity equation, given that $\partial u_z/\partial z = 0$. To make the equations tractable, we also make the assumption that

$$\tau_{xx} + \tau_{yy} \simeq 0, \qquad (9.6.16)$$

which is exact for inelastic liquids, reasonably accurate for viscoelastic liquids of low

Deborah number, and poor for high Deborah numbers, De $\gg$ 1. An equation for the normal stress difference results:

$$(\tau_{xx} - \tau_{yy}) + \lambda u_x \frac{d}{dx}(\tau_{xx} - \tau_{yy}) = 4\eta \frac{du_x}{dx}. \tag{9.6.17}$$

This equation combines with the governing stress-momentum equation, Eqn. (9.6.1), to produce the governing equation,

$$cu_x + \lambda u_x \frac{d}{dx}(cu_x) = 4\eta \frac{du_x}{dx}, \tag{9.6.18}$$

which is integrated by

$$\int_{u_0}^{u_x} \frac{\lambda u_x - 4\eta/c}{u_x} \, du_x = -\int_0^x dx, \tag{9.6.19}$$

to yield

$$\lambda(u_x - u_0) - c \ln \frac{u_x}{u_0} = -x. \tag{9.6.20}$$

At $x = L$, $u_x = u_L$, and thus $c = \lambda(u_L - u_0)/\ln(u_L/u_0)$, and Eqn. (9.6.20) becomes

$$\lambda(u_x - u_0) - \frac{\lambda(u_L - u_0) + L}{\ln(u_L/u_0)} \ln \frac{u_x}{u_0} = -x, \tag{9.6.21}$$

which can be solved for the velocity profile $u_x(x)$.

In the limit of zero relaxation time, $\lambda = 0$, Eqn. (9.6.21) gives the Newtonian velocity profile. In the limit of high elasticity, Eqn. (9.6.21) yields

$$u_x = u_0 + \lambda x, \tag{9.6.22}$$

which is a linear velocity profile.

The thickness of the film in the latter case is

$$H(x) = \frac{\dot{V}}{Wu_x} = \frac{\dot{V}}{W}\left(\frac{1}{u_0 + \lambda x}\right) = H_0\left(\frac{1}{1 + \lambda x/u0}\right), \tag{9.6.23}$$

and the tension required to draw the film is

$$F = W[H(\tau_{xx} - \tau_{yy})]_{x=L} = W\left[H_0\left(\frac{1}{1\lambda L/u_0}\right)(4\eta\lambda)\right]$$

$$= \frac{4\eta\dot{V}\lambda}{u_0 + \lambda L} = 4\eta\frac{\dot{V}}{L}\left(\frac{\lambda L}{u_0 + \lambda L}\right) = 4\eta\frac{\dot{V}\lambda L}{L\, u_L}. \tag{9.6.24}$$

In conclusion, Problem 9.5 demonstrates that extension-thinning behavior results in a thinner film and less required tension. Elasticity also results in a thinner film, but higher required tension. *Fiber spinning* is analyzed by an identical procedure.

The factor 4, appearing in the film-casting equations, becomes 3 in the fiber-spinning equations, due to the fact that the stress difference, $\tau_{zz} - \tau_{rr}$, is different:

$$\tau_{zz} - \tau_{rr} = 2\eta \frac{du_z}{dz} - 2\eta \frac{du_r}{dr} = 2\eta \frac{du_z}{dz} - 2\eta \left(-\frac{1}{2} \frac{du_z}{dz}\right) - 3\eta \frac{du_z}{dz}. \qquad (9.6.25)$$

Another similar flow is *film blowing,* which is used to produce plastic bags of cylindrical shape for food-wrapping purposes.

### 9.5.4 Lubrication Flows of Non-Newtonian Liquids

The lubrication equations derived and applied to Newtonian liquid in Section 8.1 are now extended to liquids with power law viscosity and with relaxation time. Resulting differences are due to a different constitutive equation that replaces the shear stress in the stress–momentum equation. Converging and diverging flows of shear thinning and thickening liquids are analyzed in Example 9.7, and a lubrication model flow for viscoelastic liquid is analyzed in Example 9.8.

**Example 9.7: Non-Newtonian Flow in Converging Pipe**

Study the behavior of shear-thinning and shear-thickening liquids in a converging pipe of length $L$ of inlet radius $R_0$, outlet radius $R_L$, and wall inclination angle $\theta \simeq \tan \theta$. Calculate and plot the resulting pressure distribution and compare it to those of rectilinear and diverging geometries.

The governing momentum equation is

$$-\frac{dp}{dz} + \frac{1}{r} \frac{\partial}{\partial r} (r\tau_{rz}) = 0, \qquad (9.7.1)$$

which is integrated to

$$\tau_{rz} = \frac{r}{2} \frac{dp}{dz} \qquad (9.7.2)$$

for any incompressible fluid. For a power-law liquid, this equation becomes

$$\eta \left|\frac{\partial u_z}{\partial r}\right|^n \frac{\partial u_z}{\partial r} = \frac{r}{2} \frac{dp}{dz}. \qquad (9.7.3)$$

The shear rate is negative everywhere; therefore, Eqn. (9.7.3) is conveniently rearranged in the form

$$-\eta \left(-\frac{\partial u_z}{\partial r}\right)^n \left(-\frac{\partial u_z}{\partial r}\right) = -k \left(-\frac{\partial u_z}{\partial r}\right)^{n+1} = \frac{r}{2} \frac{dp}{dz}, \qquad (9.7.4)$$

which is integrated to

$$u_z = -\int_0^{R(z)} \left(-\frac{r}{2\eta} \frac{dp}{dz}\right)^{1/(n+1)} dz + c,$$

where $R(z) = R_0 - (R_0 - R_L)z/L$ is the wall-line equation. Thus

$$u_z = \left(-\frac{1}{2\eta} \frac{dp}{dz}\right)^{1/(n+1)} \frac{(R(z)^{1+1/(n+1)} - r^{[1/(n+1)]+1}}{[1/(n+1)] + 1}. \qquad (9.7.5)$$

The unknown pressure gradient is estimated by using the known flow rate as follows:

$$\dot{V} = 2\pi \int_0^{R(z)} r u_z \, dr = 2\pi \frac{[-(1/2\eta)(dp/dz)]^{1/(n+1)}}{[1/(n+1)]+1} \int_0^{R(z)} r(R(z)^{1+1/(n+1)} - r^{1+1/(n+1)}) \, dr$$

$$= \frac{\pi[-(1/2\eta)(dp/dz)]^{1/(n+1)} R(z)^{3+1/(n+1)}}{3 + 1/(1+n)}. \tag{9.7.6}$$

Thus,

$$\frac{dp}{dz} = -2\eta \left[ \frac{\dot{V}(3 + 1/(1+n))}{\pi} \right]^{n+1} \frac{1}{R(z)^{3n+4}}, \tag{9.7.7}$$

$$p(z) = p_0 - 2\eta \left[ \frac{\dot{V}(3 + 1/(1+n))}{\pi} \right]^{n+1} \frac{1}{3n+3} \left[ \frac{1}{R(z)^{3n+3}} - \frac{1}{R_0^{3n+3}} \right] \frac{L}{R_0 - R_l} \tag{9.7.8}$$

and

$$u_z = \frac{\dot{V}(3 + 1/(1+n))}{\pi R^2(z)} \left[ 1 - \left( \frac{r}{R(z)} \right)^{1+1/(n+1)} \right]. \tag{9.7.9}$$

The preceding analysis assumes that the flow rate $\dot{V}$ is known and the exit pressure is unknown. In lubrication situations, the inlet and outlet pressures are usually known and the flow rate is unknown. In the latter case, the analysis proceeds as follows. Equation (9.7.8) is evaluated at $z = L$ to yield

$$p_L - p_0 = -2\eta \left[ \dot{V} \frac{(3 + 1/(1+n))}{\pi} \right]^{n+1} \left[ \frac{1}{3n+3} \right] \left[ \frac{1}{R_L^{3n+3}} - \frac{1}{R_0^{3n+3}} \right] \frac{L}{R_0 - R_L}. \tag{9.7.10}$$

The flow rate is calculated as

$$\dot{V} = \frac{\pi(p_L - p_0)^{1/(1+n)}[(3n+3)]^{1/(1+n)}}{(-2\eta)^{1/(1+n)}[3 + 1/(1+n)]} \left[ \frac{1}{1/R_L^{3n+3} - 1/R_0^{3n+3}} \right]^{1/(1+n)} \left[ \frac{R_0 - R_L}{L} \right]^{1/(1+n)} \tag{9.7.11}$$

and the velocity profile becomes

$$u_z = \left[ \frac{(p_0 - p_L)(R_0 - R_L)(3n+3)}{L2\eta R(z)^{2n+2}} \right]^{1/(1+n)} \left[ \frac{1}{1/R_L^{3n+3} - 1/R_0^{3n+3}} \right]^{1/(1+n)}$$
$$\times \left[ \frac{1 - (r/R(z))^{1+1/(1+n)}}{1 + 1/(1+n)} \right]. \tag{9.7.12}$$

The pressure profile, under fixed inlet and outlet pressures $p_0$ and $p_L$, becomes

$$p(z) = p_0 + \frac{p_L - p_0}{1/R_L^{3n+3} - 1/R_0^{3n+3}} \left[ \frac{1}{R(z)^{3n+3}} - \frac{1}{R_0^{3n+3}} \right]. \tag{9.7.13}$$

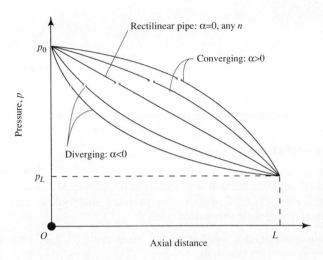

**Figure E9.7.1** Pressure profiles of shear-thinning liquid in converging and diverging pipe, under fixed inlet and outlet pressures and variable flow rate.

Equation (9.7.13) predicts a monotonically decreasing nonlinear pressure profile, as shown in Figs. E9.7.1 and E9.7.2.

The area under the curves of Fig. E9.7.2 provides an indication of the ability of viscous inelastic liquids to support a load under lubrication conditions. It is shown that shear-thinning behavior reduces load capacity, which is enhanced by shear thickening.

The friction encountered by the wall of the pipe under fixed inlet and outlet pressures is obtained by differentiating Eqn. (9.7.12):

$$
\begin{aligned}
\tau_{rz}^{W} &= \eta \left( -\frac{\partial u_z}{\partial r} \right)^{n+1}_{r=R(z)} \\
&= \frac{(R_0 - R_L)(p_0 - p_L)(3n + 3)}{2R(z)^{3n+3}L} \frac{1}{1/R_L^{3n+3} - 1/R_0^{3n+3}}.
\end{aligned}
\tag{9.7.14}
$$

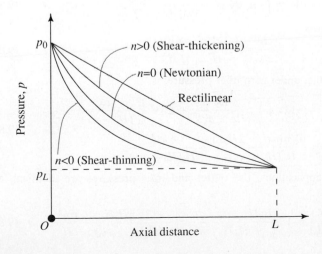

**Figure E9.7.2** Pressure profiles of power-law liquids in converging pipe under fixed inlet and outlet pressures and variable flow rate.

The total friction is

$$F = 2\pi \int_0^L R(z)\tau_{rz}^W dz$$

$$= \frac{\pi(R_0 - R_L)(p_0 - p_L)(3n + 3)}{L(3n + 1)} \cdot \frac{1}{1/R_L^{3n+3} - 1/R_0^{3n+3}} \left( \frac{1}{R_L^{3n+1}} - \frac{1}{R_0^{3n+1}} \right), \tag{9.7.15}$$

which decreases with diminishing negative $n$, and equivalently, with increasing shear thinning.

### Example 9.8: Lubrication Flow of Maxwell Liquid

Study the flow of a Maxwell liquid in the lubrication geometry of Fig. G.3b, for a rectilinear upper wall of angle inclination $a \simeq \tan a$, inlet opening $H_0$, and outlet opening $H_L$ over axial distance $L$. Calculate the pressure distribution and the resulting load capacity and friction at the wetted wall.

**Solution** Approximate analyses that appear in the literature[3,25-27] start with the lubrication form of the momentum equation,

$$-\frac{\partial p}{\partial x} + \frac{\partial \tau_{xy}}{\partial y} = 0, \qquad \frac{\partial p}{\partial y} = 0, \tag{9.8.1}$$

which in most cases are validated a posteriori to hold under typical lubrication conditions. The constitutive equation for the shear stress may be approximated by

$$\tau_{xy} \approx \eta \frac{\partial u_x}{\partial y} \tag{9.8.2}$$

for the shear stress, and

$$\tau_{yy} + \lambda u_x \frac{\partial \tau_{yy}}{\partial x} = -2\eta \frac{\partial u_x}{\partial x} \tag{9.8.3}$$

for the stress normal to the wall. The solution to Eqns. (9.8.1) and (9.8.2) is obtained by the approach outlined in Chapter 8. The velocity profile is

$$u_x = \frac{1}{2\eta} \frac{dp}{dx} (y^2 - Hy) + V \left( 1 - \frac{y}{H} \right), \tag{9.8.4}$$

and the flow rate is

$$\dot{V} = \int_0^H u_x \, dy = -\frac{H^3}{12\eta} \frac{dp}{dx} + \frac{VH}{2}. \tag{9.8.5}$$

The pressure gradient is

$$\frac{dp}{dx} = -\frac{12\dot{V}\eta}{H^3} + \frac{6\eta V}{H^2}, \tag{9.8.6}$$

and the pressure profile, under zero inlet pressure, is

$$p(x) = -12\dot{V}\eta \int_0^x \frac{dx}{(H_0 - az)^3} + 6\eta V \int_0^x \frac{dx}{(H_0 - ax)^2}$$

$$= \frac{6\dot{V}\eta(ax^2 - 2H_0x)}{H_0^2(H_0 - ax)^2} + \frac{6\eta Vx}{H_0(H_0 - ax)}. \tag{9.8.7}$$

In most lubrication applications, the outlet and inlet pressures are identical, and therefore

$$\dot{V} = VH_0 \frac{(aL - H_0)}{(aL - 2H_0)} = V \frac{H_0 H_L}{H_L + H_0}, \qquad \frac{dp}{dx} = -\frac{12V\eta H_0 H_L}{H^3(H_L + H_0)} + \frac{6\eta V}{H^2}, \tag{9.8.8}$$

and

$$p(x) = \frac{6\eta Vx}{H_0(H_0 - ax)}\left[\frac{H_L(ax - 2H_0)}{(H_0 + H_L)(H_0 - ax)} + 1\right]. \tag{9.8.9}$$

The velocity profile becomes

$$\frac{u_x}{V} = 3\left[1 - \frac{2H_0 H_L}{H(H_0 + H_L)}\right]\left(\frac{y^2}{H^2} - \frac{y}{H}\right) + \left(1 - \frac{y}{H}\right). \tag{9.8.10}$$

Equations (9.8.4) to (9.8.10) are common for Newtonian and Maxwell liquids. However, the normal stress $\tau_{yy}$ and the load supported are expected to be different. For Newtonain liquids,

$$\tau_{yy} = 2\eta\frac{\partial u_y}{\partial y} \simeq -2\eta\frac{\partial u_y}{\partial y} \simeq -\frac{2\eta V}{H}\frac{dH}{dx} \simeq 0 \tag{9.8.11}$$

is small since the inclination is small. For Maxwell liquids, the constitutive equation, Eqn. (9.8.3) yields

$$\tau_{yy} = 2\eta\frac{\partial u_y}{\partial y} - \lambda u_x\frac{\partial \tau_{yy}}{\partial x} \simeq \lambda u_x\frac{\partial \tau_{yy}}{\partial x} \tag{9.8.12}$$

with solution,

$$\tau_{yy} = C\exp\left[-\frac{1}{2}\int\frac{dx}{u_x}\right], \tag{9.8.13}$$

which enhances the load capacity $W$, when superimposed on pressure:

$$W = \int_0^L |p + \tau_{yy}|\,dx. \tag{9.8.14}$$

Equation (9.8.13) indicates that the load capacity increases with the elasticity, represented by the relaxation time, $\lambda$. It must be kept in mind that due to the approximate nature of viscoelastic lubrication models, any assumption made a priori must be validated a posterion, and if possible, by numerical modeling and experiments, too. More consistent and therefore complicated approximations not amenable to analytic solutions can be found elsewhere.[3,28]

## PROBLEMS

1. *Cone-and-plate rheometer.* A polymeric melt of unknown viscosity and elasticity is tested by means of a cone-and-plate viscometer of radius $R = 2.5$ cm and cone angle $\phi = 0.1$ rad. The torque $T$ and the total normal force $F$ required are recorded at several anglular velocities $\Omega$. Complete Table P9.1 and plot the viscosity $\eta$, the first normal stress coefficient $N_1$, and the *recoverable shear* $S_R = \tau_{12}/N_1$ as functions of the shear rate $\dot{\gamma}$. Also, characterize the melt.

2. *Parallel-plate rheometer.* The polymeric melt of Problem 9.1 is placed in a parallel-plate rheometer of radius $R = 2.5$ cm and gap $H = 1$ mm. What would be the appropriate recorded torque $T$ and normal force $F$ at the tabulated values of angular speed $\Omega$ in Table P9.1?

3. *Capillary rheometer.* A viscous inelastic liquid is tested in a capillary rheometer of radius $R = 5$ mm and length $L = 0.5$ m. The flow rates $Q$ achieved under selected pressure gradients, $\Delta p/\Delta L$, are recorded. Complete Table P9.3, plot the viscosity vs. shear rate, and characterize the liquid.

**TABLE P9.1** RHEOLOGICAL DATA OBTAINED WITH A CONE-AND-PLATE VISCOMETER

| Run | $\Omega$ (rad/s) | $\dot{\gamma}$ (s$^{-1}$) | $T$ (Pa m$^3$) | $\eta$ (Pa s) | $F$ (Pa m$^2$) | $N_1$ (Pa s$^2$) | $S_R$ |
|-----|------------------|---------------------------|----------------|---------------|----------------|------------------|-------|
| 1 | 0.0001 | | 0.003 | | 0.01 | | |
| 2 | 0.001 | | 0.033 | | 0.7 | | |
| 3 | 0.01 | | 0.26 | | 10 | | |
| 4 | 0.1 | | 1 | | 100 | | |
| 5 | 1 | | 2.2 | | 300 | | |
| 6 | 10 | | 3.3 | | 400 | | |
| 7 | 100 | | 66 | | 700 | | |
| 8 | 1000 | | 660 | | 1000 | | |

4. *Viscosity laws.* Fit the data of Tables P9.1 and P9.3 by the appropriate viscosity law. What is the physical significance of the parameters in the viscosity law?

5. *Flow of viscoplastic liquids.* Sketch the qualitative behavior (i.e., no flow at all, yielded and unyielded regions, or complete flow) of Bingham plastic liquids of yield stress $\tau_y$, in pipes of radius $R$, under pressure gradient $\Delta p/\Delta L$. Construct the following plots: $\tau_y$ vs. $\Delta p/\Delta L$ under constant $R$; $R$ vs. $\Delta p/\Delta L$ under constant $\tau_y$; and $\tau_y$ vs. $R$ under constnat $\Delta p/\Delta L$. Discuss the physical significance of your findings.

6. *Different flow behavior based on rheology.* What quantitative and/or qualitative differences are expected among Newtonian, viscous inelastic, viscoplastic, viscoelastic, and viscoplastic liquids in the following operations and processes? Explain why.
   (a) Spinnability (i.e., ability to form thin fibers under tension);
   (b) friction and wear;
   (c) load capacity;
   (d) coating (i.e., ability to form thin films);
   (e) swelling (i.e., ability to expand on exiting confined flows);
   (f) memory (i.e., ability to reverse deformation and flow);
   (g) wettability (i.e., ability to spread under gravity).

7. *Poiseuille channel flow of a power-law liquid.* Analyze Poiseuille channel flow under pressure gradient $\Delta p/\Delta L$, in a channel of width $2H$, of a power-law liquid of given material parameters $k$ and $n$: Find the velocity and the stress distribution, the flow rate, and the total friction force. Plot and compare the cases: $n < 0$ (shear thinning), $n = 0$ and $k = \eta$ (Newtonian), and $n > 0$ (shear thickening).

**TABLE P9.3** RHEOLOGICAL DATA OBTAINED WITH A CAPILLARY RHEOMETER

| Run | $\Delta p$ (dyn/cm$^2$) | $Q$ (cm$^3$/s) | $\dot{\gamma}$ (s$^{-1}$) | $\tau_{12}$ | $\eta$ | $N_1$ |
|-----|-------------------------|----------------|---------------------------|-------------|--------|-------|
| 1 | 5 | 0.0013 | | | | |
| 2 | 15 | 0.015 | | | | |
| 3 | 50 | 0.14 | | | | |
| 4 | 150 | 1.45 | | | | |
| 5 | 500 | 14.5 | | | | |
| 6 | 1600 | 150 | | | | |
| 7 | 5000 | 1400 | | | | |

**8.** *Pipe flow of viscoplastic liquid.* Analyze the flow of a Bingham plastic liquid of yield stress $\tau_y$ and viscosity $\eta$, in a pipe of diameter $2R$ and length $L$, under constant pressure gradient $dp/dx = \Delta p/\Delta L$. Find the velocity and shear stress distributions and the flow rate. Plot the achieved flow rate vs. pressure drop and find when the flow ceases.

**9.** *Film flow of a viscoplastic liquid.* Analyze the film flow of a Bingham plastic liquid of yield stress $\tau_y$ and viscosity $\eta$, down an inclined plane at angle $\phi$ with the horizontal, at flow rate $Q$ per unit width. Find the velocity and stress distributions across the film. Then find the relation between the flow rate $Q$ and the thickness $H$, and compare it with that of Newtonian liquid of the same density $\rho$ and viscosity $\eta$. Plot the velocity and the stress and explain their distribution.

**10.** *Viscous liquid in gear pump.* Repeat the analysis of Example 6.5 for viscous liquids following the power law of viscosity, $\eta = k\dot{\gamma}^n$ for $n = -0.5$, $n = 0$, and $n = 2$. Compare the results to those of Example 6.5 and justify the differences.

**11.** *Viscous liquid in screw extruder.* Repeat the analysis of Example 6.6 for viscous liquids of power law viscosity, $\eta = k\dot{\gamma}^n$, for $n = -0.5$, $n = 0$, and $n = 2$. Compare your results to those of Example 6.6 and justify the differences.

**12.** *Ketchup flow.* Ketchup is a Bingham liquid of density $\rho = 1.1\,\text{g/cm}^3$, viscosity $\eta = 500\,\text{cP}$, and yield stress $\tau_y = 500\,\text{dyn/cm}^3$. For an upside-down bottle of radius $R = 4\,\text{cm}$ and neck of length $L = 5\,\text{cm}$ and diameter $D = 3\,\text{cm}$, find the amount of ketchup necessary to have a supply of ketchup. What happens for $m > m^*$ and $m < m^*$? For $m < m^*$ find the additional external stress required for flow initiation. How is this stress supplied?

**13.** *Viscoelastic agents as friction reducers.* It has been suggested that the addition of small amounts of polymers in Newtonian liquids reduces friction on solid surfaces in contact with the flowing liquid. Justify or deny this suggestion. Similarly, it has been suggested that the addition of small amounts of polymers in lubricants results in friction reduction and load capacity increase. Justify this observation.

**14.** *Can drainage.* Find which liquid drains faster under gravity from a cylindrical can of volume $V$, radius $R$, and height $H$ through its neck of diameter $d$ and length $L$: Newtonian, shear-thinning, shear-thickening, or Maxwell viscoelastic.

**15.** *Radial flow of power law liquid.* Analyze plane sink flow of a power-law liquid of consistency $k$ and index $n$, at flow rate $Q$. Calculate velocity, stress, and pressure distributions for shear-thinning $(n < 0)$, Newtonian $(n = 0,\ k = \eta)$, and shear-thickening $(n > 0)$ liquids. Comment on the differences and similarities.

**16.** *Radial flow of Maxwell liquid.* Analyze plane sink flow of a Maxwell liquid of density $\rho$, viscosity $\eta$, and elasticity $\lambda$, under flow rate $Q$. Calculate the velocity, stress, and pressure distributions along a radial streamline. Comment on the differences and similarities.

**17.** *Thin-film lubrication flows of non-Newtonian liquids.* Study the prototype thin-film coating flow of power law, along the lines of Example 9.7, of confined lubrication flows. Solve the resulting Reynolds equations for selected limiting cases of the involved parameters.

**18.** *Thin film of Maxwell liquid.* Repeat Problem 9.17 for a Maxwell liquid along the lines of Example 9.8 on confined lubrication flow.

**19.** *Bubble growth in viscoelastic liquid.* A bubble of initial radius $R_0$ and pressure $p_0$ starts growing due to a pressure increase $p(t) = p_0 e^{kt}$ in a Maxwell liquid of viscosity $\eta_0$ and relaxation time $\lambda$.

**(a)** Find the velocity profile and the extension rate within the liquid.

**(b)** Find the viscoelastic stress distribution in the liquid.

**(c)** What are the values of parts (a) and (b) at the bubble–liquid interface?

**(d)** How would you utilize such a device to measure elongational viscosity?

20. *Pipe flow of power-law liquids.* A viscous liquid, which has a viscosity $\eta = |du/dy|^n$ cP, flows in a pipe of radius $R = 5$ cm under pressure gradient $\Delta p/\Delta L = -0.1$ atm/m. Calculate the flow rates $Q$ achieved for two shear-thinning liquids of $n = -0.8$ and $n = -0.3$, for a Newtonian liquid of $n = 0$ and for two shear-thickening suspensions of $n = 1$ and $n = 2$. Plot $Q$ vs. $n$ and comment on the ability of viscous liquids to deform and flow under pressure gradients.

21. *Film flow of power-law liquids.* A power-law viscous liquid of viscosity $\eta = |du/dy|^n$ cP flows at flow rate per unit width $Q/W = 1$ cm$^2$/s down an inclined plane at angle $\phi = 30°$ with the horizontal. Calculate the resulting film thickness for two shear-thinning liquids of $n = -0.8$ and $n = -0.3$, for a Newtonian liquid of $n = 0$ and for shear-thickening liquids of $n = 1$ and $n = 2$. Plot $H$ vs. $n$ and comment on the ability of viscous inelastic liquids to deform and spread under gravity.

22. *Film flow of viscoplastic liquids.* A Bingham liquid of density $\rho = 1$ g/cm$^3$, viscosity $\eta = 1$ P, and yield stress $\tau_y = 10$ dyn/cm flows down an inclined plane at an angle $\phi = 45°$ with the horizontal, at a flow rate per unit width $Q/W = 5$ cm$^2$/s. Calculate the resutling film thickness. Then plot the thickness vs. inclination angle to determine the angle $\phi$ at which the flow ceases.

23. *Elongational viscosity by fiber spinning.* In a fiber-spinning device, the recorded tension to yield the fiber pictured in Fig. P9.23 is $F = 1000$ dyn, for flow rate $Q = 1$ mm$^3$/s. Under these spinning conditions, the velocity and stress are uniform over the cross section of the fiber shown, and inertia, gravity, and surface tension effects are negligible.

**(a)** Fit the shape of the thickness profile, $r = r(z)$.

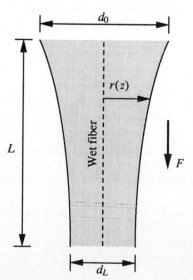

**Figure P9.23** Rheological data from fiber spinning.

(b) Find the velocity profile, $u(z)$, and the extension rate, $du(z)/dz$, along the fiber. What draw ratio is achieved?

(c) Calculate and plot the elongational viscosity vs. extension rate (along the filament).

(d) Characterize the liquid tested.

24. Answer *true* or *false* and justify your answer:

(a) Shear-thinning liquids produce thicker thin films than do their Newtonian counterparts.

(b) Rheometers utilize constitutive equations to estimate the viscosity.

(c) Shear-thickening liquids result in higher shear rates near solid walls than do their Newtonian counterparts.

(d) Viscoelasticity reduces friction and wear.

(e) Shear-thinning liquids develop normal stresses.

(f) A high degree of inclination enhances lubrication efficiency.

(g) In torsional flows, shear thinning creates rod climbing.

(h) Polymeric additions to lubricants reduce friction alone.

(i) Bingham liquids flow when the stress everywhere exceeds their yield stress.

(j) There may be stationary regions in a moving Bingham liquid.

(k) A liquid may exhibit both shear thinning and viscoelasticity.

(l) A liquid may exhibit both yield stress thinning and shear thinning.

25. *Rheological characterization.* The following data have been obtained by means of a cone-and-plate viscometer for a silicon-based coating solution.

| Stress (Pa) | Normal force (dyn) | $\eta$ (Pa s) | Torque (dyn cm) | Rate (s$^{-1}$) |
|---|---|---|---|---|
| 3.862e+01 | 9.761e+00 | 3.862e−01 | 9.665e+00 | 1.000e+02 |
| 9.290e+01 | 1.194e+01 | 3.918e−01 | 2.325e+01 | 2.371e+02 |
| 2.204e+02 | 1.599e+01 | 3.920e−01 | 5.516e+01 | 5.623e+02 |
| 2.928e+02 | 1.735e+01 | 3.920e−01 | 7.329e−01 | 7.498e+02 |
| 3.874e+02 | 2.025e+02 | 3.875e−01 | 9.697e+01 | 9.998e+02 |
| 8.721e+02 | 2.834e+01 | 3.678e−01 | 2.183e+02 | 2.371e+03 |
| 1.738e+03 | 4.779e+01 | 3.091e−01 | 4.350e+02 | 5.622e+03 |
| 2.045e+0.3 | 6.378e+01 | 2.727e−01 | 5.117e+02 | 7.497e+03 |

(a) Plot the viscosity vs. shear rate on appropriately scaled paper and determine the material parameters of the viscosity law.

(b) Repeat for the normal stress difference. Is the liquid viscoelastic?

26. *Parallel-plate viscometer.* This device consists of two parallel circular plates of radius $R$ placed horizontally a distance $H$ apart, like a cone-and-plate rheometer with the upper cone replaced by a plate. The liquid fills the cylindrical gap and the upper plate starts rotating at angular speed $\Omega$. A velocity profile,

$$u_\theta = \frac{\Omega r Z}{H},$$

with respect to the center of the bottom plate, at $r = 0$ and $z = 0$, results. With $R = 25$ mm, $H = 0.5$ mm, the following data were recorded.

| Stress (Pa) | Normal force (dyn) | $\eta$ (Pa s) | Torque (dyn cm) | Rate ($s^{-1}$) |
|---|---|---|---|---|
| 4.357e+00 | 2.315e+00 | 4.357e−01 | 1.090e+00 | 1.000e+01 |
| 9.504e+00 | 4.995e+00 | 4.009e−01 | 2.379e+00 | 2.371e+01 |
| 2.341e+01 | 4.911e+00 | 4.163e−01 | 5.859e+00 | 5.623e+01 |
| 3.132e+01 | 5.328e+00 | 4.177e−01 | 7.838e−00 | 7.498e+01 |
| 4.174e+01 | 5.527e+00 | 4.174e−01 | 1.045e+01 | 9.998e+01 |
| 9.890e+01 | 7.306e+00 | 4.171e−01 | 2.475e+01 | 2.371e+02 |
| 2.352e+02 | 1.069e+01 | 4.183e−01 | 5.885e+01 | 5.622e+02 |
| 3.112e+02 | 1.242e+01 | 4.151e−01 | 7.788e+02 | 7.497e+02 |
| 4.112e+02 | 1.478e+01 | 4.113e−01 | 1.029e+02 | 9.997e+02 |
| 9.297e+02 | 3.030e+01 | 3.921e−01 | 2.327e+02 | 2.371e+03 |
| 1.203e+03 | 4.294e+01 | 3.806e−01 | 3.011e+02 | 3.161e+03 |
| 1.412e+03 | 6.374e+01 | 3.350e−01 | 3.535e+02 | 4.216e+03 |

**(a)** What was the corresponding angular speed $\Omega$?

**(b)** Is the liquid Newtonian, shear thinning, or viscoelastic?

**(c)** Plot the viscosity and the normal stress difference vs. shear rate. Fit the resulting curves by appropriate material functions (e.g., power-law viscosity).

**27.** *Viscoelastic characterization.* The following data have been obtained by means of a cone-and-plate viscometer under sinusoidal oscillation of the upper cone.

| $G'$ (Pa) | $G''$ (Pa) | $\eta^*$ (Pa s) | Torque (g cm) | Frequency (rad/s) |
|---|---|---|---|---|
| 3.044e−01 | 4.649e+00 | 4.659e−01 | 3.519e−01 | 1.000e+01 |
| 7.497e−02 | 1.097e+01 | 4.629e−01 | 8.313e−01 | 2.371e+01 |
| 2.989e−01 | 2.598e+01 | 4.621e−01 | 1.957e+00 | 5.623e+01 |
| 5.800e−01 | 3.465e+01 | 4.622e−01 | 2.601e−00 | 7.498e+01 |
| 1.014e+00 | 4.626e+01 | 4.627e−01 | 3.464e+00 | 9.998e+01 |
| 3.289e+00 | 8.268e+01 | 4.654e−01 | 6.207e+00 | 1.778e+02 |
| 1.090e+01 | 1.486e+02 | 4.712e−01 | 1.149e+01 | 3.162e+02 |
| 1.933e+01 | 2.010e+02 | 4.790e−01 | 1.593e+01 | 4.216e+02 |

**(a)** Given the fact that

$$G'\left(\frac{1}{\omega}\right) = G(t) = \alpha \exp\left(-\frac{t}{\lambda}\right),$$

where $\lambda$ is relaxation time and $\alpha$ its coefficient, plot $G(t)$ vs. $t$ on appropriately scaled paper and determine the set $\{\alpha, \lambda\}$ of the measured liquid.

**(b)** By appropriately plotting $\eta^*$, determine the viscosity given that

$$\eta(\dot{\gamma}) = \eta^*(F).$$

**28.** *Deborah and Weissenberg number.* Based on the definitions given by Eqns. (9.8) and (9.9), show that for pressure channel flow of Maxwell viscoelastic liquid, the two numbers are related by Ws = 3De and for pressure pipe flow by Ws = 4De.

**29.** *Quick rheological characterization.* List ways to find out if an unknown liquid is Newtonian, shear thinning, shear thickening, viscoplastic, viscoelastic, or elastomeric without the use of a rheological instrument. Assume that the liquid is viscoelastic. How could you quantify its rheological properties with rheological instruments?

# REFERENCES

1. P. S. Virk, Drag reduction fundamentals, *AIChE J.*, 21, 625 (1975).
2. R. B. Bird, R. C. Armstrong, and O. Hassager, *Dynamics of Polymeric Liquids,* Vol. 1, *Fluid Mechanics,* Wiley, New York, 1977.
3. R. I. Tanner, *Engineering Rheology,* London Press, Oxford, 1985.
4. Z. Tadmor and G. Gogos, *Principles of Polymer Processing,* Wiley, New York, 1979.
5. R. B. Bird and C. F. Curtiss, Fascinating polymeric liquids, *Phys. Today,* 37, 36 (1984).
6. H. Merkovitz, The emergence of rheology, *Phys. Today,* 21, 23 (1968).
7. H. A. Barnes, J. F. Hutton, and K. Walters, *An Introduction to Rheology,* Elsevier, New York, 1989.
8. Z. Chen and T. C. Papanastasiou, Elongational viscosity by fiber spinning, *Rheol. Acta,* 29, 385 (1990).
9. J. Meissner, Development of a universal extensional rheometer for the uniaxial extension of polymer melts, *Trans. Soc. Rheol.,* 16, 405 (1972).
10. F. T. Trouton, On the coefficient of viscous traction and its relation to that of viscosity, *Proc. Roy. Soc.,* A77, 426 (1906).
11. T. C. Papanastasiou, Flows of materials with yield, *J. Rheol.,* 31, 385 (1987).
12. J. Billingham and J. W. J. Ferguson, Laminar unidirectional flow of a thixotropic fluid in a circular pipe, *J. Non-Newtonian Fluid Mech.,* 47, 21 (1993).
13. J. L. White and A. B. Metzner, Development of constitutive equations for polymeric melts and solutions, *J. Appl. Polym. Sci.,* 7, 1867 (1963).
14. M. Doi and S. Edwards, Dynamics of rod-like macromolecules in concentrated solutions, *J. Chem. Soc. Faraday Trans.,* 74, 918 (1979).
15. R. B. Bird and C. F. Curtiss, A kinetic theory for polymer melts: The stress tensor and the rheological equation of state, *J. Chem. Phys.,* 74, 2026 (1981).
16. B. Bernstein, E. A. Kearsley, and L. J. Zapas, A study of stress relaxation with finite strain, *Trans. Soc. Rheol.,* 7, 397 (1963).
17. M. H. Wagner, Zur Netzwerktheorie von Polymer-Schmelzen, *Rheol. Acta,* 18, 33 (1979).
18. A. C. Papanastasiou, L. E. Scriven, and C. W. Macosko, An integral constitutive equation for mixed flows: Viscoelastic characterization, *J. Rheol.,* 27 387 (1983).
19. J. G. Oldroyd, On the formulation of rheological equations of state, *Proc. Roy. Soc.,* A200, 523 (1950).
20. N. Phan-Thien and R. I. Tanner, A new constitutive equation for network theory, *J. non-Newtonian Fluid Mech.,* 2, 353 (1977).
21. R. G. Larson, *Constitutive Equations for Polymer Melts and Solutions,* Butterworth, Boston, 1988.
22. S. M. Dinh and R. C. Armstrong, A rheological equation of state for semiconcentrated fiber suspensions, *J. Rheol.,* 28, 207 (1984).
23. G. K. Batchelor, *An Introduction to Fluid Dynamics,* Cambridge University Press, Cambridge, 1979.
24. M. J. Crochet and K. Walters, Numerical Methods in non-Newtonian Fluid Mechanics, *Annu. Rev. Fluid Mech.,* 15, 241 (1983).

25. S. V. Pennington, N. D. Waters, and E. W. Williams, The numerical simulation of an Oldroyd liquid draining down a vertical surface, *J. non-Newtonian Fluid Mech.,* 34, 221 (1990).

26. R. Keunings and D. W. Bousfield, Analysis of surface tension driven leveling in viscoelastic films, *J. Non-Newtonian Fluid Mech.,* 22, 219 (1987).

27. D. W. Bousfield, Long-wave analysis of viscoelastic film leveling, *J. non-Newtonian Fluid Mech.,* 40, 47 (1991).

28. A. Beris, R. C. Armstrong, and R. A. Brown, Perturbation theory for viscoelastic fluids between eccentric rotating cylinders, *J. Non-Newtonian Fluid Mech.,* 13, 109 (1983).

# 10

# *Turbulent Flow and Mixing*

## 10.1 CHARACTERISTICS OF TURBULENT FLOW

The preceding chapters of this book dealt primarily with laminar flow (Chapters 5 to 9) and occasionally with turbulent flow (Chapters 3 and 4). Recall the original experiment by Reynolds[1] on the laminar–turbulent transition and the definitions of Chapter 3: *Laminar flow* occurs at low Reynolds numbers, where adjacent fluid layers move parallel with respect to each other without vertical convective exchange of mass, momentum, or energy, in such a way that a thin stream of dye or smoke injected in the flowing fluid travels with the fluid as a straight line. As the velocity of the fluid increases, the dye line becomes thinner and then begins a wavy or sinusoidal motion in such a way that at any point in the flow field the velocity appears to oscillate randomly about an average value. The latter motion is characterized as *turbulent* to signify chaotic random motion as opposed to the well-behaving, layerwise, laminar motion. The *transition* from laminar to turbulent flow is not an abrupt one nor fully defined: It depends on externally induced vibrations,[2] on whether the velocity increases or decreases,[3] and, of course, on the flow geometry.[4] For these reasons, flow is commonly split into laminar and turbulent regions separated by a rather extensive transition regime. Table 10.1 lists the three regimes in terms of the Reynolds number for selected flow situations.

Confined turbulent flow is difficult to visualize, unless injecting dye or smoke in a flow of transparent wall. Several free-stream turbulent flows are easy to

**TABLE 10.1**   LAMINAR, TRANSITION, AND TURBULENT REGIMES FOR
SELECTED GEOMETRIES

| Flow geometry | Laminar | Transition | Turbulent |
|---|---|---|---|
| Channel | $0 < \mathrm{Re} < 2100$ | $2100 < \mathrm{Re} < 3000$ | $\mathrm{Re} > 3000$ |
| Pipe | $0 < \mathrm{Re} < 2100$ | $2100 < \mathrm{Re} < 3000$ | $\mathrm{Re} < 3000$ |
| Film flow: | $0 < \mathrm{Re} < 4$ | $1000 < \mathrm{Re} < 2000$ | $\mathrm{Re} > 2000$ |
| Boundary layer: | | | |
|    Over plate | $0 < \mathrm{Re} < 3 \times 10^5$ | $3 \times 10^5 < \mathrm{Re} < 3 \times 10^6$ | $\mathrm{Re} > 3 \times 10^6$ |
|    Over cylinder | $0 < \mathrm{Re} < 150$ | $200 < \mathrm{Re} < 10^4$ | $\mathrm{Re} > 3 \times 10^5$ |
|    Over sphere | $0 < \mathrm{Re} < 450$ | $450 < \mathrm{Re} < 2 \times 10^4$ | $2.5 \times 10^5$ |

witness by mere observation, such as smoke lines emitted by stacks or left behind aircraft that expand and disappear quickly as contrasted with the same smoke lines persisting for a long period in laminar flow, three-dimensional violent waves under heavy seas, Niagara Falls, and "dancing flames," which often reveal temperature-induced turbulent flow.

Quantitative measurements of local velocity components in turbulent flow are done by hot-wire anemometry[5] based on the resulting heat loss from the wire to the surrounding turbulent flow; and by laser-doppler velocimetry[6] based on the shift of frequency of laser beam by scattering caused by particle seeds or impurities traveling with the local velocity. Turbulent fluctuations of free surfaces and interfaces are recorded by parallel-wire conductance propes[7] based on conductance changes induced by fluctuating wetting of the prope. In pipes and channels the characteristic parabolic velocity profile of laminar flow persists up to $\mathrm{Re} \simeq 2100$, with the instantaneous local velocity at a particular distance from the solid boundary is constant, independent of time. For $\mathrm{Re} > 2100$, the instantaneous local velocity fluctuates about a time-averaged velocity.

The velocity fluctuations decay significantly near solid boundaries that retard relative motion by friction, as shown in Fig. 10.1. As a result, the flow is virtually laminar within a thin film adjacent to the solid boundary, called the *laminar sublayer*.

Figure 10.2 illustrates a case of free turbulence that accompanies a jet at a high Reynolds number beyond 20,000, surrounded by potential flow. The free boundaries of the jet are distorted significantly due to local turbulence and mixing, whereas laminar jets form a smooth, well-defined free surface (e.g., Section 3.5). The distortion of a turbulent free surface demonstrates the ongoing chaotic fluid motion underneath.

Figure 10.3 shows the velocity that a stationary observer fixed at a specific point $y$ would record under laminar, transition, and turbulent steady conditions, as time progresses, at the same flow rate, defined by

$$\dot{V} = 2 \int_0^H \bar{u}_x(y)\, dy = 2 \int_0^H \left[ \frac{1}{\Delta t} \int_t^{t+\Delta t} u_x(y, t)\, dt \right] dy, \qquad (10.1)$$

where $\Delta t$ is time period larger than the time period of fluctuations $T$. Experimental evidence[9] suggests that the transition from laminar flow and the

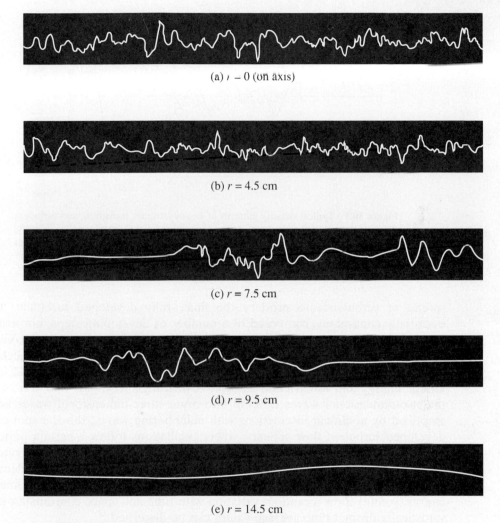

(a) $r = 0$ (on axis)

(b) $r = 4.5$ cm

(c) $r = 7.5$ cm

(d) $r = 9.5$ cm

(e) $r = 14.5$ cm

**Figure 10.1**  Measured velocity fluctuations in confined turbulent flow in a pipe from center ($r = 0$) to wall ($r = 15$). (From Ref. 8, by permission.)

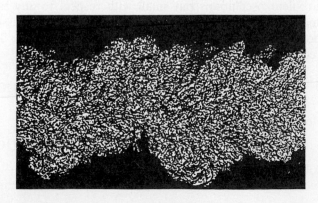

**Figure 10.2**  Turbulent jet surrounded by potential flow. (From Ref. 8, by permission.)

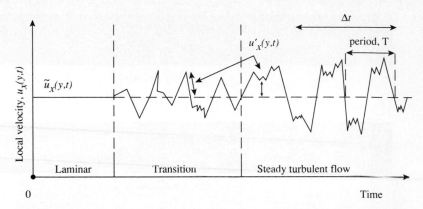

**Figure 10.3** Typical velocity patterns of steady laminar, transition, and turbulent flows, and $x$-velocity component decomposition: $u_x(y, t)$ is instantaneous velocity, $u_x'(y, t)$ is velocity fluctuation, and $\tilde{u}_x(y)$ is time-smoother velocity, such that $u_x(y, t) = \tilde{u}_x(y) + u_x'(y, t)$, at distance $y$ from the symmetry line, under common flow rate defined by Eqn. (10.1).

spread of turbulence to produce the final, fully developed turbulent flow is essentially continuous, composed of a number of developing steps, not a sudden single catastrophic change. In general, the transition process can be decomposed into four steps.[10] First, small two-dimensional waves develop due to mechanical vibrations or flow-rate fluctuation, which are then linearly amplified due to high kinetic energy and low friction associated with high Reynolds numbers; second, the two-dimensional waves develop into finite three-dimensional waves and are amplified by nonlinear interactions with neighboring waves; third, a spot of fully developed turbulent flow appears, where oscillations follow a certain periodicity with time $T$ as shown by Fig. 10.3; and finally, the turbulent spot propagates and infects the entire flow field. Under conditions of fully developed turbulent flow, the various quantities associated with the fluid motion (e.g., mass, momentum, energy, solute) show a random variation with time and space coordinates, so that statistically distinct time average values can be discerned.[11]

A time-smoothed average value, $q(y)$ of a quantity at point $y$ is defined by averaging $q(y, t)$ over a time interval $\Delta t$. This time interval must be large with respect to the time of turbulent oscillations but small with respect to any time period of the driving mechanism (e.g., time-dependent velocity or pressure gradient). This time smoothing is shown in Fig. 10.3 and discussed below. The *time-smoothed $i$-velocity* component under the conditions shown is

$$\bar{u}_i(y) = \frac{1}{\Delta t} \int_t^{t+\Delta t} u_i(y, t)\, dt \tag{10.2}$$

and the instantaneous velocity is

$$u_i(y, t) = \bar{u}_i(y) + u_i'(y, t) \tag{10.3}$$

Similar expressions can be written for all velocity components and for the pressure, which is also fluctuating. An *intensity of turbulence* can also be defined as

$$I = \sqrt{\frac{\sum_i [u_i'(y, t)]^2}{\bar{u}_i}}, \tag{10.4}$$

where $\bar{u}_i$ is the cross-sectional average velocity of the time-smoothed turbulent velocity, defined by

$$\bar{u}_i = \frac{1}{A} \int_A \bar{u}_i(y)\, dA \tag{10.5}$$

and

$$\overline{\sum_i (u_i')^2} = \frac{1}{\Delta t} \int_t^{t+\Delta t} \overline{\sum (u_i')^2}\, dt. \tag{10.6}$$

The intensity of turbulence, $I$, provides an estimation of the magnitude of the turbulent fluctuations, $u_i'$, relative to the average velocity of the flow, $\bar{u}_i$. In general, fluctuations in the direction of flow are greater than those in the perpendicular directions, but there is a tendency for the two to be the same away from solid boundaries. For turbulent flow between flat plates, the intensity $I$ may be as high as 20% in the direction of flow and 10% in the perpendicular direction.

## 10.2 THEORY OF TURBULENT FLOW

The unsteady conservation equations and the Navier–Stokes equations hold for turbulent flow, since the eddies generated are of much greater lengths than the mean free path of molecules. However, it is difficult to apply these equations directly to turbulent flow problems, because the instantaneous values of the variables involved in these equations (e.g., pressure, velocity, stress) vary with time to such a degree that their calculation would be practically impossible with the computational means available; they would even be of little use in practical applications. Thus, some modifications are needed to make these equations applicable to turbulent flow. Reynolds modified the Navier–Stokes equations so that the variables would be the time averages introduced in Section 10.1. The transformation of the unsteady Navier–Stokes equations to time-smoothed equations in terms of time-smoothed variables is highlighted below. Two main areas of theoretical study of turbulent flow have been evolved and are still among the most active research areas of fluid mechanics. The first investigates the unsteady Navier–Stokes equations in order to predict the *stability of laminar flow* with respect to perturbations that can determine the onset of transition to turbulence. The second is to determine necessary modifications to the Navier–Stokes equations to apply under turbulent conditions once a turbulent flow is established; this area is divided into *phenomenological and statistical theories of turbulent flow.* Phenomenological theories are summarized below. Statistical theories are more involved and are treated elsewhere.[4,10] Mixing promoted by turbulence is highlighted in Section 10.3 and detailed in relative publications.[12]

## 10.2.1 Stability of Laminar Flow and Onset of Turbulent Flow

*Hydrodynamic stability* theory deals with predicting whether or not a given flow is stable. A steady flow is *stable* if having been removed from its steady state by a fluctuation or perturbation (e.g., a mechanical vibration), it returns to its steady state as time proceeds. A steady flow is *unstable* if, having been removed from its steady state, it moves away to catastrophe as time proceeds after the elimination of the perturbation. A fluctuation or perturbation that neither grows nor decays with time signifies a *neutrally stable* flow.

A laminar flow in a pipe at low Reynolds numbers is stable. However, above a critical Reynolds number, the corresponding steady laminar flow is no longer stable; a small velocity or pressure fluctuation or perturbation causes the flow to move away from its steady state toward the turbulent flow. This critical Reynolds number characterizes the *onset of the transition* and is calculated as follows.

The Navier–Stokes equations are solved to calculate the steady state, which is generally a function of the Reynolds number, of steady velocity $\bar{u}^{ss}(x, y, z, \mathrm{Re})$ and steady pressure $p^{ss}(x, y, z, \mathrm{Re})$. Then the steady-state variables are perturbed to time-dependent velocity $u(x, y, z, \mathrm{Re}, t)$ and pressure $p(x, y, z, \mathrm{Re}, t)$, respectively, according to

$$u(x, y, z, \mathrm{Re}, t) = \bar{u}^{ss}(x, y, z, \mathrm{Re}) + f(x, y, z)e^{ct} \tag{10.7}$$

and

$$p(x, y, z, \mathrm{Re}, t) = p^{ss}(x, y, z, \mathrm{Re}) + g(x, y, z)e^{ct}, \tag{10.8}$$

where the sign of the exponent $c$ determines growth, decay, or nonevolution of *disturbances or instabilities*. These time-dependent variables are substituted in the time-dependent Navier–Stokes equations. Given the fact that $u^{ss}$ and $p^{ss}$ must satisfy the steady part of the Navier–Stokes equations, a relation between $c$ and Re is found. For values of Re for which $c < 0$, the flow is stable. The critical value at Re at which $c$ changes sign signifies the onset of the transition to turbulent flow. Simplified stability analysis cases are summarized in Panton[13] and are detailed in Drazin and Reid.[14]

### Example 10.1: Linear Stability Analysis of Creeping Flow

To highlight the underlying ideas of linear stability analysis, consider an idealized Couette creeping flow of zero Reynolds number between parallel plates. This flow is expected to be stable to any disturbance, because the infinitely large viscosity implied quickly dissipates the mechanical energy of any disturbance to heat, and forces velocity fluctuations to decay. The governing equations are

$$\frac{d^2 u_x}{dy^2} = 0, \quad u_x(y = 0), \quad \text{and} \quad u_x(y = H) = V, \tag{10.1.1}$$

with steady solution,

$$u_x^{ss}(y) = V\frac{y}{H}. \tag{10.1.2}$$

The flow is disturbed by

$$u_x(y, t) = u_x^{ss}(y) + \sum_n \varepsilon_n \sin\left(v\pi\frac{y}{H}\right)e^{c_n t}, \tag{10.1.3}$$

where $\varepsilon_n \to 0$ and the disturbance

$$u_x'(y, t) = \sum_n \varepsilon_n \sin\left(n\pi \frac{y}{H}\right) e^{c_n t} \qquad (10.1.4)$$

represents a generalized fluctuation made up by superposition of sinusoidal waves of wavelengths $H/n$. The fluctuation(s) satisfy the boundary conditions of zero-velocity fluctuations at the wall, and grows or decays exponentially, depending on the sign of $c_n$. Equation (10.1.3) is substituted in the corresponding transient equation,

$$\rho \frac{\partial u_x}{\partial t} = \eta \frac{\partial^2 u_x}{\partial y^2}, \qquad (10.1.5)$$

to yield

$$\rho \left[ \frac{\partial u_x^{SS}}{\partial t} + \sum_n c_n \varepsilon_n \sin\left(n\pi \frac{y}{H}\right) e^{c_n t} \right] = \eta \left[ \frac{\partial^2 u_x^{SS}}{\partial y^2} - \sum_n \left(\frac{n\pi}{H}\right)^2 \varepsilon_n \sin\left(n\pi \frac{y}{H}\right) e^{c_n t} \right], \qquad (10.1.6)$$

which, given that $\partial u_x^{SS}/\partial t = \partial^2 u_x^{SS}/\partial y^2 = 0$, simplifies to

$$\sum_n \rho c_n \varepsilon_n \sin\left(n\pi \frac{y}{H}\right) e^{c_n t} = -\sum_n \eta \left(\frac{n\pi}{H}\right)^2 \varepsilon_n \sin\left(n\pi \frac{y}{H}\right) e^{c_n t}. \qquad (10.1.7)$$

Because the functions $\sin(n\pi y/H)$ are linearly independent, Eqn. (10.1.7) implies that

$$\rho c_n \varepsilon_n \sin\left(n\pi \frac{y}{H}\right) e^{c_n t} = -\eta \left(\frac{n\pi}{H}\right)^2 \varepsilon_n \sin\left(n\pi \frac{y}{H}\right) e^{c_n t} \qquad (10.1.8)$$

for any $n$. Equation (10.1.8) reduces further to

$$c_n = -\frac{n}{\rho} \left(\frac{n\pi}{H}\right)^2 = -\nu \left(\frac{n}{H}\right)^2 < 0, \qquad (10.1.9)$$

which predicts an always stable flow to the disturbances assumed, the more so the larger the kinematic viscosity $\nu$ and wavenumber $n$ are. Also, velocity waves of small wavelength $H/n$ decay fastest. The overall conclusion is that *this* Couette flow is able to dissipate *this* type of instability by itself, without external intervention, i.e., the flow is self-controlled. Otherwise, an external intervention in the form of *process control* is required to maintain steady operation. The stability characteristics are expected to be different for nonzero Reynolds numbers and for different types of disturbances, as denonstrated elsewhere by more rigorous stability analyses.[14,15] If an unstable process must be stabilized, a *process control* scheme is required to maintain steady operation.

## 10.2.2 Turbulent Flow Equations and Reynolds Stresses

Once a laminar flow becomes unstable transition to turbulence occurs and the Navier–Stokes equations are practically inapplicable to further analysis of the flow. The same equations are made applicable by means of the time-smoothed quantities introduced in Section 10.1 [with arguments omitted hereafter, i.e., $u_i$ stands for $u_i(y, t)$, $\tilde{u}_i$ for $\tilde{u}_i(y)$, $u_i'$ for $u_i'(y, t)$, and so on]:

$$u_i = \tilde{u}_i + u_i', \qquad i = x, y, z. \qquad (10.9)$$

$$p = \tilde{p} + p'. \qquad (10.10)$$

These expressions are substituted in the original Navier–Stokes equations, which are therefore transformed to

$$\frac{\partial \tilde{u}_x}{\partial x} + \frac{\partial \tilde{u}_y}{\partial y} + \frac{\partial \tilde{u}_z}{\partial z} = 0 \tag{10.11}$$

$$\rho\left(\frac{\partial \tilde{u}_x}{\partial t} + \tilde{u}_x \frac{\partial \tilde{u}_x}{\partial x} + \tilde{u}_y \frac{\partial \tilde{u}_x}{\partial y} + \tilde{u}_z \frac{\partial \tilde{u}_x}{\partial z}\right) = -\frac{\partial \tilde{p}}{\partial x} + \eta\left[\frac{\partial^2 \tilde{u}_x}{\partial x^2} + \frac{\partial^2 \tilde{u}_x}{\partial y^2} + \frac{\partial^2 \tilde{u}_x}{\partial z^2}\right] + \rho g_x$$

$$+ \rho\left[\frac{\partial}{\partial x}(\overline{u'_x u'_x}) + \frac{\partial}{\partial y}(\overline{u'_y u'_x}) + \frac{\partial}{\partial z}(\overline{u'_z u'_x})\right] \tag{10.12}$$

$$\rho\left(\frac{\partial \tilde{u}_y}{\partial t} + \tilde{u}_x \frac{\partial \tilde{u}_y}{\partial x} + \tilde{u}_y \frac{\partial \tilde{u}_y}{\partial y} + \tilde{u}_z \frac{\partial \tilde{u}_y}{\partial z}\right) = -\frac{\partial \tilde{p}}{\partial y} + \eta\left[\frac{\partial^2 \tilde{u}_y}{\partial x^2} + \frac{\partial^2 \tilde{u}_y}{\partial y^2} + \frac{\partial^2 \tilde{u}_y}{\partial z^2}\right] + \rho g_y$$

$$+ \rho\left[\frac{\partial}{\partial x}(\overline{u'_x u'_y}) + \frac{\partial}{\partial y}(\overline{u'_y u'_y}) + \frac{\partial}{\partial z}(\overline{u'_z u'_y})\right] \tag{10.13}$$

$$\rho\left(\frac{\partial \tilde{u}_z}{\partial t} + \tilde{u}_x \frac{\partial \tilde{u}_z}{\partial x} + \tilde{u}_y \frac{\partial \tilde{u}_z}{\partial y} + \tilde{u}_z \frac{\partial \tilde{u}_z}{\partial z}\right) = -\frac{\partial \tilde{p}}{\partial z} + \eta\left[\frac{\partial^2 \tilde{u}_z}{\partial x^2} + \frac{\partial^2 \tilde{u}_z}{\partial y^2} + \frac{\partial^2 \tilde{u}_z}{\partial z^2}\right] + \rho g_z$$

$$+ \rho\left[\frac{\partial}{\partial x}(\overline{u'_x u'_z}) + \frac{\partial}{\partial y}(\overline{u'_y u'_z}) + \frac{\partial}{\partial z}(\overline{u'_z u'_z})\right]. \tag{10.14}$$

By comparing Eqns. (10.11) to (10.14) with the corresponding equations for laminar flow, Eqns. (G.5) to (G.7) of Appendix G, it is obvious that besides replacement of the laminar instantaneous variables by the time-smoothed turbulent variables $\tilde{u}_x$, $\tilde{u}_y$, $\tilde{u}_z$, and $\tilde{p}$, additional terms appear in Eqns. (10.11) to (10.14), made up by dual combinations of the velocity fluctuations $u'_x$, $u'_y$, and $u'_z$. These terms, commonly called *Reynolds stresses,* are represented by the notation

$$\tilde{\tau}_{xx}^{(t)} = \rho \overline{u'_x u'_x}, \qquad \tilde{\tau}_{xy}^{(t)} = \rho \, \overline{u'_y u'_x}, \tag{10.15}$$

The Reynolds stresses are components of turbulent momentum flux, representing turbulent convective transfer of momentum due to the local fluctuations of the velocity about its average value, which are absent in laminar flow. Indeed, Eqns. (10.12) to (10.14), under the definitions of Eqn. (10.15), can be cast in a form identical to the stress momentum equations [e.g., Eqns. (G.6) to (G.8), Appendix G], with the total stresses given by

$$\tau_{ij} = \tilde{\tau}_{ij}^{(l)} + \tilde{\tau}_{ij}^{(t)} = \eta\left(\frac{\partial \tilde{u}_i}{\partial x_j} + \frac{\partial \tilde{u}_j}{\partial x_i}\right) + \rho(\overline{u'_j u'_i}) \tag{10.16}$$

[i.e., as a summation of a time-smoothed, laminar-like part, $\tilde{\tau}_{ij}^{(l)}$, and a fluctuating turbulent part, $\tilde{\tau}_{ij}^{(t)}$].

To solve Eqns. (10.11) to (10.14), the Reynolds stresses $\tilde{\tau}_{ij}^{(t)} = \rho \, \overline{u'_i u'_j}$, must be expressed in terms of the time-smoothed velocity components, $\tilde{u}_i$. To do so, several semiempirical relations have been developed, as summarized below for channel flow in the $x$-direction.

**1.** *Boussimesq's eddy viscosity (BEV):*[16]

$$\tilde{\tau}_{yx}^{(t)} = \eta^{(t)} \frac{d\tilde{u}_x}{dy} = (\eta + \varepsilon\rho) \frac{d\tilde{u}_x}{dy} \tag{10.17}$$

**2.** *Prandtl's mixing length (PML):*[17,18]

$$\tilde{\tau}_{yx}^{(t)} = \rho \, l^2 \left| \frac{d\tilde{u}_x}{dy} \right| \frac{d\tilde{u}_x}{dy} \tag{10.18}$$

**3.** *Von Kármán's similarity hypothesis (KSH):*[19,20]

$$\tilde{\tau}_{yx}^{(t)} = \rho \, k^2 \left| \frac{(d\tilde{u}_x/dy)^3}{(d^2\tilde{u}_x/dy^2)^2} \right| \frac{d\tilde{u}_x}{dy} \tag{10.19}$$

**4.** *Deissler's formula for near wall region (DFW):*[21]

$$\tilde{\tau}_{yx}^{(t)} = -\rho \, n^2 \tilde{u}_x \, y \left[ 1 - \exp\left(-\frac{n^2 \tilde{u}_x y}{v}\right) \right] \frac{d\tilde{u}_x}{dy} \tag{10.20}$$

BEV is analogous to Newton's law of viscosity with $\eta^{(t)}$ a *turbulent viscosity made up by the laminar, $\eta$, and eddy, $\varepsilon$, viscosities.* PML rests on the approximation that a fluid particle in fluctuating distance $l$ experiences change in velocity identical to $l \, d\tilde{u}_x/dy$, which is equivalent to a $\rho(l \, d\tilde{u}_x/dy)(l \, d\tilde{u}_x/dy)$ transfer of momentum. KSH is based on the statistical similarity of local flow patterns, which relates higher-order derivatives to the first-order derivative and the mixing length. The BEV, the PML, and the KSH give reasonable results only when the parameters involved, $\eta^{(t)}$, $l$, and $k$, are made position dependent. Even in that case, they appear to hold only away from solid boundaries, where they are complemented by the DFW. The latter takes into account the dominant dissipative action of viscosity, or by a linear velocity profile in the laminar sublayer, adjacent to the solid wall.

In trying to formulate a velocity profile that is continuous throughout the flow field, Pai[22] developed the expression

$$\frac{\tilde{u}_x}{\tilde{u}_{x,\max}} = 1 + a_1 \left(\frac{y}{H}\right)^2 + a_2 \left(\frac{y}{H}\right)^{2m} \tag{10.21}$$

for channel flows, and pipe flows (with $\tilde{u}_x$, $\tilde{u}_{x,\max}$ replaced by $\tilde{u}_z$, $\tilde{u}_{z,\max}$, and $y$ by $r$). The coefficients $a_1$ and $a_2$ and the exponent $m$ are calculated by the boundary conditions and by symmetry considerations. Figure 10.4 illustrates the ability of Eqns. (10.17) to (10.21) to fit or predict data in channel and pipe flows.

It is obvious from Fig. 10.4 that none of the equations above provide a valid approximation throughout the channel width or pipe diameter. The power series of Eqn. (10.21) appears to perform most satisfactorily at Reynolds numbers less than $10^5$, where the exponent $m$ is well approximated by[10]

$$m = -0.617 \times 8.211 \times 10^{-3} (\text{Re})^{0.786}. \tag{10.22}$$

It turns out that the most accurate representation is obtained by splitting

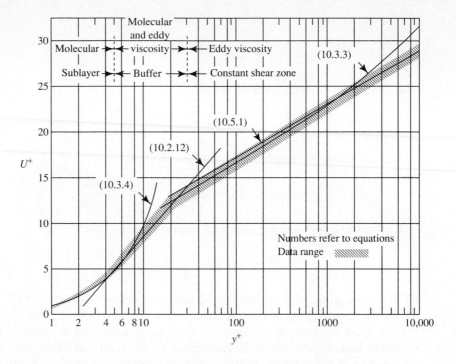

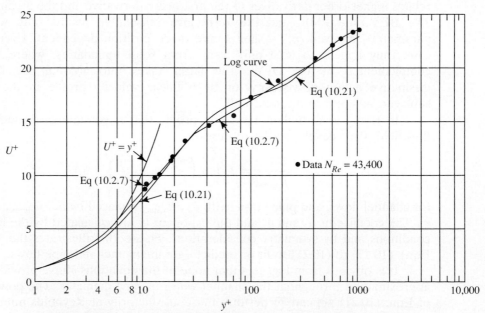

**Figure 10.4** Universal turbulent velocity distribution diagram. (From R. S. Brodkey, *The Phenomena of Fluid Motions,* Addison-Wesley Series in Chemical Engineering, Ohio State University, Columbus, Ohio, 1959; by permission.)

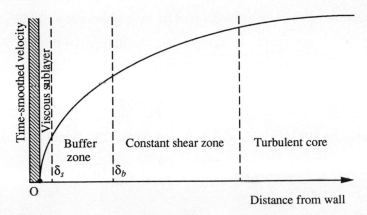

**Figure 10.5**  Velocity distributions and zones of turbulent flow in channels and pipes.

the entire width into subsections, as shown in Fig. 10.5, and applying the most appropriate equation among Eqns. (10.17) to (10.21). In the viscous sublayer, turbulence is negligible, and the shear stress is nearly constant, identical to that of the wall, $\tau_W$. The resulting velocity profile is linear, changing from zero at the wall to a finite value, $u_\delta$, where the buffer zone starts. In the constant shear zone, viscous forces become negligible, and the zone is close enough to the boundary that the shear stress is essentially that of the wall, $\tau_W$. These assumptions conform to Prandtl's mixing-length hypothesis, and therefore Eqn. (10.18) is the most appropriate in the constant shear zone. In the buffer zone, between the viscous sublayer and the constant shear zone, both viscous and turbulent stresses are important and Eqn. (10.20), which is often extended to the viscous sublayer, is the most applicable. In the turbulent core, Eqn. (10.18) still applies, but Eqn. (10.19) is a better approximation.

### 10.2.3 Solution of Turbulent Flow Equations

Equations (10.11) to (10.14) are solved for the channel (and pipe) flow, shown in Fig. 10.6. Other turbulent flow situations amenable to analytic solutions are

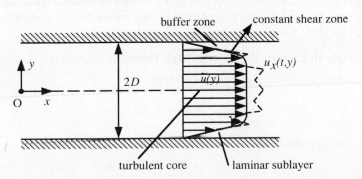

**Figure 10.6**  Turbulent channel flow.

analyzed by the same procedure. The time-smoothed velocities are

$$\tilde{u}_x = f(x, y), \qquad \bar{u}_y = \bar{u}_z = 0. \tag{10.23}$$

Furthermore, for fully developed flows,

$$\frac{\partial \tilde{u}_x}{\partial x} = 0. \tag{10.24}$$

Equation (10.16) yields $\overline{u'_y u'_z} = 0$, and Eqns. (10.12) and (10.13) become

$$-\frac{\partial \tilde{p}}{\partial x} + \eta \frac{d^2 \tilde{u}_x}{dy^2} + \rho \frac{d}{dy} (\overline{u'_x u'_y}) = 0 \tag{10.25}$$

and

$$-\frac{\partial \bar{p}}{\partial y} + \rho \frac{d}{dy} (\overline{u'_y u'_y}) = 0. \tag{10.26}$$

By differentiating Eqn. (10.26) with respect to $x$ and then interchanging the order of the resulting second-order derivatives, it turns out that $\partial^2 p / \partial^2 y = 0$, and therefore the final form of Eqn. (10.25) is

$$-\frac{d\tilde{p}}{dx} + \frac{d}{dy} \left( \rho \, \overline{u'_x u'_y} + \eta \frac{d\tilde{u}_x}{dy} \right) = 0, \tag{10.27}$$

which is integrated to yield

$$-\frac{d\tilde{p}}{dx} y + \rho \, \overline{u'_x u'_y} + \eta \frac{d\tilde{u}_x}{dy} + c_1(x) = 0. \tag{10.28}$$

At $y = 0$ (the plane of symmetry), $\overline{u'_x u'_y} = 0$ and $d\tilde{u}_x / dy = 0$; therefore, $c_1(x) = 0$. Thus, the equation becomes

$$-\frac{d\tilde{p}}{dx} y + \rho \, \overline{u'_x u'_y} + \eta \frac{d\tilde{u}_x}{dy} = 0. \tag{10.29}$$

## 10.2.4 Pipe and Channel Turbulent Flows

Equation (10.29) can be solved, given an appropriate expression for the Reynolds stresses, by Eqns. (10.17) to (10.20). This procedure is demonstrated in Examples 10.2 to 10.4, for channel and pipe turbulent flows.

**Example 10.2: Channel Flow with PML Theory**

The governing momentum equation,

$$-\frac{d\tilde{p}}{dx} y + \rho \, \overline{u'_x u'_y} + \eta \frac{d\tilde{u}_x}{dy} = 0, \tag{10.2.1}$$

is combined with Eqn. (10.18) to yield

$$-\frac{d\tilde{p}}{dx} y + \rho \, l^2 \left| \frac{d\tilde{u}_x}{dy} \right| \frac{d\tilde{u}_x}{dy} + \eta \frac{d\tilde{u}_x}{dy} = 0. \tag{10.2.2}$$

By considering the flow between the midplane of symmetry and the upper wall where $d\bar{u}_x/dy < 0$, Eqn. (10.2.2) becomes

$$-\frac{d\bar{p}}{dx}y - \rho l^2\left(\frac{d\bar{u}_x}{dy}\right)^2 + \eta\frac{d\bar{u}_x}{dy} = 0. \tag{10.2.3}$$

The mixing length is usually taken as proportional to the distance from the wall, i.e.,

$$l = k(D - y). \tag{10.2.4}$$

For fully developed flow, $d\bar{p}/dx = \Delta p/\Delta L = \lambda$, and therefore Eqn. (10.2.3) becomes

$$\lambda y + \rho k^2(D - y)^2\left(\frac{d\bar{u}_x}{dy}\right)^2 - \eta\frac{d\bar{u}_x}{dy} = 0, \tag{10.2.5}$$

with roots

$$\frac{d\bar{u}_x}{dy} = \frac{\eta}{2\rho k^2(D - y)^2} \pm \sqrt{\left[\frac{\eta}{2\rho k^2(D - y)^2}\right]^2 - \frac{\lambda y}{\rho k^2(D - y)^2}}, \tag{10.2.6}$$

which is integrated to

$$\bar{u}_x = \frac{\eta}{2\rho k^2(D - y)} \pm \frac{\eta}{2\rho k^2}\int_0^y\left(\sqrt{\frac{1}{(D - y)^4} - \frac{2\rho^2 k^2 \lambda y}{\eta^2(D - y)^2}}\right)dy. \tag{10.2.7}$$

As stated in Section 10.3.2, the PML theory applies to the constant shear stress zone of shear stress $\tau_w$. Under these conditions, Eqn. (10.2.3) yields

$$\tau_w + \rho l^2\left(\frac{d\bar{u}_x}{dy}\right)^2 - \eta\frac{d\bar{u}_x}{dy} = 0. \tag{10.2.8}$$

Furthermore, since in the shear stress zone laminar stresses are significantly less than the turbulent ones, Eqn. (10.2.8) reduces further to

$$\tau_w + \rho k^2(D - y)^2\left(\frac{d\bar{u}_x}{dy}\right)^2 = 0, \tag{10.2.9}$$

which is integrated to

$$\bar{u}_x = A - \left(\frac{\tau_w}{k^2\rho}\right)^{1/2}\ln(D - y), \tag{10.2.10}$$

with the constant $A$ determined by matching this velocity with the velocity of the adjacent zones, and $\tau_w = (dp/dx)D$. If the thickness of the buffer layer including the viscous sublayer is $\delta_b$ (Fig. 10.5), where the velocity is $(\bar{u}_x)_\delta$, Eqn. (10.2.10) reduces to

$$\bar{u}_x - (\bar{u}_x)_\delta = \left(-\frac{\tau_w}{k^2\rho}\right)^{1/2}\ln\frac{D - y}{D - \delta}. \tag{10.2.11}$$

Equation (10.2.11) is usually nondimensionalized by defining the dimensionless variables

$$u^* = \frac{\bar{u}_x}{(\tau_w/\rho)^{1/2}}, \quad y^* = \frac{y(\tau_w/\rho)^{1/2}}{\nu}, \quad \text{and} \quad D^* = \frac{D(\tau_w/\rho)^{1/2}}{\nu},$$

which transform Eqn. (10.2.11) into

$$u^* - u_b^* = \frac{1}{k}\ln\frac{D^* - y^*}{D^* - y_b^*}. \tag{10.2.12}$$

Deissler[21] found that Eqn. (10.2.12), with constants taken as $k = 0.36$, $y_b^* = 26$, and $u_b^* = 12.85$, approximates well the velocity profiles at Re $> 20,000$, except near the solid walls, where the analysis of Example 10.3 provides a better approximation.

### Example 10.3: Turbulent Velocity Distribution by DFW

Deissler's formula is applicable in the near-wall region of turbulent flow. The governing momentum equation for pipe flow is

$$\frac{\Delta p}{\Delta L}\frac{R}{2} = \tau_w = \eta\frac{d\bar{u}_z}{dr} + \rho\, n^2\bar{u}_z(R - r)(1 - \exp[-n^2\bar{u}_z(R - r)/v])\frac{d\bar{u}_z}{dr}, \qquad (10.3.1)$$

which is integrated to

$$\bar{u}_z = \int_{R-\delta_b}^{R}\frac{\tau_w\,dr}{\eta + \rho\,n^2\bar{u}_z(R - r)(1 - \exp[-n^2\bar{u}_z(R - r)/v])}, \qquad (10.3.2)$$

and in dimensionless form (see Example 10.4) to

$$u^* = \int_{R^*-\delta_b^*}^{R^*}\frac{dr^*}{(1 + n^2u^*(R^* - r^*)(1 - \exp[-n^2u^*(R^* - r^*)]))}, \qquad (10.3.3)$$

which appears to hold in $0 \leq R^* - r^* \leq 26$ with $n = 0.124$ for long, smooth tubes. For $R^* - r^* < 5$, Eqn. (10.3.3) reduces to

$$u^* = R^* - r^*, \qquad (10.3.4)$$

which is identical to the velocity profile obtained by integrating Newton's law of viscosity over the viscous laminar sublayer. Solutions of Eqn. (10.3.3), obtained by numerical integration and iteration, are included in Fig. 10.4.

### Example 10.4: Channel Turbulent Flow with KSH Theory

The governing equation becomes

$$-\frac{d\bar{p}}{dx}y + \rho\,k^2\left|\frac{(d\bar{u}_x/dy)^3}{(d^2\bar{u}_x/dy^2)^2}\right|\frac{d\bar{u}_x}{dy} + \eta\frac{du_x}{dy} = 0. \qquad (10.4.1)$$

Between the midplane of symmetry and the upper wall, the velocity gradient is negative, and therefore Eqn. (10.4.1) becomes

$$-\frac{d\bar{p}}{dx}y - \rho\,k^2\frac{(d\bar{u}_x/dy)^4}{(d^2\bar{u}_x/dy^2)^2} + \eta\frac{d\bar{u}_x}{dy} = 0, \qquad (10.4.2)$$

and equivalently,

$$\left(\frac{d\bar{p}}{dx}y\right)\left(\frac{d^2\bar{u}_x}{dy^2}\right)^2 + \rho\,k^2\left(\frac{d\bar{u}_x}{dy}\right)^4 - \eta\frac{d^2\bar{u}_x}{dy^2}\frac{d\bar{u}_x}{dy} = 0. \qquad (10.4.3)$$

By substituting $f = d\bar{u}_x/dy$, Eqn. (10.4.3) becomes

$$\left(\frac{d\bar{p}}{dx}y\right)\left(\frac{df}{dy}\right)^2 + \rho\,k^2f^4 - \eta\,f\left(\frac{df}{dy}\right)^2 = 0, \qquad (10.4.4)$$

which is highly nonlinear and cannot be solved analytically. However, since the KSH is most valid for the turbulent core where viscous stresses are minor, Eqn. (10.4.4) is approximated by

$$\frac{d\bar{p}}{dx} y \left(\frac{df}{dy}\right)^2 + \rho k^2 f^4 = 0, \tag{10.4.5}$$

with solution

$$\frac{d\bar{u}_x}{dy} = f = -\frac{\sqrt{-(dp/dx)/\rho \, k^2}}{c - 2\sqrt{y}}. \tag{10.4.6}$$

At the solid wall, $d\bar{u}_x/dy \rightarrow \infty$, and therefore $c = 2R^{1/2}$. Thus

$$\frac{d\bar{u}_x}{dy} = -\frac{\sqrt{-(d\bar{p}/dx)/\rho \, k^2}}{(2\sqrt{R} - 2\sqrt{y})}, \tag{10.4.7}$$

which is integrated further to

$$\frac{\bar{u}_x}{\sqrt{-(d\bar{p}/dx)/\rho \, k^2}} = +\sqrt{y} + \sqrt{R} \ln(2\sqrt{R} - 2\sqrt{y}), \tag{10.4.8}$$

and equivalently to

$$\frac{\bar{u}_{u,\max} - \bar{u}_x}{\sqrt{\tau_w/\rho}} = -\frac{1}{k}\left[\ln\left(1 - \sqrt{\frac{y}{H}}\right) + \sqrt{\frac{y}{H}}\right] \tag{10.4.9}$$

for channel flow. A corresponding expression for pipe flow is

$$\frac{\bar{u}_{x,\max} - \bar{u}_x}{\sqrt{\tau_w/\rho}} = -\frac{1}{k}\ln\left(1 - \frac{r}{R}\right). \tag{10.4.10}$$

Experimental work by Nikuradse[23] suggests that the best values of $k$ are 0.36 and 0.4 for channel and pipe flow, respectively, used in Fig. 10.4.

### Example 10.5: Friction in Turbulent Flow

The logarithmic velocity profile developed in Example 10.2, with $k = 0.36$ and $y_b^* = 26$ (as proposed by Deissler[21]), can be rearranged in the form

$$u^* = 2.5 \ln y^* + 5.5, \tag{10.5.1}$$

where

$$u^* = \frac{\bar{u}_x}{(\tau_w/\rho)^{1/2}} \quad \text{and} \quad y^* = \frac{y(\tau_w/\rho)^{1/2}}{\nu},$$

and $y$ distance from solid wall. The ratio $\tau_w/\rho$ enters the definition of friction coefficient $f_F$ described in Section 4.1 [e.g., Eqn. (4.8)], quantified later by Colbrook's Eqn. (4.32) and Blasius's Eqn. (4.33).

(a) By combining these equations, find the interrelation between the Reynolds number Re and the friction coefficient $f_F$.

(b) How does this relation compare to Eqns. (4.32) and (4.33)?

(c) Is there any relation to the thickness of the laminar sublayer $\delta$?

### Solution

(a) *Friction coefficient vs. Reynolds number.* Both the Reynolds number and the friction coefficient are based on the cross-sectional average velocity $\bar{u}$, which is

$$\bar{u} = \frac{1}{\pi R^2} \int_0^R 2\pi r \bar{u}_x \, dr = \sqrt{\frac{\tau_w}{\rho}} \left(2.5 \ln\left[\frac{R\bar{u}}{\nu}\sqrt{\frac{\tau_w}{\rho\bar{u}^2}}\right] + 1.75\right). \tag{10.5.2}$$

The friction factor, which is also a dimensionless stress according to the definition in Eqn. (4.8),

$$f_F = 2 \frac{\tau_w}{\rho u_m^2}, \tag{10.5.3}$$

yields

$$\sqrt{\frac{1}{f_F}} = 1.77 \ln\left(0.354 \, \mathrm{Re}\sqrt{f_F}\right) + 1.24, \tag{10.5.4}$$

which compares well with that observed experimentally[23] in smooth pipe for $5000 < \mathrm{Re} < 3.4 \times 10^6$.

**(b)** *Coolbrook equation, Eqn. (4.32).* Equation (10.5.4) is identical to Eqn. (4.32) for smooth pipe of $\rho/D = 0$. Equation (4.33) is a limiting case of Eqn. (4.32).

**(c)** *Thickness of laminar sublayer.* The thickness $\delta_s$ is expected to follow (see Example 10.8)

$$\delta_s = c_1 \frac{\nu}{\sqrt{\tau_w/\rho}} \tag{10.5.5}$$

and therefore,

$$\delta_s = c_1 = \frac{\nu}{\sqrt{\tau_w/\rho}} = c_2 \frac{\nu}{\bar{u}\sqrt{f_F}} \frac{D}{D} \tag{10.5.6}$$

and

$$\frac{\delta_s}{D} = c_3 \frac{1}{\mathrm{Re}\sqrt{f_F}}, \tag{10.5.7}$$

where the $c_i$ are known constants of proportionality. The latter equation is rearranged in the form

$$\sqrt{f_F} = \frac{c_3}{\delta_s \, \mathrm{Re}}. \tag{10.5.8}$$

Equations (10.5.6) and (10.5.7) predict that $\delta_s$ is inversely proportional to $\sqrt{\mathrm{Re}}$, low Reynolds number, and becomes nearly inversely proportional to Re at $\mathrm{Re} > 10^5$. By the same arguments, $f_F$ is proportional to $\delta_s^2$ at low Reynolds numbers and to $\delta_s^{2/7}$ for $\mathrm{Re} > 10^5$.

## Example 10.6: Universal Turbulent Velocity Profile in Pipe Flow

The constants of Eqn. (10.21) are calculated as follows: At $r = R$ (solid wall), $\bar{u}_z = 0$; therefore,

$$0 = 1 + a_1 + a_2. \tag{10.6.1}$$

The governing momentum equation for pipe flow is

$$-\frac{\Delta p}{L}\frac{r}{2} + \rho \overline{u_z' u_r'} + \eta \frac{d\bar{u}_z}{dr} = 0, \tag{10.6.2}$$

which satisfies the symmetry conditions since at $r = 0$, $\overline{u_z' u_z'} = 0$ and $d\bar{u}_z/dr = 0$. Equation (10.5.2), applied at the solid wall where $\overline{u_z' u_r'} = 0$, yields

$$-\frac{\Delta p}{\Delta L}\frac{r}{2} + \eta \left(\frac{d\bar{u}_z}{dr}\right)_{r=R} = 0, \tag{10.6.3}$$

which combines with Eqn. (10.21) to yield

$$-\frac{\Delta p}{\Delta L}\frac{R}{2} + \eta\bar{u}_{z,\max}\left(\frac{2a_1}{R} + \frac{2ma_2}{R}\right) = 0. \qquad (10.6.4)$$

The flow rate expression provides the third equation needed to determine the three unknowns $a_1$, $a_2$, and $\bar{u}_{z,\max}$.

$$\dot{V} = \int_0^R 2\pi r\bar{u}_z\, dr = 2\pi\bar{u}_{z,\max}\int_0^R r\left[1 + a_1\frac{r}{R} + a_2\left(\frac{r}{R}\right)^{2m}\right] dr$$

$$= \frac{R^2}{2} + a_1\frac{R^2}{3} + a_2\frac{R^2}{2m+2}. \qquad (10.6.5)$$

Equations (10.6.1), (10.6.4), and (10.6.5) determine $a_1$, $a_2$, and $\bar{u}_{z,\max}$. The remaining constant, $m$, is determined by Eqn. (10.22). Alternative relations that may replace some of these equations are

$$\dot{V} = \pi R^2\bar{u}_{z,\text{ave}} \qquad (10.6.6)$$

and

$$\frac{\Delta p}{\Delta L}\frac{R}{2} = \tau_w, \qquad (10.6.7)$$

where $\tau_w$ is the wall shear stress.

### 10.2.5 Laminar Sublayer and Friction Factor

Estimations of the thickness of the laminar sublayer and the resulting friction factor in channel (and pipe) turbulent flows are obtained by matching the velocity profiles of zones, as demonstrated in Examples 10.7 and 10.8.

**Example 10.7: Flow Zones in Turbulent Channel Flow**

As explained in Section 10.2.2, it is difficult to apply a unique velocity profile throughout the cross-sectional area of a channel or a pipe turbulent flow. Instead, the channel width $2D$ is split into zones where a different type of equation applies. With the dimensionless definitions,

$$u^* = \frac{\bar{u}_x}{(\tau_w/\rho)^{1/2}}, \quad y^* = \frac{y(\tau_w/\rho)^{1/2}}{v}, \quad D^* = \frac{D(\tau_w/\rho)^{1/2}}{v}, \quad \delta_i^* = \frac{\delta_i(\tau_w/\rho)^{1/2}}{v},$$

the zones are as follows:

**(i)** Viscous or laminar sublayer: $u^* = y^*$, $0 < y^* < \delta_s^*$

**(ii)** Buffer zone: $u^* = 5\ln y^* - 3.05$, $\delta_s^* < y^* < \delta_b^*$

**(iii)** Constant shear stress zone: $u^* = 2.5\ln y^* + 5.5$, $\delta_b^* < y^* < \delta_c^*$

**(iv)** Turbulent core: $u^* = 8.74(y^*)^{1/7}$, $\delta_c^* < y^* < D^*$

    **(a)** By matching velocities of adjacent zones, calculate the thickenss $\delta_i$ of each zone in terms of the characteristic length $v\sqrt{\rho/\tau_w}$.

    **(b)** What are the resulting thicknesses for $v\sqrt{\rho/\tau_w} = 0.001$, 0.01, and 0.1 cm?

**Solution**

**(a)** Matching of velocity profiles:

    **(i)** *Laminar sublayer.* Matching the velocities of the laminar sublayer and the buffer zone at the common line $y^* = \delta_s^*$ results in

$$\delta_s^* = 5 \ln \delta_s^* - 3.05 \qquad (10.7.1)$$

with solution $\delta_s^* = 5$ and therefore,

$$\delta_s = 5 \frac{\nu}{\sqrt{\tau_w/\rho}}. \qquad (10.7.2)$$

    **(ii)** *Buffer zone.* By the same procedure, the results are

$$5\delta_b^* - 3.05 = 2.5 \ln \delta_b^* + 5.5, \quad \text{thus, } \delta_b^* = 30 \qquad (10.7.3)$$

and therefore,

$$\delta_b = 30 \frac{\nu}{\sqrt{\tau_w/\rho}}. \qquad (10.7.4)$$

    **(iii)** *Constant shear stress zone.* The resulting thickness is

$$\delta_s = 50 \frac{\nu}{\sqrt{\tau_w/\rho}}. \qquad (10.7.5)$$

    **(iv)** *Turbulent core.* The difference between the half-width $D$ and the thicknesses (i) to (iii) is the thickness of the turbulent core:

$$\delta_t = D - [\delta_s + \delta_b + \delta_c] = D - 85\left(\frac{\nu}{\tau_w/\rho}\right). \qquad (10.7.6)$$

**(b)** The resulting values of the thicknesses are:

For $\dfrac{\nu}{\tau_w/\rho} = 0.001$, $\delta_s = 0.005$ cm, $\delta_b = 0.03$ cm, $\delta_c = 0.085$ cm,

and $\delta_t = 14.88$ cm.

For $\dfrac{\nu}{\tau_w/\rho} = 0.01$, $\delta_s = 0.05$ cm, $\delta_b = 0.3$ cm, $\delta_c = 0.85$ cm,

and $\delta_t = 13.8$ cm.

For $\dfrac{\nu}{\tau_w/\rho} = 0.1$, $\delta_s = 0.05$ cm, $\delta_b = 3$ cm, $\delta_c = 8.5$ cm,

and $\delta_t = 3.0$ cm.

Notice that the outer layers expand and the turbulent core shrinks as the factor $\nu/(\tau_w/\rho)$ increases, or equivalently, as the Reynolds number decreases (Example 10.8).

### Example 10.8: Thickness of Laminar Sublayer

Show that the thickness of the laminar sublayer diminishes with the Reynolds number. Is there any relation between the laminar sublayer and the boundary layer? Is this a valid statement for laminar and turbulent flow?

**Solution**  The thickness of the laminar sublayer as given by Eqn. (10.7.1) depends on the ratio $v\sqrt{\rho/\tau_w}$ , that is,

$$\delta_s = cv\sqrt{\frac{\rho}{\tau_w}},$$  (10.8.1)

with $c$ a proportionality constant. The ratio $\tau_w/\rho$ also enters the definition of the friction coefficient [e.g., Eqn. (4.18)]:

$$f = \frac{\tau_w/\rho}{\bar{u}^2/2}.$$  (10.8.2)

The two equations combine to yield

$$\frac{\delta_s}{D} = \frac{c\sqrt{2}}{\mathrm{Re}\,\sqrt{f}}.$$  (10.8.3)

Given that

$$f = 0.0395\,\mathrm{Re}^{-1/4}$$  (10.8.4)

at $\mathrm{Re} > 10^5$ [e.g., Eqn. (4.33)], Eqn. (10.8.3) becomes

$$\frac{\delta_s}{D} = \frac{k}{\mathrm{Re}}.$$  (10.8.5)

Equation (10.8.5) shows that the laminar sublayer shrinks as the Reynolds number increases, as does the boundary layer over a plate. The two situations are similar given the fact that both are dominated by vorticity, which is prevented from spreading away the higher the Reynolds number. Preventing vorticity from diffusing away is equivalent to shrinkage of these two layers. Laminar flow in channels and pipes and creeping Stokes flow around submerged bodies are limiting cases of low Reynolds number, where these layers grow and cover the entire field.

## 10.3 MECHANISMS OF FLUID MIXING

The original experiment by Reynolds on the onset of turbulent flow in a long pipe demonstrates the most elementary mechanism of liquid mixing: In laminar flows, mixing may occur by streamline convection in the flow direction in an accelerating or decelerating flow field, and by diffusion in directions normal to streamlines. In turbulent flows, in addition, mixing is induced by the oscillating velocity. This velocity convects fluid, heat, and solute in all possible directions. The fundamentals of mixing are extremely important in handling, transferring, and processing liquids and are examined in several theoretical[12] and applied[24] publications. Mixing is a process or mechanism that promotes uniformity of concentration of any kind: solute, heat, dispersed particles. In static and flowing gases and low-viscosity liquids, diffusion is the primary cause of mixing. However, in high-viscosity liquids such as polymeric solutions and melts, convection is the dominant mechanism of mixing. Convection in general enhances mixing by increasing the interfacial area between regions of low and high concentration, which in turbulent flow is augmented by the velocity fluctuations. Once such regions are brought in contact by convection, diffusion takes over, finalizing *mixing at the molecular level, where each molecule of solute is surrounded by solvent molecules.*

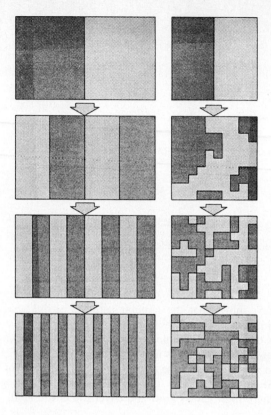

**Figure 10.7** Convective mixing in turbulent and laminar flow. (From Z. Tadmor and C. G. Gogos, *Principles of Polymer Processing,* Wiley, New York, copyright © 1979; reprinted by permission of John Wiley & Sons, Inc.)

Figure 10.7 shows the progression of mixing by convection in laminar and turbulent flow fields. The former, commonly called *distributive mixing,* is due to bulk rearrangement of material by turbulence. The latter, commonly called *streamline or laminar mixing,* is due to the deformation, shear or elongational, induced by a laminar flow. Thus, in laminar flows, convective mixing takes place only if a permanant ongoing deformation or strain is imposed on the fluid, which is not necessarily true of convective mixing in turbulent flow. The patterns shown in Fig. 10.7, achieved by either type of convective mixing, do not constitute stages of complete mixing. However, at these stages there is an extended interfacial area between the black cells of high concentration and the white cells of low concentration, such that diffusion takes over and molecular mixing is achieved, denoted by a gray color. Thus convection enhances molecular mixing in two ways: (1) by building large concentration gradients between black and white cells; and (2) by increasing the interfacial area across whch diffusion occurs.

Heat, as well as any other property associated with liquid molecules, spreads and mixes with its environment by similar mechanisms of diffusion, laminar convection, and turbulence. However, the final stage of mixing always occurs by conduction, whereas the dominant role of convection is to rearrange contacts of regions of different temperature, which is enhanced further by turbulence:

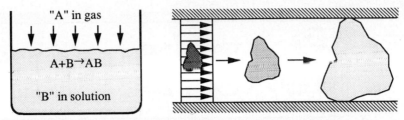

**Figure E10.9.1**  Diffusive mixing in tank reactor and in tubular plug reactor.

Convection and turbulence give rise to temperature gradients, $dT/dx$, by decreasing the distance $dx$ among regions of different temperature. Then conduction takes place, completing heat mixing at the molecular level.

Topics of turbulent mixing by promoted by impinging jets, jet mixing in tanks, micromixing phenomena in stirred reactors, industrial mixing equipment, solid agitation and mixing in aqueous systems, and use of helical ribbon blenders for non-Newtonian materials are some of the applications of these mechanisms and principles.

### Example 10.9: Laminar and Turbulent Mixing in Pipe Flow

Describe the mechanisms of diffusive and convective mixing when blending two viscous liquids.

**Solution**  *Diffusive mixing.* This is the only mechanism of mixing under no-flow or plug-flow conditions, as shown in Fig. E10.9.1. In the tank reactor, the reactant A is absorbed at the interface and then diffuses away (i.e., it mixes with the solution, due to a decreasing concentration from interface to bottom). For the same reason, the produce AB diffuses away from its formation region, spreads throughout, and mixes with the solution. In the tubular plug reactor, due to diffusion, the island of species A spreads out as it advances downstream, while its color weakens until complete mixing occurs, where almost every single molecule of A is surrounded by solvent molecules.

*Convective laminar mixing.* Consider laminar flow in a pipe. An island of a colored species is injected at the inlet. Its particles are big enough that diffusion is negligible but small enough that gravity is also negligible. The configuration of the island as it advances downstream is shown in Fig. E10.9.2. As can be seen, as the carrier liquid travels downstream, the colored island is strained more and more to a degree far downstream, where its continuity breaks and every particle is confined among solvent molecules. Notice also that the mixing in Fig. E10.9.2 is due entirely

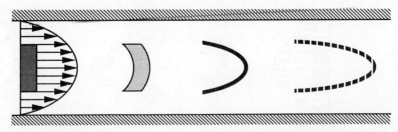

**Figure E10.9.2**  Convective laminar mixing with zero diffusion.

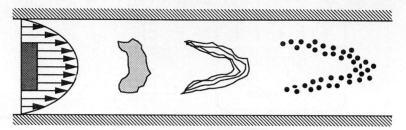

**Figure E10.9.3**   Convective laminar mixing with diffusion.

to the shearing deformation of material elements. When diffusion is not negligible, the situation of Fig. E10.9.2 changes to that of Fig. E10.9.3. Convective mixing due to deformation brings regions of low concentration in contact with regions of high concentration (i.e., it increases the interface of mass exchange). Thus, the potential of diffusion is enhanced, which, when superimposed to the convective macromixing already achieved, drives to complete micromixing. In other words, ideal micromixing occurs only due to diffusion, which is the only mechanism available to dissolve single molecules, and which is therefore the last stage of any kind of mixing.

   *Turbulent convective mixing.* Velocity fluctuations cause regions of high concentration to switch position continually with regions of low concentrations, in addition to the convective mixing of laminar flow, as shown in Fig. E10.9.4.

### 10.3.1 Atmospheric Turbulence and Mixing

Mixing, and therefore dispersion of atmospheric pollutants and heat in the atmosphere, are enhanced significantly by atmospheric turbulence. The more important fluctuations to mixing are on the order of 0.01 cycle per second and are due to either atmospheric heating, which causes density gradients, $d\rho/dz$, and therefore natural convection; or wind shear effects due to velocity gradients, $du/dz$. The first, called *thermal turbulence,* has fluctuations of periods on the order of minutes and prevails on sunny days when light winds occur and the temperature gradient is highly negative. The second, called *mechanical turbulence,* has fluctuations of periods on the order of minutes, and prevails on

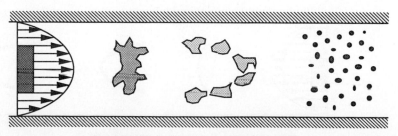

**Figure E10.9.4**   Turbulent convective mixing.

**TABLE 10.2**   TURBULENCE CHARACTERISTICS ACCORDING
TO RICHARDSON NUMBER, Ri

| Richardson number | Turbulent characteristics |
| --- | --- |
| $0.25 < \text{Ri}$ | No vertical mixing |
| $0 < \text{Ri} < 0.25$ | Mechanical turbulence |
| $\text{Ri} = 0$ | Mechanical turbulence |
| $-0.03 < \text{Ri} < 0$ | Mechanical and thermal turbulence |
| $\text{Ri} < -0.04$ | Thermal turbulence and mixing |

*Source:* Ref. 27.

windy nights with a small temperature gradient. The *Richardson number,* defined as[26]

$$\text{Ri} = \frac{(g/T)(dT/dz)}{(du/dz)^2} , \qquad (10.30)$$

where $T$ is the temperature, $u$ the velocity, $z$ the elevation, and $g$ the acceleration due to gravity, characterizes atmospheric turbulence according to Table 10.2. The table indicates that negative Richardson numbers occur with thermal turbulence of weak winds and vertical mixing. Smoke (and pollutants) emitted by a source spreads rapidly, both vertically and laterally, due to the wind and the thermal turbulence, respectively. Dispersion decreases as the Richardson number approaches zero. Positive Richardson numbers occur with mechanical turbulence of horizontal eddies and negligible vertical mixing.

Quantitatively, the time-smoothed square fluctuations introduced in Eqn. (10.30), usually taken over a 1-hour period, can be converted into estimates of the horizontal and vertical parameters, $\sigma_y$ and $\sigma_z$, of the dispersion equation,[28]

$$c(y, z) = k \exp\left[ -\frac{1}{2}\left( \frac{y^2}{\sigma_y^2} + \frac{(z - H)^2}{\sigma_z^2} \right) \right], \qquad (10.31)$$

which governs the distribution of concentration $c$ with altitude $z$ and horizontal distance $y$ from an emitting source at elevation $H$ over ground level (e.g., emission of industrial gases by stack). The concentration of such pollutants at ground level, $c(y, z = 0)$, is obtained by substituting $z = 0$ in Eqn. (10.3). It is easy to show that this concentration is maximum at $y = 0$, where the emitting source stands, and decays to zero at distances $y = 4\sigma_y$ away. Equations similar to Eqn. (10.31) are used to determine environmental pollution by industrial stacks and traveling vehicles.

## 10.4 AGITATION AND MIXING OF FLUIDS

Mixing of two or more fluids (e.g., water with alcohol, gas bubbles with water, monomers for polymerization, polymer melt blending, etc.) to produce a uniform mixture of uniform concentration of each of the involved ingredients is one of the most common operations in fluid processing. Intuitively, mixing is expected to be enhanced by (1) low interfacial tensions(s) that inhibit formation of interfaces, keeping fluids apart; (2) density ratios close to unity that prevent separation by stratification in gravity and centrifugal fields; and (3) low viscosities that promote fluidity, penetration of one ingredient by another, and mixing. Mixing of fluids that fulfill these requirements is nearly spontaneous (e.g., gas mixing, alcohol–water mixing). Mixing of other fluids is not spontaneous (e.g., water–oil, polymer blends, suspensions, and emulsions) and does not proceed without the supply of external power, usuallly in the form of *agitation.*

Supply of external power by agitation promotes mixing in several ways:

1. By mechanically increasing the interfacial area between ingredients, otherwise driven to minimum by interfacial tension. Agitation breaks the continuity of each ingredient and any interface into "pieces" that are brought into multiple contacts by the flow induced by agitators.

2. By increasing fluid circulation in multiple ways; generation of vorticity and strong convection currents; reduction of viscosity of shear-thinning liquids; breaking of yield stress of viscoplastic liquids; reduction of viscosity in general by heating due to viscous dissipation.

3. By reducing the inhibiting effects of large density ratios by appropriate equipment design that allows flexible patterns of recirculation: on vertical planes that work against gravity stratification; and multiple horizontal recirculations that work against radial stratification with centrifugation.

4. Some liquid pairs have high *heat of solution* to overcome, and the power required is provided by mechanical agitation. Similarly, the free energy of an interfacial configuration indicates its stability relative to a homogeneous (well-mixed) fluid, and the differences in free energy provide some clues as to when and how much agitation is necessary.

Figure 10.8 illustrates the fundamental agitation equipment, promoting mixing according to the principles discussed. The process of mixing assisted by supply of external power to a rotating *impeller* (e.g., Fig. 10.8) is called *agitation,* and the corresponding equipment, *agitators.* Common recirculation patterns induced by agitators are shown in Fig. 10.8. These patterns may be altered depending on the nature of the fluids, by appropriate alterations to the impeller, the *baffles,* or both. Some common impeller designs are shown in Table 10.3.

Agitation is also necessary to *stabalize suspensions and emulsions* by inducing circulation and viscous and centrifugal forces that keep particles and droplets apart, preventing coalescence and mixing of the disperse phase. The turbulent velocity fluctuations may both hinder and promote coalescence: Fluctuations bring droplets together; however, if the time required for coalescence is long, the same fluctuations take droplets apart. Therefore, agitators and impellers for this purpose are designed and operate at conditions controlling

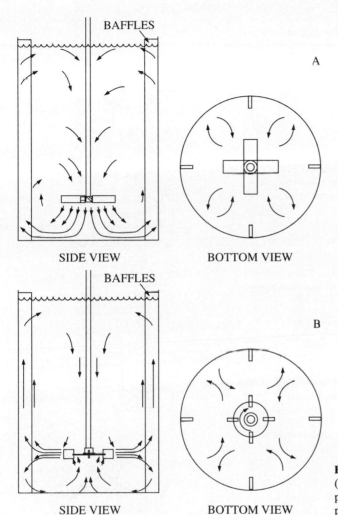

**Figure 10.8** Flow patterns produced by (A) axial flow, and (B) radial flow impellers. (From Refs. 29 and 30, by permission.)

*coalesence frequency.*[30] Profound applications are in suspension and emulsion polymerizations to stabilize the reacting disperse monomer droplets to completion of the degree of desired polymerization, and then to coalescence if desirable for the resulting liquid polymer droplets.[31] Stabilization of suspensions and emulsions is also enhanced by the presence of surfactants containing and keeping droplets apart.[32]

### 10.4.1 Design Equations of Agitation

Experimentation, dimensional analysis, and more recently, computer-aided analysis of the agitation process, are used to construct operating diagrams, as summarized below. These diagrams define the characteristics of the equipment

**TABLE 10.3**  BASIC IMPELLER DESIGNS IN COMMON USE

| Number | Name | Description | |
|--------|------|-------------|---|
| R-1 | Flat blade | Vertical blades bolted to support disk | L=1/4D<br>W=1/5D<br>DISC DIA.=2/3D |
| R-2 | Bar turbine | Six blades bolted/welded to top and bottom of support disk | L=1/4D<br>W=1/2D<br>T=1/2D<br>DISC DIA.=2/3D |
| R-3 | Anchor | Two blades with or without cross arm | W=1/10D |
| A-1 | Propeller three blades | Constant pitch, skewed-back blades | 1.5 PITCH RATIO |
| A-2 | Axial flow four blades | Constant angle at 45° | W=1/5D<br>∡=45° |
| A-3 | Axial flow three blades | Variable blade angle, near constant pitch | BLADE ANGLE AND WIDTH DECREASES HUB TO TIP |
| A-4 | Double spiral | Two helical flights, pitch = $\frac{1}{2}D_0$ | $D_0$=(OUTER)<br>$D_1$=1/3$D_0$<br>W=1/6$D_0$ |

*Source:* Ref. 29, by permission.

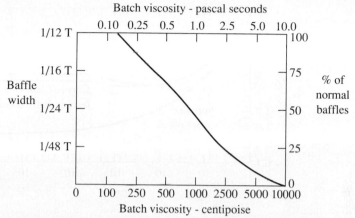

**Figure 10.9** Baffle width as a function of viscosity. (From Ref. 29, by permission.)

required, given the nature of the involved fluids as well as the final goal of agitation.

The most important feature of an impeller is flow and turbulence production, which are functions of both the impeller and the vessel design. A rotating impeller with vertical blades (e.g., $R - 1$ and Fig. 10.8) produces radial flow. Inclining the blades with respect to horizontal (e.g., $A - 2$ and Fig. 10.8) produces axial flow, which is necessary to eliminate vertical stratification and sedimentation or flotation of particles in suspensions or emulsions. Addition of baffles to the vessel produces radial velocity components, which in turn give rise to vertical streams promoting blending. *This combination produces the highest shear rates.*

These flow patterns provide the first indications as to selection of the appropriate combination of vessel–propeller. Figure 10.9 shows baffle-width variations with the viscosity of the mixture, where "normal baffles" refer to low-viscosity agitation, as shown in Table 10.3, under the same conditions of propeller design and rotation.

Due to the complex flow patterns, analytic solutions are impossible and numerical modeling cumbersome and often not featuring all aspects of agitation. Therefore, most knowledge of agitation comes from dimensional analysis (e.g., Section 1.7). Such analyses performed on geometrically similar agitators of fixed vessel diameter $T$ identify as controlling parameters the Reynolds number,

$$\text{Re} = \frac{D^2 N \rho}{\eta}, \tag{10.32}$$

and the power number,

$$N_p = \frac{P}{\rho N^3 D^5}, \tag{10.33}$$

where $D$ is the propeller diameter, $N$ the frequency of rotation, and $P$ the power

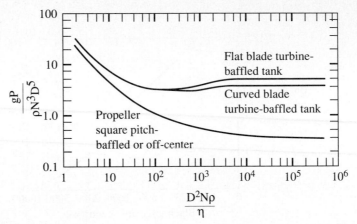

**Figure 10.10**  Power number vs. Reynolds number of agitation. (From Ref. 29, by permission.)

supply. Then appropriate experimentation guided by Eqns. (10.32) and (10.33) unravels the quantitative relation between the two, as shown in Fig. 10.10.

If the impeller is thought of as a (centrifugal) pump that increases the kinetic energy per kilogram of fluid $k/\rho$ N m/kg, flowing at the flow rate $\dot{m}$ kg/s, the power consumed is

$$P = \frac{\hat{E}_k}{\rho}\dot{m} = \hat{E}_k\dot{V}. \qquad (10.34)$$

The flow rate is cast in the form

$$\dot{m} = k_1 N D^3, \qquad (10.35)$$

where $k_1$ is a design parameter characteristic of the impeller design. These equations combine with Fig. 10.10 to provide one of the most important characteristics of an agitation equipment: The ratio of its "pumping capacity" $\dot{V}$ (i.e., at what rate fluid is processed) to the kinetic energy of the processed fluid $\hat{E}_k/\rho$ for a given amount of external power $P$ follows.

The curves of Fig. 10.10 can be divided into laminar, transition, and turbulent agitations. Under turbulent conditions, the power depends only on the impeller design, i.e.,

$$P = k_2 N^3 D^5, \qquad (10.36)$$

which combines with Eqns. (10.34) and (10.35) to yield the ratio

$$\frac{\dot{V}}{\hat{E}_k/\rho} = k_3 D^{8/3}, \qquad Re > 10^3. \qquad (10.37)$$

In the laminar and transition regions, the same ratio varies between $D^{9/3}$ and

$D^{7/3}$. These equations show that propellers with large diameters are to be preferred for applications that demand high pumping and delivery rates. Small diameters are best for required high velocities.

**Scaleup of agitation runs.**    Due to the many uncertainties entering in an agitation process, extensive experimentation is required at a pilot scale before installation at full scale is recommended. For scaling-up pilot-scale results to actual scale process, the vessel diameter $T$ must be considered as an additional variable. Then the resulting equation of dynamic similarity, obtained by dimensional analysis is

$$\frac{gP}{\rho N^3 D^5} = k \frac{ND^2\rho}{\eta} \frac{N^2 D}{g} \frac{D}{T}, \tag{10.38}$$

which includes Eqns. (10.32) and (10.33) for constant $T$. The fluid characteristic usually cannot be changed from pilot to full scale, and therefore $P$, $N$, $D$, and $T$ are the scalable variables, according to the relation

$$\frac{P_T T_p}{D_p^9 N_p^5} = \frac{P_f T_f}{D_f^p N_f^5}, \tag{10.39}$$

where the subscripts $p$ and $f$ denote pilot and full scales, respectively. According to Eqn. (10.39), if $P_p$ is power consumed with a vessel of diameter $T_p$ and an impeller of diameter $D_p$ rotating at frequency $N_p$ at pilot scale, the power required for a full-scale agitation of corresponding variables $T_f$, $D_f$, and $N_f$ will be

$$P_f = \frac{D_f^9 N_f^5 T_p}{D_p^9 N_p^5 T_f} P_p = (\text{SF})P_p. \tag{10.40}$$

In Eqn. (10.40), SF is a known *scaling factor*.

It should be noted that the summary provided in this section for agitation is rather generalized, mostly qualitative material aimed primarily at the principles of agitation. Detailed information on industrial mixing equipment can be found in a diversity of publications dealing explicitly with the theory, design, and operation of general-purpose as well as specialized agitators [e.g., Refs. 12, 24, 29, 33].

**Example 10.10: Design and Scaleup of Agitation Unit**

An agitation unit is to be designed to mix viscous liquids of viscosities around 1000 cP and densities around $1 \text{ g/cm}^3$, at low frequency and shear rates such that heating by viscous dissipation is avoided. This is achieved by design as well as by a low power supply not exceeding $P_{max} = 2.5 \times 10^{-4} \text{ W}$. The unit must process liquid at rate $\dot{V} = 1 \text{ m}^3/\text{min}$ in a vessel of diameter $T = 1$ m.

(a) By utilizing Eqns. (10.30) and (10.31), or the diagram of Fig. 10.10, select an appropriate propeller–vessel combination and calculate the maximum allowed frequency $N$.

(b) After selection of a system, you want to make sure that it will indeed perform as required. To find out, pilot-scale tests were done with the same liquid and with a similar propeller–vessel system of 50% the size of the actual one. These tests showed that with the same frequency $N$, the power consumed was $P_p = 5 \times 10^{-7}$ W. What conclusions can be drawn for the actual agitation unit?

**Solution**

(a) *Selection of large-scale unit.* The sensitivity of the liquids to heating requires a vessel with small baffles. From Fig. 10.9, the baffle width must be approximately

$$W_b = \frac{T}{24} \approx 4.2 \text{ cm.} \tag{10.10.1}$$

Also, to avoid high shear rates and heating, a flat-blade propeller or turbine of characteristic curve shown in Fig. 10.10 is selected, with diameter

$$D = \frac{T}{3} = 0.33 \text{ m.} \tag{10.10.2}$$

To determine the frequency $N$, Eqns. (10.32) to (10.35) are not helpful since the values of the design parameters are not known at this stage. Direct use of Fig. 10.10 is more convenient as follows: Assume an initial $N^{(0)} = 0.5$ Hz. The Reynolds number defined by Eqn. (10.30) is $\text{Re}^{(0)} = 33^2 \times 0.5/10 = 50$. The power number of a flat-blade turbine corresponding to this Reynolds number is found from Fig. E10.10 to be $Np^{(0)} = 38$. For this power number, Eqn. (10.33) predicts $N^{(1)} = 0.128$ Hz. Since $N^{(1)} \neq N(0)$, a new iteration starts with $N^{(1)} = 0.128$ and by the same sequence of calculations it is found that $\text{Re}^{(1)} = 128$, $Np^{(1)} = 7$, and $N^{(2)} = 0.107$ rps. A third iteration starting with $M^{(2)} = 0.107$ Hz ends with $N^{(3)} = 0.107$ Hz. Thus, the maximum allowed frequency is

$$N = 0.1 \text{ Hz} = 6 \text{ rpm.} \tag{10.10.3}$$

(b) *Pilot-scale tests and scaleup.* Equation (10.40) applies here with $T_p = 0.5T_f$,

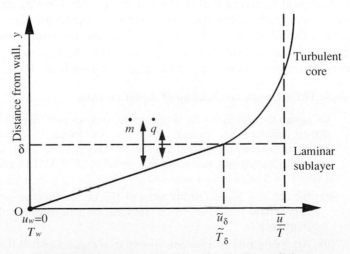

**Figure E10.11** Reynolds analogy for transfer coefficients.

$D_p = 0.5D_F$, and $N_p = N_f$. Thus,

$$P_f = [2^9 \times 1^5 \times 0.5]P_p = 512P_p = 512 \times 5 \times 10^{-7}\,\mathrm{W} = 2.56 \times 10^{-4}\,\mathrm{W},$$

(10.10.4)

which is about the allowed maximum power supply. Thus the chosen combination of vessel–propeller is appropriate.

## 10.5 ANALOGIES BETWEEN TRANSFER COEFFICIENTS

The analogies between momentum, heat, and mass transfer played an important role in the development of mass and heat transfer theories and mechanisms of turbulent flow. Furthermore, heat and mass transfer coefficients may be predicted from pressure drop data, which is relatively easy to measure. The existence of these analogies was first recognized by Reynolds in the form of a relationship between heat transfer and skin friction, which was refined further by Prandtl.

For the system of Fig. (10.11), relations for the heat flux and wall shear stress under laminar flow conditions are

$$\dot{q} = \dot{m}c_p(\bar{T} - T_W)$$

(10.41)

and

$$\tau_W = \frac{\dot{m}}{A}(\bar{u} = 0),$$

(10.42)

where $\dot{q}$, $\dot{m}$, and $\tau_W$ are heat, mass, and momentum fluxes, respectively, across area $A$. The mass flux $\dot{m}$ is eliminated between the two equations, resulting in

$$\frac{\dot{q}}{c_p(\bar{T} - T_W)} = A\frac{\tau_W}{\bar{u}}$$

(10.43)

Given that also,

$$\dot{q} = hA(\bar{T} - T_W),$$

(10.44)

where $h$ is a heat transfer coefficient, Eqn. (10.45) reduces to

$$\frac{h}{\rho\bar{u}c_p} = \frac{\tau_W}{\rho\bar{u}^2}.$$

(10.45)

Therefore,

$$\mathrm{St} = \frac{f}{2},$$

(10.46)

where St is the Stanton dimensionless number for heat transfer and $f$ is the familiar friction coefficient.

**TABLE 10.4**  VON KÁRMÁN'S EXTENSION OF PRANDTL–TAYLOR'S ANALOGY

| Laminar sublayer | Buffer zone | Turbulent core |
|---|---|---|
| $\dot{q} = -k \dfrac{dT}{dy}$ | $= -\rho c_p (\alpha - \varepsilon_T) \dfrac{d\tilde{T}}{dy}$ | $= \rho c_p (\tilde{T} - T_w)$ |
| $\dot{m}_c = -\mathscr{D} \dfrac{dc}{dy}$ | $= -(\mathscr{D} + \varepsilon_s) \dfrac{d\tilde{c}}{dy}$ | $= \tilde{u}(\tilde{c} - c_w)$ |
| $\tau_w = \eta \dfrac{du}{dy}$ | $= \rho(v + \varepsilon) \dfrac{d\tilde{u}}{dy}$ | $= \rho \tilde{u}$ |

In this way, transfer coefficients of momentum, heat, and solute are interrelated. This is extremely important in transport phenomena, because measurement of one of those often allows determination of the rest. Also, fluid behavior and solution of problems of different transfer are often similar [e.g., (momentum) boundary layer vs. thermal boundary layer vs. solute boundary layer], and conclusions from one can be utilized to study the rest.

The general approach to the problem of obtaining a relation between rates of momentum and of heat and mass transfer in turbulent flow is to obtain the eddy viscosity as a function of the distance from the wall. Next, a relationship between the eddy viscosity and the eddy diffusivity is formed. The latter, combined with the heat or mass flux equations and integrated, leads to these analogies. The combined equations are shown in Table 10.4.

In this way, transfer coefficients of momentum, heat, and solute are interrelated. This is extremely important in transport phenomena, because measurement of one of those often allows determination of the rest. Also, fluid behavior and solution of problems of different transfer are often similar [e.g., (momentum) boundary layer vs. thermal boundary layer vs. solute boundary layer], and conclusions from one can be utilized to study the rest. Table 10.4 indicates similarities among laminar transfer coefficients of heat, $k$, solute $\mathscr{D}$, and momentum $\eta$, as well as among turbulent transfer coefficients, which are the eddy viscosity $\varepsilon$, the eddy diffusivity $\varepsilon_s$, and the eddy thermal diffusivity $\alpha_T = \varepsilon_T \rho c_p$.

Further discussion on these analogies and interrelations among momentum, heat, and solute transfer is outside the scope of this book. A systematic approach to these derivations with engineering applications can be found elsewhere.[34]

## PROBLEMS

1. *Turbulent flow equations in cylindrical coordinates.* Derive the time-smoothed equations for turbulent flow in pipe with respect to the cylindrical system of coordinates. Identify the resulting Reynolds stresses and express them in terms of time-smoothed velocity derivatives.

2. *Eddy viscosity.* By assuming a logarithmic velocity distribution law, determine the

ratio of eddy viscosity to the laminar viscosity, $\eta^{(t)}/\eta$, of water of density $\rho = 62.4\,\text{lb}_m/\text{ft}^3$ and viscosity $v = 1.1 \times 10^{-5}\,\text{ft}^2/\text{s}$, flowing in a pipe of diameter $d = 4\,\text{in.}$ and wall stress $\tau_w = 10^{-6}\,\text{lb}_f/\text{in.}^2$. What is the pressure drop and the resulting flow rate under these conditions?

3. *Transition to turbulent flow.* A Newtonian liquid of viscosity $\eta = 1\,\text{cP}$ and density $\rho = 1\,\text{g/cm}^3$ flows in a horizontal pipe of internal diameter $d = 4\,\text{in.}$ Plot the curve of the pressure drop $\Delta p/\Delta L$ vs. the Reynolds number as the flow rate changes from zero to infinity. Identify the pressure drop of transition to turbulence.

4. *Laminar, transition, and turbulent flow.* For Reynolds numbers between $10^4$ and $10^5$, it has been shown experimentally that the time-smoothed velocity in a pipe of radius $R$ can be approximated by

$$\tilde{u}_z = \tilde{u}_{z,\text{max}}\left(1 - \frac{r}{R}\right)^{1/7}$$

and

$$\frac{\bar{u}}{\bar{u}_{z,\text{max}}} = \frac{4}{5}.$$

(a) Find the corresponding relations for laminar flow.
(b) Plot the pressure drop, $\Delta p/\Delta L$, for laminar and turbulent flow as a function of the flow rate and the Reynolds number.
(c) Identify the transition from laminar to turbulent flow.
(d) Construct qualitative velocity profiles for turbulent vs. laminar flow and comment on their form.

5. *Universal velocity profile.* Pai's universal velocity profiles by Eqn. (10.21) are often used to estimate the mixing length assumed or predicted by the theories of Section 10.2.2. For flow of water at flow rate $0.1\,\text{m}^3/\text{s}$ in a pipe of $10\,\text{cm}$ radius, the values of the parameters involved are $m = 0.33$, $a = -0.338$, and $a_2 = -0.662$.
(a) Calculate the ratio $\bar{u}/\bar{u}_{z,\text{max}}$.
(b) Calculate the time-smoothed velocity and its first and second derivatives at the point $r/R = 0.5$.
(c) Which of the parameters of Eqns. (10.17) to (10.20) can be estimated at the solid wall?
(d) How could you calculate the same parameters away from the wall?

6. *Turbulent flow in pipe.* Calculate and plot velocity distributions in water of density $\rho = 1\,\text{g/cm}^3$ and viscosity $\eta = 1\,\text{cP}$, flowing in a pipe of internal diameter $d = 10\,\text{cm}$ at total flow rate $\dot{m} = 3.14\,\text{kg/s}$ as driven by a pressure gradient of $1\,\text{dyn/cm}^3$, by means of:
(a) The PML and KSH theories, which yield a logarithmic velocity profile $\tilde{u}_z = A + B\ln(R - r)$.
(c) The DFW theory, which yields $\tilde{u}_z = A(R - r) + B$.
(d) Pai's formula, $\tilde{u}_z/\tilde{u}_{z,\text{max}} = 1 + a_1(r/R)^2 + a_2(r/R)^{2m}$, with $a_1 = -0.338$, $a = -0.662$, and $m = 0.33$.

By assuming that Pai's formula is accurate throughout the cross section and that the maximum velocities of the three expressions are identical, evaluate the performance of the first two equations and the portion of the cross section where they apply.

7. *Turbulent pipe flow.* Consider turbulent flow of water of viscosity $\eta = 0.01\,\text{cP}$ and density $\rho = 1\,\text{g/cm}^3$ in a pipe of internal diameter $d = 4\,\text{in.}$ driven by a pressure gradient of $0.8\,\text{dyn/cm}^3$.

**(a)** What is the wall shear stress?

**(b)** By assuming a logarithmic velocity distribution

$$\bar{u}_z = 2.5\left(-\frac{\tau_w}{\rho}\right)^{1/2} \ln\left[\frac{(-\tau_w/\rho)^{1/2}}{\nu}(R-r)\right],$$

what is the resulting thickness of the laminar sublayer?

**(c)** What is the velocity profile of the laminar sublayer?

**(d)** What is the resulting total flow rate in the laminar sublayer and the rest of the cross section?

**(e)** What is the relation between the flow rate and the pressure drop compared to the linear relation

$$\dot{V} = \frac{\pi}{8\eta}\frac{\Delta p}{\Delta L}R^4$$

of laminar flow?

**8.** *Turbulent velocity profile.* Comment on the potential of the expression

$$\bar{u}_z = A\{1 - \exp[-n(R-r)]\}$$

to fit turbulent velocity profiles in channels and pipes. What is the significance of parameters $A$ and $n$? How would you estimate these parameters? Can this velocity profile be used over the entire cross section?

**9.** *Pressure drop vs. flow rate.* By using the integrated momentum equation for turbulent flow in channels and pipes, show that the linear relation between flow rate and pressure drop cannot hold in turbulent flow. Use selected velocity profiles of turbulent flow (e.g., Problem 10.4) to show that the relation is indeed nonlinear.

**10.** *Stability of laminar flow.* Consider steady laminar flow in a channel of width $2H$ governed by the equation

$$0 = -\frac{dp}{dx} + \eta\frac{d^2u_x}{dy^2},$$

which is the steady, fully developed flow version of the transient equation,

$$\rho\left(\frac{\partial u_x}{\partial t} + u_x\frac{\partial u_x}{\partial x} + u_y\frac{\partial u_x}{\partial y}\right) = -\frac{\Delta p}{\Delta L} + \eta\left[\frac{\partial^2 u_x}{\partial y^2} + \frac{\partial^2 u_x}{\partial x^2}\right].$$

Examine the stability of the flow with respect to infinitesimal velocity disturbances of the form

$$u'_x(y, t) = \varepsilon \sin\left[\left(n + \frac{1}{2}\right)\frac{\pi y}{H}\right]e^{cn t}, \qquad u'_y(y, t) = p'(y, t) = 0$$

under the idealized assumptions that the pressure and normal velocity $u_y$ are not disturbed. Explain why the flow appears to be stable to any disturbance that conforms to these conditions.

**11.** *Stability of laminar flow.* Repeat Problem 10.10 with the additional feature that the pressure is also disturbed according to

$$p'(x, t) = -\left(\frac{\eta\pi}{H}\right)^2 \varepsilon x \sin\left[\left(n + \frac{1}{2}\right)\frac{\pi y}{H}\right]e^{cn t}.$$

**12.** *Turbulent velocity profile.* By assuming the following velocity distribution in a pipe,

$$\bar{u}_z = \bar{u}_{z,\max}\left(1 - \frac{r}{R}\right)^n,$$

calculate
   **(a)** the resulting flow rate,
   **(b)** the average velocity $\bar{u}$ and its ratio to $\bar{u}_{z,\max}$,
   **(c)** the radius where the actual velocity is equal to the average velocity, and
   **(d)** the shear stress distribution and the wall shear stress.

**13.** *Logarithmic velocity profile.* The logarithmic velocity profile

$$\frac{\bar{u}_z}{u^*} = \frac{\bar{u}_{z,\max}}{u^*} + \frac{1}{k}\ln\frac{r}{R}, \qquad u^* = \left(\frac{\tau_W}{\rho}\right)^{1/2},$$

where $\tau_W$ is the wall shear stress, identical to $(\Delta p/\Delta L)(R/2)$, and $\bar{u}_{z,\max}$ is the maximum time-smoothed velocity, applies to turbulent flow outside the laminar sublayer of thickness $\delta$ adjacent to the wall.
   **(a)** What is the flow rate outside the laminar sublayer, and how does it relate to the pressure drop, $\Delta p/\Delta L$?
   **(b)** What is the relation between the maximum and the average velocity?

**14.** *Reynolds stresses.* Use Prandtl's one-seventh power law,

$$\frac{\bar{u}_z}{\bar{u}_{z,\max}} = \left(1 - \frac{y}{H}\right)^{1/7},$$

to find approximate expressions for mixing-length distributions for Prandtl's hypothesis,

$$\bar{\tau}_{ij}^{(t)} = \rho\, l^2 \left|\frac{d\bar{u}_x}{dy}\right|\frac{d\bar{u}_x}{dy},$$

with $l = k(H - y)$ and with $l = k(d\bar{u}_x/dy)/(d^2\bar{u}_x/dy^2)$, which are the mixing lengths according to Prandtl and von Kármán, respectively. Assume that the shear stress is everywhere constant, identical to that at the wall and of the laminar sublayer of thickness $\delta$.

**15.** *Transfer analogies.* Consider the equations of Table 10.4.
   **(a)** Justify their form, check consistency of dimensionality, and comment on the physical meaning of the involved transfer coefficients.
   **(b)** By following the procedure summarized in Eqns. (10.41) to (10.46), find relations between the Reynolds number and the friction coefficient and dimensionless numbers controlling heat and solute transfer (e.g., Appendix E).

**16.** *Agitator functioning.*
   **(a)** Justify the patterns of flow shown in Fig. 10.8. Which of the two designs is better for suspension, emulsion or solution of biological cells?
   **(b)** Justify the curves of Figs 10.9 and 10.10.
   **(c)** Justify the similarity of the curves of Fig. 10.10 to the Moody diagram for friction presented in Fig. 4.3.

**17.** Agitator scaleup. A pilot-scale agitator runs in a vessel of diameter $T = 1$ ft filled to height $H = 2$ ft with the actual industrial liquid agitated by a propeller of diameter $D = 10$ cm. The relation $P_p = 0.01N_p$, with $N_p$ in rpm and $P_p$ in watts, was found. What conclusions can be drawn for full-scale agitation in a tank of diameter $T_f = 1$ m, height $H = 2\,\text{m}$, and propeller diameter $D_f = 0.33\,\text{m}$? How would things change for an industrial liquid of viscosity 10 times that of the pilot-scale liquid?

18. *Agitation frequency.* Use Fig. 10.10 to solve the following indirect agitation problem: A Newtonian mixture of viscosity $\eta = 1\,\text{cP}$ and density $\rho = 1\,\text{g/cm}^3$ is stirred in the three types of agitator and turbines of Fig. 10.10, with a common turbine–propeller of diameter $D = 20\,\text{cm}$. What angular velocity is achieved by each of them under $P = 0.01\,\text{W}$ power supply? Justify your answer. Which is the most appropriate type for solid–particle suspension, for emulsion stabilization, and for bubble suspension? What would the required power be if the angular velocities were to be doubled?

19. *Propeller design.* Position the curves corresponding to the following propeller blade designs on the diagram of Fig. 10.10:
   (a) cylindrical blade turbine;
   (b) spherical blade turbine;
   (c) flat-blade propeller;
   (d) vertical blade propeller.

20. *Turbulent agitation.* Explain in which ways turbulence promotes or hinders the following agitation processes:
   (a) mixing of gases;
   (b) mixing of low viscosity liquids;
   (c) mixing of highly viscous liquids;
   (d) mixing of gas with liquid;
   (e) suspension stabilization;
   (f) emulsion stabilization;
   (g) bubble suspension stabilization;
   (h) heat transfer;
   (i) polymerization reaction;
   (j) biological cell suspension;
   (k) powder solution in viscous liquid.

21. *Turbulent gas flow.*[35] Turbulent flow of gas in porous medium is modeled by the Forscheimer equation,

$$-\frac{dp}{dx} = \frac{\eta}{k} u + \rho \beta u^2,$$

   where $\eta$ is the viscosity, $k$ the permeability, $\rho$ the density, and $\beta$ the turbulent factor.
   (a) What is the significance of the last term? What is the limit $\rho \beta u^2 \to 0$?
   (b) Solve for the velocity $u$ and justify the resulting expression on physical ground.
   (c) What is $u$ for radial sink flow; how does it relate to the corresponding laminar form?

22. *Wall boundary condition of turbulent flow.*[36] In numerical computations of channel (and pipe) turbulent flow, the domain of laminar sublayer to the wall is supressed, being replaced by a boundary condition to the remaining turbulent core domain. One of the used boundary conditions is

$$\tau_w = \frac{u_s}{y_s} \frac{\eta y_s^*}{\ln(c_1 y_s^*)}, \qquad y_s^* = c_2 k_s^{1/2} y_s, \qquad \int_0^{y_s} \cdot\, dy = c_3 k_s^{3/2} \ln(c_1 y_s^*),$$

   where the $c_i$ are empirical constants and $(u_s, y_s)$ parallel velocity and normal distance from the wall of the line $y = s$ that is the boundary of the numerical domain.
   (a) Justify the use and the physical significance of the boundary condition.
   (b) What are the implied kinematics and stresses in the omitted laminar sublayer?

(c) Why is this omission necessary with numerical calculations?

(d) Can you propose a better boundary condition?

# REFERENCES

1. O. Reynolds, An experimental investigation of the circumstances which determine whether the action of water shall be direct or sinuous, and of the law of resistance in parallel channels, *Trans. Roy. Soc. (London)*, A174, 123 (1883).

2. G. I. Taylor, *Mécanique de la turbulence*, Editions du Centre National de la Recherche Scientifique, Paris, 1962.

3. S. J. Davies and C. H. White, An experimental study of the flow of water in pipes of rectangular section, *Proc. Roy. Soc. (London)*, 119A, 92 (1928).

4. H. L. Dryden, 'A review of the statistical theory of turbulence,' *Quart. Appl. Math.*, 1, 7 (1943).

5. G. Comte-Bellot, 'Hot wire anemometry,' *Annual Rev. Fluid Mech.*, 8, 209 (1976).

6. P. Buchhave, W. K. George, and J. L. Lumley, 'The measurement of turbulence with the LDA,' *Annual Rev. Fluid Mech.*, 11, 443 (1979).

7. M. Miya, D. E. Woodmansee, and T. J. Hanranty, 'A model for roll waves in gas–liquid flow,' *Chem. Engrg. Sci.*, 26, 1915 (1971).

8. S. Corrsin, Investigation of flow in an axially asymmetric heated jet-air, *NACA Rep. 3L23*, 1943.

9. G. I. Taylor, Stability of a viscous liquid contained between two rotating cylinders, *Philos. Trans. Roy. Soc. (London)*, 223A, 289 (1923).

10. R. S. Brodkey, *The Phenomena of Fluid Motions*, Addison-Wesley Series in Chem. Eng., Ohio State University, Columbus, Ohio, 1967.

11. J. O. Hinze, *Turbulence*, McGraw-Hill, New York, 1959.

12. J. M. Ottino, *The Kinematics of Mixing: Stretching, Chaos and Transport*, Cambridge University Press, Cambridge, 1989.

13. R. L. Panton, *Incompressible Flow*, Wiley, New York, 1984.

14. P. G. Drazin and W. H. Reid, *Hydrodynamic Stability*, Cambridge University Press, New York, 1981.

15. R. Jordinson, The flat plate boundary layer: Part 1, Numerical integration of the Orr–Somerfeld equation, *J. Fluid Mech.*, 43, 801 (1970).

16. T. V. Boussinesq, *Mem. Pres. Acad. Sci.*, 3rd ed., Paris, 23, 46 (1877).

17. L. Prandtl, Uber die augebildete Turbulenz, *Z. Angew. Math. Mech.*, 5, 136 (1925).

18. H. Schlichting, *Boundary Layer Theory*, McGraw-Hill, New York, 1955.

19. T. von Kármán, *Collected Works*, Butterworth, London, 1956.

20. G. I. Taylor, *Proc. Roy. Soc. London*, A135, 685 (1932).

21. R. G. Deissler, Theory of decaying homogeneous turbulence, *NACA Rep. 1210*, 1955.

22. S.-I. Pai, *Viscous Flow Theory*, Vol. II, D. Van Nostrand, Princeton, N.J., 1957, p. 41.

23. J. Nikuradse, Flow laws in rough tubes, *VDI Forschungsh.*, 356 (1932).

24. J. Y. Oldshue, *Fluid Mixing Technology* McGraw-Hill, New York, 1983.

25. Z. Tadmor and C. G. Gogos, *Principles of Polymer Processing*, Wiley, New York, 1979).

26. I. Prandtl, *Essentials of Fluid Dynamics*, Hafner, Publishing Co., New York, 1949.

27. O. G. Sutton, *Micrometeorology: A Study of Physical Processes in the Lowest Layers of Earth's Atmosphere*, McGraw-Hill, New York, 1953.

28. K. Wark and C. F. Warner, *Air Pollution: Its Origin and Control,* Harper & Row, New York, 1976.

29. J. Y. Oldshue, Industrial mixing equipment, in *Encyclopedia of Fluid Mechanics,* Vol. 2, *Dynamics of Single-fluid Flows and Mixing* (ed. N. P. Cheremisinoff), Gulf Publishing, Houston, Tex., 1986.

30. D. H. Napper, *Polymeric Stabilization of colloidal Dispersions,* Academic Press, London, 1983.

31. E. G. Chatzi, A. Gavrielides, C. Boutris, and C. Kiparissides, On-line monitoring of drop size distribution in agitated vessels, *Ind. Engrg. Chem. Res.,* 28, 1704 (1989); 30, 536 (1991).

32. D. A. Edwards, H. Brenner, and D. T. Wesen, *Interfacial Transport Processes and Rheology,* Butterworth-Heinemann, Boston, 1991.

33. R. H. Perry, *Perry's Chemical Engineer's Handbook,* 6th ed., McGraw-Hill, New York, 1984.

34. C. T. Geankoplis, *Transport Processes and Unit Operations,* Prentice Hall, Englewood Cliffs, N.J., 1983.

35. R. A. Wattenbarger and H. J. Ramey, Jr., Gas well testing with turbulence, damage and wellbore storage, *Trans. AIME,* 243, 877 (1968).

36. B. E. Launder and D. B. Spalding, *Mathematical Models of Turbulence,* Academic Press, London, 1972.

# Appendix A
## *Fluid Mechanics Mathematics*

## A.1 SCALARS, VECTORS, AND SYSTEMS OF COORDINATES

A *scalar* quantity or variable or function is fully defined by a numerical value accompanied by an appropriate unit of measurement. Such scalars in fluid mechanics are the temperature, the pressure, the density, the viscosity, the surface tension, the volume, the mass, and so on. For example, no further information than the statement "the fluid pressure at the center of a horizontal pipe is 2 atm" is required to define a unique fluid pressure there.

A *vector* quantity or variable or function is not fully defined by its numerical value and unit of measurement alone. A *direction* with respect to a *frame of reference* is required in addition. Indeed, the statement that "the velocity of a flowing fluid at the center of the pipe is 1 m/s" does not uniquely define the velocity there, since there may be several possibilities of such velocity in all possible directions. A unique velocity is defined by the statement "the velocity at the center of the pipe is 1 m/s, parallel to the wall, and from left to right (with respect to an *observer or reference frame*) fixed at the wall."

Vectors of the three-dimensional space are fully defined if described with respect to a frame of reference made up by an origin point and three mutually perpendicular directions around the origin. By induction, vectors of a two-dimensional space or plane, are fully defined with respect to a frame of reference made up by an origin and two mutually perpendicular directions; and vectors of a

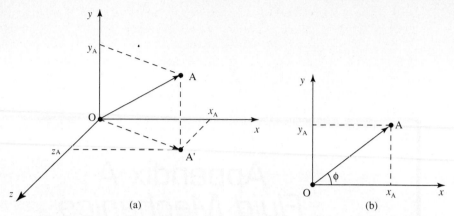

**Figure A.1** Cartesian system of coordinates in three- and two-dimensional spaces.

one-dimensional space, or line, are fully defined by an origin and a unique direction along the line.

The mathematical representation of the position and direction of a vector with respect to a reference frame yields systems of coordinates. The simplest and most commonly used is the Cartesian system of coordinates, shown in Fig. A.1 for three- and two-dimensional spaces, respectively.

Consider the vector $\mathbf{v} = OA$ of the two-dimensional space of Fig. A.1. Its direction is fully defined by its two projections or components, $x_A$ and $y_A$, onto the two mutually perpendicular directions, or axes $x$ and $y$, departing from the origin $O$. In other words, given the magnitudes $Ox_A$ and $Oy_A$, and directions $+$ or $-$, of the two components, the vector $\mathbf{v}$ is constructed by the intersection of the projecting lines $Ax_A$ with $Ay_A$. By following the reverse procedure, any given vector $\mathbf{v}$ can be decomposed to its unique components with respect to a specified system of coordinates. That is why vectors are customarily introduced by specifying their components in an appropriate way: For the example of the pipe flow cited earlier, the statement "parallel to the wall, and from left to right" defines the magnitude and direction of velocity with respect to a coordinate system fixed at the wall and with its $x$-axis parallel to the wall. By the same statement, all other velocity components are zero, and the velocity in this case is identical to its $x$-component.

These geometric relations between the vector $\mathbf{v}$ and its components $v_x$ and $v_y$ are represented by (in reference to Fig. A.1b)

$$|\mathbf{v}| \equiv v = \sqrt{v_x^2 + v_y^2} \tag{A.1}$$

$$\tan \phi = \frac{v_y}{v_x} \tag{A.2}$$

$$v_x = v \cos \phi \tag{A.3}$$

$$v_y = v \sin \phi. \tag{A.4}$$

If the vector $\mathbf{v}$ represents fluid velocity, the amount of fluid content transferred by convection through a line or surface perpendicular to the $x$-axis is proportional to $v_x$ alone, since $v_y$ directs tangentially and never passes through the line or surface. Similarly, any vector force $\mathbf{F}$ results in a normal force (component) $F_x$ onto a surface perpendicular to the $x$-axis, and in a shear force (component), $F_y$, onto the same surface. These force components divided by the area of the surface yield the corresponding normal, $\tau_{xx}$, and shear, $\tau_{xy}$, stresses on that $x$-surface: The first stress subscript characterizes the surface and the second the direction of action. By the same arguments, the same force results in a normal force $F_y$, normal stress $\tau_{yy}$, shear force $F_x$, and shear stress $\tau_{yx}$, on a $y$-surface that is perpendicular to the $y$-axis.

Position $\mathbf{r}$ is also a vector, determined by its components or projections onto the reference directions. The point $A$ on the plane $xOy$ of Fig. A.1b is fully determined by the projections or distances of its position vector $OA$ from the two axes $Ox_A$ and $Oy_A$. Thus, the statment "a point at distance $y_A$ from the $x$-axis and distance $x_A$ from the $y$-axis" defines point $A$ on the plane. The magnitude of the position vector is the distance of the point from the origin, which is not a vector:

$$|\mathbf{r}| = r = \sqrt{r_x^2 + r_y^2} = r(\cos^2 \phi + \sin^2 \phi) = r. \tag{A.5}$$

Position (but not distance), velocity, force, normal and shear stress, acceleration, momentum, vorticity, and torque are important vectors in fluid mechanics. Their study is facilitated and done with respect to each of the fundamental directions, by means of their component. For example, at equilibrium the summations of forces and torques in each of the three fundamental directions vanishes, as demonstrated in Chapter 3. Similarly, the equations of motion of fluids are derived by considering momentum balances in each of the three fundamental directions, as demonstrated in Chapter 5.

Choices of coordinate systems other than the Cartesian are possible and convenient for situations that conform better to a curvilinear than a rectangular geometry. Such systems are the cylindrical and its two-dimensional analog polar, and the spherical system of coordinates. These two important systems are shown in Fig. A.2. In the cylindrical system, the position of a point is fully defined by its

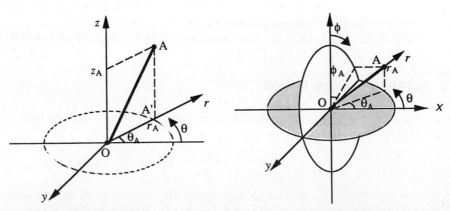

**Figure A.2** Cylindrical and spherical system of coordinates, sharing the same origin with a Cartesian system of coordinates.

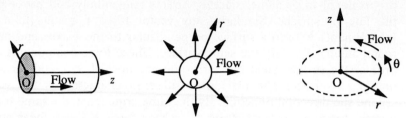

**Figure A.3** Convenient choices of cylindrical system of coordinates to analyze pipe (a), radial (b), and torsional (c) flows.

distance from an axis (the $z$-axis in Fig. A.2), by its elevation from a reference plane (the circle in Fig. A.2), and by its angular deviation from a reference line (the $\theta$-line in Fig. A.2) on the plane of the circle that is perpendicular to the $z$-axis. In two dimensions, the $z$-axis is eliminated to yield the polar system of coordinates.

The cylindrical system is placed in the most convenient configuration, the one that most facilitates analysis of the flow. For flow in pipes, the cylindrical system has its $z$-axis parallel to the flow and its base circle coincident with the cross section of the pipe. For radial sink or source flows and for torsional flows, the base circle is parallel to the flow, which is directed toward the $z$-axis (for radial flows) or around the $z$-axis (for torsional flows). These convenient choices are shown in Fig. A.3 and demonstrated in Example 5.9.

In the spherical system of coordinates, the position of a point is fully defined by its distance from the center of a sphere and by its angular deviations from two mutually perpendicular reference planes through the center of the sphere, as shown in Fig. A.2b.

A vector **v** is decomposed and represented with respect to any of the three coordinate systems, as

$$\mathbf{v} = \mathbf{e}_1 v_1 + \mathbf{e}_2 v_2 + \mathbf{e}_3 v_3. \tag{A.6}$$

Here $v_1$, $v_2$, and $v_3$ are the components or projections on the three mutually perpendicular directions defined by the directional unit vectors $\mathbf{e}_1$, $\mathbf{e}_2$, and $\mathbf{e}_3$, as shown in Fig. A.4 for the Cartesian system and tabulated in Table A.1.

## A.2 DIFFERENTIAL CALCULUS

A function $f(x)$ is differentiable with derivative $f'(x) = df/dx$, at a point $x$, if the limit

$$\lim_{\Delta x \to 0} \frac{f(x + \Delta x) - f(x)}{\Delta x} = \frac{df}{dx} = f'(x) \tag{A.7}$$

exists. Thus, the derivative represents the variation of the dependent variable, $f(x)$, per variation of the independent variable, $x$. The geometric analog of differentiation is shown in Fig. A.5.

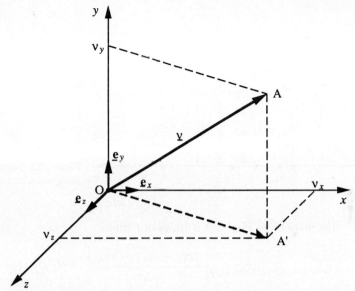

**Figure A.4**   Decomposition of vector **v** into components $v_x$, $v_y$, and $v_z$.

**TABLE A.1**  CHARACTERISTICS OF ORTHOGONAL SYSTEMS
OF COORDINATES

|  | Cartesian | Cylindrical | Spherical |
|---|---|---|---|
| Position coordinates | $x, y, z$ | $r, \theta, z$ | $r, \theta, \phi$ |
| Convenient for flows | Channel | Pipe | Spherical |
|  | Film | Radial |  |
|  |  | Torsional |  |
| Vector, **v**, components | $v_x, v_y, v_z$ | $v_r, v_\theta, v_z$ | $v_r, v_\theta, v_\phi$ |
| Directions, $\mathbf{e}_i$, vectors | $\mathbf{e}_x, \mathbf{e}_y, \mathbf{e}_z$ | $\mathbf{e}_r, \mathbf{e}_\theta, \mathbf{e}_z$ | $\mathbf{e}_r, \mathbf{e}_\theta, \mathbf{e}_\phi$ |

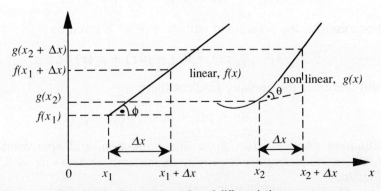

**Figure A.5**  Geometric analog of differentiation.

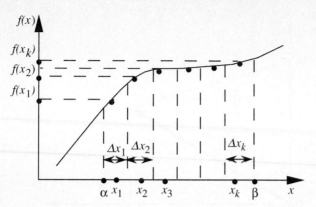

**Figure A.6** Geometric analog of integration.

By inspecting Fig. A.6 it turns out that

$$\tan \phi = \frac{f(x_1 + \Delta x) - f(x_1)}{\Delta x} = f'(x_1)\big|_{x=x_1} \tag{A.8}$$

and

$$\tan \theta \approx \frac{g(x_2 + \Delta x) - g(x_2)}{\Delta x} = \lim_{\Delta x \to 0} \frac{g(x_2 + \Delta x) - g(x_2)}{\Delta x} = g'(x)\big|_{x=x_2}. \tag{A.9}$$

Thus, the derivative is identical to the tangent of the angle formed between the graph of the function and the direction of increasing independent variable. The practical significance of Eqns. (A.8) and (A.9), and therefore of the derivative, too, is that they can be used to find an estimate of the value of a function at a point $x + \Delta x$, given its value $f(x)$ at a neighboring point $x$. Indeed, by these equations,

$$f(x + \Delta x) \approx f(x) + f'(x)\,\Delta x, \tag{A.10}$$

and in the limit of $\Delta x \to dx \to 0$,

$$f(x + dx) = f(x) + f'(x)\,dx. \tag{A.11}$$

For example, the pressure at altitude $z + dz$ is given by

$$p(z + dz) = p(z) + p'(z)\,dz, \tag{A.12}$$

and therefore the pressure gradient there is

$$dp = p(z + dz) - p(z) = p'(z)\,dz. \tag{A.13}$$

Equation (A.9) is exact for a linear function, and approximate for nonlinear functions with accuracy improving in the limit of $\Delta x \to dx \to 0$, as illustrated by Fig. A.5.

Derivatives of common functions are tabulated in math handbooks and need not be recalculated each time a case arise. For example,

$$\frac{d}{dx}(x^2 + \alpha x) = 2x + \alpha,$$

$$\frac{d}{dx}(\sin \beta x) = \beta \cos \beta x, \tag{A.14}$$

$$\frac{d}{dx}[f(x)g(x)] = f(x)\frac{dg(x)}{dx} + g(x)\frac{df(x)}{dx},$$

and so on. Higher-order derivatives are obtained by consecutive differentiations:

$$\frac{d^2}{dx^2}(x^2 + \alpha x) = \frac{d}{dx}\left[\frac{d}{dx}(x^2 + \alpha)\right] = \frac{d}{dx}(2x + \alpha) = 2,$$

$$\frac{d^2}{dx^2}(\sin \beta x) = \frac{d}{dx}(\beta \cos \beta x) = -\beta^2 \sin \beta x. \tag{A.15}$$

## A.3 INTEGRAL CALCULUS

The integral of a function $f(x)$ over an interval $\alpha \le x \le \beta$ is defined by

$$I = \int_\alpha^\beta f(x)\, dx = \lim_{\Delta x_i} \sum_{i=1}^k f(x_i)\, \Delta x_i, \tag{A.16}$$

where $\Delta x_i$ are $k$ subdivisions of the interval such that

$$\sum_{i=1}^k \Delta x_i = \beta - \alpha, \tag{A.17}$$

and $x_i$ is at the center of $\Delta x_i$, as illustrated in Fig. A.6. It is obvious from Fig. A.6 that the integral $I$ is also the area between the $x$-axis and the graph of $f(x)$, bounded by the straight lines $x = \alpha$ and $x = \beta$, respectively.

By combining Eqns. (A.7) and (A.16) it results in

$$\int_0^x \left[\frac{d}{dx}[f(x)]\right] dx = \lim_{\Delta x_i \to 0} \sum_{i=1}^k \left[\frac{df}{dx}\bigg|_{x=x_i}\right] \Delta x_i$$

$$= \lim_{\Delta x_i} \sum_{i=1}^k \frac{1}{\Delta x_i}\left[f\left(x_i + \frac{\Delta x_i}{2}\right) - f\left(x_i - \frac{\Delta x_i}{2}\right)\right] \Delta x_i = f(x) - f(0). \tag{A.18}$$

Therfore, integration is the reverse of differentiation, and in fact the integrals of the last terms of Eqn. (A.14) are the corresponding parentheses of their first terms. Integrals of commonly used functions are tabulated in standard math handbooks and need not be recomputed each time they arise.

## A.4 PARTIAL DIFFERENTIAL AND INTEGRAL CALCULUS

A function of two independent variables, $f(x, y)$, has partial derivatives with respect to each of them:

$$\left(\frac{\partial f}{\partial x}\right) = \frac{df}{dx}\bigg|_{y=\text{const.}} = \lim_{dx\to 0}\frac{1}{dx}[f(x + dx, y) - f(x, y)]$$

and

$$\frac{\partial f}{\partial y} = \frac{df}{dx}\bigg|_{x=\text{const.}} = \lim_{dy\to 0}\frac{1}{dy}[f(x, y + dy) - f(x, y)]. \qquad (A.19)$$

In other words, a partial derivative with respect to one of the independent variables is identical to the common derivative, by considering the rest of the independent variables as constants. Thus,

$$\frac{\partial}{\partial x}(x^2 + xy) = 2x + y,$$

$$\frac{\partial}{\partial y}[\sin(xy)] = x\cos(xy), \qquad (A.20)$$

$$\frac{\partial}{\partial z}(xf(y)z) = xf(y),$$

and so on. The total derivative of a multivariable function is given by the summation of its partial derivatives,

$$\frac{d}{dx}f(x, y, z) = df(x, y, z) = \frac{\partial f}{\partial x}dx + \frac{\partial f}{\partial y}dy + \frac{\partial f}{\partial z}dz, \qquad (A.21)$$

and represents the variation of the function per distance $ds = \sqrt{(dx)^2 + (dy)^2 + (dz)^2}$ of the three-dimensional space.

The same guidelines apply to integration of a function of two or more variables, with respect to any of them, by treating the rest as constants:

$$I(y) = \int_\alpha^\beta (2x + y)\, dx = (x^2 + yx)\big|_\alpha^\beta = (\beta^2 - \alpha^2) + y(\beta - \alpha)$$

$$I(x) = \int_\gamma^\delta (2x + y)\, dy = \left(2xy + \frac{y^2}{2}\right)\bigg|_\gamma^\delta = 2x(\delta - \gamma) + \tfrac{1}{2}(\delta^2 + \gamma^2)$$

$$\qquad\qquad\qquad\qquad\qquad\qquad\qquad\qquad\qquad\qquad (A.22)$$

$$I = \int_\gamma^\delta \int_\alpha^\beta (2x + y)\, dx\, dy = \int_\gamma^\delta I(y)\, dy = \int_\alpha^\beta I(x)\, dx$$

$$= (\beta^2 - \alpha^2)(\delta - \gamma) + \left(\frac{\delta - \gamma}{2}\right)(\beta - \alpha).$$

## A.5 PHYSICAL PROCESSES AND DIFFERENTIAL EQUATIONS

A differential equation is a relation between a function, any of its derivatives, and any of its independent variables, for example,

$$\frac{\partial^2 f}{\partial x^2} + \frac{\partial^2 f}{\partial y^2} + \frac{\partial f}{\partial y} + f(x, y) + x^2 + y^2 = \sin(x + y). \tag{A.23}$$

Equation (A.23) is a partial differential equation (PDE), because it involves derivatives with respect to more than one independent variable. An ordinary differential equation involves derivatives with respect to one independent variable, for example,

$$\frac{d^2}{dx^2} - 3\frac{df}{dx} + 2f = 0. \tag{A.24}$$

Methods of solution of many, but not all, PDEs and ODEs are tabulated in math handbooks. These sources provide the general solution to a differential equation. For the ODE considered here, the general solution is

$$f(x) = c_1 c^{2x} + c_2 e^x, \tag{A.25}$$

where $c_1$ and $c_2$ are arbitrary constants of integration. The general solution satisfies the differential equation no matter what the values of $c_1$ and $c_2$ are.

The general solution is not very useful, and in fact $f(x)$ given by Eqn. (A.25) cannot be evaluated for a given value of $x$, which would be the ultimate goal in seeking the solution to Eqn. (A.24). What is sought is the particular solution that satisfies not only the differential equation alone, but certain boundary or initial conditions that must accompany a differential equation in order to make a well-posed mathematical problem, as summarized below.

Consider the differential equation,

$$\frac{d^2 u}{dy^2} = 0, \qquad 0 < y < H, \tag{A.26}$$

with the two required boundary conditions

$$u(y = 0) = 0 \tag{A.27}$$

and

$$u(y = H) = V \tag{A.28}$$

that make a well-posed mathematical problem. Equation (A.26) governs the velocity distribution with respect to $y$, in any horizontally flowing incompressible fluid, in the $x$-direction, under uniform pressure. There may be many such flow situations, and this fact is reflected by the form of the general solution,

$$u(y) = c_1 y + c_2, \tag{A.29}$$

which may represent these different flow situations, depending on the value of $c_1$

and $c_2$. What the boundary conditions by Eqns. (A.27) and (A.28) do is to single out the particular solution that corresponds to the unique flow situation they represent: horizontal, incompressible flow between two parallel horizontal plates, spaced a distance $y = H$ apart, the lower of which is stationary and the upper traveling horizontally with velocity $V$, as depicted in Fig. 6.2.

The unique solution sought to the system of differential equation (A.26) *and* boundary conditions equations (A.27) and (A.28) is obtained by requiring that the general solution Eqn. (A.29) that satisfies the differential equation satisfy the two boundry conditions as well. Thus,

$$u(y = 0) = c_1 0 + c_2 = 0 \qquad\qquad (A.30)$$

and

$$u(y = H) = c_1 H + c_2 = V, \qquad\qquad (A.31)$$

which determine the specific values,

$$c_1 = 0, \qquad c_2 = \frac{V}{H}. \qquad\qquad (A.32)$$

These values are substituted in the general solution, which produces the particular solution:

$$u(y) = \frac{V}{H} y. \qquad\qquad (A.33)$$

It is easy to check that the particular solution satisfies both the differential equation and the boundary conditions.

In steady-state flows, a differential equation describes how a variable or a function—for example, velocity, pressure, stress—at any point in the interior of a fluid and away from its boundaries is influenced by its adjacent continuum, through fluxes represented by the derivatives involved. The boundary conditions describe how the fluid interacts with its surroundings at its natural boundaries. Both kinds of information are required to provide a complete picture of the flow. In time-dependent or transient flows, the differential equation involves time derivatives that describe how a current variable or a function interacts with its earlier or later states, through fluxes represented by these time derivatives. Then the accompanied initial conditions describe a reference state of the entire fluid, from which the fluid evolves with time.

In most cases there is a relation between the time-dependent and the steady solution to a flow situation. Consider the partial differential equation

$$\rho \frac{\partial u_x}{\partial t} = \eta \frac{\partial^2 u_x}{\partial y^2}, \qquad t > 0, \quad 0 \le y \le H \qquad\qquad (A.34)$$

with the initial condition

$$u_x(t = 0, y) = 0 \qquad\qquad (A.35)$$

and the boundary conditions

$$u_x(t > 0, y = 0) = 0, \qquad u_x(t \geq 0, y = H) = V, \qquad (A.36)$$

which describe how a fluid at rest in time $t$, i.e., $u_x(t = 0, y)$, between two stationary plates at distance $H$ apart, starts to flow at times $t > 0$, when the upper plate at distance $y = H$ from the $x$-axis (chosen to coincide with lower plate of $y = 0$) is set to motion with constant velocity $V$ for times $t > 0$. The solution of Eqns. (A.34) to (A.35) yields a velocity,

$$u_x = f(t, y), \qquad (A.37)$$

which describes the developing fluid velocity at each position $y$ at any time $t$ under the constraints $u_x(t, y = 0) = f(t, 0) = 0$ and $u_x(t, y = H) = f(t, H) = V$. For large time, or as $t \to \infty$, the fluid attains a steady velocity given by

$$u_x(t \to \infty) = f(t \to \infty, y) = g(y) = u_x(y) \qquad (A.38)$$

with $u_x(y = 0) = g(y = 0)$ and $u_x(y = H) = g(y = H) = V$. The solution $u_x(t \to \infty) = u_x(y)$ is also found by solving the steady-state equations, obtained from Eqns. (A.34) to (A.36) by dropping the time derivative in Eqn. (A.34) and the entire initial condition, Eqn. (A.35):

$$\frac{\partial^2 u_x}{\partial y^2} = 0, \qquad 0 \leq y \leq H \qquad (A.39)$$

$$u_x(y = 0) = 0, \qquad u(y = H) = V \qquad (A.40)$$

The solution of Eqns. (A.39) and (A.40) yields directly the steady velocity $u_x(y)$. Thus the steady-state solution is obtained directly and easily by solving the steady-state equations (A.39) and (A.40), or by solving the time-dependent Eqns. (A.34) to (A.36), and then taking the long time limit $u_x(t \to \infty, y)$. In other words, the long time limit $u_x(t \to \infty, y)$ of the time-dependent solution $u_x(t, y)$ is the steady-state solution $u_x(y)$, *if indeed there is such a steady solution*: There are flow situations that never reach a steady solution (e.g., turbulent flow), and others where the velocity grows with time to infinitely large values (e.g., shock waves).

# Appendix B
## Units and Conversions

| Property | S.I. Unit | C.G.S. Unit | C.G.S. SI units equivalent | English Unit | English SI units equivalent | Common Unit | Common SI units equivalent |
|---|---|---|---|---|---|---|---|
| **Fundamental** | | | | | | | |
| Length | m | cm | $10^{-2}$ | ft | 0.305 | mile | 1609 |
| Mass | kg | g | $10^{-2}$ | lb | 0.454 | ton (2000 lb) | 907.2 |
| Time | s | s | 1 | s | 1 | min | 60 |
| Temperature | K | °C | 1 | °F | 1.8 | °C | 1 |
| | | K | 1 | °R | 1.8 | °F | 1.8 |
| **Secondary** | | | | | | | |
| Area | $m^2$ | $cm^2$ | $10^{-4}$ | $ft^2$ | 0.093 | acre | 4047 |
| Volume | $m^3$ | $cm^3$ | $10^{-6}$ | $ft^3$ | 0.028 | U.S. gallon | $3.785 \times 10^{-3}$ |
| Velocity | m/s | cm/s | $10^{-2}$ | ft/s | 0.305 | mile/h | 0.445 |
| Density | $kg/m^3$ | $g/cm^3$ | $10^{-3}$ | $lb/ft^3$ | 16.02 | $slug/ft^3$ | 515.4 |
| Momentum | kg m/s | g m/s | $10^{-5}$ | $lb_f$ s | 4.448 | N · m | 1 |
| Force | N | dyn | $10^{-5}$ | $lb_f$ | 4.448 | poundal | 0.138 |
| Frequency | Hz | Hz | 1 | $s^{-1}$ | 1 | rpm | 0.0166 |
| Surface tension | N/M | dyn/cm | $10^{-3}$ | $lb_f/ft$ | 14.59 | Pa · m | 1 |
| Pressure | Pa | $\mu$bar | $10^{-1}$ | $lb_f/ft^2$ | 47.78 | atm | $1.013 \times 10^5$ |
| Stress | Pa | $\mu$bar | $10^{-1}$ | $lb_f/ft^2$ | 47.78 | psi | $6.895 \times 10^3$ |
| Viscosity | Pa s | poise | $10^{-1}$ | $lb_f s/ft^2$ | 47.78 | cP | $10^{-3}$ |
| Kin. viscosity | $m^2/s$ | stoke | $10^{-4}$ | $ft^2/s$ | 0.093 | cS | $10^{-6}$ |
| Work | J | erg | $10^{-7}$ | Btu | 1054 | kWh | $3.6 \times 10^3$ |
| Power | W | erg/s | $10^{-7}$ | Btu/s | 1055 | hp | 745.7 |

*Temperature scales:* $T(K) = T(°C) + 273.15 = 0.555[T(°F) + 459.67] = 0.555T'(°R)$.

*Ideal gas constant:* $R = 8.314 \, J/(mol\,K) = 8.314 \, (N \cdot m)/(mol\,K) = 8.314 \, (Pa \cdot m^3)/(mol\,K) = 0.083 \, (bar \cdot m^3)/(kmol \cdot K)$.

*Gravity acceleration:* $g = 9.81 \, m/s^2 = 32.17 \, ft/s^2 = 7.9 \times 10^4 \, miles/h^2 = 981 \, cm/s^2$.

*Speed of sound:* In air: $C \simeq 344 \, m/s = 0.214 \, mile/s$; in water: $c = 1484 \, m/s = 0.922 \, mile/s$; in soil: $c \simeq 600 \, m/s = 0.373$ mile/s; in steel: $c \simeq 1525 \, m/s = 0.948 \, mile/s$.

*Speed of light:* $c_l = 2.998 \times 10^8 \, m/s = 1.87 \times 10^5 \, mile/s = 187 \, mile/ms$.

# Appendix C
## *Physical and Mathematical Constants*

| Constant | Symbol | Value | Application |
|---|---|---|---|
| Natural base | $e$ | 2.7183 | Science, mathematics |
| Natural log | ln 10 | 2.3026 | Science, mathematics |
| Radian unit | $\pi$ | 3.1416 | Science, mathematics, angular kinematics |
| Speed of light in air | $c_l$ | $2.998 \times 10^8$ m/s | Optics, relativity theory |
| Speed of sound in air | $c$ | 340 m/s | Acoustics, compressible flow |
| Ideal gas constant | $R$ | $1.987 \dfrac{\text{cal g}}{\text{mol K}}$ | Ideal gas law |
| | | $82.05 \dfrac{\text{cm}^3 \text{ atm}}{\text{mol K}}$ | Thermodynamics |
| | | $8.314 \dfrac{\text{N m}}{\text{mol K}} = \dfrac{\text{J}}{\text{mol K}}$ | Gas mechanics |
| | | $1.544 \dfrac{\text{ft lb}_f}{\text{lbmol } °\text{R}}$ | Heat-work machines |
| Gravity acceleration | $g$ | 981 cm/s$^2$ | Mechanics |
| | | 32.17 ft/s$^2$ | Kinematics |
| Joule's constant | $\alpha$ | $788 \dfrac{\text{ft lb}_f}{\text{Btu}}$ | Thermodynamics |
| | | 4.184 J/cal | Heat–work machines |
| Planck's constant | $h$ | $6.62 \times 10^{-27} \dfrac{\text{erg}}{\text{s}}$ | Quantum mechanics |
| Avogadro's number | $N_A$ | $6.02 \times 10^{23} \dfrac{\text{molecules}}{\text{mol}}$ | Chemistry, thermodynamics |
| Boltzmann's constnat | $k$ | $R/N_A$ | Statistical mechanics |

# Appendix D
## Some Property Values of Natural Matter

| Property | Matter | p = 0.1 atm | | p = 1 atm | | | p = 10 atm | |
|---|---|---|---|---|---|---|---|---|
| | | 4°C | 20°C | 4°C | 20°C | 40°C | 20°C | 40°C |
| Molecular weight | Air | 29 | 29 | 29 | 29 | 29 | 29 | 29 |
| | Water | 18 | 18 | 18 | 18 | 18 | 18 | 18 |
| Density (kg/m$^3$) | Air | 0.129 | 0.120 | 1.29 | 1.20 | 1.13 | 12 | 11.3 |
| | Water | 1000 | 998 | 1000 | 998 | 992 | 998 | 992 |
| | Soil | | | 2500 | 2500 | 2500 | | |
| Viscosity (cP) | Air | 0.0158 | 0.0175 | 0.0165 | 0.0181 | 0.0195 | 0.0184 | 0.0198 |
| | Water | 1.792 | 1.001 | 1.792 | 1.002 | 0.656 | 1.002 | 0.657 |
| Surface tension with air (dyn/cm) | Air | 0 | 0 | 0 | 0 | 0 | 0 | 0 |
| | Water | 75.6 | 73 | 75.6 | 73 | 69.6 | 73 | 69.6 |
| Elasticity modulus ($10^9$ Pa) | Air | | $2.2 \times 10^5$ | | $2.2 \times 10^6$ | | $2.2 \times 10^7$ | |
| | Water | | | 2.02 | 2.10 | 2.28 | | |
| | Soil | | | | 0.02 | | | |
| Speed of sound (m/s) | Air | 332 | 344 | 332 | 344 | 355 | 344 | 355 |
| | Water | | | 1407 | 1484 | 1528 | | |
| | Soil | | | | 600 | | | |
| Permeability of water (darcy) | Air | ∞ | ∞ | ∞ | ∞ | ∞ | ∞ | ∞ |
| | Water | 0 − ∞ | 0 − ∞ | 0 − ∞ | 0 − ∞ | 0 − ∞ | 0 − ∞ | 0 − ∞ |
| | Soil | | | | 0.001–0.1 | | | |

# Appendix E
# Dimensionless Numbers in Fluid Mechanics

| Name (page) | Number definition | | Physical significance | |
|---|---|---|---|---|
| | Definition | Variables | Ratio | Application |
| Euler (p. 33) | $Eu = \dfrac{\Delta p}{\rho u^2}$ | $\Delta p$ = pressure drop<br>$u$ = velocity<br>$\rho$ = density | $\dfrac{\text{frictional loss}}{\text{kinetic energy}}$ | Friction in pipes<br>Potential flow |
| Fanning (p. 147) | $f_F = \dfrac{D(\Delta p/\rho)}{2u^2 L}$ | $D$ = pipe diameter<br>$\Delta p/\rho$ = pressure drop<br>$L$ = length of flow | $\dfrac{\text{frictional loss}}{\text{kinetic energy}}$ | Friction in pipes<br>Friction in channels |
| Froude (p. 37) | $Fr = \dfrac{u^2}{gL^2}$ | As above | $\dfrac{\text{inertia force}}{\text{gravity force}}$ | Surface waves<br>Flows under gravity |
| Mach (p. 193) | $\mathcal{M} = \dfrac{u}{c}$ | $u$ = fluid velocity<br>$c$ = speed of sound | $\dfrac{\text{fluid velocity}}{\text{sound speed}}$ | Sonic phenomena<br>Compressible flows |
| Reynolds (p. 37) | $Re = \dfrac{\rho D u}{\eta}$ | As above<br>$D$ = diameter<br>$\eta$ = viscosity | $\dfrac{\text{inertia force}}{\text{viscous force}}$ | Any fluid flow |
| Weber (p. 30) | $We = \dfrac{u^2 \rho D}{\sigma}$ | As above | $\dfrac{\text{inertia force}}{\text{capillary force}}$ | Atomization<br>Liquid droplets<br>Gas bubbles<br>Liquid meniscous |
| Capillary (p. 33) | $Ca = \dfrac{\eta u}{\sigma}$ | As above | $\dfrac{\text{viscous force}}{\text{capillary force}}$ | Interface flows<br>Thin-film flows |
| Stokes (p. 387) | $St = \dfrac{\eta u}{\rho g D^2} = \dfrac{Re}{Fr}$ | As above<br>$Re$ = Reynolds number<br>$Fr$ = Froude number | $\dfrac{\text{viscous force}}{\text{gravity force}}$ | Flows under gravity<br>Natural convection |
| Deborah (p. 421) | $De = \dfrac{\lambda}{L/\bar{u}}$ | $\lambda$ = relaxation time<br>$L/\bar{u}$ = residence time | $\dfrac{\text{relaxation period}}{\text{residence period}}$ | Viscoelastic flow<br>Polymer processing |
| Weissenberg (p. 421) | $Ws = \dfrac{\tau_{11} - \tau_{22}}{\tau_{12}}$ | $\tau_{11}, \tau_{22}$ = normal stress<br>$\tau_{12}$ = shear stress | $\dfrac{\text{elastic force}}{\text{viscous force}}$ | Viscoelastic flow<br>Rod climbing<br>Extrudate swell |

# Appendix F
## Stress Components of Newtonian Fluid

**TABLE F**  VISCOUS STRESS COMPONENTS FOR INCOMPRESSIBLE NEWTONIAN FLUID: $\sum\limits_{i=1}^{3} \tau_{ii} = 0$

| System coordinates | Shear stress component | Normal stress component |
|---|---|---|
| Cartesian | $\tau_{yx} = \tau_{xy} = \eta\left(\dfrac{\partial u_x}{\partial y} + \dfrac{\partial u_y}{\partial x}\right)$ | $\tau_{xx} = 2\eta\,\dfrac{\partial u_x}{\partial x}$ |
|  | $\tau_{zx} = \tau_{xz} = \eta\left(\dfrac{\partial u_x}{\partial z} + \dfrac{\partial u_z}{\partial x}\right)$ | $\tau_{yy} = 2\eta\,\dfrac{\partial u_y}{\partial y}$ |
|  | $\tau_{zy} = \tau_{yz} = \eta\left(\dfrac{\partial u_y}{\partial z} + \dfrac{\partial u_z}{\partial y}\right)$ | $\tau_{zz} = 2\eta\,\dfrac{\partial u_z}{\partial z}$ |
| Cylindrical | $\tau_{zr} = \tau_{rz} = \eta\left(\dfrac{\partial u_z}{\partial r} + \dfrac{\partial u_r}{\partial z}\right)$ | $\tau_{zz} = 2\eta\,\dfrac{\partial u_z}{\partial z}$ |
|  | $\tau_{r\theta} = \tau_{\theta r} = \eta\left[r\dfrac{\partial}{\partial r}\left(\dfrac{u_\theta}{r}\right) + \dfrac{1}{r}\dfrac{\partial u_r}{\partial \theta}\right]$ | $\tau_{rr} = 2\eta\,\dfrac{\partial u_r}{\partial r}$ |
|  | $\tau_{\theta z} = \tau_{z\theta} = \eta\left(\dfrac{\partial u_\theta}{\partial z} + \dfrac{1}{r}\dfrac{\partial u_z}{\partial \theta}\right)$ | $\tau_{\theta\theta} = 2\eta\left[\dfrac{u_r}{r} + \dfrac{1}{r}\dfrac{\partial u_\theta}{\partial \theta}\right]$ |
| Spherical | $\tau_{r\theta} = \tau_{\theta r} = \eta\left(\dfrac{\partial u_\theta}{\partial r} + \dfrac{1}{r}\dfrac{\partial u_r}{\partial \theta} - \dfrac{u_\theta}{r}\right)$ | $\tau_{rr} = 2\eta\,\dfrac{\partial u_r}{\partial r}$ |
|  | $\tau_{\theta\phi} = \tau_{\phi\theta} = \eta\left[\dfrac{\sin\theta}{r}\dfrac{\partial}{\partial \theta}\left(\dfrac{u_\phi}{\sin\theta}\right) + \dfrac{1}{r\sin\theta}\dfrac{\partial u_\theta}{\partial \phi}\right]$ | $\tau_{\theta\theta} = 2\eta\left(\dfrac{1}{r}\dfrac{\partial u_\theta}{\partial \theta} + \dfrac{u_r}{r}\right)$ |
|  | $\tau_{\phi r} = \tau_{r\phi} = \eta\left[\dfrac{1}{r\sin\theta}\dfrac{\partial u_r}{\partial \phi} + r\dfrac{\partial}{\partial r}\left(\dfrac{u_\phi}{r}\right)\right]$ | $\tau_{\phi\phi} = 2\eta\left(\dfrac{u_r}{r} + \dfrac{u_\theta}{r}\cot\theta + \dfrac{1}{r\sin\theta}\dfrac{\partial u_\theta}{\partial \phi}\right)$ |

# Appendix G
## Navier–Stokes Equations

**TABLE G.1** CONSERVATION EQUATIONS FOR AN INCOMPRESSIBLE NEWTONIAN FLUID IN CARTESIAN COORDINATES

Continuity

$$\frac{\partial u_x}{\partial x} + \frac{\partial u_y}{\partial y} + \frac{\partial u_z}{\partial z} = 0 \tag{G.1}$$

Monentum equations in terms of stresses

  *x*-component

$$\rho\left(\frac{\partial u_x}{\partial t} + u_x\frac{\partial u_x}{\partial x} + u_y\frac{\partial u_x}{\partial y} + u_z\frac{\partial u_x}{\partial z}\right) = -\frac{\partial p}{\partial x} + \frac{\partial \tau_{xx}}{\partial x} + \frac{\partial \tau_{yx}}{\partial y} + \frac{\partial \tau_{zx}}{\partial z} + \rho g_x \tag{G.2}$$

  *y*-component

$$\rho\left(\frac{\partial u_y}{\partial t} + u_x\frac{\partial u_y}{\partial x} + u_y\frac{\partial u_y}{\partial y} + u_z\frac{\partial u_y}{\partial z}\right) = -\frac{\partial p}{\partial y} + \frac{\partial \tau_{xy}}{\partial x} + \frac{\partial \tau_{yy}}{\partial y} + \frac{\partial \tau_{zy}}{\partial z} + \rho g_y \tag{G.3}$$

  *z*-component

$$\rho\left(\frac{\partial u_z}{\partial t} + u_x\frac{\partial u_z}{\partial x} + u_y\frac{\partial u_z}{\partial y} + u_z\frac{\partial u_z}{\partial z}\right) = -\frac{\partial p}{\partial z} + \frac{\partial \tau_{xz}}{\partial x} + \frac{\partial \tau_{yz}}{\partial y} + \frac{\partial \tau_{zz}}{\partial z} + \rho g_z \tag{G.4}$$

Navier–Stokes equations for a Newtonian fluid

  *x*-component

$$\rho\left(\frac{\partial u_x}{\partial t} + u_x\frac{\partial u_x}{\partial x} + u_y\frac{\partial u_x}{\partial y} + u_z\frac{\partial u_x}{\partial z}\right) = -\frac{\partial p}{\partial x} + \eta\left(\frac{\partial^2 u_x}{\partial x^2} + \frac{\partial^2 u_x}{\partial y^2} + \frac{\partial^2 u_x}{\partial z^2}\right) + \rho g_z \tag{G.5}$$

  *y*-component

$$\rho\left(\frac{\partial u_y}{\partial t} + u_x\frac{\partial u_y}{\partial x} + u_y\frac{\partial u_y}{\partial y} + u_z\frac{\partial u_y}{\partial z}\right) = -\frac{\partial p}{\partial y} + \eta\left(\frac{\partial^2 u_y}{\partial x^2} + \frac{\partial^2 u_y}{\partial y^2} + \frac{\partial^2 u_y}{\partial z^2}\right) + \rho g_y \tag{G.6}$$

  *z*-component

$$\rho\left(\frac{\partial u_z}{\partial t} + u_x\frac{\partial u_z}{\partial x} + u_y\frac{\partial u_z}{\partial y} + u_z\frac{\partial u_z}{\partial z}\right) = -\frac{\partial p}{\partial z} + \eta\left(\frac{\partial^2 u_z}{\partial x^2} + \frac{\partial^2 u_z}{\partial y^2} + \frac{\partial^2 u_z}{\partial z^2}\right) + \rho g_z \tag{G.7}$$

---

**TABLE G.2** CONSERVATION EQUATIONS FOR AN INCOMPRESSIBLE NEWTONIAN FLUID IN CYLINDRICAL COORDINATES

Continuity

$$\frac{1}{r}\frac{\partial}{\partial r}(ru_r) + \frac{1}{r}\frac{\partial u_\theta}{\partial \theta} + \frac{\partial u_z}{\partial z} = 0 \tag{G.8}$$

Momentum equations in terms of stresses

 $r$-component

$$\rho\left(\frac{\partial u_r}{\partial t} + u_r\frac{\partial u_r}{\partial r} + \frac{u_\theta}{r}\frac{\partial u_r}{\partial \theta} - \frac{u_\theta^2}{r} + u_z\frac{\partial u_r}{\partial z}\right)$$

$$= -\frac{\partial p}{\partial r} + \frac{1}{r}\frac{\partial}{\partial r}(r\tau_{rr}) + \frac{1}{r}\frac{\partial \tau_{r\theta}}{\partial \theta} - \frac{\tau_{\theta\theta}}{r} + \frac{\partial \tau_{rz}}{\partial z} + \rho g_r \tag{G.9}$$

 $\theta$-component

$$\rho\left(\frac{\partial u_\theta}{\partial t} + u_r\frac{\partial u_\theta}{\partial r} + \frac{u_\theta}{r}\frac{\partial u_\theta}{\partial \theta} + \frac{u_r u_\theta}{r} + u_z\frac{\partial u_\theta}{\partial z}\right)$$

$$= -\frac{1}{r}\frac{\partial p}{\partial \theta} + \frac{1}{r^2}\frac{\partial}{\partial r}(r^2\tau_{r\theta}) + \frac{1}{r}\frac{\partial \tau_{\theta\theta}}{\partial \theta} + \frac{\partial \tau_{\theta z}}{\partial z} + \rho g_\theta \tag{G.10}$$

 $z$-component

$$\rho\left(\frac{\partial u_z}{\partial t} + u_r\frac{\partial u_z}{\partial r} + \frac{u_\theta}{r}\frac{\partial u_z}{\partial \theta} + u_z\frac{\partial u_z}{\partial z}\right) = -\frac{\partial p}{\partial z} + \frac{1}{r}\frac{\partial}{\partial r}(r\tau_{rz}) + \frac{1}{r}\frac{\partial \tau_{\theta z}}{\partial \theta} + \frac{\partial \tau_{zz}}{\partial z} + \rho g_z \tag{G.11}$$

Navier–Stokes equations for a Newtonian fluid

 $r$-component

$$\rho\left(\frac{\partial u_r}{\partial t} + u_r\frac{\partial u_r}{\partial r} + \frac{u_\theta}{r}\frac{\partial u_r}{\partial \theta} - \frac{u_\theta^2}{r} + u_z\frac{\partial u_r}{\partial z}\right)$$

$$= -\frac{\partial p}{\partial r} + \eta\left[\frac{\partial}{\partial r}\left(\frac{1}{r}\frac{\partial}{\partial r}(ru_r)\right) + \frac{1}{r^2}\frac{\partial^2 u_r}{\partial \theta^2} - \frac{2}{r^2}\frac{\partial u_\theta}{\partial \theta} + \frac{\partial^2 u_r}{\partial z^2}\right] + \rho g_r \tag{G.12}$$

 $\theta$-component

$$\rho\left(\frac{\partial u_\theta}{\partial t} + u_r\frac{\partial u_\theta}{\partial r} + \frac{u_\theta}{r}\frac{\partial u_\theta}{\partial \theta} + \frac{u_r u_\theta}{r} + u_z\frac{\partial u_\theta}{\partial z}\right)$$

$$= -\frac{1}{r}\frac{\partial p}{\partial \theta} + \eta\left[\frac{\partial}{\partial r}\left(\frac{1}{r}\frac{\partial}{\partial r}(ru_\theta)\right) + \frac{1}{r^2}\frac{\partial^2 u_\theta}{\partial \theta^2} + \frac{2}{r^2}\frac{\partial u_r}{\partial \theta} + \frac{\partial^2 u_\theta}{\partial z^2}\right] + \rho g_\theta \tag{G.13}$$

 $z$-component

$$\rho\left(\frac{\partial u_z}{\partial t} + u_r\frac{\partial u_z}{\partial r} + \frac{u_\theta}{r}\frac{\partial u_z}{\partial \theta} + u_z\frac{\partial u_z}{\partial z}\right)$$

$$= -\frac{\partial p}{\partial z} + \eta\left[\frac{1}{r}\frac{\partial}{\partial r}\left(r\frac{\partial u_z}{\partial r}\right) + \frac{1}{r^2}\frac{\partial^2 u_z}{\partial \theta^2} + \frac{\partial^2 u_z}{\partial z^2}\right] + \rho g_z \tag{G.14}$$

**TABLE G.3** CONSERVATION EQUATIONS FOR AN INCOMPRESSIBLE NEWTONIAN FLUID IN SPHERICAL COORDINATES

Continuity

$$\frac{1}{r^2}\frac{\partial}{\partial r}(r^2 u_r) + \frac{1}{r\sin\theta}\frac{\partial}{\partial\theta}(u_\theta\sin\theta) + \frac{1}{r\sin\theta}\frac{\partial u_\phi}{\partial\phi} = 0 \tag{G.15}$$

Momentum equations in terms of stresses

r-component

$$\rho\left(\frac{\partial u_r}{\partial t} + u_r\frac{\partial u_r}{\partial r} + \frac{u_\theta}{r}\frac{\partial u_r}{\partial\theta} + \frac{u_\phi}{r\sin\theta}\frac{\partial u_r}{\partial\phi} - \frac{u_\theta^2 + u_\phi^2}{r}\right)$$

$$= -\frac{\partial p}{\partial r} + \frac{1}{r^2}\frac{\partial}{\partial r}(r^2\tau_{rr}) + \frac{1}{r\sin\theta}\frac{\partial}{\partial\theta}(\tau_{r\theta}\sin\theta) + \frac{1}{r\sin\theta}\frac{\partial\tau_{r\phi}}{\partial\phi} - \frac{\tau_{\theta\theta} + \tau_{\phi\phi}}{r} + \rho g_r \tag{G.16}$$

θ-component

$$\rho\left(\frac{\partial u_\theta}{\partial t} + u_r\frac{\partial u_\theta}{\partial r} + \frac{u_\theta}{r}\frac{\partial u_\theta}{\partial\theta} + \frac{u_\phi}{r\sin\theta}\frac{\partial u_\theta}{\partial\phi} + \frac{u_r u_\theta}{r} - \frac{u_\phi^2\cot\theta}{r}\right)$$

$$= -\frac{1}{r}\frac{\partial p}{\partial\theta} + \frac{1}{r^2}\frac{\partial}{\partial r}(r^2\tau_{r\theta}) + \frac{1}{r\sin\theta}\frac{\partial}{\partial\theta}(\tau_{\theta\theta}\sin\theta) + \frac{1}{r\sin\theta}\frac{\partial\tau_{\theta\phi}}{\partial\phi} + \frac{\tau_{r\theta}}{r} - \frac{\cot\theta}{r}\tau_{\phi\phi} + \rho g_\theta \tag{G.17}$$

φ-component

$$\rho\left(\frac{\partial u_\phi}{\partial t} + u_t\frac{\partial u_\phi}{\partial r} + \frac{u_\theta}{r}\frac{\partial u_\phi}{\partial\theta} + \frac{u_\phi}{r\sin\theta}\frac{\partial u_\phi}{\partial\phi} + \frac{u_\phi u_r}{r} + \frac{u_\theta u_\phi}{r}\cot\theta\right)$$

$$= -\frac{1}{r\sin\theta}\frac{\partial p}{\partial\phi} + \frac{1}{r^2}\frac{\partial}{\partial r}(r^2\tau_{r\phi}) + \frac{1}{r}\frac{\partial\tau_{\theta\phi}}{\partial\theta} + \frac{1}{r\sin\theta}\frac{\partial\tau_{\phi\phi}}{\partial\phi} + \frac{\tau_{r\phi}}{r} + \frac{2\cot\theta}{r}\tau_{\theta\phi} + \rho g_\phi \tag{G.18}$$

Navier–Stokes equations for a Newtonian fluid

r-component

$$\rho\left(\frac{\partial u_r}{\partial t} + u_r\frac{\partial u_r}{\partial r} + \frac{u_\theta}{r}\frac{\partial u_r}{\partial\theta} + \frac{u_\phi}{r\sin\theta}\frac{\partial u_r}{\partial\phi} - \frac{u_\theta^2 + u_\phi^2}{r}\right)$$

$$= -\frac{\partial p}{\partial r} + \eta\left[\frac{1}{r^2}\frac{\partial}{\partial r}\left(r^2\frac{\partial u_r}{\partial r}\right) + \frac{1}{r^2\sin\theta}\frac{\partial}{\partial\theta}\left(\sin\theta\frac{\partial u_r}{\partial\theta}\right) + \frac{1}{r^2\sin^2\theta}\frac{\partial^2 u_r}{\partial\phi^2}\right.$$

$$\left. -\frac{2}{r^2}u_r - \frac{2}{r^2}\frac{\partial u_\theta}{\partial\theta} - \frac{2}{r^2}u_\theta\cot\theta - \frac{2}{r^2\sin\theta}\frac{\partial u_\phi}{\partial\phi}\right] + \rho g_r \tag{G.19}$$

θ-component

$$\rho\left(\frac{\partial u_\theta}{\partial t} + u_r\frac{\partial u_\theta}{\partial r} + \frac{u_\theta}{r}\frac{\partial u_\theta}{\partial\theta} + \frac{u_\phi}{r\sin\theta}\frac{\partial u_\theta}{\partial\phi} + \frac{u_r u_\theta}{r} - \frac{u_\phi^2\cot\theta}{r}\right)$$

$$= -\frac{1}{r}\frac{\partial p}{\partial\theta} + \eta\left[\frac{1}{r^2}\frac{\partial}{\partial r}\left(r^2\frac{\partial u_\theta}{\partial r}\right) + \frac{1}{r^2\sin\theta}\frac{\partial}{\partial\theta}\left(\sin\theta\frac{\partial u_\theta}{\partial\theta}\right) + \frac{1}{r^2\sin^2\theta}\frac{\partial^2 u_\theta}{\partial\phi^2}\right.$$

$$\left. +\frac{2}{r^2}\frac{\partial u_r}{\partial\theta} - \frac{u_\theta}{r^2\sin^2\theta} - \frac{2\cos\theta}{r^2\sin^2\theta}\frac{\partial u_\phi}{\partial\phi}\right] + \rho g_\theta \tag{G.20}$$

φ-component

$$\rho\left(\frac{\partial u_\phi}{\partial t} + u_r\frac{\partial u_\phi}{\partial r} + \frac{u_\theta}{r}\frac{\partial u_\phi}{\partial\theta} + \frac{u_\phi}{r\sin\theta}\frac{\partial u_\phi}{\partial\phi} + \frac{u_r u_\phi}{r} - \frac{u_\phi^2\cot\theta}{r}\right)$$

$$= -\frac{1}{r\sin\theta}\frac{\partial p}{\partial\phi} + \eta\left[\frac{1}{r^2}\frac{\partial}{\partial r}\left(r^2\frac{\partial u_\phi}{\partial r}\right) + \frac{1}{r^2\sin\theta}\frac{\partial}{\partial\theta}\left(\sin\theta\frac{\partial u_\theta}{\partial\theta}\right) + \frac{1}{r^2\sin^2\theta}\frac{\partial^2 u_\theta}{\partial\phi^2}\right.$$

$$\left. -\frac{u_\phi}{r^2\sin^2\theta} + \frac{2}{r^2\sin\theta}\frac{\partial u_r}{\partial\phi} + \frac{2\cos\theta}{r^2\sin^2\theta}\frac{\partial u_\theta}{\partial\phi}\right] + \rho g_\phi \tag{G.21}$$

# Index